Höll, Water

Water

Examination · Assessment · Conditioning
Chemistry · Bacteriology · Biology

Karl Höll

with the collaboration of

Helmut Peter

and

Dietrich Lüdemann

Walter de Gruyter · Berlin · New York 1972

Translated from the fifth German edition by Professor M. R. F. Ashworth,
Chairman, Department of Organic Analysis of the University of the Saar, W. Germany

Karl Höll, Prof. Dr. phil., Dr.-Ing. E. H.
Food Chemist (Diploma), Hamelin, W. Germany
Helmut Peter, Dr. med., Medical Specialist of Laboratory Diagnostics, Hamelin, W. Germany
Dietrich Lüdemann, Prof. Dr. rer. nat., Scientific Director of the Institute for Water-, Soil- and Atmospheric Hygiene of the Federal Health Office, Berlin-Dahlem

ISBN 3 11 003728 9
Library of Congress Catalog Card Number 78 – 176322

Printed in Germany by Georg Wagner, Nördlingen

Foreword

In accordance with many requests, this book is being published in English translation. There is no similar specialist work which is devoted to water investigation and assessment in chemical, bacteriological and biological laboratories, and in institutes of hygiene and water works. It is hoped that the book will thereby fill a gap in the literature and be received as cordially as the five editions in German have been. We are consequently grateful to a chemist colleague, Professor Ashworth, who has undertaken the translation into his English mother-tongue. The book is devoted primarily to drinking water. Since drinking water, service water, baths water, boiler feed water, river water and wastewater, as well as mineral and medicinal water are so closely interrelated, these other themes are also treated appropriately. The drinking water specialist must be acquainted with, and respect, the quality requirements of all types of consumer; and the expert occasionally has to decide with new tappings whether a water has physiological or beneficial effects on health and may be used as mineral or medicinal water.

The book is based on practical experience over several decades and is written in easily understandable form for practical use. Theoretical discussions and chemical equations have thus been omitted where possible. It is thereby hoped to facilitate the cooperation between all water specialists, hygienists, chemists, engineers, biologists, etc. Only through this cooperation of all specialists can a solution be found in the future to the great drinking water problems, such as water supply from distant reservoirs and other surface waters.

The procedures of examination have been largely adapted to the German Standard Procedures for Investigation of Water ("Deutsche Einheitsverfahren zur Wasseruntersuchung"); and the bases of assessment to the Standard Requirements of Drinking Water Quality in Europe ("Einheitliche Anforderungen an die Trinkwasserbeschaffenheit in Europa").

Many water laboratories now have improved equipment, yet it is not possible to renounce simple methods, e. g., visual colorimetry, which can also be carried out by personnel of more limited chemical qualifications. These simple methods will also still be useful in future in field work and on expeditions.

In recent years great progress has been made in all domains of water hygiene, supply and conditioning. The literature has accordingly been covered in this book in general up to the end of 1971.

Karl Höll

Special Note on the Translation

The German edition was of course intended primarily for the German-speaking areas of Europe. References to firms, places, events, laws, etc., were thus largely restricted to these areas or their immediate neighbourhood. Practically all of these have, nevertheless, been included in the translation. The value of quoting German firms for supplying apparatus and chemicals may perhaps be questioned; the readers' home countries (USA, Britain, etc.) can probably provide these more conveniently. However, this is not necessarily certain. Moreover, those associated with the vast subject of water will surely be interested in the similarities to and differences from the methods, apparatus and equipment employed, and the ideas and regulations prevailing in their own country. This comparison can yield only profit and, it is hoped, additionally justifies the publication of a translation in English.

Table of Contents

Part 1

Chemistry of Water · Examination · Assessment · Conditioning
by Karl Höll

Foreword V
Special Note on the Translation VI

I. Investigations of Drinking Water 1
A. General Tests on the Sampling Spot 1
1. Inspection of the Site 1
a) Various Types of Well 1
b) The Three Drilling Procedures 3
c) Horizontal Wells (Collector Wells) 3
2. The Pumping Test 4
3. Points for Consideration in the Inspection of the Site 6
a) General Enquiries 6
b) Special Hygienic Enquiries 6
4. Dye Test and Other Tests for Contamination of Ground Water 9
5. Sampling 9
6. Sample Size 13
7. Preservation of Water Samples 13
8. Sent-In Samples 14
9. Necessary Tests and Investigations on the Spot 14
B. General Tests in the Laboratory 16
1. Organoleptic Examination of the Drinking Water 16
2. Clarity and Transparency 17
3. Turbidity and Settleability 18
4. Colour of the Water 19

II. General Remarks about Chemical Examination of Water 20

A. General Principles of Chemical Work 20
1. Quotation of the Results 21
B. Procedures of Investigation 22
C. Colorimetric Working Techniques 22
1. General 22
2. Apparatus for Colorimetry 23
3. Working Rules for Colorimetry 24
4. Work with Hehner Cylinders 26

III. Minor Drinking Water Analysis 27

A. Hygiene-Chemical Investigation of Drinking Water 27
1. Ammonium Ion Determination 27
a) Colorimetric Determination of Ammonium Ion with Nessler's Reagent 28
b) Determination of Protein Nitrogen as Ammonium Ion 30
c) Determination of Albuminoid Nitrogen as Ammonium Ion 30
2. Nitrite Ion Determination 31
a) Colorimetric Determination of Nitrite according to *Zambelli* 31
b) Indole Method (for small nitrite ion amounts) 32
3. Nitrate Ion Determination 33
a) Colorimetric Determination of Nitrate according to *Noll* 33
b) New Specific Nitrate Determination with Sodium Salicylate 34
c) Colorimetric Determination of Very Small Amounts of Nitrate according to *Kuisel* 35
d) Hellige Method 36
e) Nitron Method 36
f) Titrimetric Determination after Reduction to Ammonium 36
g) Serial Analyses with Ion-specific Electrodes 36
4. Potassium Permanganate Consumption (determination of organic matter in water) 37
a) Rough Field Determination 37
b) Determination of the Permanganate Number according to *Kubel* (oxidisability) 38
c) Determination of Permanganate Number according to *Schulze-Trommsdorf* 39
5. Determination of Chlorine Number 40
6. Determination of Chloride 40
a) Titrimetric Determination of Chloride according to *Mohr-Winkler* 41
b) Interferences in the Chloride Determination 42
c) Determination of Chloride according to *Schneebeli* and *Staub* 43
d) Mercuric Nitrate Method 43
e) Determination of Chloride with Ion-specific Electrodes 43
7. Determination of Phosphate Ion 43
8. Determination of Polyphosphate 45
9. Determination of Urochrome 46
10. Faecal Reaction 47
11. Determination of Sulphate Ion 47
a) Gravimetric Determination of Sulphate as Barium Sulphate 48
b) Titrimetric Determination of Sulphates in Waters of low Sulphate Content, according to *Winkler* 49
c) Determination of Sulphate according to *Ohle* 50
d) Titrimetric Determination of Sulphate according to *Sijderius* 50

IV. Assessment of Drinking Water 51

A. General Requirements of a Drinking Water concerning External Nature and Temperature 51

1. Coloration of the Water and its Improvement 52
2. Turbidity and Clarification of the Water 54
3. Odour and Taste 55
B. Hygienic Judgment of Drinking Water, based on the Chemical Analysis 55
1. General 55
2. Significance of the Chemical and Bacteriological Examinations of the Water . 56
3. Hygienic Measures when Drinking Water is Polluted 58
4. Ground Water Supply 58
5. River Water Supply 59
6. Reservoir Water Supply 59
7. Spring Water Supply 60
8. Rainwater Supply 60
C. Sterilisation (Disinfection) of the Water 60
1. Incomplete Sterilisation by Filtration 61
2. Sterilisation through Addition of Chemicals 62
3. Sterilisation with Ozone 62
4. Chlorination 63
a) Disadvantages of Chlorination 63
5. Sterilisation with Chlorine Dioxide 64
6. Other Sterilisation Procedures 65
7. On which Chemical Features is based Objection to a Well from the Point of View of Hygiene? 67
D. Pollution Indicators 68
1. Ammonia or Ammonium Ion 68
2. Nitrous Acid or Nitrite Ion 69
3. Nitric Acid or Nitrate Ion 69
4. Potassium Permanganate Number and Chlorine Number 71
5. Chloride Ion 73
6. Phosphate Ion 75
7. Sulphate Ion 76
8. Potassium and Sodium 77
9. Residue on Evaporation 77
10. Screenable Material 77
11. Carcinogenic Materials in Water 78
a) Removal 79
b) Regulations 79
E. Normal Values for Drinking Water 80
F. Examples of Drinking Water Analyses for Individual Wells and their Evaluation . 80
1. Dug Wells, strongly polluted by cow manure 80
2. Water from a drilled well which is not suspect 81
3. Water from a shallow well 81

V. Major Drinking Water Analysis of Tap Water (Chemical Examination of Drinking and Service Water from a Central Water Supply) 82

A. Working Procedure 82
B. Evaluation of Aggressivity 83

1. Reaction 83
2. pH-Determination, Hydrogen Ion Concentration 83
   a) Approximate pH-Determination 84
   b) More Accurate Colorimetric Determination of pH 84
   c) Accurate Electrometric pH-Determination 85
   d) Measurement of the Redox-Potential 86
3. Carbon Dioxide 86
   a) Determination of Total Free Carbon Dioxide 86
   b) Titrimetric CO_2-determination according to *Trillich* 86
   c) Acidimetric Determination of Free Carbon Dioxide according to *U. Hässelbarth,* with Electrometric End-point Indication 87
   d) Heyer Test with Temperature Modification of the Sample according to *U. Hässelbarth* 88
   e) Heyer Test to establish the Specific Temperature- and Salt Effects with the Water under Investigation 89
4. Acid-Base Titration 90
   a) General 90
   b) Definition of the End-point of a Titration 90
   c) Correction for the m-value 91
   d) Sphere of Application and Interferences 91
5. Determination of the Acid Consumption (m- or p-value with positive sign). 91
6. Determination of the Base Consumption (m- or p-value with negative sign) . . 92
   a) Direct Titration of Base Consumption 92
   b) Back Titration of an Excess of Base 93
7. Determinations derived from m-value, p-value and pH 93
8. Calculation of the Aggressivity to Lime of Water 94
9. Calculation of the Free Excess Carbon Dioxide according to *U. Hässelbarth* . 94
10. Lead Corrosion Experiment 101
11. Oxygen Determination 101
   a) Iodometric Oxygen Determination according to *Winkler-Bruhns* 102
   b) Oxygen Determination in Surface Waters and Contaminated Waters according to *W. Ohle;* Differential Iodine Method 104
   c) Determination of Very Small Amounts of Oxygen 105
   d) Determination of Oxygen Consumption of River Water and Wastewater . . 105
12. Oxygen Deficiency, Oxygen Saturation Deficiency 106
13. Oxygen Saturation Index 106

C. Technical-chemical Water Analysis 108
1. Determination of Iron 108
   a) Colorimetric Determination of Total Iron 109
2. Determination of Manganese 110
   a) Colorimetric Determination of Manganese 111
   b) Manganese Determination in Waters rich in Chloride 111
3. Determination of Hardness 112
   a) Accurate Determination of the Hydrogen Carbonate Ion (and carbonate hardness according to *Lunge*) 113
   b) Accurate Measurement of Total Hardness according to *Blacher* 114
   c) Calculation of Total Hardness from Calcium and Magnesium Hardnesses . 115
   d) Rapid Method for Determining Total Hardness with EDTA (Titriplex) . . 115

e) Titrimetric Ca- und Mg-determination with Complexone; Method based on the New German EV (German Standard Methods) 116
4. Determination of Calcium and Magnesium 117
a) Gravimetric Calcium-determination 117
b) Titrimetric Calcium-determination 118
c) Titrimetric Magnesium-determination 118
d) Colorimetric Magnesium-determination 118
e) Determination of Magnesium and Calcium Chlorides 119
5. Hydrogen Sulphide and Sulphides 119
a) Colorimetric Determination of Small H_2S-amounts 119
b) Iodometric H_2S-determination 120
c) Direct Iodometric Titration of H_2S 120
6. Residue on Evaporation and Ignition; Loss on Ignition 120
7. Electrolytic Conductivity 121
8. Interferometer Value 122
9. Density (Specific Gravity) 122
10. Radioactivity of Drinking Water and its Measurement 123
a) Measurement of the Radioactivity 124
b) Decontamination 125
D. Artificial Water Additives 126
1. Free Chlorine and Combined Available Chlorine 126
a) Free Available Chlorine 126
b) Combined Available Chlorine 126
c) Total Chlorine (Available Chlorine) 126
2. Determination of Free Available Chlorine 126
a) Colorimetric Procedures 126
b) Titrimetric Determination of Free Chlorine with DPD 127
3. Titrimetric Determination of Total Chlorine 127
4. Determination of Combined Available Chlorine 127
5. Chlorine Demand (Chlorine combination capacity, breakpoint) 128
6. Ozone 128
7. Hydrogen Peroxide 129
8. Silver 129
9. Determination of Aluminium 130
a) Colorimetric Aluminium Determination 130
b) Aluminium Determination according to *Gad* 130
10. Determination of Polyphosphates 130
11. Determination of Sulphite 131
12. Determination of Hydrazine 131
E. Determination of the Metal Ions, etc. 131
1. Determination of Lead 131
a) Sampling and Preparation of the Sample for Lead Investigation 131
b) Colorimetric Determination of Lead 132
2. Determination of Copper 133
a) Colorimetric Determination of Copper according to *Winkler* 133
b) Copper and Lead together 133
3. Determination of Zinc 133
4. Determination of Potassium 134

a) Titrimetric Determination of Potassium 135
5. Determination of Sodium 136
6. Determination of Chromium 136
7. Determination of Arsenic 137
8. Determination of Selenium according to *Quentin* 138
9. Determination of Tin 138
F. Determination of Fluoride Ion (according to *Sanchis*) 138
1. New Working Procedure 139
2. Significance of Fluoride Ion for Health 140
G. Determination of Iodide Ion (according to the new EV) 141
H. Determination of Silica 142
1. Colorimetric Determination according to *Winkler* 142
I. Detection of Ground Water contamination through Mineral Oil 143
1. Chemical Investigation 143
2. Removal of Mineral Oil from the Water 144
3. Seepage of Mineral Oil 145
Mineral Oils and Carcinogenic Substances 146

VI. Assessment of the Drinking Water and Service Water for Central Water Supply 148

A. Treatment and Conditioning of the Water 148
1. General 148
a) Basic Demands on Tap Water 148
b) Drinking Water Conditioning Regulation (foreign matter decree) of 19/12/59 (W. Germany) 148
2. Corrosion 151
a) Corrosive Power of the Water (cold water) 151
b) Increase of Corrosive Power on Mixing Two Waters 152
c) Pipe Corrosion and Water Hygiene 153
d) Behavoiur of Metal Materials towards Water 154
e) Water Pipes 155
f) Drinking Water Containers 160
3. Hydrogen Ion Concentration, pH 160
4. Carbon Dioxide 161
5. Free Carbon Dioxide 162
a) Technical Significance of the Carbon Dioxide Content of the Water 162
b) "Associated" Carbon Dioxide 163
c) Carbonate Balance and Anti-rust Protective Layer 163
d) Excess Carbon Dioxide 164
6. Deacidification of the Water 166
a) Mechanical Deacidification 166
b) Chemical Deacidification (marble, ignited dolomite, $Ca(OH)_2$) 168
c) Protection of the Pipes against Corrosion by means of Phosphate Dosage . 172
d) Protection against Corrosion by means of Treatment with Silicates 174
7. Iron and Manganese 174
a) Iron and Manganese in Tap Water 174

b) Removal of Iron (open, closed, chemical) . . . 175
c) Removal of Manganese . . . 177
8. Hardness of the Water . . . 178
a) Concept of Hardness . . . 178
b) Occurrence of the Hardness-yielding Agents . . . 180
c) Hardening of Ground Water through Refuse Tips . . . 180
d) Sanitary Significance of Water Hardness . . . 181
e) Significance of Water Hardness for Service Water and Technical Purposes . 182
f) "Fur" or "Scale" Deposit . . . 183
g) Softening of Service Waters . . . 183
h) Ion Exchange Procedures . . . 185
j) Addition of Polyphosphates . . . 187
9. Metals, etc. . . . 189
a) Lead . . . 189
b) Copper . . . 190
c) Arsenic . . . 191
d) Zinc . . . 192
e) Tin . . . 192
f) Aluminium . . . 192
g) Chromium, Cadmium, Antimony, etc. . . . 193
10. Oxygen . . . 193
11. Hydrogen Sulphide . . . 194
B. Examples of Analysis of Tap Water and its Assessment . . . 195
1. Survey of the Analysis Values Found . . . 195
a) Examination of the Water from the Iberg spring . . . 195
b) Waterworks L. . . . 197
c) Water from a Deep Well in C. . . . 198
d) Water from the Deep Well of the Firm of B. . . . 199
e) Favourable and Unfavourable Values for Tap Water . . . 201

VII. Examination and Assessment of Swimming Bath Water, Water for Building Purposes, Surface (Receiving) Water and Wastewater . . . 202

A. Swimming Bath Water . . . 202
1. Chemical Examination of the Swimming Bath Water . . . 205
2. Hygiene-Chemical Assessment of the Swimming Bath Water . . . 206
3. Detection of Urine Components in Swimming Bath Water . . . 206
B. Water for Building Purposes . . . 208
1. Assessment of the Harmfulness to Concrete . . . 208
2. Concentration Limits for Harmfulness to Concrete . . . 209
3. Sulphate-resistant Cements . . . 210
4. Water for Concrete Mixing . . . 211
5. Concentration Limits for Water in Contact with Iron and Steel . . . 211
C. Surface and Receiving Waters . . . 212
1. Contamination of Waters . . . 212
2. Determination of the Biochemical Oxygen Demand . . . 215
3. Biomass Titration . . . 216

4. Determination of the Phenol Content 217
5. Quantitative Determination of Detergents (Tensides) 217
D. Wastewater (Sewage) 219
1. Sampling 219
2. Some Special Tests of Waste Water 220
3. Assessment of the Wastewater 223
4. Sewage Clarification 224
5. Biological Purification Procedures 225
6. The Third Sewage Purification Stage 227
7. Trade Wastewaters 227
E. Some Comments on Boiler Feed Water, Boiler Water and Cooling Water 230
1. Boiler Feed Water 230
a) General 230
b) Silica in Boiler Feed Water 231
c) Oxygen in Boiler Feed Water 232
d) Hydrazine Determination 232
2. Cooling Water 233
F. Water for Brewing Purposes 234

VIII. Mineral Water and Medicinal Water 237
A. Mineral Water 237
B. Medicinal Water 238
1. Standard Analyses 241
a) Major Medicinal Water Analysis 241
b) Minor Medicinal Water Analysis 242
c) Control Analysis 243
d) Hygienic Investigation 243
2. Explanatory Comments on the Medicinal Water Analysis 243
1. Determinations of Cations 244
a) Potassium and Sodium 244
b) Lithium 245
c) Flame-photometric Determination of Lithium 245
d) Calcium and Magnesium 246
e) Iron 246
2. Determinations of Anions 246
a) Chloride 246
b) Sulphate 247
c) Boric Acid 247
d) Iodide and Bromide 248
3. Determinations of Trace Elements 248
a) Dithizone Method 248
b) Dithizone Analysis 250
c) Example of a Minor Medicinal Water Analysis 253
C. Water as a Material 254

IX. Reagents for Investigation of Drinking Water 256

X. Covering Document for Water Samples 262

XI. Bibliography of the Chemical Part 264

Part 2

Bacteriology of Drinking Water
by Helmut Peter

I. General Remarks about Bacteriological Investigation Procedures 273

II. The Bacteriological Working Place 276
III. Water and Disease 278
IV. Inspection of the Site 283
V. Sampling 286
VI. Determination of the Bacterial (Plate) Count (Number) 290
VII. Determination of *E.coli* and of Coliform Bacteria 297
VIII. Pathogenic Coli Strains 309
IX. Detection of Enterococci and *Clostridium Welchii* 310
X. Detection of Viruses and Bacteriophages 314
XI. The Membrane Filter Procedure 317
XII. Bacteriological Examination of Baths Water 324
XIII. Detection of Pathogenic Bacteria in the Water 327
XIV. Bibliography of the Bacteriological Part 333

Part 3

Biology of Water
by Dietrich Lüdemann

I. Biological Water Investigations 345

II. Methods 346

A. Sampling 346
1. Biological Investigation of Waters 346
B. Apparatus 347
1. Membrane Filter Procedure 351
2. Investigations of Bottom Materials and Sludges 353
C. Ecological Procedures of Biological Water Analysis 355
1. Saprobic Systems 356
2. Indicator Organisms 358
3. Assessment of the Results 360

III. Biological Investigation of Drinking Water 365

A. Organisms in Plants for Obtaining and Conditioning Water 366
B. Shortcomings in the Tappings of Springs 368
C. Water Works Plants 368

1. Slow Filters 369
2. Rapid Filters 372
3. Activated Carbon Filters 372
4. Water Distribution System 372

IV. Methods of Combating Organisms in Drinking Water Supply Plants 373

V. Bibliography of the Biological Part 378
Index of Contents of Parts 1 to 3 381

Abbreviations

DIN	=	*D*eutsche *I*ndustrie-*N*ormen (German Industrial Standards) (Beuth-Verlag, Berlin)
*	=	Reagent of the German Pharmacopoeia (DAB = *D*eutsches *A*rznei*b*uch)
S. u. S.	=	Schleicher & Schüll (Filter)
GWF	=	*G*as- und *W*asser*f*ach (Journal, published by Verlag Oldenbourg, Munich)
Lit. Ber.	=	Literaturberichte über Wasser, Abwasser, etc. (Reports on articles in the literature concerning water, wastewater, etc)
DVGW	=	*D*eutscher *V*erein von *G*as- und *W*asserfachmännern (German society of specialists in subjects concerning gas and water)
Ges. Ing.	=	Gesundheitsingenieur (Journal, "Sanitary Engineer")
Vom Wasser	=	Yearbook, published by Verlag Chemie, Weinheim, W.Germany
IVL	=	*I*nternationale *V*ereinigung für *L*imnologie

Part 1

Chemistry of Water · Examination ·
Assessment · Conditioning
by
Dr. phil. Karl **Höll,** Hamelin, W. Germany

I. Investigations of Drinking Water

A. General Tests on the Sampling Spot

1. Inspection of the site

A drinking water plant cannot be judged without accurate knowledge of the local situation. This necessary knowledge must be acquired by the expert in a previous local inspection before he gives his opinion. *Spitta* says pointedly that whoever ventures a hygienic assessment of a drinking water without having seen and tested the place of water intake behaves with few exceptions little differently from a quack doctor attempting a cure at a distance.
Definite specialist knowledge is required for inspecting a site. First, the different types of well must be distinguished.

a) Various Types of Well

One distinguishes: Dug wells (Schacht-, Kessel-, Senk-, Sodbrunnen); hollow ram pump or abyssinian wells (Ramm-, Schlag-, Abessinier-, Stossbrunnen); and tube, drilled or bore wells (Bohr-, Rohr-, Röhrenbrunnen).
A dug well consists in its simplest form of a walled, more or less deep basin in which the ground water collects from below and from which the water is withdrawn by means of a bucket. This is the open dug well (cistern) which is the least satisfactory from a hygienic point of view and which is to be rejected nowadays (see Arch. f. Hygiene *148,* 40 [1964]). It should have at least a metal roofing, saddle-shaped with a lateral flap for the bucket. Dug wells of unrabbeted field stones, wood or unmortared bricks are especially unsuitable since surface water can seep in laterally; this is recognisable through dark stripes below the joints. The walls of the dug well should at least be made of well rabbeted and hard-fired bricks. As a minimum requirement, one should aim at well joined concrete rings which are let into the earth when the well basin is being built and to a depth of at least 7 metres (23 feet).
The well walls should be watertight in any case to a depth of at least 3 metres (10 feet) below the surface of the earth. The walls of the dug well are best given a crushed clay lining of 30–50 cm (12–20 inches) thickness over the complete depth of the well while it is being built.
If water is drawn by bucket, always only one and the same bucket must be used, which is to be kept in a clean state and held by a chain.
There should be a bench alongside the well, on which the bucket can be stood which is filled by that used for drawing.
The closed dug well can be accorded a more favourable hygienic judgment. It is closed with

a lid of iron or concrete which should overlap the sides of the well. Older wells of this type still have wooden lids which are, however, insufficiently tight and consequently accumulate fungi and become carriers of infection. No wooden parts at all should be used in dug wells for this reason.

The pump is situated on the cover of the well or beside it. Its lift tube passes through the cover almost to the bottom of the well or, depending on the reserve of water, more suitably to a point 0.5 to 1 metre (20–40 inches) above the bottom. The well pit should extend to at least 50 cm (20 inches) above the surface level so as to hinder entry of rainwater and dirty water to the well.

Dug wells can be externally distinguished from tube (bore or drilled) wells by the pump's being on or beside a lid. Dug wells are mostly shallow wells and therefore hygienically less favourable than deep wells (drilled wells). Dug wells nevertheless have the advantage of always possessing a larger reserve of water, from which a larger amount of water is rapidly available at any time (p. 3–4).

Cisterns (rainwater tanks): Collection of rainwater has been since immemorial times the sole way of procuring drinking water in some areas, e. g. on the Friesian coast or on Heligoland. Roof surfaces and the air carry contamination which can be easily washed along. The first rainwater collected when it rains should thus be diverted with a swivel-mounted gutter and that collecting later be guided into the cistern (see *K. Höll,* Arch. f. Hygiene *113,* 283–295 [1935] concerning water supply through cisterns on Heligoland).

A surface roof area of 150 m^2 (ca 1500 sq. feet) is regarded by *R. Meyer* as a standard for a 6-membered family ("Zur Trinkwasserversorgung aus Zisternen", Arch. f. Hygiene *137,* 454–476 [1953]).

Drilled (bore, tube) wells: Drilled wells are divided into shallow (3 to 7 metres deep) and deep (7 to 50 metres and more).

Shallow wells are, for example, the abyssinian driven wells, which are constructed by ramming a pointed well pipe into the ground and mounting a hand pump on it. They are thus also called hollow ram pump wells. The steel point of the well pipe has slits or holes (from 3 to 6 mm diameter, spread over 1 metre) clad with bronze or brass sieves (lace or twill material). The soil and ground water must not be aggressive. Plastic points are now available, with slits and plastic-lace material. The lace tissue is fixed to the filter pipe with wires and clips.

This well filter developed from wooden filters via steel, copper and earthenware filters to plastic filters on the one hand and to artificial resin-compressed wood (Schönebecker Filter Ges., Hannover) on the other. The three last-mentioned have the advantage of being insensitive to attack. The corrosion-resistant OBO-filters of the Schönebecker Brunnenfilter GmbH are especially suitable for aggressive waters, including strongly aggressive mineral waters. Nowadays, polyamide filters and rising pipes are used with submersible pumps. V_2A and V_4A filters are used in the rare cases of high aggression.

In the course of the years, filters can become encrusted and coated with a yellow deposit ("verockern" in German) and have to be cleaned by washing. Flushing with acid is often needed. The firm of Petzold, GmbH, Munich 9 specialises in this (Steintod-Verfahren).

Deep wells are drilled wells. They are made by drilling a lining pipe into the earth and the filter and making-up pipe are then let down into the hollow lining pipe. The filter or the

strainer are embedded in a gravel wall. The gravel packing is controlled by the sieve analysis of the soils in the vicinity, so that the well does not become choked with sand (gravel wall well). The well is ready for use as soon as the lining pipe is "drawn" and the well sealed.
Several wells are combined to farm a battery of wells in larger water works, and united with a common suction or siphon conduit, which may lead to a common collecting tank. Suction and siphon conduits must be laid especially water- and airtight. Drainage pipes are also united to form infiltration galleries (Eternit (asbestos cement) or earthenware). "Layer wells" ("Tellerbrunnen") have been developed in the last few years for medium to large water works and have evidently proved very good in practice; they do not become so quickly coated with yellow deposit (*C. Abweser,* Kommunalwirtschaft No. 2/3 [1955] and *G. Bohrisch,* ibid. No. 7 [1956]).
Deep wells are operated with sub-aqueous pumps, abyssinian driven or hollow ram pump wells with pumps on the surface.
Shallow wells, especially dug wells, should no longer be built for drinking water supply. The deeper they are, the better they are protected from penetration of bacteria (p. 284) and nitrates (p. 69).
Further information about wells and well hygiene is to be found in "Hygienische Leitsätze für die Trinkwasserversorgung" (Hygiene guiding principles for drinking water supply) (Veröffentlichungen auf dem Gebiet der Medizinalverwaltung, Berlin [1932]; in *H. Beger,* Brunnenhygiene, Berlin [1947]; in "Brunnenordnung", published by Carl Heymann, Berlin [1935]; and in *E. Bieske,* Bohrbrunnen, Munich [1953]). DIN 2001 (*18*) contains guiding principles for individual drinking water supply.

b) The Three Drilling Procedures

Formerly, one worked by hand with the rotary drilling procedure; nowadays, work is carried out by the flat chisel (pitching borer) procedure (the cutting chisel is mechanically allowed to fall intermittently) or hydraulic boring (rotary boring, with rolling cutter-type bits and tool washing). Dry and wet drilling procedures are thus distinguished. Wet drilling requires far less working time but does not yield unaffected soil samples which are often desired.
DIN 18301 and DIN 18302 give instructions about carrying out drilling and acquisition of ground water. DIN 18302 also prescribes thorough cleaning of the lining pipes, making up pipes and filter pipes before they are built in; this is often neglected.

c) Horizontal Wells (Collector Wells)

Horizontal filter wells are constructed nowadays for larger water supplies. They combine horizontal feeder pipes with vertical collecting shafts. In the Ranney procedure, these feeder pipes are forced into the ground water carrier by hydraulic pressure from a shaft. In the Fehlmann procedure, a drilling pipe is first driven horizontally and the feeder pipe introduced. The Preussag procedure, which can be used for fine sand also, is similar; the horizontal feeder pipes are surrounded with gravel packed layers. Horizontal wells are much more efficient than vertical wells.
Horizontal wells also provide a certain water reserve in the collecting shafts; they do not

upset the water-bearing strata and the ground as much as do vertical wells. The large gathering surface in horizontal wells is in itself more favourable than the point abstraction of water in vertical wells. This applies especially to regions where more intensive demand brings a danger of salting up and hardening (coning effect). In a 4-metre shaft, there is a reserve of 12.5 cubic metres (ca 3500 gallons) per metre depth.
A small horizontal well has already been introduced in Austria (*L. D. György*, Gas, Wasser, Wärme *17*, No. 11 [1963]) and has now obtained general acceptance.
The Mannesmann "Schräginfiltrator" (Inclined "infiltrator") serves for enriching ground water with surface water (firm of Mannesmann, Düsseldorf).
Especially thorough preliminary studies are necessary before a horizontal well is installed, so as to prevent planning errors (*G. Keller*, Bohrtechnik-Brunnenbau *7*, [1956]).

2. The Pumping Test

A pump test is performed to ascertain the capacity of a newly drilled well, usually with a provisional pump. The tests should last 3 days with smaller drilling enterprises and 1–3 weeks with larger. During this time, the fall of the natural ground water level is measured at definite time intervals as the pumping rate is increased. This measurement is done with a cable plumb line which is hung down in the bore hole alongside a graduated strip. The lowering of the water level is cone-shaped ("cone of depression" or "pumping depression cone"). As soon as the water level remains constant for some time at constant pumping rate, the "steady state" is attained. This means that as much water flows in from the ground water carrier as is taken out. If further pumping causes the ground water level to sink further, the pumping rate must be reduced appropriately.
The subsequent rise of the ground water level must be measured frequently since this is important for assessment of the capacity of a well. Slow depression of the water level during the pumping test and a rapid rise after pumping is stopped indicate that the well has a good yielding capacity and that the reserve of ground water is large.
The DVGW working sheet W 111 (1963) (p. 269) and DIN 18301 and 18302 (1958) give further particulars for the well constructor. W 111 contains also a scheme for hydrochemical analysis during the pumping test.
Several careful chemical investigations must be made by an experienced water chemist during the pumping test, at least at the beginning and end of the test and, if necessary, at various delivery stages.
A tap on the riser pipe or on its pipe bend is prescribed in the DVGW working sheet W 111, so that sampling is possible without objection. It is, namely, important to establish if there is an increase or decrease of the oxygen- and CO_2-contents. When new water acquisitions are made, it must in all cases be established whether the water has aggressive properties and needs to be processed. This is important for planning the building of the water works and the financing of a processing plant. It is likewise important for the choice of the well screen.
It is more reasonable to measure the temperature at this tap than at the end of the often long discharge pipe. The exposed part of the pipe can often cause cooling during frosty weather

or warming during high summer. Fluctuations of tenths of a degree are of significance for the assessment, however. The water temperature rises when deep water is drawn, and also in summer when surface water is drawn (p. 15). For filling the sample bottles, the plastic sampling tube is drawn over the tap and one proceeds as on p. 11. Great care must be taken that no air bubbles gain access to the sample bottles.
Very fine bubbles often originate from the eddy of the subaqueous pump and the absence of oxygen in the samples is seen from the white manganese precipitate in the oxygen test; this is regularly the case with deep water. Change in oxygen content is important for judging whether surface water has entered. Later, it is a clue to the tight fit of the well lining.
If the riser pipe bend carries no tap, the delivery pipe must be designed at the end so that it rises slightly. The emerging water is then dammed up in it. A discharge pipe which slopes downwards does not fill up completely, Air then has access to the water and leads to an incorrect result. Carbon dioxide also can then escape from waters which are richer in CO_2, especially with mineral waters and medicinal waters (acidulous waters).
After the delivery pipe has been set up, the water samples can be taken from the end of the pipe after a brief wait, using the plastic sampling tube (p. 11). The delivery pipe should, however, be as short as possible and, if necessary, perhaps shortened for the sampling; this means that the first pipe connection after the pipe bend is opened for a short time.
The sample is best taken, however, with a tap at the pipe bend of the riser pipe, as said above. Waste pipes are often rusty and release iron in dissolved and undissolved form into the water. In short pumping tests, one has to reckon with this sometimes with the riser pipe, too.
Important for the first assessment of the newly tapped water are questions of aggressivity, the iron- and manganese contents, the degrees of hardness and their fluctuations during the pumping test. Soluble or undissolved iron from the series of pipes can sometimes be detected in the first investigation at the start of the pumping test. Before the pumping yields a completely clear water (which can be tested with the help of a viewing tube 1 or 1.5 metre long), the iron content found is problematic and an iron investigation is premature. With some soils (sand and gravel), pumping to clarity lasts only a few hours; with others (loam, loess, especially with very fine material), many days. The best moment for sampling should be agreed on between the drilling chief and the chemical laboratory, since change to a higher pumping output usually brings about renewed turbidity formation.
Through measurement of the redox potential, it is possible to determine if a water tends to suffer from yellow deposit. It is not to be feared if the redox potential E_{OH} is 10 ± 20 mV at pH values of ca 7.0 or the rH value 14.5 ± 0.5. With higher values, especially higher rH values, it is probable, however, if more than 0.2 mg Fe is constantly present for some days in the pumped water.
The chloride and sulphate contents are of prime importance in some regions, e.g. when salt masses or lenses are to be expected in the soil formations concerned.
Bacteriological investigation of the water is inadvisable before the well is finally built, or else it may lead to erroneous conclusions. This happens, for example, when the water overflows out of the bore hole before the pumping test is begun and returns into it during pumping. Before pumping off, the water in the bore hole does not allow the surface water access to it. After the pressure ceases, the surface water easily enters the empty bore hole and brings microorganisms with it–yet it will be assumed that only deep water is being drawn.

When the pump has been finally built into the ready well installation, the bacteriological investigation, after at least three days of pumping, yields a better picture of the bacterial content of the water and also of how watertight the installation is. Even then, however, unfavourable bacteriological results are to be regarded with reservation, depending on the cleanliness of the series of pipes and of the gravel packing used.

On the other hand, favourable bacteriological data prove that the pipes are clean, that surface water is not entering the well and that the series of pipes is watertight.

High bacterial counts are sometimes due to water from upper ground water levels descending through the bore hole and becoming mixed with the profound ground water. The bacteriological and chemical results are not parallel in such a case.

The ground water levels must be measured in the sounding pipes which are set up around the boring; the cable plumb line, mentioned previously, is used for this.

New formation of ground water should be included also in the pumping test. Care must be taken that the pumped-off water cannot infiltrate in the region of the well. It must be led away through pipes or gutters to the nearest receiving water. In no case may the pumped-off water remain in the cone of depression of the well.

3. Points for Consideration in the Inspection of the Site

a) General Enquiries

1. Is it a) a dug well (open or closed), or b) a bore well (see above)?
   If b), is it a shallow or deep well?
2. How deep is the well?
3. Or is it a spring tapping, and how is it covered and tapped?

Concerning 1: How deep is the ground water level[1]?

Does the water level fall in the well when considerable water is taken from it[1]?

Does the ground water level change after prolonged heavy rain or when the level of neighbouring waters rises or falls? In which direction is the gradient of the ground water flow? (towards or away from the well?) This can, if necessary, be ascertained through measurements with sounding pipes or of the water level in wells in the vicinity.

How far from the well is the house to be supplied?

Are the filter pipes deep enough for surface water not to be taken in?

b) Special Hygienic Enquiries

Individual water supply

1. Are there in the neighbourhood of the water catchment area, stables, effluents from stables, lavatories, sewers, waste water drains, leaching trenches, seepage pits, rubbish

1 For measurement, German apparatus includes: the Rang well measurer or the more accurate "Lichtlot" (firm of H. C. Spohr, Frankfurt/Main); or the new measuring instrument of the Hydrom. Werkstätten, Dr. Bovensiepen, Kaufbeuren, Allgäu, Bavaria.

tips, sewage farms, cemeteries, camping sites, clarification plants, gravel pits, stored mineral oil, plants with radioactive wastes, petrol stations, oil heating plants etc.?
If so, how far away from the well and in which direction does the ground water flow? The distance must not be less than 10 to 20 metres (32–64 feet) from shallow wells in sandy soils and not less than 50 metres in gravelly soils, provided that the contamination sites are downhill from the well. The distance should be correspondingly greater from cesspits and waste water seepages. Sometimes lesser distances suffice in heavy clay soils, and also with deep wells, depending on the nature of the ground (an exception is coarse gravel, especially in steeply sloping terrain, such as the Molassegebiete of Lower Bavaria).
The well should be at least 50 metres from the river in bank filtration of river water. Experience of the author has shown that river and stream beds are usually well sealed; wells which are situated directly on rivers of varying salt content rarely show that they are influenced by this. This has been confirmed by recent investigations (*J. Holluta*, Kommunal-Wirtschaft [1955]).
A larger protective area is prescribed for central water supplies: 1st order protective zone, the catchment area of the well; 2nd order protective zone, the immediate protective zone; and the 3rd order protective zone, the water intake area. The DVGW working sheets (p. 269) W 101 (for ground water) and W 102 (for reservoir water, camping sites, dumping areas, etc.) contain directives for drinking water protective areas.

2. Do domestic animals graze all the time or occasionally in the vicinity of the well? Is there any chance at all of pollution at a smaller or greater distance? (Especially important in mountain areas with broken up ground).
3. Are there any salt deposits, brine springs, swamps or moors near the well?
4. Does the ground slope towards or away from the well?
5. Can the water withdrawn from the well by pumping run away?
   Or does it seep through by the well?
   There should be no depressions in the ground near the well, in which water is standing and can infiltrate; these must be filled up with loam. A walled gutter should take away overflowing water or rinsing water. If possible, a zone 1 to 2 metres round a well should be covered with thick concrete and the well pipe cemented in firmly. The area around the concreted part should be sealed with clay.
6. How old is the well system?
7. Is the well in continual use?
8. Is the water temperature always constant or does it change appreciably with the time of year?
   Or does the water temperature change frequently with the air temperature and precipitation?
9. Does the water show turbidity or colour from time to time?
10. Are there trees or bushes near the well? If so, it would be advisable to remove them or dead trees on account of loosening of the soil and formation of cracks through the roots.
11. Dug wells must be uncovered for the inspection and the walls of the well must be looked at. An experienced man can tell whether the walls are watertight by illuminating them.

Darker or lighter streaks, running down from the joints or cracks in the well, show that bacteria-containing or even contaminated surface water has entered. In that case, the well must be emptied, thoroughly cleaned and disinfected and newly joined internally. The rim of the well must project about 0.5 metres (20 inches) above ground level and the well footing must be well cleared. For a depth of at least 3 metres the well sides must be watertight. The rubber rings in the grooves of the iron cover must be intact.

12. It must be established for surface waters whether any contaminating influents are present or are temporarily present. It may be necessary to find out more about the intake area of reservoirs. It should be ascertained with flowing waters whether there is a possibility of pollution above the point of intake of water. An overflow is made with springs and small streams by building a weir of wood or stones. Sampling is then carried out after a few days.
13. Have there been any illnesses in recent years which can be transmitted by drinking water? (e.g. typhoid, paratyphoid, p. 278).
    DIN 2001 and DVGW working sheets W 101–102 (p. 269) provide forms for objections against wells.

Central water supply

In central water supply plants the questions below have to be answered in addition to those above.

1. When sampling from water mains, are there lead pipes or zinc galvanised pipes in the house connections? If so, a sample must be taken after the water has stood for a longer time in the pipe; this sample is for lead or zinc determination.
2. Are there defects, e.g. leaks, in the mains, or installation defects? From what materials have the connections been made? (see *H. Beger,* "Leitfaden der Brunnenhygiene", Berlin [1952] for further information).
3. If a filtration plant is being used, samples must be taken before and after filtering; these are best corresponding samples, taken at a time interval governed by the rate of flow through the filtration plant.
4. Have individual wells an unfavourable water composition at the moment of sampling? According to the guiding principles of DIN 2000 for drinking water supply, it should be possible to take a separate sample from each single well!
5. In what sort of condition is the construction of the well intake? Is the manhole watertight or does water rise in it sometimes?
6. Is the pipe line network of the parish or town in order? Pipes with marked incrustations are dangerous in that they cause local resistance to water flow. The ensuing fall in pressure may bring about subpressure in the system, so that foul water or waste water can enter by suction through leaky places in the pipe (*H. Wittmann* and *W. Bleines,* GWF *98,* 556–562 [1957]).
7. Are the aeration pipes secure, the well tops tight and is the rubber washer still intact?
8. Are there enough protective zones?
    Immediate protective zone = first order (catchment zone of the well or wells) and indirect protective zone = second or third order (water intake area).
9. Has an observation service for ground water been organised with several sounding pipes? Also the enquiries as in b), p. 6–7. Forms for complaint are in DIN 2000.

4. Dye Test and Other Tests for Contamination of Ground Water

A dye test is carried out if there is suspicion that a neighbouring source of contamination (lavatory, manure heaps, sewers, etc.) is influencing a well. In this, a solution of the dye is allowed to percolate into the source of contamination, e.g. the pit of the lavatory. Uranine (sodium fluorescein)[1] is the most suitable dye. Fluorescein can be used also but not with acidic waters ($pH < 7$). However, fluorescein must be dissolved in a little alcohol and then added to dilute sodium hydroxide solution. Eosin is thus usually preferred if uranine is not available. Both these dyes have only to be ground with a little alcohol and added to a full bucket of water. 10 to 20 g of dye suffice for testing single wells. If the dye appears in the well water, the well is open to objection and must be closed. One must not abandon observation too soon. The dye is destroyed in chlorinated water. A dug well must therefore not be chlorinated before the test. Uranine can be detected with a fluorescence lamp in amounts down to 1 part per 10^{11}.

A certain caution is nevertheless necessary when the result of the dye test is negative. Some soils, especially acidic humus soils, can adsorb dyes, notably uranine. Accordingly, a similar test is performed with common salt or phenol, two materials which are not adsorbed by any soil and which are easily detected in the well water. The *Gibbs* reagent, 2,6-dibromoquinonechloroimide, is the best for detecting phenol (blue coloration). From 10 to 800 kg of crude salt is needed, however, for the test, depending on the distance away of the suspected contamination site and the depth of the well (*F. Schwille,* GWF *93,* No. 8 [1952]; *G. Ebeling,* Deutsch. Gewässer-Mitt. *1,* No. 3 [1957]; *K. Eissele* reported the deficit in larger salt tests in GWF *104,* 1158–1160 [1963]).

The recently recommended "Saprol" (firm of Nördlinger, Flörsheim, W. Germany) is also not absorbed; in the positive test, it is shown even through its odour in the well water.

Short-life isotopes also can be introduced at the contamination site for testing. In some cases it is enough to add a paste of pollen grains or *Prodigiosus* bacteria, particularly in karst areas (*J. Zötl,* Steir. Beiträge zur Hydrogeologie [1959]; and *J. Mairhofer,* Steir. Beiträge zur Hydrogeologie [1963/64] for the tracer methods with ^{131}I and ^{82}Br).

A geological expert should be asked in cases of doubt concerning the nature of the soil.

5. Sampling

Correct sampling is a condition for obtaining results which are above suspicion. Basically, the investigating expert himself or at least a qualified person should therefore take the sample. When there are questions of corrosion, as nearly always in plants for central water supply, the chemist must anyhow determine the free CO_2 and pH-value, carry out the Heyer test and oxygen tests, and perhaps fix any H_2S which is present, all at the site.

Samples of both unpurified and purified water must be taken if there are no plants for filtering or other treatment (p. 8). The rate of flow of the water through the plant must be ascer-

1 Uranine AP conc. for testing subterranean waters, firm of Merck (harmless up to 100 mg/l for fish in interceptors, *H. Bandt,* Deutsch. Fischerei-Z. *170* [1957]). The firm of AC-Chemie. Euskirchen, W. Germany now supplies uranine in liquid form.

tained before sampling. If this is not seen in the records of the plant, a dye or salt test must be carried out (see above). The identity of the samples must be especially watched with sewage purification plants.

With larger waterworks, samples must be taken from every single well, since these often vary considerably in their contents of iron, manganese, oxygen and carbon dioxide and their pH-values. The total mixed water can sometimes be improved by eliminating any well which is particularly rich in iron, thereby saving use of an iron removal plant. Waters vary considerably particularly in their iron and manganese contents.

Pollution can concern only one of the wells, or part of one, while the mixed water appears in order.

The following points are to be observed accurately when sampling:

Sampling bottles: Only absolutely clean bottles of colourless glass may be used, preferably of Jena glass; the silica content of some waters increases in ordinary glass bottles and the pH value changes. Polyethylene flasks also are used and are unbreakable. It is, however, more difficult to check their cleanliness. They are prescribed for radioactivity measurements. Transparent plastic flasks are available (e.g., from the Frohn-Plastik GmbH, Deisenhofen near Munich).

It must be remembered, however, that plastic bottles can adsorb some water constituents which then evade determination. This applies to all hydrocarbons, thus including mineral oils; pesticides; also inorganic constituents, such as silica and phosphate ions, are rapidly adsorbed on the walls. This is important in the investigations of silica in boiler feed water and mineral water.

Oxygen can gain access to some plastic flasks; this can lead to iron-containing deposit which may adsorb substances present in traces.

The sample bottles should be closed with glass stoppers in perfect order or, if necessary, with corks which are wrapped with tinfoil or have been coated by dipping into molten paraffin. Untreated corks, especially old corks but also new ones, cannot be used. Rubber stoppers and rings and bakelite stoppers also cannot be used since they give off interfering odours. The bottle must not be sealed directly on the cork, which is often encountered with sent-in samples. Best for glass-stoppered bottles is the Lübbert-Schneider device for closing flasks and made of galvanised iron wire.

As a protection against contamination of the neck of the flask, the stoppers are wrapped round with parchment paper and tied up. The necessary identification marks must be written on the bottle immediately after sampling.

The sample flasks should be cleaned by shaking with pure sea sand and water, then rinsing out with water and distilled water. Soda and acids should be avoided for cleansing purposes since they adhere very firmly to the glass wall and can then be removed only by longer standing and washing out many times. Brushes also should be avoided. Sample bottles which are frequently used should, however, be cleaned at definite time intervals with chromic acid/sulphuric acid or with a washing powder, after which they are left to stand for a longer time containing water which is often renewed.

Preparations for sampling: Before samples are taken, the water which has stood (not only in the house pipe connections but in the mains) must first be run off. Hence it is best to let the water run slowly for 15 to 30 minutes, during which the tap is opened and closed several times. See p. 131 for sampling for the test of lead solubilty.

Hand pumps must have been operated beforehand for 10 to 20 minutes, slowly and regularly. The water must not flow back into the well during this. Any approach to emptying the well must be avoided, however. If the well has not been used for a longer period, water must be pumped from it for 10 to 20 minutes on several days preceding sampling.
A water sample is taken before and after a pumping spell from wells which are under suspicion. Should the composition of the water differ, particular hygienic dangers are present. With dug wells a second sample should be taken after completely emptying the well and refilling with ground water; further samples may be needed after repeated refilling of the well.

Filling the sample bottles under exclusion of air: A cleaned rubber tube or, better, transparent plastic tube is pulled over the end of the tap (if necessary, the plastic tube may be warmed). The tube should be rather longer than the flask and is introduced into the flask to reach its bottom (water has been previously allowed to run through the tube for a while). Water is then left running slowly through the flask until it has been renewed several times. Formation of air bubbles and bubbling of the water must on no account occur during this. The tube is then slowly drawn out or the flask lowered. The bottle is filled right to the brim and the stopper is introduced so that no air bubbles remain in the bottle (test by inclining it).
The water sample should always be taken in principle under exclusion of air; this prevents iron deposition and consequent loss of phosphate and also loss of carbon dioxide with consequent calcium precipitation. If the sample cannot be investigated straight away, a special sample should be taken at the site and treated with acid (HCl and HNO_3).

Sampling with the help of the submerged funnel; With large taps or tubes of pumps, the water is allowed to run into a funnel, from which a rubber tube leads to the bottom of the sampling vessel. The funnel must be permanently full, accomplished by pinching the rubber tube if necessary. The end of the tap or pump tube is below the level of the water in the funnel. After the water has run in and out of the sampling vessel for a longer period, a water sample which has not been in contact with the air is thus obtained here also.

This type of sampling is indispensable for study of oxygen, carbon dioxide and hydrogen sulphide with ordinary pumps and very wide taps. It is advantageous also for the pH determination since there is scarcely any loss of CO_2.

For sampling bore holes or sounding pipes, a vane-type pump is fitted or screwed to the pipe and pumping carried out until clear water emerges, or for at least 15 minutes. The sample is collected by the funnel method (see above); the funnel is held immediately below the exit of the pump and the rubber tube attached to the funnel extends to the bottom of the sampling bottle.

In water tapping from new boreholes, water is pumped off with a motor pump until it is clear; this takes 1 to 5 days. An accurate study of the water at the beginning and end of the "pump test" must be made with larger water acquisition schemes, in order to ascertain if there is any change in water composition during the test. This yields important information for the geological specialist (p. 14).
If the ground water can no longer be obtained from a bore hole with a pump, or if samples from different depths are required for reasons of geological interest, the water-drawing ap-

paratus of Friedinger is used (manufactured by firm of Hydrobios, Kiel-Holtenau, W. Germany). This apparatus is open above and below while letting down and water passes through it during this. This is not so with certainty with the Ruttner apparatus, especially in narrow bore pipes.

Prof. *Kruse* has developed an improved Ruttner apparatus (firm of Franz Bergmann KG, Berlin-Zehlendorf).

Gad's water-drawer is suitable for narrow bore pipes e.g., 8 cm (ca. 3″) sounding pipes, since the water streams through it (firm of Franz Bergmann KG, Berlin-Zehlendorf). The half-litre apparatus has a diameter of 7 cm; the litre-apparatus, of 9 cm.

The firm of H. C. Spohr, Frankfurt/Main has recently brought out a sampling tube of 4 cm diameter and 47 cm (ca. 18″) length for very narrow sounding pipes. The firm of Pumpen-Wart, Erkrath, W. Germany supplies a special sampling pump for pipes of diameter about 2 inches for ground water observation. It functions with compressed air and does not disturb the ground water (Luwa-Pump). The firm of Fichtel & Sachs has a special pump also.

Sampling of surface water: Special apparatus is needed for taking samples from open waters. The Friedinger apparatus mentioned above is suitable for this, as is also the useful Ruttner apparatus. The firm of Franz Bergmann of Berlin-Zehlendorf supplies the latter apparatus and also that of Spitta-Imhoff. *H. Hölzl* has recently improved the Ruttner apparatus. The Ekman-Nansen water-drawer is good for sampling deep lakes. A water sample can be taken without its entering into contact with air by introducing into the water a cylinder with plane ground glass ends and closing it under water. The ends must close tightly of course. After taking out and removing the upper lid, the water can be drawn off into bottles from below with a siphon; this water has not entered into contact with the atmosphere.

R. Liepolt's profundal water-drawer is recommended for sampling waters at a depth. With it, one can determine accurately the microstratification above the mud region (p. 59).

Sampling from lakes, reservoirs and flowing waters: Lakes often have a pronounced stratification, which must be borne in mind when sampling. The surface layer (pelagial) has a completely different composition from the deep layer (the profundal) below the thermocline. This may be highly important in the evaluation, particularly for the oxygen and carbon dioxide situation and for iron and manganese (*I. Grim,* GWF *98,* 234–239 [1957]) (p. 60).

Meyer's drawing flask can be used as a makeshift for *taking deep samples from waters.* It consists of a thickwalled bottle (a champagne bottle, weighted below with a lead weight which is held in wire plaited round the bottle). A cord or cable is attached to a loop of the wire and the bottle lowered into the water. The sample cannot be used for determination of gaseous constituents, (CO_2, O_2, H_2S) however. For this, the Ruttner or Friedinger apparatus is necessary (see above). The *Schilling sampling bottle* (GWF *79,* 229 [1936]) is suitable for lesser depths. It carries a cork with two holes, through which pass a short and a long glass tube, the latter extending to the bottom of the bottle. A rubber tube is joined to the short glass tube and is kinked through a weight. The kink is removed by lifting the weight at the desired depth in the water.

Special "oxygen bottles" (p. 102) are employed for taking samples of water from the surface layers of waters for *oxygen determination.* Their stopper carries one glass tube reaching almost to the bottom of the bottle but emerging only slightly from the stopper; and a second glass tube, in contrast, just enters the flask but has a longer limb outside, through which air escapes from the bottle.

When this device is dipped into the water, a sample is obtained which has not or only scarcely entered into contact with the air. In the Spitta-Imhoff equipment, water is forced through two oxygen bottles.

This apparatus cannot be used with shallow waters, especially smaller brooks, since it cannot be immersed enough. *Höll's sampling apparatus for streams* is then used (Ber. d. I. V. Limnologie *12,* 360–372 [1955]). An exact determination of oxygen in small streams cannot be performed without such an apparatus. The same applies to CO_2 and H_2S determination.

Water samples should be taken from the thread of the current of small streams, and also from near the banks, where there is little water movement. If necessary, samples are taken from various depths and from both the concave and convex slopes of river bends.

See p. 219 for *sampling of wastewaters.*

The bottles should be filled to the brim so that when the stopper is inserted into the flask, no air remains in the bottle.

If there is a *danger of frost,* the stoppers should be briefly removed from time to time during transport or else the bottles might burst on freezing. A similar danger is present in the height of summer, when very cold water warms up in a firmly closed flask.

The sampling flasks must not be quite filled up in very cold weather-in so far as such a time of the year is at all suitable for sampling! (see below)

6. Sample Size

1 to 2 litres of water suffice for the minor hygiene-chemical well analysis (p. 27). For total analysis of a drinking and service water, at least 2 to 3 litres are needed, together with the samples for oxygen (p. 102), carbon dioxide (p. 86), silica (p. 142), phenols (p. 217), lead (p. 131) and possibly also for hydrogen sulphide (p. 119).

It is best to fill several litre flasks under exclusion of air so that there is always water available for investigation which has not been in contact with the air (p. 11). Many determinations are thereby facilitated (e.g., of ammonium, iron and manganese).

Far larger amounts of water are required for medicinal water analyses (p. 244).

7. Preservation of Water Samples

It is advisable to preserve water samples with a few ml of chloroform per litre for despatch or longer transport periods. 3 ml of dilute sulphuric acid per litre water are used to preserve samples for the $KMnO_4$-determination.

Unsuitable moment for examination: Water examination should, if possible, not be carried out during or just after a prolonged period of frost. This might give a too favourable impres-

sion since access of contaminated water to the ground water is hindered by frozen ground. Over-favourable results would be obtained also in summer after a long dry spell, since the dried ground retains the surface water and adsorbs impurities better.
One must wait 2 to 3 weeks before sampling a well that has been newly constructed, repaired or cleaned: during this time, it should be pumped out considerably.
A single investigation at an unsuitable moment may give an incorrect picture. A definitive judgment of a central water supply can thus be arrived at only after several examinations at different times of the year, especially after longer periods of rain and of dry weather.

8. Sent-in Samples

Before the samples are sent in, the investigating laboratory should hand out instructions for sampling according to the above exposition, and information for facilitating assessment of the results of the investigation should be obtained through a printed form of the type given in the appendix (p. 262). The samples should be brought to the laboratory by the quickest way (personal delivery) in times of hard frost or intense heat.
Only exceptionally should investigation of sent-in samples be tolerated and limited then only to components which change little on standing (e.g. chloride, nitrate, hardness).

9. Necessary Tests and Investigations on the Spot

Temperature: The temperature of the water and the atmosphere are measured at the time of sampling. The former must be measured only with a dry-bulb thermometer, and in the shade. If the thermometer bulb is not dried first, the latent heat of evaporation gives rise to too low a value. It is thus advisable to measure the air temperature first. In prolonged pumping tests, the water temperature is measured often–at least at the beginning and end of the test. Fluctuations of temperature indicate that surface water has penetrated the source or that deep water has been withdrawn which has a higher temperature (1° C per 33 metres, ca. 100 feet; see below).
The temperature of tap water and well water is best measured in a vessel of at least 1 litre content, through which the water is run for some minutes. The thermometers are graduated in 0.1° but there are special water thermometers, of range from 0° to 30° C and graduated in 0.02°.
The thermometer must not be removed from the water before it has been read. The mercury bulb of the "well thermometer" (firm of Spohr, Frankfurt/Main) remains in a beaker with the well water when the apparatus is drawn up.
Thumm's flow thermometer permits more accurate measurements on tap water. The so-called reversing thermometer of Negretti and Zambra is used for temperature measurements in dug wells, boring pipes and spring pipes (e.g., riser pipes from medicinal springs). With narrow boring pipes, a special thermometer is used which hangs in a drawing bucket.
Maximum thermometers are recommended with warm mineral springs and medicinal springs (thermal springs).

In the investigation of shallow waters and springs, the temperature is measured on the water which has been filled into a bucket. The reversing thermometer, mentioned above, is employed for studies in the deep zone of streams.

Accurate temperature measurement (to 0.1° C) is needed to evaluate the degree of saturation of oxygen of a water (p. 106).

Very accurate temperature measurements are important too in questions of water acquisition of ground water and medicinal springs. In central Europe, the average yearly temperature at 10 m. depth is 9.5° C. If ground water is cooler than this in winter or warmer than this in summer, it must be assumed that surface water has had access to it; surface water is warmer than ground water in summer and cooler in winter.

According to observations of the author, the average yearly temperature of ground water is appreciably less at higher altitudes, especially on exposed mountain summits (e.g., 6.5° C on the Wasserkuppe, W. Germany). Even on exposed hills of 200 to 300 m (600 to 1000 feet) above sea level in N. Germany, water temperatures of only 7–8° C are found even in summer.

In central Europe, the water temperature increases 1° C with every 33 metres of depth (geothermal gradient) (*R. Kampe,* Schrift.-Reihe des Deutschen Bäderverb., No. 8 [1952]).

The external character (e.g. odour and taste) must always be tested immediately after sampling since some odours, such as those of hydrogen sulphide and chlorine, can disappear during transport of the water sample. The appearance of the water sample can change rapidly on standing, too. Water which was clear and colourless on sampling can soon become coloured and cloudy through iron deposition, for example. These observations are important in assessing the processes of iron removal of a water.

As already mentioned, *the test for free carbon dioxide* must in any case be performed or begun on the sampling spot. The Heyer marble solubility test must likewise be begun there and shaking of the samples taken care of during longer transport. This applies also to determination of the saturation index (p. 96).

If the *smell of hydrogen sulphide* is noticed during sampling, it must be reacted immediately with cadmium acetate or zinc acetate for quantitative determination according to *K. E. Quentin* (p. 120).

The pH value should also be measured on the spot in more accurate determinations (p. 84). The oxygen tests must be begun (p. 101). If the uptake of lead from the pipes is suspected, a special sample is taken of water which has stood for a longer time in the pipe, and is treated with acetic acid (p. 131).

According to the author's experience, the most necessary field examination can be carried out on the sampling spot with the help of the water examination cabinet of the firm of Franz Bergmann KG of Berlin-Zehlendorf. Schilling's "rapid burette" (from the same firm) has also proved useful in field investigations.

B. General Tests in the Laboratory

1. Organoleptic Examination of Drinking Water

(to be carried out immediately on every drinking water)

Odour: When the tester himself takes the sample, the odour is noted immediately at the sampling spot, as mentioned. Some odours, e.g. that of hydrogen sulphide, soon vanish. For sent-in samples or more thorough study of samples which have been brought to the laboratory, it is best to take 100 to 150 ml of the water in an erlenmeyer flask of 200 to 300 ml capacity. It is closed with a glass stopper and warmed with a small flame or on the water-bath to about 40–60° C, swirled round, the stopper removed and the odour immediately tested. General odours can be termed:

earthy	fishy
peaty	putrefying
musty	faecal
mouldy	chemical (see below)
manure-like	aromatic

Chemical odours include:

hydrogen sulphide	ammonia
carbon dioxide	phenol
chlorine	chlorophenol ("chemists' shop smell")
mineral oil	xylenol
benzine	tar
benzene	

Typical smells sometimes arise in surface waters as a result of plankton growth, e.g.:

earthy odour, through Cyanophyceae (blue algae)
fishy odour through *Asterionella, Melosira* and other diatomaceous algae
cucumber odour, through *Synura uvella* (Flagellate)
blubber odour, through *Uroglena volvox* (Flagellate)
seaweed odour, through *Dinobryon sertularia* (Flagellate)

Waters from very deep wells, especially those rich in iron, often have a peculiar metallic or musty smell. If the smell of hydrogen sulphide is weak or in doubt, a parallel odour test is carried out on a sample treated with some cadmium acetate.

Odour intensities are:

1. very weak, detectable only by experienced testers, especially in industrial control of one and the same water, and only through comparison with an odourless reference sample.
2. weak but detectable by every tester through comparison with an odourless reference sample.
3. distinct, detectable by every tester.
4. strong, unpleasantly noticeable to every tester and impairing consumption.

The sense of smell is ten times more sensitive than the sense of taste.

Smokers have a less sensitive sense of smell than nonsmokers.

Threshold Odour Number: The threshold odour number (T.O) (in German, Geruchs-schwellenwert, GSW) is the degree by which the water under investigation must be diluted by odourless water until its smell is still just noticeable.

According to the American standard procedures [2], T.O. $= \frac{A + B}{A}$, where B = volume of the water used for dilution and A = the volume of the sample (ml).
(In the German procedure, GSW $= \frac{B}{A}$ i.e., one neglects A in comparison with the usually much larger B).
The determination is carried out in odourless erlenmeyer flasks. A dilution series is made up with absolutely odourless water which has been filtered through active charcoal. The existence of odour is tested on the series, starting with the least dilution. If the sample is diluted 1:10, the threshold odour number is 10; if 1:100, then it is 100 etc. (Standard procedures [1] B1/2 and *F. Malz* and *J. Gorlas,* Vom Wasser *34* [1967]).
The threshold odour concentration (in German, GSK = Geruchsschwellenkonzentration) is the concentration of a compound in mg/l at which its odour is still just detectable. Chlorophenol has an threshold odour concentration of 0,001 mg/l and mineral oils of 1 mg/l.
Further information may be found in *J. Holluta,* Geruch- und Geschmacksbeeinträchtigung des Trinkwassers, Ursachen und Bekämpfung (Impairment of odour and taste of drinking water, origins and control), GWF *101,* 1018–1023 and 1070–1076 [1960]. *H. Liebmann* has reviewed the specific interferences with odour and taste by various plankton algae [55].
Taste: Taste is tested by warming about 200 ml of the water in a covered beaker to 30–40° C and then tasting (caution with water suspected of being infected!).
Taste impressions are:

1. insipid ("flat") (soft waters)
2. salty (high content of sodium chloride)
3. bitter (high content of magnesium salts)
4. alkaline (high pH-value)
5. acidic (low pH-value)
6. astringent, inky (high iron or aluminium content)
7. metallic (high content of lead, copper or zinc: especially noticeable as "after-taste")
8. putrefying, offensive (polluted: care ir suspicion of infection)
9. peaty, musty (water from moorland areas)
   and, finally, the normal case:
10. no special features

Taste is classified as weak, distinct or strong.
"Water-tasters" in London take a piece of apple into the mouth between each test.
According to the Standard Methods [1965], the taste threshold of 5 should not be exceeded. This means that when the sample is diluted by a factor of 5, no taste should be noticeable.

2. Clarity and Transparency

(to be carried out immediately on every water sample)
Preliminary test on the spot: As has been said already, a water should be tested for clarity and transparency straight after sampling, or at least very soon afterwards; waters, rich in iron, soon become coloured and cloudy due to separation of iron. This is less to be feared where the water has been sampled with the rubber tube procedure. One always expects tur-

bid waters to yield a deposit and thus clarify on standing; delay before testing would thus give incorrect results.
For the test, the water is filled into cylinders of 50 cm length or Hehner cylinders of 100 ml content, and inspected in diffuse daylight against a white background. The following degrees of clarity are encountered:

white-clear	opalescent
clear	weakly turbid
almost clear	strongly turbid
weakly opalescent	

The transparency of the sample is compared with that of distilled water.
If the water is cloudy, the procedure in the next section is adopted.

3. Turbidity and Settleablity

The Snellen test writing (DIN No. 3.5) is laid beneath the cylinder or Hehner cylinder of 100 ml.
Water is removed until the writing can be read distinctly. The depth of the water column in cm gives the degree of turbidity (the writing standard is quoted).
The degree of turbidity can be measured also directly through reference turbidities. Best for this is an accurately standardised kieselguhr suspension.
Comparison is made in Hehner cylinders or in glass tubes of 50 cm length and 2.5 cm width, with flat bottoms, using the kieselguhr standard suspensions (Reagent No. 17). The amount of kieselguhr needed to prepare the reference suspension, in mg/l, yields the degree of turbidity. E.g. degree of turbidity of 5 = 5 mg/l kieselguhr with 50 cm depth of water column.
Mastic (resin) solutions are more suitable for very weak turbidities (*J. Rodier* and *M. Faivre-Duboz*, L'eau *44*, 263–271 [1957]).
The Jackson candle turbidimeter is used as standard method in the USA, especially for turbidities exceeding 25 units (Standard Methods, 12th Edition, New York, 1965, p. 312–316).
The step photometer has also been used for accurate turbidity measurements, and nowadays also the Elko II with frosted glass standards and neutral wedge.
The *Lange photoelectric colorimeter* or turbidimeter can be used to determine rapidly a turbidity, independently of the colour of the water once a calibration curve has been prepared. The Lange colorimeter can be used also as a field apparatus by running it from the accumulator of a vehicle.
Lange has now introduced a colorimeter with long cuvette tube for continuous measurements at the water works. It has a pen recorder and is reasonably priced (firm of Dr. Lange, Berlin-Zehlendorf). Turbidities can be recorded continuously through a photoelectric device. The new turbidimeter for continuous measurements (firm of Metrawatt Nuremberg and Munich) is based on the principle of Rayleigh scattering.

Assessment of Water Turbidity: The sudden appearance of turbidity in a ground water can mean that surface water has entered it.
The *settleability* of a turbid water is tested by comparing the turbidity value of the freshly

taken sample with that after 24 hours. It plays a part in the silting up of pipes through untreated water. If turbidities are intense, the undissolved matter and its loss on ignition ought to be determined quantitatively as on p. 120.

Depth of visibility: The transparency and colour of surface waters are determined with the help of the "sighting (or visibility) disc". This is a white porcelain disc of 20 cm diameter or length of side which is lowered into the water by means of a graduated surveyors' chain or cable. The depth of visibility is the length of the chain when the visibility disc just ceases to be visible. It can be fairly accurately determined by withdrawing and introducing the disc several times.

4. Colour of the Water

The colour of the water also must be recorded immediately after sampling since iron and manganese compounds may separate and, especially if they are in colloidal form, they impart a yellowish appearance to the water.

Rough field determination: The water under investigation is placed in a Hehner or other cylinder of 50 cm height and observed against a white background. The colour is noted as

colourless	yellowish-brown
weakly yellowish	brown
yellowish	yellowish-green
yellow	greenish

If the coloration is extremely weak, comparison is made with very pure distilled water (Reagent No. 10) in a second cylinder, or the sample is designated 'almost colourless'. Alternatively, the *Ostwald colour standard atlas* can be used and its specified colours compared with those of the 50 cm water columns.

Accurate determination: 50 cm cylinders are used for the accurate field determination. Cloudy waters must be centrifuged before the colour determination, or the reference solutions must be adjusted to the same turbidity values by adding kieselguhr (p. 18). The American standard procedure [2] distinguishes between "apparent" and "real" colour of a water which is turbid and coloured (i.e., before and after centrifuging).
Colour determinations under field conditions can be performed with colour glasses in the Hellige apparatus (firm of Hellige & Co., Freiburg i. Breisgau), if necessary by removing the opalescent glass disc from the comparator and using the Nessler tube auxiliary. Colour values from 5 to 50 mg/l Pt are read on two colour discs.
The potassium chloroplatinate reagent (Reagent No. 13) or caramel solution (Reagent No. 15) are used for more accurate colour comparison. Methyl orange solution is suitable for some waters (10 µg as unit, i.e., colour 1 = 0,01 mg/l of methyl orange) (see *W. Ohle,* Archiv f. Hydrobiologie p. 386 [1934]). The use of potassium chloroplatinate and copper sulphate (Reagent No. 16) together is recommended for yellow-green shades; the copper is expressed as mg Cu/l along with the platinum value in the result.
The colour values are reported as mg Pt/l or mg caramel/l for water columns of 50 cm height. Using a photometer, the colour value can be determined more accurately with the 460 nm filter.

II. General Remarks about Chemical Examination of Water

The analysis of water requires scrupulously clean work since the constituents to be determined are present in extraordinarily small amounts in the water and some of them could be introduced through apparatus which has not been especially well cleaned.
Attention must be therefore paid from the beginning to cleanliness of vessels for sampling and investigation. Special chemical cleaning and preparation of the glass apparatus is unavoidable for some determinations, such as that of potassium permanganate consumption (p. 37), of manganese (p. 110) and of trace elements (p. 21).
The chemicals and reagents used in investigation of water must be especially pure, since the ordinary commercial qualities can contain impurities which are then introduced into the water samples and interfere.

A. General Principles of Chemical Work

1. *Blank determinations* must always be carried out under the same conditions as in the analysis itself, in order to establish the *purity* of the reagents employed before they are used.
2. The full *effectiveness* of the reagents must be controlled from time to time with the help of reference solutions.
3. Constituents are separated in only a few of the determinations in the examination of water. The constituents present interfere with one another in numerous investigations.
   The possible interferences must thus be ascertained before each individual determination. The interferences which occur most often are listed specially in each section of the book and means of evading them are briefly described. It is therefore advisable to adhere to the sequence of procedures for drinking water, given for the first time in this book.
4. The samples of water must be investigated as soon as possible after being taken, at the latest after 10 to 12 hours, since they can undergo changes during standing. This applies particularly to polluted water, the permanganate consumption of which may change within a few hours. The sequence of procedures is consequently described in this book so that the inexperienced worker can see how he has to begin an examination of water and which determinations must be carried out immediately. It can be seen from the sequence which determinations must be made for the individual cases (minor investigation of a well, waterworks control, service water, etc.)

Some lime separates from hard waters, especially those containing iron, when the samples stand for a longer time in half-filled bottles; this leads to incorrect results. Some determinations need to be begun or made immediately after a special sampling procedure on the spot (e.g. oxygen, free carbon dioxide, hydrogen sulphide, ferrous compounds).

5. Many determinations cannot be carried out at all with samples which are sent and are often 2 or 3 days old or more (e.g., determination of CO_2,O_2,H_2S,Cl_2,pH-value and odour). Other determinations would be too inaccurate according to what was said under 4. ($KMnO_4$-consumption). In such cases, it is best to take fresh samples oneself or to arrange for them to be fetched. The chemist whose opinion is sought should, however, really take the samples himself.
6. Turbidities and precipitates should not necessarily be removed by filtration before the investigation; these may sometimes have adsorbed important components (e.g., lead, iron, manganese and arsenic compounds). Moreover, filter paper often contains traces of ammonia, phosphate and other materials which influence the accuracy of the analysis. It is thus better to centrifuge turbid waters. "Glass fibre papers" are preferable to paper filters (*G. Giebler* and *T. Kempf*, Z. Anal. Chem. *199*,23–24 [1964]).
7. The bottles containing the water must be shaken before a sample is removed. If iron and manganese are to be investigated, the bottom and sides of the bottle must be rubbed with a rubber policeman (p. 110), should these determinations not be carried out very soon after the sample is bottled.

1. Quotation of the Results

The results are quoted as prescribed in the standard procedures for examination of water [1]. Cations and anions are expressed as mg/l, e.g. mg Fe^{++} or mg SO_4^{--}/l; or as millival per litre (mval/l), the value of which is directly given by the titration volume of a 0.1 N reagent solution per 100 ml water sample. Undissociated materials are given as millimole per litre (mmole/l). Expression as mval/l permits the analysis to be checked; total cation and anion equivalents must be the same.

The quotations in parts per million (ppm), customary in the English-speaking countries, correspond to the German values in mg/l. The committee of experts of the WHO has resolved that, in future, analysis values shall be given as mg/kg in all countries.

Values of over 10 mg/l are generally quoted to the nearest unit; those between 10 and 1 mg/l, to the nearest tenth; and those below 1 mg/l, to the nearest hundredth. This enables the degree of accuracy of the method to be expressed.

Results in the analysis of medicinal springs are still quoted as mg/kg, however, since these waters, which contain much carbon dioxide, are difficult to pipette and the complete contents of a flask must be used for some determinations.

The statement "traces" or "zero" must be avoided at all costs. The lowest limiting concentration of the determination must be seen, e.g. iron: 0 ($<$ 0.01 mg Fe/l). Qualitative data are valueless in water analysis.

Phosphate and nitrogen values are still quoted as P and N in limnological work, but the form of the compound is mentioned, e.g. 10.0 mg/l $N(NO_3)$.

B. Procedures of Investigation

Knowledge of the gravimetric and volumetric methods of quantitative analysis is assumed. These are treated in detail in standard works, such as *W. D. Treadwell* "Kurzes Lehrbuch der analytischen Chemie", Vienna [1949]; *W. Autenrieth-A. Rojahn* "Quantitative chemische Analyse", Dresden [1939]; *W. Autenrieth-O. Keller,* Dresden [1958]; and *E. Brennecke, K. Fajans,* etc., "Neuere massanalytische Methoden", Stuttgart [1951].
A wholly new analytical method for examination of water is that using the ion-specific electrodes of the Orion Research Inc., Cambridge, Mass., USA. Calcium-, fluoride-, chloride- and other ions, in the amounts in which they occur in water, can be determined with the help of these electrodes.
Colorimetric methods are perhaps generally less well known than the gravimetric and volumetric procedures, since these have found little mention in the standard works quoted (except in *Autenrieth-Rojahn*) and have been somewhat neglected in higher education.
Since colorimetric methods are important in water analysis, their essentials are summarised here. Further details are to be found in: *Freund,* "Leitfaden der kolorimetrischen Methoden", Wetzlar [1938]; *G. Kortüm,* "Kolorimetrie und Spektralphotometrie", Springer, Berlin, (4th Edition 1964); *B. Lange,* "Kolorimetrische Analyse", Weinheim, 6th Edition [1964]; and *B. Kakáč,* "Handbuch der Kolorimetrie", Jena [1962 and 1963].

C. Colorimetric Working Techniques

1. General

Principle of colorimetry: Colorimetry depends on the fact that many compounds enter into colour reactions with certain reagents and that the intensity of the colour yielded is proportional to the amount of the compounds (Beer's law). Conversely, the intensity of the colour given in the reaction permits conclusions to be drawn about the amount of a compound (to be determined) in a solution. The amount of such a compound in a particular solution can be determined through comparison of the colour intensity, yielded after adding reagents to the sample of water, with that given by reference solutions of the compound which are treated in the same way. The comparison can be made in viewing glasses, colorimeter cylinders, cuvettes or photometers (p. 23 et seq).
The maximum colour intensity from reaction is not reached immediately after addition of the reagent, but usually after some minutes. The colorimetric comparison must therefore be made only at the right moment or after the right time interval specified in the individual determinations. This is especially important when comparing with standard colour discs, etc.
On the other hand, one should not wait too long before comparing, since the colour intensity may fall after a certain time (see below).

Colour intensity and concentration of compound are linearly related in only a limited region of concentration. Samples of higher concentration must thus be diluted; those of lower concentration must be concentrated (both before addition of reagent [s]). Further information can be found under the various determinations.

Advantages of colorimetry: Colorimetric procedures enable extremely small amounts to be determined, which would not be possible using most gravimetric and volumetric methods. Colorimetric methods are thus particularly suitable for investigation of water because of this sensitivity. It is generally unnecessary to increase the concentration of the compound under investigation by evaporation or precipitation, filtration and washing. Colorimetric analysis thus has the advantage of simplicity and rapidity of execution.

Disadvantages of colorimetry: Subjective errors sometimes impair the results of colorimetric analysis. Consequently, visual colorimetry has not always been accepted as an exact method. Colorimetric methods are more accurate nowadays since the apparatus has been greatly bettered.

Extensive tests have shown that 60% of male experimentors have impaired colour vision, yet only 10% of females.

Capacity of colour distinction of the experimentor: Visual colorimetry naturally requires the experimentor to possess normal colour vision. Every investigator has particular difficulties when the colour tint of the solution under examination does not correspond with that of the reference solution. Experience is needed to be able to distinguish depths of colour when the tints are not the same. The human eye can distinguish yellow-green tints best, and reds and blues less well. These subjective errors are largely eliminated when photometers are used in which light is converted into electric energy by means of photocells (p. 24).

2. Apparatus for Colorimetry

Hehner cylinders: Making up numerous reference solutions is avoided by using the 105 ml Hehner cylinders, which are graduated in ml. They have a tap at the side to enable the liquid to be drawn off. Any desired depth of column can therefore be set easily and rapidly. The cylinders are held some centimetres above a white background (opal glass, porcelain plate, paper etc.) for colour comparison. The cylinders must be made of colourless glass or at least both cylinders must be from the same glass; the tints of studied and reference solution may otherwise differ, which interferes.

Drum colorimeter (Meinck-Horn) (firm of Bergmann, Berlin-Zehlendorf and Hamburg): With this apparatus, the colours yielded in a viewing tube are compared with colour discs attached to a rotatable drum. This suffices for the minor drinking water analysis of individual wells, especially for examinations on the spot, e.g. of nitrite and ammonium ions, both of which suffer concentration changes in polluted waters. The inaccuracy of the method is partly compensated by the large layer thickness, which can be increased at will by using long cylinders; and by being able to carry out the determination immediately at the sampling site. The colour discs were newly worked out in 1966 with the help of the author.

Hellige comparator (Hellige & Co., Freiburg i. Br.): In the Hellige comparator, colour glas-

ses, mounted on a rotatable disc, are used for colour comparison. Using nine of these colour glasses, all the colorimetric determinations required for drinking water can be made. Any weak colour from the water itself can be compensated by use of the "compensation cuvette". The investigation is carried out in the "Hellige-Neo-Comparator" with cuvettes of 13 or 40 mm thickness; or in the Nessler tube comparator with a layer depth of 250 mm. The same colour disc can be used for the various layer thicknesses if a conversion factor is utilised. Dilution or evaporation of the water sample is thus virtually unnecessary; this, together with not having to make up reference solutions is a considerable advantage in field investigations. If the 40 mm cuvette is used instead of the 13 mm one, the value read from the colour disc is divided by 3; after replacement of the 13 mm cuvette by the 250 mm Nessler tube, the value is divided by 20.
The Hellige apparatus is well suited to investigations on the spot since it is easily transported and the colours of the reference glasses are fast. This applies particularly to determinations of ammonium ion and pH-value in the Nessler tube comparator, two determinations which have to be carried out quickly before any alterations occur.

None of these methods is adequate for an exact water analysis, even though, with large layer thicknesses on the spot, they are often more useful than the determinations on very thin layers with modern measuring instruments in the laboratory many hours or days later-e.g., with constituents, the amounts of which do not remain constant (pH, NH_4^+, NO_2^-)

Leitz photometer (Leifo) (firm of Leitz, Wetzlar): This is a filter photometer on which the extinction is read from a projection scale. The Leifo has cuvettes from 1 to 200 mm thick. The Leifo E is an electric photometer.

Elko II photometer (firm of Zeiss, Oberkochen): This is similar to the Leifo and has cuvettes up to 50 mm thickness.
Neo-Helcometer (firm of Hellige, Freiburg): This has extra tubes 300 mm long for very weak colour intensities.

Universal-Colorimeter of Lange: This has two photocells in a compensation circuit for measuring the photoelectric currents which arise. The colour of the water itself can be compensated and very accurate colorimetric determinations may be carried out with it, which are free of all subjective errors. The photoelectric currents are laid on to a Multiflex galvanometer, so that the determinations can be made very easily (firm of Lange, Berlin-Zehlendorf).
Nonius-Colorimeter: This is a new and larger instrument in this domain, from the Netherlands.
Spectrophotometers are used for scientific research (e.g., that of Zeiss, Oberkochen).

3. Working Rules for Colorimetry

1. The solutions studied must be absolutely clear. Even the finest turbidities simulate more intense colorations. In exact investigations, filtration is not advisable, since any filter fibres which pass through would interfere in the colorimetric comparison. Interference is, howewver, small when hardened filters and longer filtration times are used with one and the same filter. "Glass fibre papers" are more suitable than filter paper (p. 21)

Dissolved iron compounds are retained by all filters, even hardened, and escape determination.

It is thus better to centrifuge the turbid sample but it must be remembered that iron-containing and other compounds may be adsorbed by the suspended matter and thus evade determination.

Filtration or centrifugation must always be undertaken before carrying out the colour reaction, and best on both the reference solution and that under investigation.

2. The colour of the glass of the viewing cylinders is significant with very weak colorations. This can often be recognised by interchanging the cylinder used for the solution under study and that for comparison.
3. Almost every water is coloured, which can interfere in very accurate investigations. The reference solution is thus made up using not distilled water but the water of the sample from which the constituent to be determined has previously been abstracted.

   A more marked colour of the water studied can be compensated by the Walpole technique; the reference coloured solution or colour disc is viewed through the water sample to which no reagent(s) have been added.

   Dark, peaty waters must, however, be decolorised with active charcoal or sodium carbonate-sodium hydroxide; this is given in more detail in the individual precedures.
4. The solutions under examination and for reference should be at the same temperature, as far as possible; under no circumstances should the temperature difference exceed 3 to 5° C. (cf. *Krüss* "Colorimetrie", Hamburg [1909]).
5. The layer thicknesses of both solutions should be as nearly as possible the same and in no case should differ by more than 1:2 or 1:3. The best results are obtained when the concentrations and layer thicknesses of both solutions are the same.
6. Diffuse daylight and not direct sunlight should be used as illumination. Artificial light from ordinary electric light bulbs is unsuitable; the "daylight lamps" are better (fluorescence tubes).
7. The colours of studied and reference solutions must be compared several times and the mean value taken as the result.

   Female experimentors are appreciably more capable with colour comparison than are male (p. 23).
8. The colour intensity yielded after addition of reagent is not constant. It increases to a maximum, remains then constant for some time and then more or less rapidly diminishes. Colour comparison is made during the period of the maximum. It is thus clearly important that the solution under study and that for reference are treated with the reagent at the same time and the comparison undertaken within the time interval given for the maximum intensity in the individual determinations.
9. The concentration limits given for the individual determinations must be adhered to. This is facilitated by the preliminary tests.
10. The colorimetric methods are highly sensitive to all external influences. The contents of the viewing cylinders should consequently not be mixed by swirling round after placing the hand over the mouth of the vessel. Instead the "mixing bulbs", proposed by the author (see *K. Höll,* Die Chem. Fabrik [1934]), should be employed. Suitable too is the "stirring ring" as described by *R. Czensny* (Z. f. Fisch. *1,* 253–258 [1952]).

4. Work with Hehner Cylinders

Hehner cylinders (see above) have a capacity of 105 ml. 100 ml of water under investigation are always filled into the left-hand cylinder; if it is necessary to dilute the sample, 50 or 25 ml of distilled water are first introduced and then the sample. After the reagents have been added, the volume is made up to 105 ml. To the right-hand cylinder are added 80–90 ml of distilled water, standard solution and reagents, in that order, mixing after each addition; the volume is likewise made up to 105 ml with distilled water. The colours produced are compared above a white background in bright but diffuse daylight.
Calculation is as follows:
Example 1: The colour from the water sample is more intense than that from the standard; some of it must therefore be run off, say to a remaining volume of 80 ml. The reference solution contains 0.5 mg Fe/1 (i.e. 0.05 mg in 100 ml). The concentration, x, (mg/100 ml) of the solution under investigation is thus given by:

$$80 : 0.05 = 100 : x. \quad \text{Thus } x = \frac{0.05 \times 100}{80} = 0.062 \text{ mg/100 ml}$$

$$= 0.62 \text{ mg Fe/1}$$

Example 2: The standard solution yields a more intense colour than the water sample; some of the standard solution must therefore be run off, say to a remaining volume of 70 ml. The standard solution contains 0.1 mg NO_2 in 100 ml and thus 0.70 mg in 70 ml. This time,

$$100 : 0.1 = 70 : x. \quad \text{Thus } x = \frac{0.1 \times 70}{100} = 0.07 \text{ mg/100 ml}$$

$$= 0.70 \text{ mg } NO_2/1$$

III. Minor Drinking Water Analysis

A. Hygiene-chemical Investigation of Drinking Water

(Determination of the pollution indicators)

The real drinking water analysis (hygiene-chemical analysis) is treated in this chapter. Its purpose is to establish whether the water is without fault from a hygiene point of view (in combination with the bacteriological examination).

Smell, taste and external qualitites are investigated and, as soon as possible after taking (at latest, 12 to 24 hours later) the sample (kept cool), the following are determined:

1. Ammonium ion
2. Nitrite ion
3. Nitrate ion
4. Permanganate consumption
5. Chlorine number
6. Phosphate ion
7. pH and reaction.

The determinations below can be made after a longer delay if necessary:

8. Chloride ion
9. Sulphate ion
10. Urochrome
11. Hardness
12. Iron
13. Manganesc

(11.–13.: see Chapter E on "Major Drinking Water Analysis")

1. Ammonium Ion Determination

(To be carried out straight after sampling; otherwise, the sample must be conserved with chloroform. see p. 13)

Occurrence: Up to 0.1 mg NH_4/1 in pure water. Up to 1 mg/1 and more (especially in peaty water), depending on the geological area[1]. From 0.1 up to nearly 10 mg/1 in contaminated water, but usually then from 1 to 3 mg NH_4/1 (assessment of the ammonium ion content, p. 68). Ammonia may change rapidly into nitrite and nitrate in contaminated waters as a result of biogenic processes; the reverse can also take place.

1 As a result of the mineralogical composition of the soil. This includes all occurrences in soil which are related to natural processes in soil subjected to no contaminating influences. See the relevant section in "Assessment of the water".

Rough field determination: 20 ml of the water to be tested are treated in a test tube with 10 drops of Rochelle salt solution (Reagent No. 28) and 3 to 5 drops of Nessler's reagent. If, after mixing, a yellowish or brownish colour is seen against a white background, ammonium ion is present. Waters containing 0.2 to 1 mg NH_4/1 yield a yellowish colour; those with ca 2 mg/1, a yellowish-brown colour; and with 3 mg/1 and over, a brownish colour. Waters which are very rich in ammonium ion (5 mg/l and more) give a brownish turbidity at once. If the waters are themselves yellowish, a portion of the untreated water sample must be used for comparison in a second test tube. This is advisable in most cases; the colour of the water itself is often not recognisable in the sampling bottles especially as they often have opaque sides, hindering examination in transmitted light. Most errors arise through not paying attention to the colour of the water itself.
An estimate of the NH_4-content is possible by using reference solutions based on Reagent No. 4.

a) Colorimetric Determination of Ammonium Ion with Nessler's Reagent

2 ml of Rochelle salt solution (Reagent No. 28) are added to a 100 ml sample of the water and to 100 ml amounts of reference solutions. After mixing, 2 ml of Nessler's reagent (Reagent Nr. 21b) are added to each solution. After 5 to 10 minutes, the solutions are transferred to Nessler or Hehner cylinders (which are filled up to 105 ml-see p. 26). or to the Meinck-Horn colorimeter. The colours are compared with the standards (NH_4-standard, Reagent No. 4) made up at the same time, or in the last-mentioned case, with the colour scale on a drum. Using a column of liquid of 21,5 cm, amounts of ammonium ion from 0.05 to 2 mg/1 can be determined. If values below 0.2 mg/1 are found, the determination is repeated using 4 ml of Nessler's reagent.

Should more than 1.5 mg NH_4/1 be found, the water sample is diluted with genuinely ammonia-free water. If the amount found exceeds 5 mg NH_4/1, the determination is replaced by a titrimetric method. 500 ml of water are distilled with 1 to 2 g of magnesium oxide and the evolved ammonia is absorbed in 10 ml of 0.1 N sulphuric acid. 1 ml of 0.1 N H_2SO_4 consumed = 1.8 mg NH_4.

Hellige comparator: Comparison in the Hellige comparator is especially suitable with the coloured and cloudy waters which often occur. Interference through this colour and turbidity can be compensated if not too excessive, but at least 0.2 mg/1 must be present in the 40 ml cuvette. Amounts down to 0.02 mg NH_4/1 can be accurately determined in the Nessler tube comparator (250 mm tube). Great attention must then be paid to the water colour itself (compensation or clarification, see under "Interferences"). The 13 mm cuvette is used with waters very rich in ammonia. The Hellige Nessler reagent No. 5013 must be used.
Nessler's reagent is stable for only a short time; this is not generally known.

Accurate, objective results are obtained by photometric determination at 425 nm in instruments such as the Elko or Leifo.

Interferences in the ammonium ion determination: Before every measurement, one should ensure by regarding the viewing cylinders that the water is not itself coloured; if it is coloured, this gives the impression that ammonia is present or that more ammonia is there than in reality.
Interferences through colour or turbidity of the water can be eliminated as follows: 2 ml of

sodium carbonate/sodium hydroxide solution (Reagent No. 29) are added to 130 ml of water sample and the mixture shaken. A few drops of 30% zinc sulphate solution are then added and the mixture is left for several hours. By tapping and turning the vessel, flakes of precipitate adhering to the walls and the surface can be dislodged so that they settle to the bottom. The clear water can then be decanted or siphoned off. Filtering through paper is best avoided since this sometimes releases ammonium ion. Filtering through paper which has been washed with hot water is satisfactory; through glass fibre papers still better (p. 21).
With *waters containing humic substances,* it is best to add 0.5 g sodium hydroxide to 200 ml water, shake with 2 g of active charcoal (Reagent No. 1) and then to filter through cotton wool or a filter which has been washed with hot water.
According to the author's experience, the colour of peaty water due to humic acids is intensified by alkali. If one does not wish to remove the humic substances by the above method, the water colour must be compensated by using it (plus some Rochelle salt and alkali) in the reference tube or cuvette. Standard solutions which have been made up with distilled water yield too high values otherwise.
Waters of over 25° hardness often yield a turbidity with Rochelle salt. In such cases, 2 ml of sodium carbonate/sodium hydroxide solution (Reagent No. 29) are used per 200 ml water instead of the Rochelle salt solution. The clear, softened water after the precipitate has settled is siphoned off. 101 ml are treated directly with 2 ml of Nessler's reagent and the procedure continued as above.
Ammonia can be determined only after distillation in the case of brines and other mineral waters, rich in salts (especially iron). Sodium hydroxide or a borate buffer (10.6 g boric acid + 75.0 ml of 1N NaOH, made up to 1 litre) is added before destillation to bring the pH to 9, and the distillate is collected in 4% boric acid solution. 1 ml of 0.05N H_2SO_4 used for titration is equivalent to 0.90 mg of NH_4. One can distil as usual into 10.0 ml of 0.1N HCl; the end of the condenser must then extend into the acid. 1 ml of 0.1N HCl = 1.8 mg NH_4.
The sensitivity of the Nessler reaction is reduced with *soft waters of low salt content.* This is also the case if too much potassium iodide is used in preparing the Nessler reagent (*Stall,* Z. Anal. Chem. [1937]). To such waters are accordingly added 0.1 g of common salt or 1 g of phosphate mixture (0.9 g sec. and 0.1 g prim. sodium phosphate) per 100 ml of sample or distillate.
According to *R. Wagner* (Vom Wasser *36* [1970]), the turbidity which occurs when the Nessler reagent itself is added can be prevented by adding hydroxyethylcellulose (e.g., Tylose H 20 of the firm of Kalle AG, Wiesbaden, W. Germany).
Hydrogen sulphide gives a colour similar to that from ammonium ion which, in contrast to the latter, does not vanish on adding sulphuric acid. If H_2S is present, as recognisable by its smell, some cadmium or zinc acetate solution is then added, precipitating cadmium or zinc sulphide. The water may be tested for ammonium ion after the precipitate has settled. Small amounts of H_2S are removed from the water beforehand by longer shaking.
Very small amounts of ammonium ion can be determined as indophenol blue through reaction with phenolate/hypochlorite (Berthelot's reaction) (*R. Wagner* Vom Wasser [1970]). This method may also be automatised.
In this procedure, 5 ml of 2% sodium polyphosphate solution, 5 ml of a sodium salicyla-

te/sodium nitroprusside solution (8.5 g and 60 mg, respectively, in 100 ml of water) and 5 ml of 0.3N NaOH are added to 25 ml of sample. After having mixed well, 5 ml of sodium dichlorocyanurate solution (100 mg in 100 ml of distilled water) are added. The solution is allowed to stand for 1 hour and is then made up to 50 ml and evaluated at 690 nm. Instead of one hour at room temperature, the solution can be warmed for 15 min at 37° C on the water-bath; more intense colorations then result. The standard ammonium ion solutions for comparison are treated in the same way (Reagent No. 4).
Calculation: Results are given as mg NH_4/l.

$$NH_3\text{-value} \times 1.059 = NH_4\text{-value}$$
$$\text{mg } NH_3/l = \text{mg } NH_4/l \times 0.944$$
$$\text{mg } N/l = \text{mg } NH_4/l \times 0.777$$

b) Determination of Protein Nitrogen as Ammonium Ion

(Carried out only in special cases of drinking water pollution, on wastewaters and receiving waters, then soon after sampling.)
100 ml of water are acidified with 2 ml of 1:1 sulphuric acid and heated for 15 min on the boiling water bath with 10 ml of 10% potassium persulphate solution (the persulphate must obviously be completely free of ammonium salts; this must be checked). The solution is made alkaline with NaOH after cooling. To it are then added 2 ml of Rochelle salt solution (Reagent No. 28) and 2 ml of Nessler's reagent (Reagent No. 21b). The mixture is filled into Hehner cylinders, made up to 105 ml, swirled round and evaluated; or made up to 100 ml and the light absorption measured at 425nm.
The coloration I which is developed is compared with that obtained in the normal ammonium ion determination, II (p. 28). If I is more intensive than II, standard NH_4-solution (Reagent No. 4) is added to II until the intensities are identical. The amount of protein nitrogen, expressed as ammonium ion, is calculated from the added amount of standard NH_4-solution.

c) Determination of Albuminoid Nitrogen as Ammonium Ion

(Performed only in particular cases, e.g. on effluents)
200 ml of water are rendered alkaline with magnesium oxide and the ammonia thus set free is expelled by boiling. The residue from distilling about 150 ml is treated with 50 ml of alkaline permanganate solution (18 g of $KMnO_4$ in 600 ml water, plus 400 ml of 50% KOH). 100 ml of this mixture is distilled into 20 ml of 0.2N sulphuric acid.
Unused 0.2N sulphuric acid is back titrated with 0.2N NaOH, using Congo red as indicator. 1 ml of consumed 0.2N H_2SO_4 is equivalent to 18 mg/l of albuminoid-NH_4 when a 200 ml sample of water is used. If the albuminoid nitrogen content is smaller, the colorimetric method is employed after adding alkali to the second distillate, as given for protein ammonia determination.

2. Nitrite Ion Determination

(to be carried out on every drinking water sample as soon as possible after sampling; otherwise it must be conserved with chloroform–see p. 13)
Occurrence: No detectable amounts in pure water. From 0.1 to 1.0 mg NO_2/l (in peaty water), depending on the soil. Sometimes of few tenths of a mg/l in rainwater. Contaminated waters may contain 0.1 to 2 mg NO_2/l, generally 0.1 to 0.5. (Assessment, p. 69)

Rough test: 5 drops of phosphoric acid[2] are added to a 20 ml sample of water in a test tube, followed by 10 drops of zinc iodide/starch solution[3] after having shaken. Nitrous acid is shown by the appearance of a blue coloration within some minutes. If the amount exceeds 0.6 mg/l, the blue colour appears immediately. Coloration comes after some 10 seconds if 0.4 to 0.5 mg NO_2/l is present; after 30 seconds with 0.25 mg/l; after 3 minutes with 0.15 mg/l; and after 8 minutes with 0.06 mg NO_2/l.
The test must not be carried out in direct sunlight; this yields a blue colour even in absence of nitrite. The solution is best kept away from the light while waiting.
Field method (test with Riegler's reagent): The test with Riegler's reagent would be especially recommendable in the field since there is no fear of interferences as with the zinc iodide/starch reagent and it is very sensitive.
1 ml of Riegler's reagent (Reagent No. 26) and 2 drops of hydrochloric acid[4] are added to 20 ml of water sample in a test tube, the mixture shaken and then 1 ml of ammonia[5] is added. The appearance of a reddish colour within a few minutes indicates the presence of nitrite ion.

> 2 mg NO_2/l → wine (Burgundy) red colour immediately
1 to 2 mg NO_2/l → raspberry red colour within 10 sec
0.5 to 1 mg NO_2/l → raspberry red colour within 30 sec
0.1 to 0.5 mg NO_2/l → pink colour within 10 sec
0.01 to 0.1 mg NO_2/l → pale pink colour within 10 sec

Larger amounts of iron and manganese interfere and must be removed with sodium carbonate/sodium hydroxide. Free chlorine also interferes. If necessary, the sample can be taken before the chlorination plant is reached.

a) *Colorimetric Determination of Nitrite according to Zambelli*

100 ml of water are treated successively with 2 ml of sulphanilic acid solution (Reagent No. 31) and 2 ml of naphthylamine solution (Reagent No. 21a), mixing after each addition. 5 ml of 25% ammonia are then added immediately. The pink or red colour yielded is compared in Hehner cylinders or a colorimeter with those from standard solutions (Reagent No. 23) which are made up at the same time. Amounts down to 0.03 mg NO_2/1 can be determined.
Photometric determination at 530 nm is also possible.

Humic substances in the water may *interfere* through the colour which they impart. To prevent this, 200 ml of water are treated with 3 ml of NaOH solution and shaken with 2 g of

2, 3, 4, 5 composition as in the German Pharmacopoeia (also denoted with an asterisk)

active charcoal (Reagent No. 1). The mixture is filtered, the first 20 ml of filtrate are discarded and 100 ml of subsequent filtrate are taken for determination as above.
Active charcoal eliminates also interference by free chlorine. If the water is being examined in the waterworks, the chlorination process can be turned off for a time before testing and then, after a suitable run-through interval, the untreated water can be investigated.
Very hard waters are recognised through the turbidity yielded when the ammonia is added. The determination is then repeated after adding 5% tartaric acid solution to the sample.

Calculation: Values are expressed in mg $NO_2/1$. N_2O_3-values are converted to NO_2-values by multiplication by 1.211.
Nitrite determination according to Griess-Ilosvay: The method of *Griess-Ilosvay* is even better than the above method since it does not require the use of ammonia.
2 ml of sulphanilic acid solution (3 g/l in 5N acetic acid) and 2 ml of α-naphthylamine solution (1 g/l in 5N acetic acid) are added to a 100 ml water sample. The colour which develops is compared after 10 minutes in the Nessler tube comparator with the colour disc of the Hellige comparator or estimated after 2 hours in a photometer. If free chlorine is present (in chlorinated water) the naphthylamine is added first and the sulphanilic acid 5 minutes later; or the water sample is treated with alkali, shaken with active charcoal (see above) and filtered to give a chlorine-free sample which is subsequently used.
Interferences: see below

b) Indole Method (for small nitrite ion amounts)

To 100 ml water are added 0.5 ml of 1:3 sulphuric acid and, after mixing, 0.2 ml of indole solution (0.2 mg of indole in 150 ml alcohol). A pink to violet colour gradually develops which, after 10 minutes, is compared in a viewing cylinder with the colour scale of the Meinck-Horn colorimeter (firm of Franz Bergmann KG, Berlin-Zehlendorf and Hamburg). The indole solution attains its full activity only after 8 days from the time of preparation. Nitrite amounts down to 0.02 mg/l can be determined.
This determination should best be carried out on the spot where sampling is made, in particular as a qualitative preliminary test using viewing cylinders of 100 cm lenght. Assessment of the water is much easier when a negative result is found with cylinders of this size; moreover, one is then sure that no nitrite has been lost during transport (to 0.002 mg/l). A more accurate colour comparison can be undertaken in Hehner cylinders using reference nitrite solutions (Reagent No. 23) made up at the same time as the sample.

Interferences: Free chlorine simulates nitrite. Chlorination must thus be switched off before sampling at the waterworks. Hydrogen sulphide also yields a pink colour similar to that from nitrite. This interference can be eliminated by aerating the water energetically before adding the reagents.
Waters containing humins are treated with alkali, then shaken with active charcoal (Reagent No. 1), filtered and acidified.

3. Nitrate Ion Determination

(to be carried out on every drinking water sample as soon as possible after sampling; otherwise it must be conserved with chloroform–see p. 13).

Occurrence: Usually only 1 to 5 mg NO_3/l in pure water. Up to 20 mg/l, rarely more, due to soil conditions. 200 mg/l and more, usually 50 to 150 mg/l, in contaminated water. See p. 69 for assessment.

Rough field determination (brucine test): 3 ml of conc., nitrate-free sulphuric acid are added with cooling to 1 ml of water sample in a test tube. 10 to 20 mg of brucine are added to the fully cooled solution and the mixture shaken. A cherry-red colour, gradually turning orange, appears at once if more than 100 mg NO_3/l are present. The solution turns pink red with 50 mg/l and light pink with 10 mg NO_3/l. The pink colour appears only after some minutes when 1 mg NO_3/l is present.

Interferences: Nitrites do not interfere in presence of the large excess of acid. Higher chloride contents (above 1000 mg/l) interfere and must be precipitated with silver sulphate (avoid a large excess of this reagent).
Too much brucine leads to formation of a pink colour even in absence of nitrate. The amount "a pinch" (eine Messerspitze voll), quoted in the former and new standard procedures for examination of water [1] in W. Germany, is too unprecise. On no account should more than 30 to 50 mg be used.
Free chlorine interferes and must be removed (see below).
The Swiss foodstuff regulations prescribe as rough tests diphenylamine/sulphuric acid for smaller nitrate concentrations (up to 3 mg/l) and brucine/sulphuric acid for larger amounts.

a) Colorimetric Determination of Nitrate according to Noll

To 10 ml of water sample and 10 ml of standard nitrate solution (Reagent No. 22) in two 100 ml erlenmeyer flasks is added exactly 0.6 ml of 5% brucine solution in acetic acid. After mixing, 20 ml of conc. sulphuric acid (d = at least 1.84 and free of nitrate–see below[6] are added to each solution in a single pouring movement. After 30 seconds, the mixtures are transferred to Hehner cylinders or viewing tubes containing 73 ml of distilled water (Reagent No. 10). To prevent loss, part of the mixture is poured back into the flasks. After mixing and volume contraction, the total volume of each mixture is 100 ml. After 4 to 6 minutes, the colours are compared against a white background (see below). Any air bubbles present must all be removed by tapping the cylinders gently before comparison; they would interfere with it and lead to incorrect results.

The determination can be made in a colorimeter[7], e.g. that of Meinck-Horn[8]. The colours are compared with those on the drum. The range of measurement is from 1 to 20 mg NO_3/1. This saves preparation of the reference solutions and also the sulphuric acid needed for them.

6 A blank is important! If necessary, nitrate-containing H_2SO_4 can be freed from nitrate by heating for 10 minutes after having added 5 g/l of common salt.
7 *With 0.5 ml of brucine solution in this case.*
8 *Firm of Bergmann & Altmann, Berlin-Zehlendorf.*

Only 5 ml samples are taken from waters containing over 20 mg/1 and still less from contaminated waters.

Conversion calculations:

$$1 \text{ mg } NO_3 = 0.871 \text{ mg } N_2O_5$$
$$N_2O_5\text{-value} \times 1.148 = NO_3\text{-value}$$

Interferences with the nitrate determination:
Free chlorine and hypochlorite interfere with the brucine method as with the diphenylamine method and must be removed (active charcoal after adding alkali) (see below).

1 ml of sodium hydroxide solution and then active charcoal (Reagent No. 1) are added to turbid waters or those containing *humic substances.* After shaking, the solution is filtered through a filter that has been previously washed through with hot water to remove any nitrate possibly contained in it. The pH of the water must be adjusted to ca. 9 by addition of alkali. Clarification and decoloration can be largely accomplished also by adding 2 ml of sodium carbonate/sodium hydroxide (Reagent No. 29) to every 200 ml of water, or 2 ml of a 10% solution of aluminium sulphate per 200 ml of water.

Nitrite-Ion: Nitrite ion in amounts exceeding 0.5 mg/l must be removed. This may be done by acidifying with dilute H_2SO_4 and adding 0.2 g of urea to a 10 ml water sample; by leaving for 24 hours; or by adding 1 drop of a 5% sodium azide solution to the water which has been acidified with acetic acid.

Chloride-Ion: A very high content of chloride ion (over 500 mg Cl/l) must be precipitated with silver sulphate, taking care to avoid using a larger excess of this reagent. If the precipitated silver chloride coagulates markedly, it must be well washed and then filtered. The filtrate is concentrated on the water-bath.

Iron: Waters containing much iron (more than 3 mg/l) are filtered with some zinc oxide; the first part of the filtrate is discarded.
Particles of mud or sludge interfere in the brucine method through the dark colour they impart to the reaction mixture. They must be removed before the determination, by filtration (use nitrate-free filters) or allowing to settle.

b) New Specific Nitrate Determination with Sodium Salicylate

The salicylate method (*R. Müller* and *O. Widemann,* Vom Wasser *22* [1955]) is used nowadays since the brucine method is highly susceptible to interferences and is not particularly accurate. It gives slightly lower values than the brucine method, however.

A mixture of 20 ml of water sample and 1 ml of a freshly prepared 0.5% sodium salicylate solution is evaporated down on the water-bath. After cooling, 2 ml of sulphuric acid (d = 1.84) are added. 10 minutes later, the solution is washed with 25 ml distilled water into a colorimeter cylinder or cuvette and 7 ml of alkali reagent (30% NaOH + 0.6% Rochelle salt) are added. After a further 10 minute period, the solution is made up to 50 ml with distilled water and the yellow colour yielded compared with those from nitrate reference solutions (Reagent No. 22) treated identically, or evaluated photometrically at 420 nm.

Interferences: Waters containing humic substances must be clarified with aluminium sulphate (see above). Chloride amounts exceeding 200 mg Cl/l must be precipitated with silver sulphate, avoiding excess of the reagent. The filtrate is treated further as above. *Wagner* has found stannic sulphate to be a masking agent for chloride ions (Vom Wasser *36* [1970]). Larger amounts of iron (concentrations from 1 to 5 mg Fe/l) must be removed by shaking with zinc oxide, as in the chloride determination (p. 42). Nitrites do not interfere unless present in larger amounts and then only slightly. If the amounts present exceed 1 to 2 mg NO_2/l, it is an advantage to add 0.05 g of ammonium sulphate before evaporating down in order to prevent the yellow coloration through nitrite.
Acidic waters must be neutralised.

Reduction of Nitrate with Devarda's Alloy
Mineral and brackish waters and *those containing humins but little nitrite* are best reduced by distilling with Devarda's alloy (1 g of Devarda's alloy and 0.5 g of magnesium oxide per 20 to 100 ml of water sample). The distillate is collected in 3% borate solution in a standard flask and the volume made up to 100 ml. The ammonia in the distillate is determined with Nessler's reagent (p. 28) and the amount calculated as nitrate; 1 mg is equivalent to 3.44 mg of NO_3; and 1 mg NH_3 to 3.64 mg of NO_3.

c) Colorimetric Determination of Very Small Amounts of Nitrate according to Kuisel
(Helv. Chim. Acta *18*, 195 [1934])

A 25 to 100 ml sample of water is evaporated to dryness in a glass dish on the water-bath after having added an amount of 0.1N sulphuric acid 0.5 ml less than the amount required for the determination of the carbonate hardness. After cooling, 1 ml of phenoldisulphonic acid[9] is added and the dish replaced for a minute on the boiling water-bath. After cooling again, the contents of the dish are washed with distilled water into colorimeter cylinders or cuvettes and 10 ml of 10% ammonia added. The solution and reference solutions which are similarly treated and used for colour comparison are all made up to the same volume. The accuracy is ca 0.1 to 0.2 mg NO_3/l. Chloride amounts exceeding 50 mg Cl/l interfere.
Atkins' method, using diphenylbenzidine and developed for µg amounts in sea water, is another possible procedure for waters of low nitrate content (J. Marine Biol. Ass. *18* [1932]). To 2 ml of water are added 6 ml of sulphuric acid[9] and, after cooling, 1.5 ml of diphenylbenzidine reagent (20 mg of diphenylbenzidine in 100 ml of H_2SO_4). After 24 hours, the blue colour yielded is compared with that from standard solutions.
Reduction with zinc dust can be recommended for determining small amounts of nitrate accurately. 0.5 ml of sulphuric acid (1 + 3) and then some zinc dust is added to 100 ml of water. As soon as the mixture has become clear, some indole is added. The nitrite which is formed on reduction gives a red-violet colour which can be evaluated colorimetrically according to the instructions on p. 32.

9 The phenoldisulphonic acid is prepared by heating 3 g of phenol with 37 g conc. H_2SO_4 for a longer period on the water-bath (see also *F. Gessner*, Int. Rev. ges. Hydrob. *43* [1943]).

d) Hellige Method

Nitrate determination with the Hellige apparatus (from the firm of Hellige & Co., Freiburg i. Br., W. Germany) is very convenient.
According to the instructions, 10 ml of water, 1 ml of brucine/sulphuric acid (see below) and 20 ml of pure conc. H_2SO_4 (d = at least 1.84) are mixed in a 50 to 100 ml flask. After 5 minutes, the mixture is cooled in tap water and the colour compared in the 13 mm cuvette with the colour disc 3060/63 in the comparator.
The range of application is between 4 and 20 mg NO_3/1. The Hellige apparatus is especially suitable for field laboratories since large amounts of sulphuric acid are not required for reference solutions. Another advantage is that the colour of the samples themselves can be compensated with the help of the compensation cuvette (2.5 ml water + 5.25 ml conc. H_2SO_4). Smaller amounts of nitrate can be determined with the phenoldisulphonic acid reagent, using the Nessler tube comparator of the same firm (see above).
The brucine/sulphuric acid reagent is prepared in a 50 ml erlenmeyer flask which has been moistened with distilled water. The small amount of water facilitates solution of the 0.2 g of brucine in the added 10 ml of sulphuric acid. The reagent is not stable and must be prepared freshly each day.
High chloride amounts *interfere* (see above).
Calculation: The values for mg N_2O_5/1 found from the Hellige method are converted to mg NO_3/1 by multiplication by 1.148.

e) Nitron Method

The nitron method is highly suitable for waters and wastewaters which are rich in nitrate (over 200 mg NO_3/1). A suitable water sample is acidified with sulphuric acid, 10 ml of nitron solution are added (10% solution of Merck product in 5% acetic acid) and the solution evaporated on the water-bath to a small volume. After leaving for 2 hours in a refrigerator, the separated nitron nitrate crystals are filtered off, washed with ice-water and finally weighed after drying at 105° C. Chlorides interfere.
1 mg nitron nitrate = 0.1653 mg NO_3.

f) Titrimetric Determination after Reduction to Ammonium

500 to 1000 ml of water plus 33% NaOH and 1 g of Devarda's alloy are distilled into a 3% boric acid solution until about 100 ml have passed over. The distillate is titrated with 0.05N H_2SO_4 to the colour change of the added mixed indicator (blank titration with 50 ml of boric acid solution). 1 ml of 0.05N H_2SO_4 is equivalent to 3.1 mg of NO_3.
If very little acid is needed, the distillate can be used for ammonium determination with Nessler's reagent (see above), usually after appropriately diluting.

g) Serial Analyses with Ion-specific Electrodes

The ion-specific electrodes of the Orion Research Inc. of Cambridge, USA enable considerable time to be saved in serial determinations of nitrate ion (see also p. 22).

4. Potassium Permanganate Consumption
(determination of organic matter in water)

to be carried out on every drinking water as soon as possible after sampling, otherwise the sample must be conserved–see p. 13).

Values encountered: Usually a few up to 12 mg/l in pure ground waters, most values falling between 3 and 8 mg $KMnO_4$/l. Up to 300 mg/l, depending on the soil[9] (e.g. peaty waters). 20 to 150 mg $KMnO_4$/l in contaminated waters.

Pure surface waters show mostly 10 to 30 mg $KMnO_4$/l. Alpine lakes and oligotrophic lakes (low in nutrient) of the plains have only 2 to 10 mg/l.

See p. 71 for assessment.

a) Rough Field Determination

1. *Colour of the water:* the water colour shows already if much or little organic matter is present. A 50 cm column of water is introduced into a 200 ml glass cylinder or viewing tube with flat bottom. The cylinder is held a little above a white background and the colour observed through the column length, comparing with a column of distilled water in another cylinder of the same type.

The following correlations can be cited:

Water with

1 to 10 mg/l	$KMnO_4$-consumption	colourless	
10 to 15 mg/l	$KMnO_4$-consumption	just detectable yellowish colour compared with distilled water	
15 to 20 mg/l	$KMnO_4$-consumption	yellowish; detectable even without comparison with distilled water	
30 to 40 mg/l	$KMnO_4$-consumption	yellowish	
60 to 80 mg/l	$KMnO_4$-consumption	markedly yellow (like white wine)	
80 to 100 mg/l	$KMnO_4$-consumption	yellow-brown	waters containing humic substances
120 to 150 mg/l	$KMnO_4$-consumption	brown	
200 to 300 mg/l	$KMnO_4$-consumption	red-brown	

When over 30 mg $KMnO_4$/l are consumed, a yellowish colour is recognisable in the water in the colourless sample bottles when viewed by transmitted light; 60 mg/l manifest a yellow colour when the bottle is regarded from the side.

These tests must be carried out only on freshly sampled water. During standing, colloidal or precipitated iron may bring about appreciable changes.

2. *Shaking test:* Freshly sampled, contaminated waters on shaking for a longer time in a test tube often yield froth which persists several seconds; peaty waters behave similarly. Added methylene blue makes this more easily visible.

3. *Test tube experiment with permanganate:* To 10 ml of water in a test tube are added 5 drops of dilute H_2SO_4 and 3 drops of 0.01N $KMnO_4$ solution. After shaking, the mixture is left for 5 minutes and then carefully warmed to boiling. The test tube may be held with a paper "collar" during this. It must be kept moving continuously since it tends to bump readily and boiling liquid then spurts out.

Waters with permanganate numbers exceeding 30 mg/l show decoloration within a few minutes at ambient temperature. 5 to 6 drops of permanganate solution are decolorised if permanganate consumption is over 50 mg/l. Waters with 20 to 30 mg/l consumption of permanganate are decolorised on boiling. Decoloration takes place when the boiled solution is allowed to stand a few minutes, if the permanganate number is 12 to 20 mg/l. With permanganate numbers below 12 mg/l, no decoloration is observed, even after boiling and longer standing.

b) Determination of the Permanganate Number according to Kubel
(oxidisability)
(with chloride contents of up to 300 mg Cl/l; see p. 41 for preliminary test for chloride)

Vessel cleansing: Specially cleaned 300 ml erlenmeyer flasks are used for this determination. The cleansing procedure involves boiling for a longer period with a ca 0.001N $KMnO_4$ solution, acidified with sulphuric acid[10]. The vessel is then carefully rinsed with distilled water (Reagent No. 10). One must take care that no spots of manganese dioxide have formed on the walls; such flasks cannot then be used for the determination. A reserve of 3 to 6 clean flasks should always be kept in continuous investigations. The flask opening is covered with an inverted beaker or cold finger. Automatic burettes are recommended for the determination when it is done in fixed laboratories. Burettes with a rubber exit tube carrying a pinchcock must be avoided.

5 ml of sulphuric acid (1 + 3 vol)[11] are added to 100 ml water and the mixture is brought to the boil. To prevent bumping (to which the permanganate solution has a great tendency) some glass beads or ignited pumice stones are added or, even better, a glass rod, hollowed from underneath. This greatly facilitates the determination.
15 ml of 0.01N $KMnO_4$ are added rapidly to the still boiling solution and the flask opening is closed with a funnel or cold finger. The mixture is again brought to the boil and steady (but not too vigorous) boiling is maintained for 10 minutes (measured with an alarm clock or stop watch).
Should the solution nearly or completely decolorise during this period, another 15 ml amount of the 0.01N $KMnO_4$ solution is added; this is allowed for in the result. Only a 25 or 50 ml sample of yellow waters containing humins is taken; it is diluted to 100 ml with distilled water (Reagent No. 10). 15 ml of 0.1N $KMnO_4$ solution are used with 50 ml samples of wastewaters.
After the 10 minutes boiling have elapsed, 15 ml of 0.01N oxalic acid are added (caution! boiling is delayed!) and boiling is continued until decoloration is complete. The boiling solution is then titrated with 0.01N $KMnO_4$ to a just visible pink. The hot flask is thereby held by a rolled paper "collar" or a leather strip round its neck. The titration is performed against a white background in good daylight. Duplicate determinations must be made.
If more than 10 ml are consumed, the determination is repeated with two 15 ml amounts of 0.01N $KMnO_4$ solution, best after diluting 1:1.

Determination of the titre of the $KMnO_4$ solution: The factor of the 0.01N $KMnO_4$ solution is determined from time to time in the pale pink end-solution from a titration. This is done by adding again 15 ml of 0.01N oxalic acid and retitrating the boiling solution with the permanganate solution to pale pink (see below for the calculation). The 0.01N oxalic acid must also be checked occasionally. The titre of the 0.01N $KMnO_4$ solution must not fall too far or incorrect results are obtained even when the factor is employed.
Not more than three samples should be heated at once in *serial investigations.* Three further flasks can be filled in the meantime. All three samples must be boiled simultaneously; if necessary, one must regulate this by adjusting the burner flame or cautiously heating over an open flame. After 10 minutes boiling, the flasks are placed in cold water for a short time (using lead rings round the flask necks).
Any hydrogen sulphide, sulphides or nitrites present *interfere.* In these cases, the sample is

10 The 1% $KMnO_4$ solution in the standard procedures is much too strong for this purpose; it yields a deposit of manganese dioxide on the walls of the vessel when strongly acidified.
11 A few drops of 0.01N $KMnO_4$ are added to the still hot, diluted sulphuric acid until a pale pink colour persists.

boiled for a few minutes after having acidified with sulphuric acid and before the permanganate addition.
If Fe^{2+} is present in larger amounts, 0.57 mg of $KMnO_4$ must be subtracted for each mg of this ferrous ion; the corresponding correction for Mn^{2+} is 0.6 mg $KMnO_4$ per mg to be added. Suspended matter consumes much permanganate, so that the values found are too high. The permanganate determination is thus carried out on both filtered and unfiltered samples from surface water of lakes with water bloom. The determination on wastewater is done after the mechanical purification stage of sedimentation.

c) Determination of Permanganate Number according to Schulze-Trommsdorf
(with chloride contents above 300 mg Cl/1)

If the amount of chloride is in excess of 300 mg/1,0.5 ml of 33% NaOH is used instead of 5 ml H_2SO_4 and, after the addition of oxalic acid, 5 ml of sulphuric acid (1 + 3 vol) are added; the procedure is otherwise the same as above.

Hydrogen sulphide and nitrites interfere during boiling of the alkaline solution. These must be previously removed by briefly boiling the acidic solution.

Calculation of the permanganate consumption: 1 ml of 0.01N $KMnO_4$ solution corresponds to 0.3161 mg $KMnO_4$. When a 100 ml sample is used, the permanganate titration in ml multiplied by 3.161 yields the permanganate consumption in mg $KMnO_4$/1. Any deviation of the titre from 0.01N must be borne in mind in the calculation (see below).
The correction mentioned above must be made if iron is present. The following table gives the multiplication factor for conversion to mg $KMnO_4$/1 when the titre of the permanganate reagent differs from 0.01N. For example, if the 15,0 ml of 0.01N oxalic acid requires only 14.7 and not 15.0 ml of 0.01N $KMnO_4$ in the above standardisation procedure, then the multiplication factor is 3.225 instead of 3.161.

Titre determination, with 15.0 ml 0.01N oxalic acid:

permanganate titration	multiplication factor
14.7 ml	3.225
14.8 ml	3.203
14.9 ml	3.182
15.0 ml	3.161
15.1 ml	3.140
15.2 ml	3.119
15.3 ml	3.100

Preparation of the 0.01N $KMnO_4$ solution: 3.161 g of purest $KMnO_4$ are dissolved in distilled water (Reagent No. 10) and the volume made up to 1 litre with this water. The 0.1N solution thus obtained is diluted by a factor of ten.

Preparation of the 0.01N oxalic acid solution: 6.698 g of sodium oxalate (dried at 240° C) are dissolved in about 200 ml of water in a 1 litre standard flask. 50 ml of sulphuric acid (1 + 3 vol) are added and the volume is made up to the 1 litre. Tenfold dilution of the resulting solution yields 0.01N oxalic acid. Purest distilled water (Reagent No. 10) is used always.

So-called oxygen consumption: The oxidisability of a water can be expressed as mg/l oxygen consumption instead of as $KMnO_4$ values. To obtain oxygen consumptions, the permanganate consumptions are divided by 4. The permanganate numbers may also be quoted approximately as mg/l of organic matter. A rough conversion factor for this is: 1 mg $KMnO_4$/l = 5.25 mg/l of organic matter.

Determination of humic substances
Humic substances are not completely determined by the potassium permanganate method. It is possible to distinguish in one of two ways between pure and contaminated waters which contain humic substances: either by colour evaluation (see p. 19); or, more accurately, by evaporating down the water sample, adjusting its pH to 4.8 with sodium acetate buffer and precipitating with 0.01M $FeCl_3$ solution. Further details are to be found in Handbuch der Lebensmittelchemie, Part 8, p. 625 (*D. Eichelsdörfer*) and in Mitt. Verein. Grosskesselbes. *73*, 276–279 [1961] *(G. Bohnsack).*

5. Determination of Chlorine Number

General: Like the permanganate number, the chlorine number informs about the content of organic matter. The chlorine number, however, determines organic compounds (especially those contaminants of faecal origin) which are not determined in the permanganate number. The chlorine number thus exceeds the permanganate number when faecal pollution has just occurred.

Procedure: 25 ml of chlorine reagent (Reagent No. 7) are added to a 300 ml erlenmeyer flask containing 100 ml of the water. The mixture is brought to the boil within exactly 5 minutes (stop watch or alarm clock). The flask is then closed with a cold finger. After exactly 10 minutes of boiling, the flask and contents are cooled rapidly in tap water.
0.5 g of KI and 1 ml of phosphoric acid[12] (or 7.5 ml of dilute HCl) are added to the cooled (room temperature) solution. After 10 minutes, the liberated iodine is titrated with 0.02N or 0.01N sodium thiosulphate solution (starch indicator added towards the end of the titration). The water under examination must be diluted if the chlorine number exceeds 100 mg/l.
At the same time in a parallel determination, 25 ml of chlorine reagent are added to 100 ml of distilled water (Reagent No. 10), boiled and treated exactly like the sample.
Calculation: When 100 ml water sample is used, 1 ml of chlorine reagent consumed corresponds to 7.09 mg chlorine/l. See p. 72 for evaluation.

6. Determination of Chloride-Ion

Occurrence: From 1 to 30 mg Cl/l in pure water. Up to 200 mg/l, depending on geological circumstances[13]. Often appreciably more in the vicinity of salt springs or Zechstein or in

12 Composition as in the German Pharmacopoeia.
13 *see footnote on p. 27.*

coastal regions (brackish water). Contaminated waters contain up to 300 mg/l and more, usually 50 to 150 mg Cl/l.
See p. 71 for assessment

Rough field determination: 20 ml of water are added to 10 drops of dilute nitric acid in a test tube, the mixture shaken and then 1 to 2 ml of 0.1N $AgNO_3$ solution are added. The following rough correlation holds:

Chloride content:	Observation:
3– 10 mg Cl/l	opalescence
10– 15 mg Cl/l	opalescent turbidity
15– 20 mg Cl/l	weak turbidity
20– 30 mg Cl/l	turbidity as DAB 6 (German Pharmacopoeia)
30– 80 mg Cl/l	marked turbidity, still transparent
80–150 mg Cl/l	marked, milky turbidity
150–300 mg Cl/l	flocculating turbidity, opaque, gradually coagulating into a cheesy precipitate
over 300 mg Cl/l	immediate cheesy precipitate

In the field examination, comparison is made with a standard KCl solution (0.063 g KCl/l = 30 mg Cl/l), treated in the same way. If the turbidity of the water investigated is more marked than that of the reference solution, the chloride content gives cause for concern (unless geologically dictated[14].

a) Titrimetric Determination of Chloride according to Mohr-Winkler

To 100 ml of water in a white glazed porcelain dish or in an erlenmeyer flask standing on a white background, 1 ml of 10% potassium chromate solution is added and the mixture titrated with 0.02N silver nitrate solution or an empirical solution (see below) to a just detectable yellowish-brown colour. This colour must persist 1 to 3 minutes or further silver nitrate must be added. The mixture is continuously stirred with a glass rod during titration.
The colour change can be more easily recognised if a second dish is placed alongside, containing the same total amount of water, plus the 1 ml of potassium chromate solution.

If the titration exceeds 10 ml of the 0.02N $AgNO_3$ solution, it must be repeated on a suitably diluted sample; the intense turbidity otherwise interferes with the end-point indication. It is an advantage in this case to take the titrated sample as a reference, after having removed the colour by addition of a crystal of common salt or some of the water being studied.
W. Ohle claims that the colour change is better recognisable if the operator wears yellow spectacles.
The end-point can be determined much more accurately with a potentiometric method, especially in mineral or brackish waters, etc. Towards the end, one can change over to 0.01N silver nitrate solution, even with brines.
Gravimetric methods are the best for brackish water, brines and waters from the neighbourhood of brine springs; or Volhard's method is used, working with 0.5N $AgNO_3$ solution and 0.5N ammonium thiocyanate solution (see under "Medicinal waters"); or the sample is titrated with 0.1N or 0.5N silver nitrate solution, using potentiometric end-point indication.

14 *see footnote on p. 27.*

Winkler's correction tables: In precise investigations, corrections have to be subtracted from the measured silver nitrate titrations. This applies to titration values below 10 ml of 0.02N silver nitrate. The corrections, based on 100 ml water sample, are tabulated below:

Titration	Correction
0.2– 0.3 ml 0.02N $AgNO_3$	0.13 ml
0.5– 0.7 ml 0.02N $AgNO_3$	0.15 ml
0.8– 1.0 ml 0.02N $AgNO_3$	0.16 ml
3.0– 4.0 ml 0.02N $AgNO_3$	0.18 ml
5.0–10.0 ml 0.02N $AgNO_3$	0.20 ml

Swedish limnologists found that this method gave too low results with waters containing only 2 to 3 mg Cl/l; the water must therefore be concentrated.
An empirical silver nitrate solution is very suitable for titrating waters low in chloride. It contains 4.791 g analytical grade silver nitrate in boiled, distilled water, made up to 1 litre. When used with 100 ml samples of water, 1 ml of this solution (which contains 1 mg Cl) is equivalent to a chloride concentration of 10 mg/l in the water.

b) Interferences in the Chloride Determination

Iron is removed from *waters rich in it* by shaking with some zinc oxide and filtering. The zinc oxide (German Pharmacopoeia, DAB 6) must naturally be free of chloride or else it must be washed with hot distilled water before use until the water from the washing is free of chloride. Interference of iron can be eliminated also by addition of sodium pyrophosphate (2 drops of 5% solution).
Peaty waters can be freed of their content of humins by shaking with freshly precipitated, chloride-free aluminium hydroxide (Reagent No. 2) or active charcoal (Reagent No. 1).
Brown waters containing humic substances are boiled with $KMnO_4$ solution and the unused permanganate is then removed with alcohol.
Hydrogen sulphide is removed by aeration or boiling the water sample. Small amounts of hydrogen sulphide or sulphide can be rendered harmless by adding some drops of H_2O_2.
Strongly acidic or alkaline waters are neutralised with the help of pH test papers. Waters rich in CO_2 must be degassed before titration, by warming.
Bromide and iodide ion are included in the titration and must be subtracted in the calculation for mineral waters.

Calculation
1 ml of 0.02N silver nitrate solution is equivalent to 0.7092 mg Cl. If a 100 ml water sample is titrated, the number of ml of 0.02N $AgNO_3$ solution, multiplied by 7.092, gives the chloride content of the water in mg Cl/l. If the empirical silver nitrate solution is employed (see above), the corresponding multiplication factor is 10.
1 ml of 0.1N $AgNO_3$ is equivalent to 3.5457 mg Cl
1 ml of 0.5N $AgNO_3$ is equivalent to 17.7285 mg Cl.
The Swiss limnologists, *W. Schneebeli* and *M. Staub* found lack of agreement amongst values found with Mohr's method for their primary rock waters of extremely low chloride content. They proposed the colorimetric method below for determination of chloride amounts between 0.4 and 4.0 mg Cl/l (Mitt. Geb. Lebensm. Untersuchg. *36,* 20–24 [1945]).

c) *Determination of Chloride according to Schneebeli and Staub*

As in Mohr's method, 1 ml of potassium chromate solution is added to 100 ml of water sample. Excess silver nitrate solution is then added, namely, 0.45 ml of a N/35.5 silver nitrate solution, dropwise with very vigorous stirring using a glass rod.
Test solutions containing 0.4, 0.8,1.2 and 2.0 mg Cl/l are treated in the same manner and compared with the sample determination after 1 to 2 minutes. If the coloration of the water sample is weaker than that of the test solution containing 2.0 mg Cl/l, a further 0.2 ml of the silver nitrate solution is added to the water sample until the coloration yielded falls within the test series. The result is calculated from the following table:

ml added Ag-solution	mg Cl/l				
0.45	0.4	0.8	1.2	1.6	2.0
0.65	2.4	2.8	3.2	3.6	4.0
0.85	4.4	4.8	5.2	5.6	6.0

d) *Mercuric Nitrate Method (for small chloride amounts)*

An empirical mercuric nitrate solution is very suitable for titrating waters of low chloride content. 1 ml of nitric acid (free from chloride!) and 1 ml of diphenylcarbazone solution (0.1% in alcohol) are added to a 100 ml sample of water and titrated with the mercuric nitrate solution to a violet colour. Blank titrations are advisable.
The mercuric nitrate solution is prepared by dissolving 4.832 g of $Hg(NO_3)_2 \cdot 2H_2O$ and 10 ml of HNO_3 in water and making up to 1 litre. The solution is standardised against a NaCl solution of 1.6486 g/l (1 ml = 1 mg Cl).

Interferences: A few drops of 5% sodium pyrophosphate solution are added to waters containing more than 5 mg Fe/l.

e) *Determination of Chloride with Ion-Specific Electrodes*

A completely new method is that based on ion-specific electrodes of the Orion Research Inc. (see p. 22).
Serial investigations can be very quickly and easily carried out, using this procedure; a special example of application concerns questions of over-salination of waters.

7. Determination of Phosphate Ion

Occurrence: Usually no detectable amounts in pure ground water. Ca. 0.05 mg/l in surface waters. No phosphate where water bloom is encountered. Sometimes 0.1 to 0.2 mg PO_4/l in peaty waters. Contaminated waters may contain 0.2 to 1.5 mg/l and more, generally 1 to 2 mg/l. Hot springs often contain large amounts of phosphate from apatite *(K. Höll)*.
See p. 75 for assessment.
Rough field determination according to *Medinger:* 10 ml of water are placed in one test tube

and 1 ml of strychnine-molybdate reagent (Reagent No. 24) in a second. The contents of both test tubes are mixed by first adding the water to the reagent and then pouring the mixture from one tube into the other and back two to three times. The table below correlates observation with phosphate content:

Immediate precipitate	5–10 mg PO_4/l
Immediate turbidity	3 mg PO_4/l
Immediate weak turbidity	1.5 mg PO_4/l
Weak turbidity after 3 seconds	1 mg PO_4/l
Weak turbidity after 10 to 20 seconds	0.5 mg PO_4/l
Opalescence after ca 1 minute	0.15 mg PO_4/l

Interferences: Hydrogen sulphide interferes and must be removed beforehand by vigorous aeration. The phosphate determination is made more difficult if large amounts of silica are present (over 25 mg/l). 2 ml of reagent solution are used with very hard waters.

Colorimetric determination of phosphate according to *Atkins, Denigès,* modified by *W. Ohle.*

1 ml of molybdate-sulphuric acid reagent (Reagent No. 20) is added to 100 ml of water. After mixing, 0.12 ml of stannous chloride solution (Reagent No. 33) is added from a 0.5 ml pipette. The blue colour which develops is compared in viewing or Hehner cylinders with those yielded by standard solutions (Reagent No. 25); this is done within 10 to 20 minutes. The colour glasses of the Hellige comparator are also suitable for comparison and permit compensation of the faint greenish tinge from waters which are very slightly yellowish.

Sensitivity: Down to 0.01 mg PO_4/l can be determined.

Accurate and objective colorimetric estimation is carried out at 690 nm in a photometer or "double-beam" photometer *(W. Ohle)*, with the help of a calibration curve. The reagents must be mixed rapidly and intensively, best by using the "mixing bulb" (p. 25).

Interferences: Waters containing large amounts of iron are treated with potassium cyanide before the molybdate-sulphuric acid and stannous chloride reagents are added. 2.0 ml of a 0.01% KCN solution are added for every 0.3 mg Fe/l. Silica does not interfere if no more than 50 mg/l is present. Nitrites interfere and must be removed by shaking or by prior addition of 10% sulphamic acid. Hydrogen sulphide and sulphides also interfere and can be removed by adding acid and shaking.

Turbid waters must be centrifuged (not filtered). Blank tests of reagent purities, especially of the sulphuric acid, are particularly important on account of the high sensitivity of the reaction. Temperature uniformity during the colorimetric evaluation must be watched.

In the rare cases where arsenic is present in the water, thiourea is added and then the other reagents 30 minutes later. This is especially necessary in the determination of total phosphate (according to *Ohle).*

The blue colour has a greenish tinge with waters which are rich in chloride (brackish waters). Reference solutions with equivalent chloride amounts should then be employed.

If the water contains large amounts of salts, only the molybdate-sulphuric acid reagent is first added to the 100 ml sample. The resulting solution is extracted with isobutanol-benzene (1 + 1). The stannous chloride reagent is added to the separated organic layer and the blue colour evaluated colorimetrically or photometrically.

Large amounts of fluoride interfere also.

According to *Thomas,* surface water should not be evaporated down, since this causes decomposition of the microorganisms in the water. Organic phosphorus is then released and displaces the ratio between organic and inorganic phosphorus. This error may be eliminated by using filtrates from Cella-filters.
Plastic flasks sometimes adsorb phosphate from water samples.

Calculation: Results are quoted as mg PO_4/l.

mg P_2O_5 = 0.7475 mg PO_4
mg PO_4 = 1.338 mg P_2O_5
mg P = mg HPO_4 × 0.3227

These values appear differently in the standard methods, namely,

1 mg PO_4 = 0.75 mg P_2O_5
1 mg P_2O_5 = 1.34 mg PO_4

The above determination gives only the content of inorganic phosphorus. If (as with surface water) organic phosphorus is also to be determined, a few ml of sulphuric acid are added to a 10–100 ml sample of water and this mixture is evaporated down in a Kjeldahl flask. The residue is reheated with 4 drops of hydrogen peroxide. Alternate careful water addition and evaporation down is carried out until excess hydrogen peroxide has been expelled. The solution is neutralised with ammonia after it has cooled.
In *studies of inland waters,* organic combined phosphorus must be considered as well as the inorganic phosphate (*R. Liepolt,* Österr. Wasserwirtschaft No. 1 [1965]; extensive work of *H. Ambühl* and *H. Schmid* on the determination of phosphate ion in inland waters, e. g. in the Schweizer. Z. für Hydrologie *27,* 172–173 [1965]; *J. Wernet,* Phosphate in Oberflächenwässern und Abwässern in "Münchner Beiträge" *12,* 1965; [56]). The content of condensed phosphates must be determined separately (see below and on p. 130). Thus three different determinations are necessary with surface water (see also *J. Markowski,* Wasser, Luft und Boden 7, No. 1 [1963]).

8. Determination of Polyphosphate

To determine the polyphosphates, which are added to drinking water to prevent corrosion, a second 100 ml water sample is boiled for 45 minutes with 1 ml of HCl. After cooling, the solution is neutralised and the above procedure used. The amount of polyphosphate is given by the difference between the results from the two determinations. 50% sulphuric acid + 0.4% nitric acid may be used instead of hydrochloric acid. The water which boils away is replaced from time to time.
According to *Lofamacar,* pyrophosphate (which comes from the detergents) must first be converted to orthophosphate by means of sodium bisulphate.
E. Naumann and *T. Kempf* (Schrift. Reihe Wasser-, Boden-, Lufthygiene *19,* 1–8 [1961]) state that the condensed phosphates can be more accurately differentiated by paper chromatography.
Determination of total phosphate in boiler water by Demberg's method: 2 ml of the water are mixed with 0.5 ml of molybdate reagent (see below) in a 26 mm cuvette; 0.3 ml of benzidine solution and 7.2 ml of saturated sodium acetate solution are then added.

Within 15–30 seconds of mixing with a glass rod, colour comparison is made in the Hellige comparator; the compensation solution of 2 ml of boiler water and 8 ml of distilled water is placed previously in the comparator.
The molybdate reagent consists of solutions A and B. The former contains 10 g of ammonium molybdate and 10 g of crystalline sodium sulphate in distilled water, made up to 100 ml. The latter contains 62 ml conc. nitric acid, diluted to 100 ml with distilled water. Solution A is poured into solution B to give the final reagent.
The benzidine reagent is prepared by dissolving 0.1 g of benzidine in 20 ml of glacial acetic acid and diluting to 200 ml with distilled water.

9. Determination of Urochrome

(direct test for urine and liquid manure in the water)

Urochromes are decomposition products of the colouring matter in blood. They gain access to the urine through metabolism. *O. Hettche* isolated urochrome A and B from urine.
According to *O. Hettche,* a faecal contamination of water, especially through urine or liquid manure, can be directly detected by determination of urochrome. 20 ml of 0.1M alum solution (4.7% potassium aluminium sulphate, 4.744 g $KAl(SO_4)_2 \cdot 12H_2O$) and 2 drops of phenolphthalein solution (1%) are added to 500 ml of the water in a litre cylinder. After mixing, 2 to 4 ml of 5% ammonia are added, swirling the solution, until a faint pink colour persists. Excess must be avoided. The pH-value should be ca. 7.8. The precipitate is allowed to settle overnight and as much as possible of the supernatant siphoned off next day. The precipitate is separated from the remaining solution by centrifuging, dissolved in 5 ml of 85% formic acid and centrifuged again if necessary. 0.5 ml of 85% phosphoric acid is added if iron is present. The solution is made up to 50 ml and its colour evaluated after 30 minutes. The following solution is used for colour comparison:

1.5 ml of 1% potassium dichromate solution
7.6 ml of 5% cobalt nitrate solution
+ distilled water up to 100 ml.

1 ml of this reference solution made up to 50 ml with distilled water corresponds to 1 mg/l of urochrome in the visual determination.
Photometric determination. The extinction value given with the S 38 filter and the Elko II apparatus, multiplied by the number of ml of solution (50) yields the "colour value". This "colour value" is multiplied further by 1.9 (for urochrome in formic acid solution).
This is expressed as the equation:

$$X = E1 \times A \times 1.9$$

where X = mg/l of urochrome
E1 = extinction in the 1 cm cuvette with filter S 38 (380 nm)
A = number of ml of formic acid coloured solution.

A second measurement with the S 53 filter (530 nm) is made when the presence of humic substances cannot be excluded.
The difference between the logarithms of the extinctions with filters S 38 and S 53, multiplied by 1000, yields the "Q-value", i.e., $1000 (\log E_{38} - \log E_{53}) = Q$.

If the Q-value is 0.9 or higher, no humins are present and the value obtained is pure urochrome. Q-values between 0.8 and 0.9 indicate an appreciable proportion of urochromes; and Q-values of 0.6 or less show that only humic acids are present.
O. Hettche reviews the importance of urochromes in drinking water and how goitre arises from water contaminated with urine (Ges. Ing. *76* [1955]).
This method for determining urochrome is not acceptued everywhere (*P. G. Sattelmacher* and *E. Fürstenau,* Ges. Ing. *82,* 16–20 [1961]). The author's experience with the method indicates that it provides reliable evidence for the occurrence of contamination with urine – when properly applied. Both the determinations given immediately below can be applied also to detect faecal contamination.

10. Faecal Reaction

Faecal contamination can be detected by adding 5 ml of diazonium salt reagent to 100 ml of water sample, swirling round and then adding 5 ml of 1N soda solution (53 g Na_2CO_3/l). Extensive faecal pollution is shown by a yellow-red to red colour.
The diazonium salt reagent is prepared by adding dropwise 2% sodium nitrite solution to a solution of 0.10459 g of sulphanilic acid in 0.1N HCl; ca 8 drops per 50 ml are needed.

Contamination with urine
See p. 206 for further details about direct detection of urine components (section on Swimming Bath Water).

11. Determination of Sulphate Ion

Occurrence: Generally 10 to 30 mg SO_4/l in pure ground water. Gypsum and salt deposit areas may have waters with 300 mg/l and more. 50 to 250 mg SO_4/l and more are usually found in contaminated waters.
See p. 76 for hygienic assessment.
Rough field determination: To 10 ml of water in a test tube, exactly 0.5 ml of HCl* is added, the mixture shaken and then 2 ml of 2% barium chloride are added. The presence of sulphates is shown by turbidity on shaking. At ambient temperature one observes:
1000 mg SO_4/l: An immediate powerful turbidity, becoming opaque in a moment and yielding a precipitate which settles within 1 to 2 minutes, leaving a clear supernatant.
500 mg SO_4/l: Almost instantaneous turbidity, gradually intensifying and becoming opaque after about 30 seconds.
200 mg SO_4/l: Solution becomes quickly turbid and opaque within 2 to 3 minutes. The precipitate settles partly only after some minutes. Cloudy, opaque turbidity on heating.
100 mg SO_4/l: Remains clear for a moment and then turbidity sets in, gradually intensifying. The cloudiness is shown when the test tube is moved, especially against a dark background. Cloudiness increases a little on warming but remains transparent. Some precipitate settles after a few minutes.

* *Composition as in German Pharmacopoeia (see p. 37).*

50 mg SO_4/l: Weak turbidity only after some seconds, becoming distinct after 3 minutes. Somewhat intensified on warming, faintly cloudy against a dark background. No precipitate separates.
30 mg SO_4/l: No change for 1 minute. Weak turbidity after some minutes, intensifying on heating.
20 mg SO_4/l: Very faint opalescence after standing for some time, becoming stronger on boiling. No precipitation.
10 mg SO_4/l: Remains clear and any turbidity is scarcely recognisable even on boiling.

Approximate field determination on waters rich in sulphate: 40 ml of water in a small erlenmeyer flask are treated with 1 ml of hydrochloric acid (25%) and with increasing amounts of 0.6% barium chloride solution. First, 1 ml is added and the mixture briefly boiled. The precipitate is filtered and the filtrate treated with a further 1 ml of $BaCl_2$ solution.
If a precipitate is formed, the initial 1 ml amount was not enough. The present mixture is boiled and filtered and the filtrate again tested with an additional 1 ml of $BaCl_2$ solution. This is continued until the filtrate no longer yields a precipitate with $BaCl_2$.
Some sodium sulphate solution is then added to another part of the final filtrate. If a turbidity is yielded, an excess of barium chloride is present and the experiment must be repeated with a smaller amount of reagent. In this way one proceeds further according to the accuracy desired, if necessary using smaller amounts of the water.
1 ml of $BaCl_2$ solution consumed is equivalent to 60 mg SO_4/l in this system. The results are mostly ca 10% too low, so that a corresponding correction should be added.

a) Gravimetric Determination of Sulphate as Barium Sulphate

250 ml of the water (completely clear, filtered if necessary) in a 400–500 ml beaker are acidified with 1 ml of hydrochloric acid* and evaporated down to about 50–100 ml. This residual volume is rinsed into a smaller beaker and further heated. Hot 10% barium chloride solution is added dropwise to the gently boiling contents of the beaker until no more precipitation occurs. A capillary dropping tube is used for this. 5 to 10 ml of reagent suffice with normal waters. The liquid should remain boiling during the addition.
A 1 ml excess of barium chloride solution is added to ensure total precipitation and the mixture is heated an additional 30 minutes over a small flame or on a boiling water-bath.
After standing for a longer time (overnight), the barium sulphate precipitate is filtered through a paper for quantitative work (e.g., Blauband 589 or 1575 of Schleicher u. Schüll, W. Germany) or through a porcelain filter crucible A1, and washed with hot water until the washings no longer show chloride reaction (test with silver nitrate solution, acidified with HNO_3). The filtrate is tested also for completeness of sulphate precipitation by concentrating and adding further $BaCl_2$; this is particulary important with mineral waters and other waters of high sulphate content.
A small rubber policeman is absolutely necessary for transferring the last traces of the precipitate.
The filter is dried, ignited in the weighed crucible and, after allowing to cool in a desiccator, the contents of the crucible are weighed. The mg of barium sulphate/250 ml water are obtained after having subtracted the weight of the empty crucible.
The filter crucible is dried for 1 to 2 hours in an oven at 110° C and weighed after cooling in a desiccator.
The determination must be repeated using a larger size if less than 10 to 20 mg of $BaSO_4$ are found. Conversely, the determination must be repeated with a smaller sized sample if over 300 mg of $BaSO_4$ are found. When heavy precipitates are formed, the test for complete precipitation on the concentrated filtrate must not be forgotten.

Calculation: Multiplication of the found weight of $BaSO_4$ by 0.4115 gives the mg SO_4 in the volume of water sample used; multiplication of this by 4 yields the mg SO_4/l. Should more or less water be used for the determination, this must be allowed for in the calculation.

The following corrections must be added to the found mg $BaSO_4$/l:

Corrections for the ignited $BaSO_4$ precipitate

Weight of precipitate at g	0°	1°	5°	10°	20°	30°	calcium 50° hardness
				added correction in mg			
0.30	3.6	5.4	8.5	10.0	11.0	13.2	14.7
0.25	2.7	4.1	6.0	7.0	8.5	9.4	10.7
0.20	1.8	3.6	3.7	4.4	5.2	6.0	7.3
0.15	1.3	1.6	1.8	2.3	2.8	3.3	4.1
0.10	0.7	0.8	0.9	1.2	1.4	1.6	1.8
0.05	0.3	0.3	0.3	0.3	0.3	0.3	0.3

The precipitation is complete if alcohol or acetone is added beforehand (*E. B. Sandell,* Col. Detn. of Traces of Metals, New York [1959]).

Humic substances can *interfere.* In such cases, the water is boiled with a little 0.1N $KMnO_4$ solution and the unused amount is destroyed by boiling with alcohol. The solution is filtered and sulphate determined in the filtrate. It is sometimes enough to shake the water with freshly precipitated aluminium hydroxide. *Larger quantities of iron* can be removed by shaking the water sample with zinc oxide or through precipitation with excess ammonia after having oxidised with H_2O_2; the ferric hydroxide is then filtered off.

b) Titrimetric Determination of Sulphates in Waters of low Sulphate Content, according to Winkler

The barium chromate procedure is suitable for serial laboratory investigations, even though it is less dependable than the gravimetric method.

Into a standard flask of 200 ml capacity are introduced 100 ml of the water and 20 ml of barium chromate solution (correspondingly more if the water is rich in sulphate i.e., over 50 mg/l) (Reagent No. 5). If nitrite is present, 0.5 ml of 5% sodium azide solution is added. The mixture is left for 30 minutes with frequent shaking.

Ammonia is then added dropwise until the colour changes from yellow-red to pure yellow. The volume is made up to 200 ml and the solution left for 5 minutes. It is filtered through fluted filter Nr. 560 (Schleicher & Schüll). The first part of the filtrate is discarded.

100 ml of the remaining filtrate are treated with 0.5 g KI and 10 ml hydrochloric acid (25%) in a ground glass stoppered flask. After 10 minutes, the iodine separated is titrated with 0.01 N sodium thiosulphate solution. Starch solution is added generously towards the end of titration. Should over 40 ml of the thiosulphate solution be required, the determination is repeated using a smaller water sample.

Larger amounts of organic matter can *interfere* (over 50 mg/l $KMnO_4$ consumption). As mentioned above, this interference can be circumvented by treatment with $KMnO_4$ and then alcohol.

Hydrogen sulphide must be removed beforehand by aeration.
Iron present in amounts exceeding 1 mg Fe/l must be precipitated by ammonia beforehand.

Calculation: Under the above conditions, 1 ml of 0.01N sodium thiosulphate solution is equivalent to 6.4 mg SO_4/l. 5.1 mg of SO_4 are to be subtracted from the end-value as a blank.

$$\text{mg/l of } SO_3 \times 1.2 = \text{mg } SO_4/\text{l}$$
$$\text{mg/l of } SO_4 \times 0.834 = \text{mg } SO_3/\text{l}.$$

c) Determination of Sulphate according to Ohle

To 100 ml of sample in a 300 ml erlenmeyer flask is added 0.1 ml of methyl orange indicator (DIN 8106). The solution is titrated with 0.1N HCl and one extra ml of the acid added, followed by 25 ml of $BaCl_2$ solution (1.3 g/l). The flask, covered with a watch glass, is heated until the contents boil and then rapidly cooled (some small glass beads are best added before bringing to the boil). After 30 minutes, 25 ml of potassium chromate solution (1.3 g/l) are added, together with one drop of 10% aluminium chloride solution and 1.6 ml of 0.1N KOH.
After standing for an hour, the solution is filtered through Schleicher & Schüll filter 595 or 560. 100 ml of filtrate are put into an iodine number flask and 10 ml of freshly prepared 10% KI solution and 5 ml of 25% HCl are added. After brief standing, the solution is titrated with an empirical sodium thiosulphate solution, containing 7.7520 g crystalline sodium thiosulphate and 1 ml of 20% NaOH per litre. Starch is used as indicator. 1 ml of this solution is equivalent to 1 mg SO_4/l.
W. Ohle's procedure is more accurate than the older *Winkler* method.

d) Titrimetric Determination of Sulphate according to Sijderius

The water is passed slowly through a cation exchange column[15]; the first eluate is discarded. 100 ml of subsequently issuing water are boiled with 25 ml of a 0.02N barium chloride solution (4.886 g $BaCl_2 \cdot 2H_2O$ in 1 l of aqueous solution). The mixture is maintained hot over a small flame or on the water-bath for about 15 minutes. After cooling and standing, 4 ml of buffer solution (5 g of magnesium-EDTA dissolved in 100 ml water, to which a solution of 35 g of ammonium chloride in 900 ml ammonia ($d = 0.910$) has been added and 7 drops of Eriochrome Black indicator are added. This is immediately titrated with 0.02M EDTA-solution (7.444 g Na-EDTA in 1 l of aqueous solution, standardised with the above $BaCl_2$ solution) to an ink-blue colour.
1 ml of the solution is equivalent to 1.92 mg SO_4.

15 *E.g. Dowex 50 W, x 4,50–100 mesh*

IV. Assessment of Drinking Water[1]

A. General Requirements of a Drinking Water concerning External Nature and Temperature

Water is indispensable and irreplaceable for life. Humans can go without food for a longer period but endure only 2 to 3 days without drink (p. 56). Apart from the requirements of sterility (p. 56) and purity, a fundamental demand of a drinking water is that it is appetising and of appetising origin (Paragraph 3 of German Foodstuff Regulations). Good drinking water should be clear and colourless and devoid of any strange taste or odour. Water which is coloured or turbid is rejected by the consumer as unappetising.

Water Requirements

The water should be available in adequate amounts. The DVGW working paper W 402 can serve for working out the water demand of a district. Provision of satisfactory drinking water in always sufficient amounts is one of the most important public duties.

The (W. German) law of 24/8/65, governing security of water supplies, provides plans for a regular water economy, also in emergencies, by utilising ground water throughout in the event of radioactive contamination of surface waters.

Nowadays, the water needs of larger towns must be estimated as at least 300 litres (ca 70 gallons) per head of population per day. The requirement is not so high in smaller towns. 1000 litres (ca 250 gallons) are used per head per day in New York and the figure in Switzerland is 300–500 litres (70–120 gallons).

Water needs often jump during the hotter periods. The local water supply must be ready for this, even when this involves financial difficulties for the municipal authorities (DIN 2000). Experience shows that too high a water pressure brings about increased consumption.

Only 2% of the water is drunk or used for food preparation. Nevertheless, separate pipe systems for drinking water and for service water can scarcely be provided. The Federal German Health Office (Bundesgesundheitsamt) estimates the minimum drinking water need for human nutrition in emergencies to be 2.5 litres (ca 0.5 gallon) per head per day. This estimate is valid for a few days and then larger amounts are required (10 to 15 litres, i.e. 2 to 4 gallons, per person per day). Diabetics drink much more water than normal consumers (*Lederer*, Münchn. Med. Wochenschr. *36* [1967]).

The consumption of water per head per day has grown also in the country in W. Germany to 100 litres (25 gallons), since almost all parts are served by central water supply and appropriate sanitary arrangements have been made. One has to reckon in the country with a further 50–80 litres (12–20 gallons) of water per day per animal for cattle.

1 This chapter is coordinated with "Hygienic directives for drinking water supply" [14 to 18].

Disproportionately large amounts of water are used by some factories. E.g., steelworks need 150 cubic metres (over 30 000 gallons) of water for cooling purposes in the preparation of 1 ton of steel (unless the cooling water is recirculated, with should be aimed at). In the industrial part of W. Germany, in Rhineland-Westphalia, 23 times as much water is used as in the remainder of the country despite efforts by the industries to limit the water consumption (e.g. in the steel industry, with cooling water, in the sugar and mineral oil industries and breweries, etc.).
The water should possess a refreshing taste; this is determined by the temperature (see below) and a particular content of free CO_2.
The water should be agreeable in every connection. A suitable temperature is vital here and it should be between 8° and 12° C as far as possible. Colder water is not enjoyed and if its temperature is 5° C or less, it can even cause stomach and intestinal complaints (*M. Rubner,* Lehrbuch der Hygiene, Leipzig, [1907] and Hygien. Leitsätze 14). Water at temperatures above 15 to 17° C is no longer refreshing.
Supplies from surface water are too warm in summer and too cool in winter. The temperature of water in deep wells is more constant, at a value corresponding to the average yearly temperature of the region. This temperature is 9.5° C in W. Germany, thus falling between the desired limits of 8 and 12° C. This provides a further incentive, in addition to hygienic considerations, for situating the well intake strainer as deep as possible.
Sudden changes in the temperature of a well water can be due to entry of surface water; this needs to be examined further.
Last but not least, the degree of hardness of the drinking water plays a certain part. Very soft water tastes insipid ("flat"); very hard waters (over 30–40° hardness) lead to intestinal disturbances and are unacceptable (p. 181)

Behaviour of the Water in the Pipe System

Water destined for drinking water systems should not attack the pipes and containers. It should also give sensibly no deposit on standing in the pipes, otherwise their diameter becomes reduced. In addition, drinking water must be free from components of the pipe material and the tanks, e.g., iron, zinc, lead, etc. Inside coats of the pipes must have no more than an initial effect on the odour and taste of the water. § 3 of the Foodstuff Statute (W. Germany) applies fully to drinking water also (p. 56).
See p. 160 for the behaviour of the water in tanks.

1. Coloration of the Water and its Improvement

The waters from moor regions of N. W. Germany, coloured yellow or brown with humic substances, do not provide good drinking water. Water with a colour value exceeding 25 (mg Pt/l) must be decolorised in central water suppley. This is best performed by *flocculation using aluminium sulphate.* About 10 to 20 g of aluminium sulphate (DIN 19600) are used per cubic metre (ca 250 gallons) of water, with the reagent in the form of a 5% solution. If sodium aluminate is used, only a half of this amount is needed.
The water temperature is an important factor in this flocculation with aluminium sulphate. The reaction time is doubled when the temperature is 10° lower, as, for example, in river water in winter.

The best pH for the treatment is 4.5 to 6. Purified water of this pH would attack the pipes (p. 54) and its pH-value must therefore be adjusted to 7.2–7.4 before introduction into the supply system. Hard, alkaline waters thus require two treatments. Sodium aluminate (DIN 19601) is consequently more advantageous as flocculating agent; one must then watch, however, that calcium does not separate from the purified water. Continuous control of pH is necessary.

For precipitation with sodium aluminate, the pH of the water should be between 7.5 and 8.5. *C. G. Hampson* reported marked acceleration of flocculation after addition of a sol of silica (procedure of *Baylis,* Paper Trade Journal *131,* 27–37 [1950]). *G. Giebler* proposed activated silica as a flocculating agent: a dilute solution of waterglass is treated with acid to bring the pH to nearly 7 (Schrift. Reihe WaBoLu, No. 19 [1961]).

"Flocculation accelerators" are being used more and more. Examples are Ferrosil (firm of Henkel & Co., Düsseldorf); Wisprofloc and Bentonite-additive after dosage with aluminium sulphate; or Separan (firm of Defac, Düsseldorf). *H. Bernhardt* (Arch. f. Hydrobiol. *61,* 311–317 [1965] reported the removal of humic substances from water in reservoirs using aluminium sulphate and Wisprofloc. The latter is a degraded starch product (see also *W. Wiederhold* and *F. Hiebenthal* GWF *98,* [1957] ("settling or alluvial filtration").

Flocculation agents based on polyacrylamide have been used with success for a longer time in America (*G. D. Geise, R. W. Pitman* and *G. W. Wells,* J. Am. Water Works Ass. *59,* 1303–1309 [1967]). The polyacrylic mixed polymer Sedipur TF2 (ca. 0.5 mg/l of raw water) has proved satisfactory in the Nuremberg Water Works at Mühlhof and in the river water works at Krefeld.

From the point of view of foodstuff regulations, flocculation agents are "technical auxiliaries"; they do not thus come under the decree for processing drinking water. After conditioning, the flocculation agents should be present only in technically unavoidable amounts and, in any case, in amounts causing no concern for health. This applies particularly to agents on a polyamide basis. *U. Hässelbarth* showed that in normal dosage of 0.5–0.7 mg/l with Sedipur, which contains polyacrylamide, the effluent from the filter contained only 0.005 mg Sedipur/l (Veröff. Wasserchemie Karlsruhe No. 3, 1969).

Lime or soda must be added to very soft waters to achieve flocculation. The lowering of the pH by the acidic aluminium sulphate also has an influence with very soft waters; the pure product is highly aggressive unless lime is added subsequently.

Schröder reported on *conditioning of river water* with aluminium sulphate (DVGW-circular No. 54 [Feb. 1951]). In the same circular, *Iwanowski* described the treatment of water from the rivers Rhine and Main with chlorine, ferric chloride and "Hydraffinkohle" (an active charcoal). Flocculation with ferric chloride is sometimes better than with aluminium, e.g. in treatment of river water (see *Iwanowski* above), especially after using the Accelator-process (Lurgi, Frankfurt/Main) or Reaktivator-process (Bran & Lübbe, Hamburg).

Stadager, however, says that aluminium sulphate has proved to be the best clarification agent for treating Main water (Wasser u. Boden *9,* No. 4 [1957]). In the Fuhrberg waterworks of the city of Hanover, water containing humic acids is decolorised successfully by adding lime and ozone in the Accelator. See p. 226 for information about Accelator, Reaktivator and Contact-Flocculator.

Difficulties often arise due to the plankton organisms present in summer when surface water is treated with aluminium sulphate; these can be circumvented only by regulating the zeta-potential (the apparatus is from the firm of Zeta-Meter, Inc. New York 28) (see *H. Bernhardt,* Vom Wasser *32,* 193–225 [1965]).

Flocculation with ferric chloride is somewhat dearer than that with aluminium sulphate. Open filter tanks soon acquire an unpleasant-looking brown colour during flocculation with ferric chloride.

Ferric chloride in concentrated form attacks the human skin and can thus endanger the personnel.

Further, ferric chloride may lead to flocculation at a later moment, especially with waters containing humins but little calcium. Its action is also dependent on temperature. Ferric chloride is far superior for purification of wastewaters.

The flocculation with aluminium sulphate is upset by synthetic detergents of the polyphosphates present in them.

Active charcoal (DIN 19603) can be used in filter vessels for removing humins.

The Scandinavian countries are dependent everywhere on waters containing humic substances. In Norway, where there is scarcely any real ground water, decoloration with large ozone amounts (6 to 8 mg/l) is now being carried out. Electricity is very cheap there (water-power), so that the process is economical.

Total sterilisation of the surface water is achieved at the same time.

Microsieves are mentioned on p. 55.

Subsequent treatment with lime, rendering the solution alkaline, causes the brown colour to return; the author recommended filtering through ignited dolomite (Akdolit or Magno filter masses) to prevent corrosion of the pipes. This latter process is satisfactory in Finland (*O. Peräkylä,* Wasser u. Boden *13,* 46–48 [1961]).

Akdolit-carbon (Akdolit GmbH, Düsseldorf) can be employed to remove humic substances and aggressive carbon dioxide, which occur together in peaty waters, at the same time.

The Krupp-Werke, Essen, provide *mobile equipment for preparing drinking water;* it comprises treatment with sodium hypochlorite, ferric chloride and active charcoal, for cases of emergency. A colour value of 10 to 20° (mg Pt/l) may be regarded as the limit for bathing purposes.

2. Turbidity and Clarification of the Water

As already said, turbidity and colour of the water reduce its appetising quality. The consumer equates them with dirt.

For central water supplies, a degree of turbidity of 8 units (mg/l of kieselguhr) is considered to be the limit. The amount of suspended matter should not exceed 1 ml/cubic metre drinking water. In small waterworks especially, clay suspensions are removed in a rapid filter after flocculation with aluminium sulphate. 10–100 g of the aluminium sulphate or half this amount of sodium aluminate (see above) are added to every cubic metre (ca 250 gallons) of water. A filter of fine-grained gravel suffices in some cases for removing turbidities or algae growths (green coloration and turbidity). Ferric chloride is used successfully also to remove very fine turbidities.

Berkefeld filters and ceramic filters are used nowadays as "settling or alluvial filters". The

kieselguhr which is thereby added has a markedly better clarification action. A mixture of kieselguhr and active charcoal is used in especially difficult cases.
Turbidity which suddenly appears in a water may indicate penetration of inadequately filtered surface water. Thorough bacteriological and chemical investigations must then be undertaken immediately. Clay particles of ca 1 μ diameter require about 18 days to settle und those of ca 20μ, about an hour.
Microsieves Plastic tissue with a mesh diameter of down to 10 μ is used successfully for removing more extensive turbidities and plankton (*H. Kriegshoff* GWF *104,* 1049–1051 [1963]).
Standards are given for filters for water conditioning in the DIN leaflet 19605.
The odour of hydrogen sulphide is removed by aerating the water (p. 195).
The transformation of a very poor raw water to a drinking water is admittedly economically possible today but should be considered only as a temporary emergency measure.

3. Odour and Taste

Filters of active carbon (fine-grained charcoal, e.g. Norite or Hydraffin charcoal[2]) are used to remove interfering odour or taste as well as some colorations from drinking water. Charcoal filters adsorb many organic compounds; these then provide a good culture medium on the filter, leading to multiplication of bacteria if the water is not very pure. The charcoal can therefore be added directly as a powder to the water (2 mg/l) (*Iwanowski,* DVGW-circular No. 54 [1951]). Especially good economically is also the *"filter layer process"*, with alternating layers of granular active charcoal and of filter sand.
Spent active charcoal filters can be regenerated by thermal treatment or by treatment with 3% hydrogen peroxide solution (*P. Koppe,* Schrift. Reihe Gewäss. Schutz *2,* 143–145 [1969] Aachen). (See further on p. 62).
Waters with marked alkaline or salty taste are excluded right away as drinking water (Hygien. Leitsätze für die Trinkwasserversorgung, 14). Waters with alkaline taste can be neutralised with acid to make them available for technical purposes. In view of corrosion dangers to the pipes, one must watch for any excessively high salt concentration.

B. Hygienic Judgment of Drinking Water, based on the Chemical Analysis

1. General

Basic demands of hygiene: A drinking water has to be not only pure and appetising (p. 51) but must be absolutely free from disease germs and materials dangerous to health. A drinking water should in fact be sensibly free from germs of all kinds. § 11 of the Fe-

2 see *Holluta,* Kommunalwirtschaft, No. 2/3 (1955) (Treatment of Rhine water with chlorine and flocculation with Al was unsuccessful).

deral (German) epidemic law of 18/7/61 regulates this as a skeleton law for the coming drinking water directive of the federation, the draft of which is ready. This covers all foodstuff concerns.
According to the German Foodstuff law of 21/12/58 and DIN 2000 of May 1959, drinking water is a foodstuff (see *E. A. Scheuermann,* GWF *103,* 231 and 447 [1962]). Water is "foodstuff No. 1" according to *Hossbach* (Komm. Wirtsch. No. 6 [1957]). Drinking water is given as the most important foodstuff in DIN 2000; it cannot be replaced by other materials. According to DIN 4046, drinking water is "a water, suitable for human pleasure and use, with qualities according to DIN 2000 and 2001".
Provision of good drinking water is said by *W. Hosang* to be a recognised legal duty of communities (Komm. Wirtsch. No. 2/3 [1955]).
According to *H. O. Hettche,* active carbon filters retain not only chlorine, odoriferous and other contaminant matter but also the urochrome. He found that water, contaminated faecally, led to goitre when continuously consumed, as a result of the urochrome (Ges. Ing. *76* [1955] and Arch. f. Hyg. u. Bakt. [1956]). He stated that urochrome is removed with an alumina filter. Ozone, also, removes urochrome (*K. H. Strackenbrock*).
Newly laid pipes must be well washed through and disinfected for a while with strongly chlorinated water before use; a bacteriological examination must show a good result before the pipe system can be used (DIN 2000).
It should be established when inspecting a source of water at the site, whether it is protected all the time from pollution and consequently from entry of disease-provoking bacteria (see p. 6–8).
Diseases which can be carried by drinking water include: various *Salmonella* illnesses, such as typhoid and paratyphoid; dysentery; amoebic dysentery; cholera; vomiting as "water illness"; Weil's disease, caused by Leptospira (capable of life for a long period in water) (see *Rankow,* Z. f. Hygiene *140* [1955]); anthrax; infectious jaundice; spinal infantile paralysis in water supplies from river and surface water. In addition there are various worm afflictions and, from swimming baths, conjunctivitis and fungus diseases of the foot ("Athlete's foot") (p. 278, 324).

2. Significance of the Chemical and Bacteriological Examinations of the Water

The principal goal of bacteriological examination of water is determination of bacterial number and the coli titre, i.e., the smallest amount of water which yields a positive test for *Bact. coli*[3].
Chemical analysis gives a clue to pollution, e.g., ammonium, nitrite and phosphate ions, as well as to any materials dangerous to health, such as lead, zinc and arsenic; this informs about whether the water can be used for household purposes.

3 Appearance of coli bacteria, normally present in the human and animal intestine, indicates faecal contamination of the water. If coli is found in 100 ml or even 1 ml of water, an adequate characteristic for contamination of the water is provided, especially when the bacterial count is greatly in excess of the limiting value of 100 bacteria per ml at the same time. The limiting value in Switzerland is 10 bacteria/ml (see p. 58 and 295 for further information).

The bacteriological and the chemical investigations complement one another. Neglect of the one or the other branch of water testing can have disadvantageous consequences. Even where waters from a depth below thick upper layers would appear to be unsusceptible to any influence, control of the iron, manganese, carbon dioxide and oxygen contents is just as important as the chemical-hygienic control. Only a pure, unaffected water possesses an absolutely constant chemical compostion: polluted ground waters which are dependent on surface waters show fluctuations all the time.

The constancy of the concentrations of several constituents of a water, e.g. of sulphate and chloride ions and of hardness constituents, may thus be a sign of an unaffected water; these values change quickly when surface water or any extraneous water enters.

L. Werner reported in "Archiv f. Hygiene, Vol. 132, that *Kisskalt* was right in stating that ammonium ion enters a well many days sooner than do bacteria. Bacteria are namely adsorbed in the soil much more easily than are soluble compounds derived from faeces. These facts have been established in recent experimental series with drainage wells in the USA (*R. G. Butler, G. T. Orlob* and *P. H. McGanhey,* J. Am. Water Works Ass. *46,* 97–111 [1954]).

Should, therefore, ground waters appear bacteriologically in order after passing through soils of high adsorptive power, positive tests for ammonium, nitrite and phosphate ions and increased chloride ion content can indicate that the water is contaminated from the start, and consequently at least unappetising. In this connection one must always bear in mind that the filtration power of the soil can diminish in the course of time or that ground water channels can form in the soil. There is thus a danger that infection-producing germs will gradually get into the water. In such cases, the bacteriological results thus appear too favourable.

In other cases, e.g. with newly drilled wells, the bacteriological findings appear sometimes too unfavourable since many harmless soil bacteria can be present. A fully different picture is then often obtained after the soil has "settled down" during 3 to 10 weeks.

Since it is now possible to detect faecal matter directly in very small amounts, chemical water analysis has recently been gaining importance. Faecal contamination is shown by the urochrome determination. (p. 46).

In such dubious cases, the chemical data in the end usually represent better the real state of affairs. Further, many germs, especially those of infections, have only a limited length of life in cold, pure water. The bacteriological findings thus interpret more the momentary condition of the water; bacteriological investigation must thus be undertaken more often (p. 286).

The chemical findings, however, yield a better insight into the permanent condition of the water. They yield, as said, an early warning – earlier than the bacteriological data; and they enable the occurrence of pollution to be recognised for a much longer time, since the changes in the organic and inorganic materials in the water remain detectable for a longer period. The chemical analysis thus has an importance for water hygiene at least as great as that of the bacteriological water examination. An exact quantitative chemical analysis is naturally a condition for this.

A final hygienic assessment must therefore be based on the results of bacteriological investigation, chemical analysis and inspection of the site.

An additional advantage of the chemical analysis is that the data, where necessary, can be much more rapidly acquired than those from the classical bacteriological examination which, for instance, needs 48 hours incubation time. However, there are now fast bacteriological methods, such as the *membrane filter method.* This last-named method has proved

useful in waterworks in the USA (*R. A. Eliassen,* Water and Sewage Works *102,* 523–524 [1955] (see p. 317)).
H. Zeiss and *E. Rodenwaldt* in "Einführung in die Hygiene und Seuchenlehre", Stuttgart 1942, write that a bacterial number greater than 100, an agreed compromise value, is unacceptable in drinking water; 30 years experience of innumerable controls of water in all civilised countries show this to be correct, under the condition that these bacteria do not include *Bacteria coli,* the surest indicator of water pollution by animal or human excrement. *B. coli* must not be detectable in 100 ml of water. Biological examination of the river or pond bed and of the plankton gives information about contamination which has occurred in surface water over a longer period (see. p. 346).

The Pyrogen test
If the soil has a suitable filtering power, all bacteria are retained in the upper soil layers; the dissolved metabolism products of the bacteria seep through into the ground water, just like ammonium, nitrite and nitrate ions. The author has suggested that, in cases of doubt and opinion of fundamental importance, the "pyrogen test", should be employed. Rabbits which are given a parenteral dose of the suspected water acquire a temperature (see *K. Höll,* Vom Wasser Vol. 35).

3. Sanitary Measures when Drinking Water is Polluted

If inspection of the site shows that human or animal waste has access to the water, so that there is a danger of transmission of infection, a more accurate investigation becomes superfluous. In this case, the well must be closed until the interference is removed. A new water source will then have to be sought or, should this prove difficult, the origin of contamination (usually a lavatory or leaching cesspool) must be removed and the well pumped out for a longer time. Dug wells have to be thoroughly disinfected after emptying, using bleaching powder, hypochlorite solution or Caporit or Perchloron. If no alternative for obtaining drinking water has been obtained up to this point, the water may be used for drinking and for preparing cold meals and drinks, only after having been boiled.
Only boiled water may be used also for washing-up, since there is a fear of huge multiplication of bacteria on small, scarcely visible food residues or newly prepared meals; this augments transmission of disease, a fact which is often valid and much too little known.

4. Ground Water Supply

According to DIN 4049, ground water is that which fills continuously the cavities of the earth's crust and is subject only to gravity. *Stini* says that one should add that it can move gently under hydrostatic pressure (according to *A. Wetzel*).
Thurner considers ground water to be genuine pore ground water, i.e., water which fills continuously the soil pores and forms a ground water level.
Ground water is that primarily considered as a source of drinking water. The water is puri-

fied and sterilised to a certain extent by filtration through the soil. The water is practically devoid of bacteria at 3–5 metres depth when soil conditions are good (i.e., fine sandy soil). Contamination may occur over longer stretches if the soil is coarse-grained and especially in broken ground.

5. River Water Supply

Many regions do not have enough ground water. It is then necessary to go over to surface water supply from rivers, streams or lakes, or to concentration of ground water (bank filtration or artificial production of ground water). Large cities are dependent on *river water* supply nowadays and have to tolerate hygienic shortcomings such as those of taste and extreme temperatures. Examples are Paris (Seine water), London (Thames), Minneapolis, St. Paul and New Orleans (Mississippi), Dortmund (Ruhr), Wiesbaden (Main), Rotterdam, Amsterdam and Krefeld (Rhine).

The interesting publications of *R. G. H. Bettaque* (Studien zur künstlichen Grundwasser-Anreicherung-Studies of artificial concentration of ground water-Hannover (1958) and of *W. H. Frank* (Die künstliche Grundwasser-Anreicherung, publication No. 9 of the Dortmund Municipal Waterworks, 1966) deal with artificial concentration of ground water with the help of infiltration galleries, drainage wells and underground percolation via ponds, tanks and submerged meadows.

Advantages of underground percolation from seepage ponds, after bank filtration, are equalisation of extreme river temperatures, the purification and the aesthetic side.

6. Reservoir Water Supply

Several reservoirs have been constructed in recent times for water supply of larger towns. In W. Germany, the oldest is the Remscheid reservoir; others are the Sösetal reservoir for Hildesheim and Bremen and, recently, the Wahnbachtal reservoir near Siegburg. Reservoir water is purer and has a lower bacterial number than other surface water[4]. Treatment with rapid filters to remove plankton, and precautionary sterilisation are usually enough. Reservoir waters are mostly very useful as service water since their degree of hardness and iron content are low. However, they generally need deacidification for central water supply (p. 166). On the other hand, deacidification must not lead to a too intense alkalisation (p. 171). The bacteriological and chemical nature of lakes and reservoirs undergoes fundamental change during the circulation periods in spring and autumn; this is often not considered during planning of surface water supplies. Diurnal fluctuations occur through the assimilation and dissimilation of phytoplankton in the upper water layers (epilimnion); this concerns especially oxygen, CO_2 and the pH-value. The carbonate hardness of lakes with harder water alters during the height of summer.

4 see *L. Minder,* Die Wasserversorgung aus Seen-Water supply from lakes (Arch f. Hydrobiologie 1954 and GWF 1951).

Distinct formation of layers is noticeable in reservoirs and lakes in summer. The sampling pump with the filter strainer must thus be displaceable vertically. Manganese and iron occur frequently in the profundal layers of deeper waters, as a result of disappearance of oxygen and reduction processes. The filter strainer must thus be temporarily positioned higher and the deep water taken off through bottom outlets as often carried out by *H. Bernhardt* at the Wahnbachtal reservoir (Vom Wasser *30,* 11–49 [1963] and DVGW-brochure "Gutes Trinkwasser").
When the sampling depth is altered, one should remember that the influent streams enter the standing reservoir water at a variable level, according to the temperature-dependent densities of stream and standing water; in spring, entry is at the upper surface; in summer, at the "thermocline"; and in autumn, at the reservoir bed. *J. Grim* studied this accurately in 1957 (GWF *98,* 234–238 [1957] and Gewässer u. Abwässer, No. 24 [1959]).
K. Haberer and *H. Bernhardt*'s work on the distribution of radioactive fallout in the Wahnbachtal reservoir is valuable in this connection (Vom Wasser *31* [1964]).
In Norway (Oslo-Baerum), Scotland and the USA, the plankton of reservoir or lake water is filtered off using rotating drums with microsieves of steel material (*H. Haberer,* J. Am. Water Works Ass. *57* [1965]). Water from the Thames in London has recently been filtered through large rotary micro-filters of stainless steel of 30 μ mesh, as in Oslo; water from the Lake of Constance has been treated likewise in the Bodensee-water works.

7. Spring Water Supply

Spring water is regarded hygienically less favourably than ground water as a result of the greater susceptibility to pollution from today's overpopulation. The opposite view was taken formerly. In 1936 still, the sentence "spring water is mostly very pure" was written by *J. Brix, H. Heyd* and *E. Gerlach* [18]. Further, it comes about that spring delivery is often too meagre or is markedly reduced in the height of summer when demand is at its greatest. This was shown to a startling extent in the dry year 1959. Bacteriological examination is especially frequently necessary, already at the planning stage.

8. Rainwater Supply

The water supply for individuals is often dependent on acquisition of *precipitation.* In some parts of N. W. Germany, where the ground water is peaty and salty over considerable areas, rainwater is caught on roofs and led into cisterns.

C. Sterilisation of the Water *(Removal of germs)*

(the terms 'sterilisation' and 'disinfection' appear to be used interchangeably and about equally in the literature on water treatment: the predominance of the former in the translation is therefore quite arbitrary) German: "Entkeimung"
Surface waters are permanently exposed to pollution and infection. Sterilisation must thus

always be undertaken as a preventive move, especially in thickly populated areas. This applies equally to ground waters from soils of poor filtering properties, to bank-filtered river water, and to spring water from fissured terrain – in other words, to all which are not flawless from the viewpoint of epidemic hygiene. Sterilisation can be performed through filtration or, much more effectively, by addition of chemicals (Cl, Ag, Cu, O_3, ClO_2 etc). See *G. Müller,* Bundesgesundh. Blatt 12, No. 16, 1969 for a discussion of the concepts of sterilisation, disinfection and removal of germs.

1. Incomplete Sterilisation by Filtration

Slow filters: Adequate sterilisation can be accomplished for example according to the principle of the "slow filter", previously known as "Hamburg filter". ("European filter"). The water is fed in to brick reservoirs containing about 1 metre (3.3 feet) of fine sand situated on a gravel bed, the grading of which increases from top to bottom. A "biological filter surface film" slowly forms on the surface of the filter which retains the contaminants and bacteria. The filter sand rests on the gravel base and the supporting layers, beneath which are the effluent cross-sections. Sterilisation is adequate after the filters have "run in"; in any case a considerable reduction of bacteria occurs. The filtrate should not have a bacterial number exceeding 100/ml and should have no detectable *B. coli* in 100 ml. The speed of filtration is kept below 1 m/h. Construction costs are high and thus slow filters (0.1 to 0.2 m/h) are now built only in exceptional cases. The largest are now being constructed for concentration of ground water through replenishing basins.
The sterilising action is better with turbid waters than with clear waters (*C. Truelsen,* Bohrtechnik-Brunnenbau, No. 11 1961).

Rapid filters: For economic reasons, so-called rapid filters are built nowadays. They have from 5 to 50 times the speed of filtration and are smaller, so that construction costs are also less. Their sterilisation action is, however, much less than that of the slow filters. The bacterial number is reduced only to about 10%. No "filter film" is formed and so the biogenic activity is replaced by the mechanical purification action of the filter grain. Rapid filters are therefore more suitable for clarification and improvement of the water. If sterilisation is necessary, it must be chemical.
Both rapid and slow filters are used as submerged filters, i.e., from 0.5 to 1 m depth of water should be present above the filter bed. High layer filters have been used recently with filtration speeds up to 100 m/h (*E. Naumann,* GWF *91,* 108–115 [1950]).
The *Immedium-filter* of the Lurgi firm permits especially high filtration speeds, since the water to be purified enters simultaneously from above and below. The space occupied by this filtration plant is thus smaller and the filtration process shortened (firm of Permutit, AG, Duisburg).
The backwashing of the filters, necessary from time to time, is always done today with a water/air mixture; the filter mass is whirled in and out and the particles of detritus and bacteria are more easily removed.
Rapid filters are used as a rule as closed filters which need pumping only once.
Rapid filters are fitted after other conditioning plants as a security measure, as e.g. in the Bo-

densee (Lake of Constance) waterworks. A primary filter of coarser grading is advisable for surface waters with strong plankton development, especially water bloom from plankton algae.

Considerable expert knowledge is required to estimate the correct dimensions of a filtration plant and errors are often made. Special reference is made in the 10th report of the WHO to the overloading of many filtration plants. *W. Gandenberger* (DVGW-brochure "Trinkwassergüte", 1954) wrote of the theory of filtering through sand and construction of sand filters.

Multilayer filters, used nowadays and in which the complete filter space acts, are based on another principle (hence called capacity filter as opposed to layer filter). Water passes through multilayer filters in the direction from the coarse to the fine grading. Larger quantities of solid material are thereby retained and yet filtration speeds are higher and running times before backwashing are longer. Multilayer filters are consequently more economical, especially in connection with flocculation plants (coagulation filtration). If the raw water is treated with ozone at the same time, the filtration running times are lengthened still further (for further information, see *H. Sontheimer,* Vom Wasser *34* [1967]).

The layers of variable grading can be combined also with layers of active charcoal or anthracite or with "Magno-Filt" (Magno-Werke, Duisburg).

Anthracite filters have been in use for a long time in the USA and, recently, in Germany too. German anthracite, a thermally treated hydroanthracite, has the advantage of appreciably greater porosity than filter sand. This gives rise to a better purifying action and higher filtration speeds than obtained with gravel filters.

2. Sterilisation through Addition of Chemicals

Chemical sterilisation; Chemical sterilisation should be undertaken when the water is not absolutely satisfactory from the point of view of hygiene. Automatic chemical feeders have been available for some years; these function quite reliably. Considerable progress has been made recently in this domain of water conditioning.

3. Sterilisation with Ozone

Sterilisation with ozone appears to be the best method from a sanitary viewpoint since no extraneous chemicals are introduced into the water (excess ozone soon decomposes into oxygen in the treated water), which is often highly desirable. No dissolved ozone reaches the consumer. Larger piped systems need a "safety chlorination".

Further, ozone has a much more powerful bactericidal action than chlorine and chlorine dioxide; smaller plants thus suffice.

If ozonisation and filtration are carried out at the same time, the filtration running times are increased (see above; *H. Sontheimer* GWF *109* [1968]).

Treatment with ozone removes any taste and smell possessed by the water, as well as effecting sterilisation.

Water which has been treated with ozone attacks rubber in washers, junctions and linings of

containers. Before fitting up an ozonisation plant, these must be replaced by suitable metal.
Sterilisation with a calibrated mixture of ozone and air, as used formerly, is difficult and costly. The air has to be absolutely dry before ozonisation; silica gel plants can be used for this, for example. The ozonisation process has become more economical since it has been possible to prepare ozone solutions of high concentration (indirect procedure with injector).
Ozonisation of water is independent of the general nature of the water. Only with very hard waters is it advisable to add polyphosphate to prevent calcium precipitation. Polyphosphate is claimed to improve the distribution of the ozone (*G. E. Kurzmann*, Wasser, Luft und Betrieb 7, No. 1 [1963]). However, the water must be free of iron or else turbidity is yielded.
0.3 to 2g/cubic metre (250 gallons) are used and a time of action of 10 minutes is enough. The cost is about 0.5 to 1 pfennig/cubic metre (ca. 0.5 to 1 U.S. cent or 0.25 to 0.5 new [British] penny per 1000 gallons), which is not financially crippling.
The firm of Berkefeld-Filter GmbH of Celle, W. Germany, supplies mobile ozonisation plants.

4. Chlorination

Chlorination is the most usual method of sterilisation, although it still has various shortcomings.
Using a special apparatus, the chlorine from steel cylinders is introduced in the gas form very accurately into a branch of the water. The chlorine/water mixture is led into the main water pipe, in amounts of 0.2 to 0.3 g chlorine/m^3 of water. The amount added is governed by the chlorine consumption of the water (chlorine combination capacity, see p. 128). This is generally 0.1 to 0.2 g/m^3 but can be very much higher. It must be ascertained experimentally. An additional amount of not more than 0.2 g/m^3 should be introduced beyond this combination capacity value. Chlorination in America is usually more intense since surface waters are used to a great extent. The consumer's interest in Germany precludes more energetic chlorination.
30 minutes action is required for a chlorine excess of 0.12 g/m^3; 15–20 minutes for an excess of 0.2 mg/l.
Considerable consumption of chlorine occurs in pipe systems with encrustation lumps. "Safety chlorination" is rejected nowadays since it does not offer adequate protection in the event of contaminated water gaining access to the system (consequent high chlorine consumption).

a) Disadvantages of chlorination

A disadvantage of chlorination is that the water tastes and smells unpleasant, really an offence against the basic requirements of a drinking water. Further, chlorination of river way yield chlorophenols with the phenols which are often present in such water; these impart a particularly disagreeable taste to the drinking water. They can be detected in a dilution of 1–20.000.000 (*E. Böhm*, Z. f. Lebensmittel-Unters. u. -Forschung *106*, 240–241 [1957]). The ADM-process (*A*dler-*D*iachlor-*M*utonit) is an intensive chlorination process followed by filtration through active carbon. It provides water which is completely sterile and has no chlorine taste; it is accompanied by separation of organic matter and removal of water components which taste and smell. The chlorine dioxide process likewise gives rise to no taste of chlorophenols (see below).

In the chlorination of waters contaminated with much organic matter, a breakpoint is obtained when the reaction between the chlorine and certain compounds is concluded. The chlorination must be aimed at this point (breakpoint chlorination). It is chiefly chlorination of ammonia to mono-, di- and trichloramines. The process takes the form of a curve, the minimum of which is termed the breakpoint.

The limiting concentration at which chlorine can be tasted is 0.5 g/m^3 (0.5 mg/l); the limiting concentration of odour is far lower.

Sterilisation is practically complete after 1 hour. As a safety factor, 0.1 to 0.2 mg Cl/l must still be present in the whole pipe system after this time; the chlorine consumption of the water, which can vary greatly, must be borne in mind in this connection (see above). An excess of 0.2 mg Cl/l is now prescribed in the German land of Hesse (Official Gazette, Hessen, No. 10 [1955]). Viruses and anthrax spores are not killed off with certainty (see also *K. Haak*, GWF *100* [1959]).

Filters of active carbon and Dechlorit (granulated calcium sulphite) are used to remove the troublesome excess of chlorine.

Chlorine is converted into hypochlorous acid by powerful agitation. Hence, if chlorine dosage is performed in a suitable way before the pump is reached, i.e., in the suction pipe a superior sterilising effect is obtained and some of the disadvantages of elementary chlorine are avoided. The Neue Contintale Baugesellschaft, Hannover-Linden, constructs suitable plants.

Chlorination reduces the carbonate hardness and increases the non-carbonate hardness of the water; this has an unfavourable effect where chlorination is repeated, e. g., in swimming bath water. Each mg Cl/l reduces carbonate hardness by 0.04° dH and raises the non-carbonate hardness by the same amount. At the same time, 1 mg of chlorine liberates 0.628 mg CO_2 of combined carbon dioxide. This interferes with the carbonate balance (p. 163). The action of the chlorine is weaker in alkaline waters than in acidic. The former must thus be chlorinated more intensively (*Holluta*, GWF *90* [1949]). It is better to use chlorine dioxide in such cases.

5. Sterilisation with Chlorine Dioxide

Chlorine dioxide has a *much less marked taste and smell* than chlorine, even though they are 2.5 times stronger than assumed theoretically. Chlorine dioxide has a more powerful action on bacteria than has chlorine; smaller amounts of it thus suffice. It has the additional advantage of forming not the unpleasant chlorophenols with phenols (p. 63), but the much less unfavourable trichloroquinone. The bactericidal superiority of chlorine dioxide over chlorine is especially marked in alkaline water. Its activity also persists longer.

Chlorite and chlorate are formed in alkaline solution, however; these are less strongly bactericidal and cause irritation in swimming bath water. Chlorine dioxide is therefore now being less used although chlorite formation can be suppressed or reversed by adding small amounts of chlorine gas.

The taste of chlorophenols created by treatment with chlorine, mentioned in the previous section, is removed by after-treatment with ClO_2. (*K. Jägers,* Städtehygiene *5,* 131 [1953]).

A 30% solution of sodium chlorite is used for dosage with chlorine dioxide. Chlorine dioxide is liberated by addition of HCl. 0.05 mg ClO_2/l generally suffices. Chlorine dioxide is yielded also by adding chlorine gas to the sodium chlorite solution. Re-formation of chlorite must be prevented (*H. Berndt,* Zentralblatt Bakt. I *191,* 365 [1963]). Hypochlorite solution is used in smaller waterworks but it readily precipitates calcium in hard waters. This applies with special force to waters with a deficit of "associated CO_2" (see p. 163). Chloramine solution is preferable in this case, although it is dearer. Chloramines possess less bactericidal activity than chlorine and hypochlorite, however.

Along with sodium hypochlorite solution, calcium hypochlorite ("Caporit") and "HTH"[5] are also used (HTH is cheaper but has a stronger smell).

The Chlorator GmbH of Grötzingen, W. Germany, has recently brought out the "Chlorinette", a small-scale chlorination apparatus.

6. Other Sterilisation Procedures

The *copper treatment process* with subsequent removal of copper *(Höll)* has proved useful in practice (e.g. in Bad Reinerz, W. Germany), especially in the conditioning of swimming bath water; algae are destroyed at the same time (see also p. 374–5).

The *Cumasin process of Dr. Kruse* and *Dr. Fischer* (Gesundheits-Ing. *22,* [1936]) is also a good sterilisation procedure. It is based on the sterilising action of silver ions (see below). The untreated water for it must be pure and clear and contain no more than 1000 microorganisms per ml. Larger containers are needed because the time of action is 2 hours. The process has recently been applied with success in plants in the Harz mountains (W. Germany). From 0.05 to 0.1 mg Ag/l are added.

The principle of the *Katadyn process (Dr. G. Krause)* is the bactericidal activity of certain metal atoms, notably silver (oligodynamic action, see *H. Freundlich* and *K. Söllner,* Zur Erklärung der oligodynamischen Wirkung, Biochem. Z. *203,* 3–13 [1928]).

In the Electro-Katadyn process, an electric direct current releases silver from special electrodes into the water which flows past. Despite the variable chemical composition of the water being treated (especially due to fluctuations of salt content), the dosage can be regulated so that the water always has the prescribed content of silver ion (0.05 to 0.1 mg Ag/l) after treatment.

A newer form of application of the Katadyn principle is dosage of the effective metal ions as a water-soluble preparation. The preparation, Micropur, suffers partial decomposition on dissolving in water, to an extent depending on the concentration; an easily displaceable

5 *High Test Hypochlorite.*

equilibrium relates the unchanged complex and the precipitated or dissolved silver chloride (*R. Hey,* Z. f. Bakt. *152,* No. 5 [1947]). Micropur contains 1% of silver. 10 g of it, added to 1 cubic metre of water, hence yields a concentration of 0.1 mg Ag/l. Argentox is a similar preparation for silver treatment.
The real advantage of metals for sterilisation of drinking water, compared to other processes, is that the bactericidal activity persists in the treated water for a long time and even acts against infections which gain access later. A further great advantage of silver treatment is that the taste of the pure water is not impaired. A disadvantage is that the oligodynamic action attains its maximum only after some hours; if the distance from source or well to the piped system is too short, storage plants must be introduced between them to delay arrival of water in the system until this "incubation period" is past. Further information is to be found in *G. A. Krause*[6], *F. Konrich*[7] and in *J. F. Braune,* Die Trinkwassersilberung, Munich 1957.
The Katadyn process has again shown its value recently with pure spring waters and is especially to be recommended for small water supply plants, notably where the water has a low chloride content and permanganate consumption.

Ultraviolet Irradiation

Irradiation with ultraviolet light is used for sterilisation in smaller water works with quite clear, colourless water. The advantage of this is that the water is not changed in any way and that no extraneous matter is introduced. However, the effect of the radiation is limited and the intensity of the mercury lamps fluctuates. A disadvantage is the immunity of microorganisms which are held within particles of dirt. The UV-process can therefore be applied only to clear water (see also *G. Müller.* Bundesgesundh. Bl. *12,* 245–251 [1969]).

Sterilisation of drinking water on a small scale

Water can be *sterilised* on a *small scale* by boiling or by various additions, such as of Para-Caporit (firm of Bayer, Leverkusen, W. Germany) or "chlorination tablets" which include HTH[8]-chlorine tablets or grains (Akdolit-GmbH, Düsseldorf); commercially available also are de-chlorinating tablets, e.g. sodium thiosulphate or calcium sulphite (Dechlorit).
Berkefeld filters are to be recommended when continuous sterilisation of household water is necessary. These consist of earthenware containers of 5 to 10 litre content, in which the water is filtered through a practically bacteria-tight filtering candle of kieselguhr. These candles can be impregnated with sparingly soluble silver compounds, so that there is no fear of bacteria growing through them. The filtering candles must be brushed clean on the outside from time to time, however. Berkefeld filters with a pre-filter are supplied for waters containing suspended mud. Berkefeld filters for pressure pipe lines have been available for some time, and, recently, mobile plants also (firm of Berkefeld-Filter GmbH, Celle, W. Germany).

6 *G. A. Krause,* "Oligodynamische Wassersterilisierung durch Katadynsilber" (Ges.-Ing. No. 27 [1929]).
7 *F. Konrich,* "Über oligodynamische Trinkwassersterilisierung vermittels des Katadynverfahrens" (Ges. Ing. No. 47 [1929]).
8 High Test Hypochlorite.

"Berkefeld-active candles" contain deposited silver with oligodynamic activity, so that microorganisms cannot grow through.
The Katadyn steriliser is used for clear water of low content of organic and suspended matter (oligodynamic silver sterilisation, Dr. Krause's process, Katadyn Ges., Munich).
The Swiss firm of Actinag AG, Zürich brought out in 1954 a UV-steriliser for individual water supply.

7. On which Chemical Features is based Objection to a Well from the Point of View of Hygiene?

The most important question in assessing the water is whether human or animal waste products have access to it.

If a water is contaminated by human or animal waste products, various organic and also inorganic materials are introduced into it, e.g., phosphate and chloride ions and, in particular, the nitrogen-containing ions, ammonium, nitrite and nitrate. Faecal contamination thus always leads to an increase in certain materials occuring in drinking water. The author terms these "*pollution indicators*". They include: ammonium, nitrite, nitrate, chloride and phosphate ions and potassium permanganate consumption. Any increase, even only small, in permanganate consumption, chlorine number, chloride and nitrate ion contents with simultaneous appearance of nitrite and ammonium ions, especially protein ammonia, justify always an objection to the water; this applies even when the bacterial count is low and *B. coli* has not been detected (p. 56).
It is, however, wholly wrong to assume faecal pollution when only one of the pollution indicators has increased in amount. A high ammonium or nitrate ion content does not always indicate contamination. This applies in particular to interpretation of the permanganate consumption but also to the other pollution indicators. As seen below, their increase can have quite other reasons. Only the full chemical analysis is decisive. The bacteriological results and, above all, those from inspection of the locality, must be considered for the final judgment: in special cases, the urochrome value also.

Information may be obtained also for assessment by comparison with neighbouring well waters, coming from the same depth and same soil formation.
Waters from very deeply situated soil layers seldom contain ammonium, nitrite, and nitrate ions. These nitrogen compounds are reduced by denitrifying bacteria to nitrogen which escapes through the soil air.
Large amounts of ammonium ion are found where ground water has been affected by refuse tips (*F. Nöring et al,* GWF *109,* 137–142 [1962]).

A distinction can be drawn between pollution indicators in the narrow sense (ammonium, nitrite and phosphate ions and permanganate consumption) and those in the broader sense (such as chloride, nitrate, sulphate and perhaps even carbonate ions).

The relevant sections contain information about pollution by mineral oils, phenols, detergents, pesticides, etc.

D. Pollution Indicators

1. Ammonia or Ammonium Ion

The content of ammonium ion is hygienically important since it can arise through decomposition of human or animal excrement. E.g., urea decomposes into NH_3 and CO_2. Pure water generally contains no detectable amounts of ammonium ion. This applies to all types of water (ground water, spring water, surface water and rainwater). The appearance of ammonium ion in water is thus usually reason for suspicion. There are, however, a number of exceptions where the presence of ammonium ion has no hygienic significance, especially in mineral springs (see below).

Contaminated waters contain 0.1 mg/l up to about 10.0 mg/l of NH_4^+, usually 1 to 3 mg/l. If ammonium, nitrite, nitrate and phosphate ions are present and permanganate consumption is increased, the water is objectionable as contaminated and hygienically (epidemiologically) suspect. In special cases, an individually high ammonium ion content (>3 mg/l) can indicate recent pollution. If NH_4^+ is lacking in water which contains nitrite and nitrate ions, organic materials and chloride ion in larger amounts, contamination cannot be ruled out; the ammonium ions may have undergone biological oxidation in the soil.

Protein ammonium must not be present in drinking water on any account, since it indicates faecal pollution. The absolute upper limit is 0.1 mg/l. This amount is yielded by decomposition of 1 ml urine in 100 l water.

Special cases: Ground waters of high iron and manganese contents, in particular those in the N. German plain, contain sometimes ammonium ion resulting from reduction of nitrate; this is due to H_2S yielded under pressure from iron sulphide and carbon dioxide in the soil; they are then almost always free of nitrite and nitrate ions. Deep ground water must therefore be judged quite differently from spring water in this connection. Peaty water and rain water in industrial regions also contain now and then a small amount of ammonium ion. Humic substances abstract all the oxygen from ground water, even from its nitrates and nitrites. These are reduced to ammonium ion which appears in the humic waters in amounts of a few tenths of a milligram.

Ammonia can enter ground water from ammonia-containing artifical manures (see *K. Höll,* Vom Wasser *30* [1963]). Surface waters can gain ammonia and phenols from the effluents of gasworks.

One must remember in the examination of tap water, that ammonium ion is sometimes added during conditioning of the water (e.g. the chlorine-ammonia-sterilising procedure, the so-called pre-ammonification procedure for avoiding the taste of chlorophenol).

The ammonium ion content in these special cases gives no cause for concern; cases are, however, imaginable where a content has a suspicious origin.

Tap water should have no more than 0.5 mg NH_4^+/l, for technical reasons (growth of organisms, difficulties in chlorination, etc.). See *W. Rummel* and *G. Werner,* Wasserchemie, Vol 4, for removal of ammonium ion.

2. Nitrous Acid or Nitrite Ion

Nitrite ion is almost never found in pure water, at the most in traces (up to 0,01 mg NO_2^-/l); amounts up to 0.3 mg/l are found in strongly iron-containing, boggy ground water and in rainwater after thunderstorms. The appearance of nitrite in water is generally evidence of faecal contamination. The nitrite content, along with that of ammonium ion, is therefore the most important pollution indicator. 0.2 to 2.0 mg NO_2^-/l, rarely more and usually about 1 mg/l, are found in polluted water, together with increased amounts of nitrate, ammonium, posphate and chloride ions and organic material (see below for exceptions).

If a water contains no nitrite, yet the other pollution indicators are present, contamination cannot nevertheless be dismissed from consideration; the nitrite ion may have been biogenically oxidised further or reduced to ammonia (see below).

If filter plants are present, the raw water must always be investigated; nitrite disappears during aeration and filtration of the water.

A slight increase in only nitrite and ammonium ion contents indicates decomposition of plant materials (0.02 mg NH_4^+/l and NO_2^-). The author has often found larger nitrite (and ammonium) amounts in lysimeter runs, since these come from 2 m depth, in grown soil (see *K. Höll,* Vom Wasser *30* [1963]). Heath soils are an exception.

The small amounts of nitrite in water are in themselves not detrimental to health, but indicate, as said, unhygienic conditions. Usually no trace of nitrite is found in medicinal and mineral waters which emerge from greater depths unless the water has become contaminated in the surface layers or in corroded pipes or tappings.

Special cases: Above hygienic suspicion is nitrite from freshly cemented dug wells, from zinc pipes or galvanised pressure boilers, provided that the other pollution indicators give no hint of contamination. Traces of nitrite ion can occur in rapid filters and in closed plants for iron removal, if aeration is inadequate. The amounts of nitrite occurring here are, however, a few tenths of a milligram/litre at the most.

As said, traces of nitrite can arise in ground waters, rich in iron, and in peaty waters.

3. Nitric Acid or Nitrate Ion

Small amounts of nitrate ion are present in all waters. The usual amounts are 5 to 10 mg/l. Up to and over 20 mg/l NO_3^- can be dictated by the soil. In North Germany (e.g. near Bremen and Emden), pure ground waters, yet of high nitrate content of over 100 mg/l, are found; the remaining composition is normal. According to *G. Schneider,* ground waters with 250 mg NO_3^-/l occur often in eastern Westphalia although no source of pollution is present. A nitrate content of up to 35 mg/l is considered normal in our region, according to *Stoof*; in drinking water, more than 50 mg/l should not occur (see below).

High amounts of nitrate alone can mean that a single contamination has taken place. This is particularly the case when the nitrate content fluctuates. At the moment, the soil then has adequate purifying power. Nitric acid is namely the end product of oxidation of nitrogen compounds in the biogenic process of mineralisation of waste products. The chloride ion content is then also usually higher, whereas ammonium ion and nitrite ion are not present. High amounts of only nitrate are found, for instance, in ground water from soils which are strongly loaded with waste materials. Objection must be made to such water; for one thing, it is unappetising, since it has extracted waste products; further, the decomposing power of the soil can gradually decrease and weaken its filtering power.

This applies especially to *water from shallow wells* which contain a lot of nitrate but little ammonium and nitrite ion. One must remember here that the nitrate oxidation may take place very rapidly. Special care is therefore necessary with shallow wells if a high content of nitrate is found.

The nitrate content is an indicator for both the degree of loading of a soil with organic and inorganic nitrogen-containing waste products, and the degree of decomposition of these materials in the soil. Higher nitrate values are generally found today than 50 years ago, due to the increased population.

Decomposition of plant material can be assumed when the water shows a small increase in nitrate, little nitrite and ammonium ion, and otherwise normal composition. Fallen leaves of deciduous trees yield a lot of nitrate to soil and water; this often plays a part with uncovered waters.

Forest soils impart higher nitrate ion concentrations to ground water; this is due to the activity of nitrifying bacteria and applies especially to pinewood soils. *K. Höll* established this in prolonged studies with lysimeter runs (seepage). "Pinewood regions are therefore not suitable for water acquisition" (*K. Höll*, Vom Wasser *30,* [1963]).

Iron-containing ground waters from deep wells usually have no nitrate but ammonium ion instead. Nitrate or nitrite is found instead of ammonium in the water after the plant for iron removal has been passed. Nitrate disappears when mineral oil seeps into the soil (p. 146).

Contaminated waters usually contain 50 to 150 mg NO_3^-/l (often 200 mg/l and more) alongside many organic materials, chloride and high nitrite and ammonium ion contents. That adequate mineralisation has occured during passage through the soil many be concluded when only the nitrate and chloride ion contents are high; one must consider here also that the filtering power of the soil may have diminished (see above).

Waters from present and past sewage farms have high contents of nitrate, chloride, sulphate and phosphate. Ammonium and nitrite contents are mostly relatively low, as is the permanganate number.

Drinking water should contain no more than 20 mg N_2O_5/l according th the Swiss Foodstuff Handbook (Lebensmittelbuch). The geological formations in Switzerland (primary rocks and calcareous formations) are quite different from those in N. Germany, for example. The limits are much lower there, since the normal contents of nitrate in ground and surface waters are only a few mg/l. This applies to the German alpine regions also.

It may be mentioned that a high nitrate content is not in itself considered to be detrimental to adults. It has been used so far only as a pollution indicator.

Damage to health of infants through nitrate-containing water.

Numerous cases have been described in the medical literature of recent years in which water, rich in nitrate, used for diluting mothers' milk, has considerably demaged to the health of babies. *H. H. Comly* was the first to describe the harm done to the blood count of infants by nitrate contents of 50 mg/l (J. Am. Water Works Ass. *41,* 147 [1949]). *H. E. Robertson* and *H. Draycott* made similar observations (Can. J. Public Health *40,* 30 [1949]). *E. F. Downs* then made a detailed account in the reports of the WHO. Attention was then directed to this problem in Germany and the harmfulness to babies was confirmed.

The damage is due to a reduction of nitrate to nitrite in the upper small intestine by a non-physiological intestinal flora, causing serious cyanosis and methaemoglobinaemia. *F. W. Wedemeyer* reported such damage when the water contained 25 to 40 mg NO_3^-/l (Archiv für Kinderheilkunde *152,* 267–275 [1956]). *G. Kittel* reported serious illness of babies with nitrate amounts of 161 to 218 mg/l a proposed 45 mg NO_3^-/l as the limit for drinking water (Öffentl. Gesundheitsdienst *18.* No. 12. [1957]). *F. Souchon* described similar cases from the Bremen region, where the water for dilution had 218.0 mg nitrate/l (Deutsche Medizin. Wochenschr. *81,* 1091–92 [1956]). *L. Teerhag* and *H. Eger* termed 50 mg NO_3^-/l the "toxic limiting value" according to their experience (Der öffentl. Ges.-Dienst *20,* 1–2 [1958]; and also *K. Horn* (Städtehygiene *9,* 21–25 [1956]). On the basis of general experience the WHO has fixed 100 mg/l as the limit for drinking water. This value was reduced to 50 mg/l by the "Study Group for Drinking Water Questions of the European Bureau of the WHO"; the coming (drafted) drinking water decree of the Federal German Republic has accepted this value.
It is little known that spinach, often figuring in baby nutrition, contains large amounts of nitrate, depending on the manuring (over 2000 mg/kg, Bund. GesundhBl. [1965]). A high nitrate amount leads to goitre in animal experiments; more information is to be found in the monograph of *H. Warmbach, A. Biwer, W. Buchsteeg* and *H. Thiele,* Düsseldorf [1962].

P. G. Sattelmacher demands that sterilised methylene blue solutions should always be available as a life-saving agent against infant cyanosis in hospitals in districts where the drinking water has over 50 mg NO_3^-/l (GWF *104,* 1321–1322 [1963]).
Twelve waterworks with water having more than 50 mg NO_3^-/l are to be found in the "Chemische Wasserstatistik der Wasserwerke in der Bundesrepublik" (*G. Giebler,* Munich [1959]). Of the remainder, 265 works have 0–5 mg NO_3^-/l; 98 have 5–10; 110 have 10–20; and 65 have 20–30. The remaining (14) have values of 30–50. In only a few exceptional cases are there thus high nitrate contents. The danger arises practically only in individual water supplies.
High nitrate amounts can be effectively removed from water according to *I. Delius,* by using the anion exchanger ES of the Permutit AG. Nitrate is undesirable in water also in some factories (e.g., for sugar works and industrial fermentation plants). See *H. Kruse,* Schriftenreihe d. Ver. f. Wasser-, Boden- und Lufthygiene, No. 14 [1959]. (Uniform Requirements of Drinking Water Composition and Procedures of Investigation in Europe); *K. Haag* GWF *100* [1959]; and *J. Delius,* Ges. Ing. *50,* 181–185 [1959]).

4. Potassium Permanganate Number and Chlorine Number

Reliable ground waters have only a very small permanganate consumption. It is usually less than 12 mg $KMnO_4$/l, generally only 3 to 6 mg/l. Surface waters, except reservoir and mountain waters, contain originally appreciably more organic materials (ca 10 to 30 mg $KMnO_4$/l).
Ground waters and surface waters in peaty areas always have much organic matter (hu-

mins). These are not damaging to health but impart to the water an unpleasant taste and yellow colour. Humins should therefore be removed in the central water supply (p. 52). The permanganate number of such peaty waters can be as high as 80 to 350 mg $KMnO_4$/l.
A water which has been contaminated by human or animal waste products can almost always be recognised through increase in its permanganate number, which can rise to 60 mg/l or more.
It is, however, incorrect to object to a water on account of an increased permanganate number, especially if it alone increases. Some handbooks (e.g., *Beythien,* "Laboratoriumsbuch für Lebensmittelchemiker", Dresden [1944]) contain the misleading information that the permanganate number must not exceed 10 to 12 mg $KMnO_4$/l. Such limits must be used with caution, particularly with the permanganate number. If one judged according to this limit, objection could be made to the greater part of the waters from the peaty parts of the N. German plain. Local and regional experience of the expert is therefore very important in the appraisal.
Humins are types of carbohydrate and consequently only partly determined by the permanganate oxidation. *B. Åberg* and *W. Rodhe* have drawn attention to this, based on their experience with the Swedish dystrophic lakes.
The situation is wholly different in Switzerland. The waters of the primary and calcareous rock regions possess only extremely small amounts of organic materials and have permanganate numbers of 1 to 5 mg/l only. The Swiss directions thus give a limit of 6 mg/l for the permanganate number. Similar considerations are to be applied to the German alpine waters, although the occurence of peaty waters must be borne in mind (see above).

Organic Matter and Water Colour
The surface waters of the alpine lands and other regions of primary rock and pure calcareous areas also have similarly low permanganate numbers. The alpine lakes show the well-known blue colour as a result of this extremely low content of organic substances. The colour of the water depends on the content of organic matter. The purest water, containing only traces of organic material, is blue in its upper layer. An increase in organic substances corresponding to a permanganate number of 10 to 15 mg $KMnO_4$/l yields greenish water (mixture of blue and yellow). Waters of permanganate number of over 25 mg/l are yellowish. This gradually intensifies with increase in permanganate number and, with 50 to 60 mg/l, the colour can be appreciated in even thin layers. Peaty waters with permanganate numbers of 80 to 100 mg/l are brown and numbers of over 200 mg/l yield red-brown colours.

Special cases: Waters which have stood in wooden pipes or containers have higher permanganate numbers which, as with waters from moor regions, are not always an indication of contamination. This shortcoming should, however, be eliminated (p. 2). The permanganate number of open waters is raised by leaves falling into them.
In all these last-named cases, a permanganate consumption of up to 25 mg/l is perhaps permissible for central water supply. Beyond this, treatment of the water must be planned (p. 52).
Chlorine number: It is possible to estimate the organic substances coming from faecal pollution by determining the chlorine number also. Urea and decomposition products of proteins respond more to chlorine than to permanganate.

If the chlorine number exceeds the increased permanganate number, the water is to be regarded as freshly contaminated, provided that the other pollution indicators indicate this also.

Organic Materials in Tap Water
Newer investigations show that the content of organic matter, in soluble and insoluble form, plays a big part in sterilisation of drinking water, as well as having the sanitary significance mentioned above. A very small increase in permanganate consumption of only 5–10 mg/l leads in long pipelines to increase in bacterial content, disappearance of anti-bacterial agents, loss of oxygen and development of odour. This is often disturbingly evident in municipal works and consumers at the ends of the pipe lines complain about brown or cloudy water.

5. Chloride Ion

Normal ground waters contain 10 to 30 mg Cl^-/l. In the alpine regions and those of the oldest rocks, surface and ground waters contain less than 10 mg/l, usually 3 to 5. On the other hand, surface waters very rich in chloride are often found in sea coastal areas (brackish water can have several hundred mg Cl^-/l). Higher chloride ion contents are often found inland also, then due to geological causes, such as proximity to salt deposits and zechstein formations as in the GDR. Standing and flowing waters (e.g. the rivers Werra and Weser) and consequently neighbouring ground waters are enriched in chloride through the waste waters of the potash industry in the GDR.
Variegated sandstone formations have waters of low chloride content (ca. 10 mg Cl^-/l).

In all cases where a high chloride ion content (over 30 mg/l, or over 10 mg/l in many mountain regions-see above) is not the result of geological-mineralogical circumstances, contamination through waste products can be assumed, provided of course that the other pollution indicators (increased permanganate number, appearance of ammonium, nitrite, nitrate and phosphate ions, etc.) indicate this at the same time.

Contaminated waters always have an increased amount of chloride; they usually contain 30 to 100 mg Cl^-/l, even 200 to 300 mg/l if pollution through waste waters is considerable. The chloride content of a contaminated water remains high and constant even after seeping through to the lowest soil layers. The chemical analysis thus yields the chance of recognising waters of unappetising origin in cases where the bacteriological investigation shows nothing (p. 57).
An objection to a water merely on the basis of an increase of chloride content alone is, however, incorrect. Chloride values of up to 250 mg/l occur not infrequently in ground waters of the N. German plain, although this cannot be related to salt deposits or impurities. Ground waters near salt deposits often have 100 to several 100 mg/l (e.g., Eschershausen, 1000.0 mg Cl^-/l so far). What was said on p. 68 about the increase in only one of the pollution indicators applies with special force to judgments based on chloride content.
If one has doubts about the assessment, the situation may be clarified through serial investigations. Pure unaffected waters have constant chloride values, whereas fluctuating values

are always obtained in cases of contamination. Water from past or present sewage farms possesses high chloride, nitrate and sulphate ion amounts but a low permanganate number.

A modest increase in chloride content in itself has no evil consequences for health. The other food consumed contains much more chloride than that which can occur in water. Amounts of over 250 mg Cl^-/l give the water a salty taste, however. According to "Leitsätze für Trinkwasser" and DIN 2000, the chloride content of drinking water must not exceed 250 mg Cl^-/l. The individual chlorides behave differently. The limits set are thus: 400 mg NaCl/l, 500 mg $CaCl_2$/l and ca. 168 mg $MgCl_2$/l. According to the author's experience, increased chloride content is especially noticeable in the taste of coffee, which becomes unpleasant if the content exceeds 250 mg/l. It is similar with tea.

The WHO has proposed a limit of 350.0 mg Cl^-/l.

Drinking waters containing permananently 1000 mg/l or more cause kidney maladies, or at least susceptibility to these, if drunk over a longer period (*K. Höll,* Vom Wasser, [1952]). Such waters are unsuitable as drinking water, apart from the salty taste, especially that of drinks prepared from it.

Chloride ion contents of 700 mg/l are permitted in North African regions for continuous consumption and 2800 mg Cl^-/l for temporary consumption (potabilité momentanée).

On the other hand, waters low in salt are dangerous to health over a longer period; they increase the internal cell pressure as a result of osmosis (see below). Waters with high mineral salt content influence the blood composition. The effect of a cure deliberately aimed at with particular medicinal waters and achieved by a "transmineralisation" in the blood circulation, is, however, not good for the health if permanently taken with any salt-containing waters.

For some uses of service water, chlorides can be removed with copper zeolites.

A high chloride content has a destructive influence on concrete, mortar and iron. These materials must be provided with firmly adhering protective paint coatings for contact with waters which are rich in salts. The softer the water, the more powerful is the influence of its chloride on iron pipes. Even 200 mg Cl^-/l in waters with less than 10° hardness bring about increased corrosion. The corrosive powers of brackish peaty waters from the Friesian and Oldenburg regions of N. Germany are the highest (see. p. 40).

Demineralisation

An economical procedure for demineralisation of brackish water and that of high mineral content has not yet been worked out in Germany. Water supply in the marshy areas is thus still an unsolved problem. Research, supported by the Federal Ministry for Water Economy, is to help in finding a solution.

The ion exchange apparatus of the firm of Permutit AG, Duisburg, W. Germany has nevertheless proved valuable for salt removal from brackish and also sea water in single water supplies. Apparatuses of this type are now used in emergency sea services. Salts are wholly removed by the ion exchange. The Atlas Werke of Bremen (Krupp A.G.) are soon to construct smaller and larger plants for demineralisation.

The firm of Becker Brothers, Beckum, W. Germany produces a "sea-water evaporator". The electrodialysis procedure is also used at sea. 1 cubic meter (ca 250 gallons) of partly

demineralised water costs 1 to 2 DM (12–25 new British pence or 30–60 U.S. cents) (Firm of Karl Klein, Mannheim).
Both these procedures are cheaper than distillation (*R. Elliassen,* GWF *96,* 54 [1955]).
The winning of drinking water from sea water is gaining increasing importance in various parts of the world, especially in hotter countries. A demineralisation plant is to be built now in Heligoland an account of the difficulties with fresh water.
Salt removal from sea water is economical nowadays in areas of water shortage, as a result of international research (WHO and FAO). A large plant for demineralisation of sea water was built already in 1961 in the Gulf of Mexico, providing 4 million litres of fresh water per day for the local region of water-shortage.
Three practical procedures have been developed: distillation, evaporation and freezing procedures. For more details, refer to the symposium "Süßwasser aus dem Meer" (Fresh Water from the Sea), to *K. Höll* and *K. W. Giesler,* Stoff-Hütte, 4th Edition, Vol. 3 [1967] and to the Dechema Monograph *47* [1962].
1.2 milliard cubic metres (ca. 3×10^{11} gallons) water are used daily in the USA and the available water reserves amount to 1.9 milliard m^3 per day. This limit will soon be reached if consumption increases, as is expected.
Fully demineralised water does not appeal to human consumption. A little non-demineralised or unsoftened water must be added.

6. Phosphate Ion

Pure waters, especially mountain waters, contain phosphate amounts not exceeding 0.1 mg PO_4/l, usually less than 0.03. Large amounts can get into water through wastewaters, refuse, scullery waste and especially urine (4.5 g phosphate per 24 hours per person). Objection must be made to waters containing over 0.1 mg PO_4/l, provided that the other pollution indicators show the same. Values of over 0.3 mg PO_4/l always give grounds for grave suspicion of contamination. Phosphate content is the surest indicator of faecal pollution.

Special cases: Phosphate amounts of up to 1 mg/l are sometimes found in moor regions. Artificial manures (phosphate or mixed manures) can raise the phosphate content of ground water.
Phosphates are strongly adsorbed in some soils; the phosphate content of deep, contaminated ground waters is thus sometimes normal. A low phosphate amount is therefore not always to be regarded favourably. Against this, high phosphate content of the water (0.1 to 0.5 mg PO_4/l) has geological origins in some regions. These waters possess, however, steadily constant values, whereas the phosphate values fluctuate in the course of time or continuously where there is contamination. Exact periodic studies enable this to be established.
Hot springs often have very high phosphate amounts (*K. Höll,* Heilbad u. Kurort *21* [1969]).
The phosphate content is significant for the pipe system with some aggressive waters; it yields a natural protective coating to the pipes. According to *B. Haupt,* the natural phosphate content of only 0.1 mg/l suffices for this (Vom Wasser VII. *225* [1939/40]). Poly-

phosphates are added nowadays to aggressive waters for this purpose of protecting the pipes (p. 172).

7. Sulphate Ion

Normal waters have from a few up to 50 mg SO_4/l, generally 10 to 30; several hundred mg/l may occur in waters from salt areas. Waters with a more or less high gypsum content (up to 500 mg SO_4/l and more) occur in some regions. Waters from brown coal regions frequently have a high sulphate content, sometimes even some free sulphuric acid (e.g. waterworks at Elsterwerda, E. Germany). Waters from the "Münder Mergel" (Münder marl) and similar geological formations contain likewise up to 1000 mg SO_4/l. Those from the periphery of petroleum regions are almost always without sulphate (*W. Carlé,* Heilbad u. Kurort *13,* No. 6 [1961]).
Variegated sandstone formations yield waters of low sulphate content (10 to 20 mg SO_4/l).

Hygienic significance: Contaminated waters usually have increased sulphate content (ca 100 to 200 mg SO_4/l and more) along with much chloride, nitrate and phosphate. This can be considered with some caution in the hygienic assessment of the drinking water (see above). Urine and liquid manure namely contain much sulphate. Water from present and former sewage farms always possesses high sulphate, chloride and nitrate ion content and a low permanganate number.
The sulphate amount is often exceptionally high in the vicinity of refuse deposits (from primarily formed H_2S).

Technical importance of the sulphate content: Gypsum-containing water is unsuitable as boiler water. Scale containing gypsum like that rich in silicate is especially dangerous in steam boilers and difficult to remove (see p. 231).
Waters containing much sulphate are harmful to concrete buildings. Attack begins already with 150 to 200 mg SO_4/l, especially when magnesium sulphate is present (p. 209). The mortar also is attacked.

Sanitary importance: Waters of high sulphate content, particularly alkali and magnesium sulphates, act as a purgative. Even small amounts such as 200 to 300 mg/l are noticeable through their disturbance of the intestinal function. The body becomes accustomed to it after a while. 250 mg SO_4/l may be regarded as the limit. A content of over 1000 mg/l gives rise to concern. Waters rich in sulphate are frequent in Lower Saxony (200 to 1000 mg/l). The sulphate content in the tropics and some subtropical lands is of great significance, since many waters there have an enormous amount of sulphate, thus acting as an aperient. Since the inhabitants in any case suffer from thirst, diarrhoea, which augments thirst, must be avoided at all costs.
Water containing more than 500 mg SO_4/l and 1000 mg Cl/l is unsuitable for consumption by domestic animals also. Pigs are especially sensitive.
A. I. Bokin carried out detailed animal studies with waters, rich in sulphate, and made observations on people (Ref. in Lit. Ber. Wasser u. Abwasser *5,* 284 [1956]). According to

his findings, amounts of 1000 mg SO_4/l and more bring about changes in the alimentary canal and subsequently disturbances in the nervous system.
The European bureau of WHO has now established 250 mg/l as the limit (see *H. Kruse, 22*).

8. Potassium and Sodium

Sodium and potassium contents, like chloride, are increased by faecal contamination; the potassium content is higher than the sodium content in such cases. Normal ground waters contain from a few to 50 mg Na/l and only 1 to 2 mg K/l. Higher values imply contamination unless in mineral water issuing from salt lenses or zechstein. This problem can often be cleared up by parallel studies on neighbouring wells. *K. Knie* and the author have found enormously high potassium values (several hundred mg/l) in the shallow wells in the "Seewinkel" of the Neusiedler See, Austria. Thermal and geyser waters also contain much potassium (*K. Höll*, Heilbad u. Kurort *16* [1964]). Occasionally increased potassium amounts arise in ground water through leaching of potassium salts in manures (*K. Höll*, Vom Wasser *30* [1963]).
See p. 134 for determination of potassium.

9. Residue on Evaporation

Contaminated waters have an increased residue on evaporation. No limiting values can, however, be given. A hint which assists judgment can be obtained only by comparison with values from neighbouring well waters which are above suspicion.
The residue on evaporation of polluted waters blackens on ignition and yields an unpleasant smell (like burnt hair). A high relative loss in weight (over a third of the total weight) is also an indication of contamination.
Waters with over 1000 mg/l residue from evaporation can be termed mineral waters. Half of the content of hydrogen carbonate can be added to the residue mineral waters of low salt content and the total used as a basis for assessing the water. The medicinal water analysis confirms the presence of a mineral water, since in it all the hydrogen carbonate is counted, whereas only a half of this appears in the residue on evaporation.

10. Screenable Material

Turbid drinking water can be submitted to the Nannoplankton net of *K. Höll* (see Part III) (manufactured by Franz Bergmann KG, Zehlendorf, Berlin); it permits microscopic examination of the sieved material, giving an insight into the nature of the substances causing the turbidity.
Good drinking water should not contain more than 0.5 ml of screenable material per cubic metre.

11. Carcinogenic Materials in Water

Many polycyclic aromatic hydrocarbons possess carcinogenic properties, acting both on the skin and on the stomach-intestinal tract, i.e. internally and externally. 3,4-Benzpyrene, 10, 11-benzfluoranthene and 1,2-benzanthracene are principal examples of such compounds.

Thanks to the fundamental work of *J. Borneff,* we are well informed about the occurrence of such carcinogenic substances in surface and ground water (Arch. f. Hygiene [1959–1963]). *J. Borneff* found these compounds everywhere in random samples of water and also in larger amounts in filter materials (gravel, active charcoal) and in the mud from surface waters.

The carcinogenic compounds mentioned are present in mineral oils of all types, in petrol (gasoline), fuels such as diesel oil, kerosine (aviation petrol) and fuel oils. As a result of leaks in containers of mineral oils, ground water acquires a disagreeable taste and smell and also the carcinogenic compounds enter the water. In addition, mineral oils take up the carcinogenic materials present in the soil since they possess good solvent power for them.

Rain helps carcinogenic compounds to enter the soil from the atmosphere of thickly populated and industrial regions. Exhaust from automobiles and motors and especially from fuel oil plants, soot from fuels and diesel oil and dust contain the carcinogenic substances mentioned. They reach ground and inland water with various industrial waste products (from the petroleum industry, coke-producing plants, chemical factories, roofing felt factories). Even the wastewaters from slate factories contain many carcinogenic compounds. Run-off water from asphalted roads and surfaces treated with bitumen brings carcinogenic materials into water. Ships releasing oil introduce into water large amounts of carcinogenic material and hence into drinking water where supplies come from rivers. Sealing with bitumen of drinking water tanks of concrete and masonry, hitherto customary, therefore now appears dubious. In future, concrete will have to be protected from attack of water by using new painting materials, such as Inertol or by sealing off with foil (p. 160).

Larger amounts of carcinogenic substances must always be reckoned with in rivers such as the Rhine, Main, Mosel, etc., on account of the loading with waste waters. Further, river waters contain nowadays detergents (p. 217) which increase appreciably the carcinogenic action through their surface activity. Maintained consumption of untreated river water represents, according to *Borneff,* a danger of cancer, since several mg carcinogenic substances are taken up per person per year. *O. Hettche* drew attention to the augmented activity of small amounts of carcinogenic materials in water containing detergents (Arch. f. Hygiene *144,* 467 [1960]). *T. Böhm-Gössl* and *R. Krüger* of Hüls, W. Germany have, however, shown experimentally that the amounts of detergent in the receiving water are insufficient to bring about an enrichment of benzpyrene (Lit. Ber. *14,* p. 50).

The more contaminated a surface water is, the greater is the danger through carcinogenic substances. Water which tastes of mineral oil should be used as drinking water only after filtration through active charcoal. The limit of safety with carcinogenic materials is said to be 0.03 mg/cubicmetre.

Even if no serious danger threatens from the drinking water alone, one must, according to *Borneff,* reckon with the additive and perhaps multiplicative effects of other foodstuffs; amounts of 1 to 10 mg/year of these already reach the danger limit for humans. *Borneff* thus says that drinking water must be free of carcinogenic compounds or else it must be purified with active carbon filters (*J. Borneff* and *R. Fischer,* Arch. f. Hygiene u. Bakt. *145* [1961]).
Studies of *Borneff* and *Fischer* have demonstrated that heath soils also contain traces of carcinogenic materials (Arch. f. Hygiene *146,* 1–16 [1963]); evidently these materials get into the soil and seepage just from the atmosphere.

Determination

To determine the carcinogenic substances, the water sample is extracted with cyclohexane, the extract evaporated down, the residue separated chromatographically and the 3,4-benzpyrene determined spectrophotometrically via fluorescence (*L. Scholz* and *H. J. Altmann,* Z. Anal. Chem. *240,* 81–91 [1968]).

a) Removal

Treatment of river water with active charcoal largely removes the carcinogenic compounds as well as taking out chlorophenols and improving taste. According to *Borneff,* a contact period of 15–30 min for 6–50 g/m^3 of water (either as filter or added powder) is needed. Treatment with carbon is advisable in all other cases of carcinogenic waters also.
Sand filters reduce the content of carcinogenic matter by 67% but mechanical filters after flocculation with iron chloride, by 98%. *Borneff* states that graphite filters are ineffective.
Chlorine dioxide is far better than chlorine for removal of carcinogenic substances (*J. Borneff,* GWF *110,* 1–6, [1969] and Arch. Hyg. *152,* 265–276 [1968]). Chlorination requires 2 hours, chlorine dioxide only 15 min.
Carcinogenic substances are largely decomposed by ozone also, according to *J. Reichert*; it must be preceded by aeration and filtration, however (GWF *110,* 474–482 [1969]).

b) Regulations

This information about the carcinogenic matter in water has given a new basis to the ban on washing cars in outside waters. So-called petrol separators in the sewers are thus additionally important for protecting waters against carcinogenic substances.
"Model regulations concerning deposition of liquids dangerous to ground waters" (Muster-Verordnung über die Lagerung grundwassergefährdender Flüssigkeiten) (from the LAWA, Länderarbeitsgemeinschaft Wasser in W. Germany) includes extensive precautions against leaks and the DVGW regulations (Deutscher Verein von Gas- and Wasserfachmännern) likewise.
S. S. Blioch reviews the literature in the Russian journal for "Hygiene and Health" *30,* 86–89 [1965].

E. Normal Values for Drinking Water

With every reservation based on what has been said above, the following values can be given as normal for good drinking water:

Ammonium ion	0.0 mg NH_4/l (except in iron-containing ground waters)
Nitrite ion	0.00 mg NO_2/l
Nitrate ion	5.0 to 30 mg NO_3/l
Phosphate ion	below 0.03 mg PO_4/l
Chloride ion	10 to 30 mg Cl/l
Sulphate ion	10 to 60 mg SO_4/l
$KMnO_4$-consumption	3 to 8 mg $KMnO_4$/l
Total hardness	5 to 15° dH
Iron	< 0.05 mg Fe/l
Manganese	< 0.1 mg Mn/l
Urochrome	0 mg/l

See p. 201 for values for tap water

F. Examples of Drinking Water Analyses for Individual Wells and their Evaluation

1. Dug Wells, strongly polluted by cow manure ("Minor drinking water analysis")

Appearance	clear, faintly yellowish
Smell	musty, mouldy
Ammonium ion	1.6 mg NH_4/l
Nitrite ion	1.2 mg NO_2/l
Nitrate ion	125.0 mg NO_3/l
Phosphate ion	0.65 mg PO_4/l
$KMnO_4$-consumption	88.0 mg $KMnO_4$/l
Urochrome	25.0 mg/l
Chloride ion	260.0 mg Cl/l
Sulphate ion	90.5 mg SO_4/l
Total hardness	26.50° dH
Carbonate hardness	19.63° dH
Non-carbonate hardness	6.87° dH
Iron	0.5 mg Fe/l
pH-value	6.83

Assessment: The water cannot be approved on grounds of the chemical analysis. Considerable contamination is shown by the high content of nitrogen compounds (ammonium, nitrite and nitrate ion), of organic matter and of phosphate and chloride ions.

The analysis demonstrates the influence on the water of the cowshed, only 15 m (50 feet) distant. The well must be moved and moved immediately. In the meantime, another well must be used and would have to be tested beforehand. If no alternative well is available, the water from the polluted well may be used after boiling (10 mins).

2. WATER FROM A DRILLED WELL WHICH IS NOT SUSPECT
("MINOR DRINKING WATER ANALYSIS")

Appearance	clear, colourless and without sediment at the time of sampling; yellowish-brown sediment of iron oxide hydrate 24 hours later
Taste and smell	no special feature
Ammonium ion	0 mg NH_4/l (below 0.03)
Nitrite ion	0 mg NO_2/l (below 0.01)
Nitrate ion	1.0 mg NO_3/l
Phosphate ion	0.01 mg PO_4/l
$KMnO_4$-consumption	6.5 mg $KMnO_4$/l
Urochrome	0
Chloride ion	23.8 mg Cl/l
Sulphate ion	98.5 mg SO_4/l
Iron	1.6 mg Fe/l
Manganese	0.3 mg Mn/l
Total hardness	33.5° dH
Non-carbonate hardness	17.5° dH
Carbonate hardness	16.0° dH
pH-value	7.26

Assessment: The water is without objection from a sanitary point of view, as based on the chemical analysis and also because the bacteriological data are similarly favourable.
The high iron content and hardness render the water unusable for service and domestic purposes, e.g. very unfavourable for preparing food (such as legumes) and drinks (tea, cocoa); for laundering purposes it also cannot be used on account of the hardness (high soap consumption) and high iron and manganese content (staining with spots).
If no other source of water supply is available, it would be advisable to soften the water and to remove the iron with a domestic filter.

3. WATER FROM A SHALLOW WELL
("MINOR DRINKING WATER ANALYSIS")

Appearance	faintly yellowish, clear, little sediment
Ammonium ion	0.15 mg NH_4/l
Nitrite ion	0.06 mg NO_2/l
Nitrate ion	25.0 mg NO_3/l
$KMnO_4$-consumption	12.0 mg $KMnO_4$/l
Chloride ion	18.6 mg Cl/l
Sulphate ion	16.5 mg SO_4/l
Phosphate ion	0 (below 0.01 mg PO_4/l)
Iron	0.3 mg Fe/l
pH-value	7.0
Urochrome	0
Total hardness	5.7° dH
Carbonate hardness	4.2° dH
Non-carbonate hardness	1.5° dH

Assessment: The water can be used as drinking water, since the bacteriological finding indicates the same. If possible, the well should be deepened.
See p. 195 for examples of the major drinking water analysis.

V. Major Drinking Water Analysis of Tap Water

Chemical Examination of Drinking and Service Water from a Central Water Supply

The following working procedure is recommended for examining water for a central supply, and piped water:

A. Working Procedure

1. *Evaluation of Aggressivity* (on the site)
   1) Reaction
   2) pH-value
   3) Total free carbon dioxide
   4) Heyer-test
   5) Oxygen tests
   6) Hydrogen sulphide
   7) Attack on lead
   8) Qualitative nitrite test
   9) Clarity and colour
2. *Hygiene-chemical analysis*
   1) Colour
   2) Degree of turbidity
   3) Ammonium ion
   4) Nitrite ion (see above)
   5) Nitrate ion
   6) Chloride ion
   7) Permanganate consumption and chlorine number
   8) Phosphate ion
   9) Sulphate ion
   10) Urochrome
3. *Evaluation of the service value of the water*
   (technical water analysis, investigation of service water)

   Total iron, iron (II) compounds, iron (III) compounds
   Manganese
   Combined carbon dioxide
   Alkalinity, acid consumption
   Acidity, alkali consumption
   Calculation of the aggressive carbon dioxide and the carbon dioxide which interferes with rust protection, from the Heyer test and the pH value
   Carbonate hardness

Total hardness
Non-carbonate hardness
Calcium content
Magnesium content
Hydrogen sulphide and sulphides
Residue on evaporation and on ignition
Electrolytic conductivity
Silicic acid and silicates
Evaluation of artificial additives (chlorine, aluminium, polyphosphate, etc.)
Evaluation of lead, copper, zinc, arsenic, aluminium, tin and potassium
If required, determinations of phenols, mineral oil, detergents and pesticides.

B. Evaluation of Aggressivity

1. Reaction

Determination of the reaction of a water with litmus paper generally has little value nowadays. A pH-determination is usually made. As a rough test of the aggressivity, determination of the reaction with an indicator may be of some use, e.g., with rosolic acid (see below), methyl red (0.1% alcoholic solution), methyl orange (0.1% aqueous solution) and phenolphthalein (0.375% alcoholic solution); an approximate idea is thereby obtained whether the water is aggressive. An acid or alkaline reaction, as shown by these indicators, points the way for subsequent investigations.
All soft waters which are rich in carbon dioxide and also harder waters with considerable excess carbon dioxide give a yellowish colour with *rosolic acid.* The reddish or yellow-red colour of the indicator remains unchanged with all other waters.

Rosolic acid reagent: 0.25 g rosolic acid is dissolved in a little alcohol by warming. 20 ml alcohol are added, the solution is neutralised with a few drops of barium hydroxide and made up to 50 ml. Five drops of this reagent are added to 50 ml of water under investigation.
Methyl orange (Reagent No. 18) yields a red coloration with waters containing free mineral acid, such as occur in brown coal regions; and an orange-red colour with waters rich in carbon dioxide. Waters showing an acid reaction to methyl orange are naturally strongly aggressive.
Waters giving a pink or reddish colour with phenolphthalein contain carbonates and perhaps even hydroxides. These attack lead pipes (p. 154).

2. pH-Determination, Hydrogen Ion Concentration

General: Hydrogen ions bring acid reaction. The number of hydrogen ions in g/l, the hydrogen ion concentration, is 1×10^{-7} for purest water. In alkaline waters it is smaller, although still measurable; in acidic waters, it is larger than 1×10^{-7}. The hydrogen ion concentration is a measure of the reactivity degree of the water. 10^{-14} is the dissociation constant of absolutely chemically pure water at 22° C; 10^{-14} g water per litre is thus dissociated into H-ions and OH-ions. Equal numbers of both must be present in this pure, neutral water, so that it contains 10^{-7} H-ions and 10^{-7} OH-ions. The indices are somewhat clumsy and consequently the hydrogen ion concentration is given as pH-value, namely, the negative logarithm of the hydrogen ion concentration. pH = 7 is therefore neutral; pH 7 to 14 is the alkaline region and pH 0 to 7, the acidic region.

Hydrogen ion concentration	10^{-1}	10^{-7}	10^{-14}
pH-value	1	7	14

pH-value	0	1	2	3	4	5	6	7	8	9	10	11	12	13	14
		strongly → acid ← weakly							weakly		→ alkaline ← strongly				

Values Encountered: Normal waters have pH-values of about 7. Processed waters can have values of over 8. Waters from peaty regions have pH-values of 5 to 6; this pH is shown also by soft waters with much free carbon dioxide, and mineral waters, rich in carbon dioxide (acidulous waters, pH 4 to 5.5).
The pH of distilled water ranges from 5.0 to 5.7. depending on the carbon dioxide and hydrogen carbonate contents of the water distilled.
K. Schwabe's monograph, 3rd Edition, Dresden, contains further information about pH.

a) Approximate pH-Determination

1. *Merck's universal indicator* permits rough pH-determination in the range 0 to 9, with an accuracy of 0.5 unit.
   A few drops of each of the two indicator solutions (for pH regions 0 to 5 and 4 to 9) are mixed in a test tube or on a porcelain spotting tile with some ml of the water under investigation. The colour(s) are compared with the scale provided by the firm of Merck.
2. *Merck's pH-papers,* specially for water investigation, can be used also.
3. *The Czensny apparatus* (Firm of Bergmann KG, Berlin-Zehlendorf and Hamburg) yields rough values for pH between 3 and 9, with an accuracy of 0.5.
4. Using the Ruhrverband pH-pocket colorimeter.

b) More Accurate Colorimetric Determination of pH

The Hellige-comparator gives fairly accurate pH values in intervals of 0.2 unit and permits estimates of intermediate values of 0.1 pH unit. For pH-values of ca 7, bromothymol blue is used as indicator; 0.5 ml is added to 10 ml water in a special cuvette of 13 mm thickness. Phenol red is a very suitable indicator for weakly alkaline waters. The colours are compared with the colour discs (stable to light) of the Hellige-comparator.

Interferences: Turbidity and the colour of the water itself are compensated with the com-

pensation cuvette. The colour is affected by chlorinated water; the pH-value is thus estimated after the chlorination plant has been switched off.
If the turbidity or water colour is more intense, the sample must be diluted with conductivity water or purest distilled water (Reagent No. 10). The compensation cuvette must always contain clear colourless water.
An indicator error of 0.1 to 0.3 pH-units arises with soft waters of very low salt content. A Nessler tube of 25 cm column height is then used and the indicator concentration is reduced (e.g. 0.25 ml for 100 ml water with bromothymol blue). The salt error resulting from very salty waters (above 0.5N) can be avoided by diluting with conductivity water.
Contaminated waters, especially wastewaters, give a "protein indicator error", except with the nitrophenol indicators.
The Hellige indicator solutions can suffer colour change on long standing. For example, bromothymol blue can change from green to blue. The correct colour can be reinstated with 0.1N HCl or NaOH. This control is important in pH measurements on weakly buffered, i.e., soft waters. The colour of the froth yielded by shaking the indicator solution shows if it is all right. Thus, the bromothymol blue indicator must have a yellow-green froth; and phenol red, bluish-red.

Temperature error: This error is about 0.1 pH unit above and below 15° to 25° C (see *Kolthoff,* Colour Indicators).

c) Accurate Electrometric pH-Determination

General: Electrometric pH-determination is performed by measuring the potential difference between the measuring electrode (quinhydrone, glass or antimony electrodes) which dips into the water under examination, and the reference electrode of known potential, the calomel electrode.
At room temperature, the temperature error is negligible but becomes significant at more extreme temperatures, e.g. in field experiments, especially with highly alkaline waters. Above 18° C, the values found are too high and below 18° C, too low. E.g. at 10° C, the deviations amount to 0.1 to 0.2 pH unit.
The pH apparatus should be checked from time to time.This is done with buffer solutions of known pH-values, e.g., standard acetate solution of pH 4.616 at 20° C (100 ml 1N NaOH + 200 ml 1N acetic acid, diluted to 1 litre with distilled water [Reagent No. 10]). The firm of Merck recently introduced buffer solutions for preparing very accurate pH-standard solutions.
The pH-apparatus can be checked also with the "pH-Wert-Geber" of the Firm WTW, Weilheim, Bavaria, W. Germany.
In recent years practically only the glass electrode has been used, since it has been brought to a state of very high perfection. In serial investigations it must be washed several times with the water until two duplicate measurements agree precisely.
All types of oil impair the glass electrode seriously. It is then difficult to clean. If necessary, colorimetric methods can be used for wastewaters, for which they nearly always suffice.
The pH must be measured on the spot as soon as sampling has been carried out. The pH-meter must thus be portable. Should such an apparatus not be available, more reliable re-

sults will be obtained in the warm summer months if a colorimetric determination is made on the spot (e.g. with the Hellige-comparator) instead of an electrometric determination another day after having carried the water samples a longer distance to the laboratory.

d) *Measurement of the Redox-Potential*

Together with the pH-value, the redox-potential is important in water chemistry, e.g., in the chlorination of swimming bath water (p. 202).
If two compounds are present in an aqueous solution which can be converted one into the other by reduction or oxidation, respectively, one speaks of a redox-medium. Strongly and weakly oxidising and reducing compounds can be present. This can be more accurately characterised by measurement of the redox-potential. The redox-state in drinking water differs from that in swimming bath water (*S. Carlson* and *U. Hässelbarth,* Vom Wasser *35,* 266–283 [1968]).
Measurement of the redox-potential requires a Knick pH amplifier, a platinum electrode, a saturated calomel reference electrode and a measuring vessel. The firm of WTW, Weilheim, Bavaria, W. Germany provides a very handy apparatus.

3. Carbon Dioxide

Occurrence: A few up to 100 and more mg/l of free carbon dioxide occur in all natural waters. Ground waters have usually 10 to 50 mg/l free CO_2, many springs much more. Surface waters generally contain only a few mg/l, especially running water (see also p. 161). Accurate CO_2-determinations are highly important for central water supply plants. The free carbon dioxide is divided into "associated" and excess CO_2 (p. 163–166).

a) *Determination of Total Free Carbon Dioxide*

Preliminary test: 100 ml of water sample are treated with a few drops of a 0.1% phenolphthalein solution, just turned pink by addition of NaOH. Decoloration shows the presence of free CO_2. If decoloration occurs also with a boiled (and then cooled) sample, it is due to free mineral acid (mine water). If the pink or red colour is intensified by addition of phenolphthalein solution, no free CO_2 is present; the pH value is over 8.0 and aggressive carbon dioxide is thus absent.

b) *Titrimetric CO_2-determination according to Trillich*

Preparation: A long, wide-necked flask of about 250 ml with a mark at 200 ml is used. The water is introduced into the flask with the help of a sampling rubber tube and long-maintained overflowing, as described on p. 11; the water does not then come into contact with the air. The volume is emptied down to the mark. 1.0 ml of an alcoholic phenolphthalein solution (0.375 g/l methanol) is immediately added and the contents titrated against a white background in bright daylight (not in sunlight) with 0.05N NaOH to a just detectable pink colour. It must persist 2 to 3 minutes. If it disappears immediately, as is usually the case, further NaOH must be added. The determination is repeated on a second water sample, to which

the total amount of alkali reagent required for the first titration is added directly; usually a higher titration is found, since the losses due to shaking in the first titration are eliminated.
The colour change is slow with hard waters; it is far sharper with soft waters.
Comparison with an untitrated sample of the water is strongly recommended, especially with waters which are coloured, e.g., through humins. A doubtful colour change, e.g. with hard water, can be made clearer by adding further indicator.
With water very rich in CO_2, the determination is repeated after having covered the water meniscus at the 200 ml mark with ether.
It is difficult to fill the usually narrow-necked standard flask with the sampling tube. These flasks are used in the standard procedure and present a problem. The standard flask would have to be filled from a litre flask, already filled in the correct way; this transfer would be with a very thin siphon tube and the water must be allowed to run over for a long time.

The 0.05N NaOH is highly sensitive to carbon dioxide and should thus be frequently prepared anew or standardised.

Calculation
For a 200 ml water sample, a 1 ml titration of 0.05N NaOH corresponds to 11.0 mg/l of free carbon dioxide.

Interferences: Rochelle salt solution (1 or 2 ml of a 33% solution, neutralised to phenolphthalein) is added to waters, rich in iron (more than 3 mg Fe/l, recognised through the shaking test (p. 108)). This is advisable also with very hard waters (30 to 50° dH and over), especially mineral waters. As an alternative, one can dilute with distilled water which has been boiled and rapidly cooled.
Rather too high values are found with very soft waters *(Strohecker).*
An appropriate value must be subtracted from the alkali consumption when hydrogen sulphide is present (sulphur springs).
According to the detailed and informative work of *U. Hässelbarth,* the titrimetric method of Trillich is no longer sufficiently accurate for practical use (*U. Hässelbarth,* DVGW-Congress, Wiesbaden 1961, GWF *104,* 90–93 and 157–160 [1963]; Z. anal. Chem. *214,* 264–280 [1965], Neue Deliva No. 8 [1963] and GWF *104,* 506–512 [1963]).
In this edition the old Trillich method is again given, along with the new method of *Hässelbarth* immediately below.

c) Acidimetric Determination of Free Carbon Dioxide according to U. Hässelbarth, with Electrometric End-point Indication

Preparation: A titration flask is required for the determination, which has a ground-glass fitment (14.5 mm) for the glass electrode and inlets for filling and for fitting in a burette exit. The water sample is introduced from the flask by means of a pump pipette which can easily be made from an ordinary 100 ml pipette, a stopper with two holes and a bent glass tube with a rubber pressure bulb. A stirring device with temperature control is necessary (e.g., the AGMOR, firm of Dr. Thiedig & Co., Berlin). Most water-works can use their own tap water for cooling, otherwise a thermostat is required or, for temperatures below 10° C, a cryostat. A pH-measuring instrument with temperature correction and accuracy of measurement of ± 0.05 pH is needed together with a glass electrode for end-point determination.

Procedure: The pump pipette is fitted on the mouth of the sample flask and rather more than 100 ml of the water is carefully forced into the pipette by means of the pressure bulb; the meniscus is then adjusted to the 100 ml mark by letting some water run out. The 100 ml are allowed to flow into the dry titration flask which contains a teflon-covered stirring bar. The tip of the pipette must touch the bottom of the flask. The flask with contents are placed in the holder in a thermostat at the required temperature and the glass electrode and burette fitted in. 0.02N NaOH is added to the titration flask from a 10 ml burette with 0.02 ml divisions. After each addition, the stirring mechanism is set in operation for 2 to 3 sec. This is repeated until the pH of the sample has attained 8.35 and remains there. A repeat titration is performed in which the amount of NaOH required for the first titration is added dropwise without interruption and then titration continued if necessary.
If more than 20 ml of reagent solution are needed, titration is carried out with 50 ml of sample water, diluted with CO_2-free, distilled water to 100 ml. 1 ml 0.02 N NaOH solution is equivalent to 0.88 mg free CO_2.

Interferences: More than 3 mg iron/l and carbonate hardness exceeding 10° dH interfere. Higher concentrations of these interferers are rendered harmless by adding 1 ml of a Rochelle salt solution. This consists of a 50% Rochelle salt solution, the pH of which has been adjusted to 8.35 with HCl or NaOH. This pH is checked by adding 1 ml of solution to 100 ml freshly demineralised water (with an ion exchanger) and measuring the pH of this diluted solution; this eliminates the salt error. Should the measured value deviate from 8.35, acid or alkali is added to the concentrated Rochelle salt solution and the dilution and measuring procedure repeated until the desired pH is established (*Hässelbarth,* GWF *104*, 88 et seq & *157* et seq [1963]).
With waters containing larger amounts of hydrogen sulphide (sulphur springs), the corresponding consumption of alkali must be subtracted from the titration.

d) Heyer-Test with Temperature Modification of the Sample according to U. Hässelbarth

This method is distinguished from the classical Heyer-test only through the fact that the sample is brought to the temperature that it had at the moment of sampling outside. This takes into account the considerable influence of temperature on the carbonate balance.

Sampling: A glass bottle of 0.5 l content with a stopper is filled on the spot as described on p. 11 (with the aid of a plastic tube or through a submerged funnel by allowing the water to flow for a longer period through it). The water must not come into contact with air and the sample must be free of bubbles. 1 g of powdered marble (Merck) and a teflon-covered magnetic stirring bar of 2.5 cm length are introduced into the water. The bottle is stoppered and stirred at 1000 rpm in a thermostat, accurate to ± 0.5° C, which has been previously set to the temperature of the water at the moment of sampling. A cryostat may be necessary here (Firm of Haake, Berlin).
The mixture is stirred for 2 hours and then filtered through a glass fibre filter. The first part of the filtrate is discarded. 100 ml of clear filtrate are pipetted out and titrated in an erlenmeyer flask with 0.1N HCl to pH 4.3 (electrometric end-point indication).

1 ml consumption of 0.1N HCl in excess of that consumed by the untreated water under identical conditions, is equivalent to 0.5 mmole/l (50.05 mg/l) $CaCO_3$ (*U. Hässelbarth*, GWF *104*, 89 [1963]).

e) Heyer-Test to establish the Specific Temperature- and Salt Effects with the Water under Investigation

Salt content and temperature have a powerful influence on the carbonate balance. Numerical values are given in DIN 50930, (Tables on p.111 et seq.) and were obtained from laboratory experiments and theoretical considerations.
The peculiarities of a particular water can, however, be determined only in a practical test with it. The Heyer-test as in d) is carried out. The analytical results yield a point in a diagram (*A. Grohmann*, Ges. Ing. *90*, No. 9 [1969]), through which the equilibrium curve is drawn in the prescribed form. It is advantageous to use a prefashioned diagram stamped on metal, together with a curved rule held with magnets (firm of G. Klinger, Berlin 45).
The diagram includes very soft waters with equilibrium pH of over 9.0 and of 7.0 and below, and utilises the "Analysis via Qc" of the carbonate balance. Qc is the sum of carbon dioxide and its anions:

$$Qc = \text{free } CO_2 + HCO_3^- + CO_3^{2-}.$$

After carrying out the Heyer-test as under d), the following four determinations are necessary:
pH-value
m-value (mval/l)
p-value (mval/l)
Ca^{2+} (mmole/l) (1 mmole/l Ca = 40.08 mg Ca/l = 56.08 mg CaO/l = 5.608° dH)
Qc = m – p (mmole/l). The concentration of Qc is given in mmole/l when the m- and p-values are quoted in mval/l. Most natural waters have a pH below 8.2. They must be titrated with NaOH to pH 8.2 and this yields a p-value with negative sign (see p. 92). One must bear this in mind when calculating Qc.
The value of log (Qc × Ca) is now plotted against the pH-value (x- and y-axes, respectively) in the diagram, giving the equilibrium point through which the curve has to be drawn. The peculiarities of a water, as far as salt content and temperature are concerned, are then fully respected.

Shortened method: Since it is immaterial which point of the equilibrium curve is obtained in the Heyer-test, 0.5 l of the water sample is deacidified with NaOH to a point in the vicinity of the expected equilibrium pH. Air exclusion is no longer necessary and it is possible to work even in an erlenmeyer flask without a stopper. 1 g marble powder and the teflon-covered stirring bar are introduced and the mixture stirred; 10 minutes suffice to establish equilibrium.
Adjusting Heyer-test: The shortened Heyer-test, just mentioned, has been given this name because it adjusts the equilibrium curve to the special features of the water being studied.

Comparative Heyer-test: The classical Heyer-test of d) compares the m-value of the untreated water with the m-value of the water after treatment with powdered marble. The continuously-functioning instruments "Kalkas" after *G. Axt* and "Conmor" after *U. Hässelbarth* can be used here.

Analysis via Qc: There is a noticeable tendency in water chemistry to ascertain the sum total (Qc) of carbon dioxide and it anions as a component of the water; this is comparable with, for example, total iron, total phosphate, total chlorine, etc. Calculations are then based on this value [66a]. This has advantages in the description of the equilibrium of very soft waters with an equilibrium pH of over 8.0 (up to 9.5); in the description of the biogenic development in eutrophic lakes and reservoirs; and particularly in calculating the lime aggressivity of mixtures of waters of varied compositions (*A. Grohmann,* Vom Wasser 1971).

4. Acid-Base Titration

a) General

A water sample with pH deviating from 7.0 contains H^+ or OH^--ions in excess, the amount of which can be calculated from the pH-value. This measures only the dissociated protons and hydroxyl ions. When weak acids and bases are not present in the water sample, the result of the acid-base titration agrees with that calculated from the pH-value. Weak acids and, or bases have, however, a so-called buffering action on the hydrogen-ions and this causes a discrepancy between acid-base titration value and data calculated from the pH.
The presence of weak acids or bases reduces the sharp jump in pH obtained during acid-base titration in their absence (thus titration of a 100 ml sample of water containing mineral acid with 0.1N NaOH gives a jump of pH from 3 to 11); the extent of the reduction depends on their nature and amount. The titration end-points of acid-base titrations in water chemistry accord with the pH jumps of the carbon dioxide. The end-points are defined as if carbon dioxide were the sole weak acid and its anions the sole weak bases in the water sample. This is true in very many cases, especially of natural waters, but it is an arbitrary assumption.
The results of acid consumption are therefore to be quoted as positive m- or p-values, without commentary; and those of base consumption likewise as negative m- or p-values. Only when conclusions are to be drawn from the titration results, e.g., about the concentration of free CO_2 in the sample, must one ask if such conclusions are possible (p. 93, 7).

b) Definition of the End-point of a Titration

Every pH-region between 1 and 14 can be buffered by judicious choice of buffer substances. One cannot therefore count on having a pH jump at all in the acid-base titration. The end-points must thus be defined and, as explained above, in connection with the two pH jumps found in the titration of carbon dioxide:
3 to 5 with <u>m</u>ethyl orange as indicator (colour change at pH 3.5 to 4.3; the result is the <u>m</u>-value).

7.8 to 9 with phenolphthalein as indicator (colour change at pH 8.3 to 9.5; the result is the p-value).

Titration to a colour change yields, however, poorly reproducible values. More accurate results are obtained by titrating to a pH end-point. For this reason, pH-determination with the help of glass electrodes is preferable to using colour indicators. Based on the titration of carbon dioxide, end-points in the determination of the p- and m-values are defined as follows:

(German Standard Procedures H_7)

pH end-point of the m-value determination: pH 4.30 (valid at exactly 20° C and at an ionic strength of 0.01 mole/l)

pH end-point of the p-value determination: pH 8.20 (valid at exactly 20° C and at an ionic strength of 0.01 mole/l).

If the pH-value is measured at the place of sampling, the pH end-point of 8.30 is taken, valid ad exactly 10° C and at an ionic strength of 0.01 mole/l.

c) Correction for the m-value

Only the components of the water sample are to be included in the determination. The acid consumption to attain pH 4.3 of water containing no buffering components must therefore be subtracted from the result. This correction can be obtained with twice-distilled, CO_2-free water and is $m_{cor} = 0.05$ mval/l; it must be used for every result where m is below 5 mval/l.

d) Sphere of Application and Interferences

The m- and p-values can be determined in all waters when the titration end-point is electrometrically indicated. Titration using indicators is always to be avoided if substances other than carbon dioxide and its anions cause buffering in the pH regions of 4.3 and/or 8.2; or if the colour of the sample masks the colour change of the indicator (cf also 3c, p. 87).

Substances such as humic acids, phosphates, citrates, tartrates, etc. buffer the hydrogen-ions in the region of pH 4.3 or 8.2, thereby causing the pH jump to be protracted. Expulsion of the free and liberated carbon dioxide facilitates recognition of the colour change with methyl orange for mineral waters rich in CO_2; this permits more accurate results to be obtained. According to *Berger* at *Lunz*, very precise results may be obtained if air is drawn through the titration vessel with the help of a glass sinter during determination of the m-value (private communication). This artifice is not necessary if the end-point is determined electrometrically.

5. Determination of the Acid Consumption (m- or p-value with positive sign)

Acid consumption is the number of milligram equivalents (mval) of acid which are needed in titrating 1 l of water to the pH value 8.20 (p-value) or 4.30 (m-value); this is that given under "Alkalität" (Alkalinity) in the DIN-norm 19641.

The positive m-value of a water is termed acid combination power (SBV in German = Säurebindungsvermögen) in limnology. A pocket apparatus for determining it in the field according to Prof. *Ohle* is supplied by the firm of Krebs, Freiburg, W. Germany.
Finely suspended calcium carbonate separates from surface waters sometimes in summer; this is established by measurements of turbidity before and after addition of acid (*W. Ohle*, Arch. f. Hydrobiologie [1952]).
100 ml of sample are introduced into the titration vessel and titrated with 0.1N HCl to the defined pH end-point. If indicators need to be used instead of electrometric end-point indication, 0.5 ml of 0.2% aqueous methyl orange solution (m-value) or 0.5 ml of 0.2% alcoholic phenolphthalein solution (p-value) is added. Should the acid titration be less than 2 ml, the titration is repeated with 0.02N acid. The same apparatus may be employed as for carbon dioxide according to *U. Hässelbarth*. The titration is carried out at room temperature, however.

Calculation
p-value: d ml water sample were titrated with S ml of fN acid to the pH-value 8.20.
p-Value = + 1000 · f · S/d (mval/l) and positive
m-value: d ml water sample were titrated with S ml of fN acid to the pH-value 4.30.
m-Value = + (1000 · f · S/d) − 0.05 (mval/l) and positive

6. Determination of the Base Consumption (m- or p-value with negative sign)

Base consumption is the number of milligram-equivalents (mval) of base which are needed in titrating 1 l of water to the pH value 4.30 (m-value) or 8.20 (p-value).
Titration with a base to pH 8.20 is possible only when the pH of the water is below this value. It is thus a determination of the volatile free carbon dioxide in the case of a natural water. To prevent loss of CO_2 during transport of the sample to the place of titration, insulated boxes are recommended containing lumps of ice enclosed in plastic. Alternatively the base consumption is measured on the sampling spot. The end-point pH of 8.30 is to be used for water sampled at lower temperature (valid for 10° and 0.01 mol/l ionic strength).

a) *Direct Titration of Base Consumption*

The same points are to be observed as in the determination of the acid consumption.

Calculation
m-value: d ml water sample were titrated with B ml of a fN base to the pH-value 4.30.
m-Value = − (1000 · f · B/d) − 0.05 (mval/l) and negative
p-value: d ml water sample were titrated with B ml of a fN base to the pH value 8.20.
p-Value = − (1000 · f · B/d) (mval/l) and negative.

b) Back Titration of an Excess of Base

Back titration is employed only when the content of weak acid (free CO_2) is large and over 10 ml 0.1N base (NaOH) are required in the direct titration according to a) to attain the end-point (*K. E. Quentin et al,* Z. Anal. Chem. *231,* 21 [1967]).
N ml of 0.25N NaOH and 5 ml of a masking solution (see below) are placed in a titration flask. d ml of water sample are introduced under the solution without entrainment of bubbles. The excess NaOH is titrated with 0.25N HCl to the pH 8.20 end-point. Let S ml acid be required.

Calculation: p-value $= -\dfrac{(N-S) \times 0.25 \times 1000}{d}$ (mval/l) and is negative.

Masking solution: 1 mol Rochelle salt + 1 mol sodium citrate in 2l distilled water, adjusted to pH 8.20 (cf the 50% Rochelle salt solution in *U. Hässelbarth*'s determination of free carbon dioxide).

7. Determinations derived from m-value, p-value and pH-value

The hydrogen ion concentration is given by log $[H^+] = -$ pH. The hydroxyl ion concentration is given by log $[OH^-] =$ pH $- 14$.
For further calculations it is a condition that the concentration of other weak acids and bases can be neglected in comparison with those of carbon dioxide and its anions. This is very often not the case with very soft waters (reservoirs).

$$\begin{aligned}
\text{m-value} &= 2\,CO_3^{2-} + HCO_3^- + OH^- - H^+ \\
\text{p-value} &= CO_3^{2-} - \text{free } CO_2 + OH^- - H^+ \\
Qc &= CO_3^{2-} + HCO_3^- + \text{free } CO_2 = \text{m-value} - \text{p-value}
\end{aligned}$$

pH-value of water below 8.0. The carbonate concentration is so small here that it plays no part in the above totals. The concentration of free CO_2 is identical with the minus p-value (free $CO_2 = -$ p-value); the concentration of hydrogen carbonate is identical with the m-value; and Qc = (m-value) − (p-value).
These derivations represent by far the most frequent applications of the p- and m-values in water chemistry. The determination of free CO_2 according to *Trillich* and to *Hässelbarth* correspond to them.
The hydrogen ion concentration is also to be considered for water of pH below 5.0.

pH-value of water above 9.0. Free CO_2 is still present in very small amounts but is without significance for the above totals. The carbonate concentration equals the plus p-value and the hydrogen carbonate concentration equals m-2p. Qc is again given by Qc = (m-value) − (p-value).
The hydroxyl ion concentration is also to be considered for water of pH above 10.0.

8. Calculation of the Agrressivity to Lime of Water

As Tillmans in his fundamental work (Ges. Ing. *35,* 669–677 [1912]), stated, the Heyer-test is ultimately the standard for characterising a water; the sample must be under thermostatic control during the experiment (p. 88). There is, however, the desire to draw conclusions about the character of the water from the analytical data. To this end, calculation is made from tables or formulae, of the equilibrium concentration of any compound which is concerned with the carbonate equilibrium; this is compared with the value analytically found. It is mostly carried out for the free carbon dioxide (see below), a method introduced by Tillmans at the time when a pH-measurement was not that easily possible! (1912)
The practical worker will become better acquainted with the remaining methods when he considers the following: The "associated" carbon dioxide (see p. 163) is not a compound in the chemical sense but rather a "desired value", as in regulation technique, with which the analytically determined "actual value" is compared. In the same way, the concepts of "excess CO_2", "lime-aggressive CO_2", etc, are ways of expressing the calculated deviation of "actual" from "desired" value. It is clear that a "desired value" can be formulated just as well for the other participants in the carbonate equilibrium (hydrogen carbonate, carbonate, calcium and also pH-value) (*G. Axt,* Vom Wasser *28* [1961] and *32* [1965]; *A. Grohmann,* Ges. Ing. *90,* No. 9 [1969]).

Analysis via the free carbon dioxide (Tillmans' method). Calcium, hydrogen carbonate and free CO_2 must be determined and also the temperature and ionic strength (salt content) as parameters. The Tillmans curve (curve according to Auerbach) and Tillmans table do without the calcium determination. They assume that calcium and hydrogen carbonate are equivalent. This condition was checked with the DVGW statistics for waterworks. Even with a generous interpretation, it applies to only about 10% of German waterworks. Free CO_2 should therefore no longer be calculated according to the Tillmans table (*K. Höll,* "Wasser", 4th Edition, p. 104) but only according to *U. Hässelbarth*'s procedure given in the following pages (Vornorm DIN 50930).

Analysis via pH-hydrogen carbonate (Saturation index according to *Strohecker* and *Langelier*). Calcium, hydrogen carbonate and the pH-value and also the parameters of ionic strength and temperature must be determined (p. 96). *J. Hallopeau* gives a graph (Terres et Eaux *35,* 4 Trim. [1960]).
Analysis via pH-Qc (Total content of carbon dioxide and its anions, p. 93). Calcium, the pH-value and the Qc-value (Qc = m-p) as well as the parameters of ionic strength and temperature must be determined. The parameters do not have to be determined specially if a suitable equilibrium curve has been adjusted to the peculiarities of a particular water (p. 89) (*A. Grohmann,* Ges. Ing. *90,* No. 9 [1969]). Analysis via Qc has its special meaning in the calculation of lime aggressivity of mixed waters (p. 152).

9. Calculation of the Free Excess Carbon Dioxide according to Hässelbarth

The following analytical data are needed to calculate the content of free, excess carbon dioxide in a water:

C_{CO_2}	(mval/l) or (mg/l)	=	concentration of the free CO_2
$C_{HCO_3^-}$	(mval/l) or (mg/l)	=	concentration of the hydrogen carbonate
$C_{Ca^{2+}}$	(mval/l) or (°dH)	=	concentration of calcium (calcium hardness)
$C_{Mg^{2+}}$	(mval/l) or (°dH)	=	concentration of magnesium (magnesium hardness)
C_{Cl^-}	(mval/l or (mg/l)	=	concentration of chloride ion
$C_{SO_4^{2-}}$	(mval/l or (mg/l)	=	concentration of sulphate ion
t	(°C)	=	temperature of the water

The content of excess carbon dioxide is given by the difference between the amount of free carbon dioxide and of "associated" carbon dioxide (see p. 163) (equilibrium concentration)
$C_{excess\ CO_2} = C_{free\ CO_2} - C_{equilibrium\ CO_2}$
Whereas the amount of free CO_2 can be determined analytically, the "associated" concentration of CO_2 is calculated according to the corrected Tillmans relation:

$$C_{equilibrium\ CO_2} = \frac{K}{f_T} C^2_{HCO_3^-} \cdot C_{Ca^{2+}} \text{ (tables 1 and 2)}$$

where K is the Tillmans constant and f_T an activity coefficient which takes into account the effect of electrolytes on the equilibrium. This is a function of the total ionic strength μ,

$$\log f_T = \frac{3\sqrt{\mu} + 1.7\mu}{1 + 5.3\sqrt{\mu} + 5.5\mu}$$

To save frequent calculation, the values of f_T against ionic strength μ are tabulated (Table 3).
The ionic strength μ total is calculated from the analytical data according to the equation
$\mu_{total} = 5 \cdot 10^{-4}\ \Sigma\ nv = 5 \cdot 10^{-4}(2C_{Ca^{2+}} + 2C_{Mg^{2+}} + C_{HCO_3^-} + C_{Cl^-} + 2C_{SO_4^{2-}})$
n is the concentration in mval/l and v = valency of the ions. Values in mg/l must be divided by the equivalent weight of the ion and values in "German degrees (of hardness)" °dH, divided by 2.8. Example:

$C_{Ca^{2+}}$	= 3.0 mval/l	$C_{SO_4^{2-}}$	= 2.9 mval/l
$C_{Mg^{2+}}$	= 1,8 mval/l	C_{Cl^-}	= 0.15 mval/l
$C_{HCO_3^-}$	= 1.9 mval/l	t	= 10° C

$\mu_{total} = 5 \cdot 10^{-4}(2 \times 3.0 + 2 \times 1.8 + 1.9 + 0.15 + 2 \times 2.9) = 8.725 \times 10^{-3}$

With this value one finds in the Table (3) the nearest f_T = 1.555. The equilibrium concentration of the free carbon dioxide can now be calculated when the value for K is used which corresponds to the water temperature of 10° C (taken from Table 1) namely 1.339×10^{-2}.

$$C_{equilibrium\ CO_2} = \frac{1.339 \times 10^{-2}}{1.555} \times (1.9)^2 \times 3.0 = 0.093 \text{ (mval/l)}$$

The K-value is taken from Table 2 if concentrations are given in mg/l and °dH. The activity coefficient remains the same.

CaO	= 8.4 °dH	combined CO_2	= 41.8 mg/l
MgO	= 5.05 °dH	SO_4^{2-}	= 139.2 mg/l
Carbonate hardness (p. 179)	= 5.3 °dH	Cl^-	= 5.25 mg/l

$$C_{\text{equilibrium } CO_2} = \frac{2.173 \times 10^{-4}}{1.55} \times 41.8 \times 8.4 = 2.05 \text{ mg/l}$$

Should the content of free carbon dioxide of the water be 0.2 mval/l or 4.4 mg/l, the the "excess" amounts are

$C_{\text{excess } CO_2} = 0.2 - 0.093 = 0.107$ mval/l, or

$C_{\text{excess } CO_2} = 4.4 - 2.05 = 2.35$ mg/l

Part of the excess carbon dioxide is lime-aggressive. It can be sufficiently accurately calculated by a method of approximation (*Hässelbarth* GWF *104*, 506 [1963]). This value is usually obtained in practice by the Heyer-test, however. (p. 87) (*Hässelbarth* GWF *104,* 89, 157 [1963]).

Saturation Index according to *Strohecker* and *Langelier*

In order to assess the aggressivity of a water, the saturation index of *Strohecker* and *Langelier* is needed as well as the data on contents of excess and lime-aggressive carbon dioxide. It is defined as the difference between the measured pH-value and the "equilibrium-pH", expected from the given calcium and magnesium hardnesses and the contents of hydrogen carbonate, sulphate and chloride ions in the carbonate equilibrium:

$I = \Delta$ pH = measured pH − equilibrium pH

Negative values of I are obtained in the presence of excess and lime-attacking carbon dioxide. Waters in the carbonate equilibrium have a saturation index of zero. Waters having a free carbon dioxide content lower than the equilibrium concentration display positive saturation indices.

The equilibrium-pH needed to calculate the saturation index is worked out from the corrected Strohecker-Langelier equation:

equilibrium pH = $pK^* - \log C_{HCO_3^-} - \log C_{Ca^{2+}} + \log f_L$

pK^* is the Strohecker-Langelier constant, the values of which are given as a function of temperature in Table 4 (for concentration data in mval/l) and Table 5 (concentrations in mg/l and °dH). Log $C_{HCO_3^-}$ is the logarithm of the hydrogen carbonate concentration, and log $C_{Ca^{2+}}$ the logarithm of the calcium concentration. Log f_L is an activity coefficient, derived from the ionic strength of the water according to the relation:

$$\log f_L = \frac{2.5\sqrt{\mu}}{1 + 5.3\sqrt{\mu} + 5.5\,\mu}$$

Table 6 contains values of log f_L as a function of total μ to simplify calculation.
The ionic strength μ is calculated as given on p. 95.

Example:

$C_{Ca^{2+}} = 3.0$ mval/l, log $C_{Ca^{2+}} = 0.47711$

$C_{Mg^{2+}} = 1.8$ mval/l

$C_{SO_4^{2-}} = 2.9$ mval/l

$C_{Cl^-} = 0.15$ mval/l

$C_{HCO_3^-} = 1.9$ mval/l, log $C_{HCO_3^-} = 0.27875$

$t = 10°$ C

$\mu_{\text{total}} = 8.725 \times 10^{-3}$

Then log $f_L = 0.151$ and $pK^* = 8.6389$
Equilibrium pH $= 8.6389 - 0.47712 - 0.27875 + 0.151 = 8.034$
pK^* is taken from Table 5 when data are in mg/l and °dH. Log f_L has the same value.
$CaO = 8.4$ °dH, log $CaO = 0.92428$; $MgO = 5.05$ °dH; KH $= 5.3$ °dH; combined $CO_2 = 41.8$ mg/l, log combined $CO_2 = 1.62118$; $SO_4^{2-} = 139.2$ mg/l; $Cl^- = 5.2$ mg/l
$\mu total = 8.725 \times 10^{-3}$
Then log $f_L = 0.151$ and $pK^* = 10.4286$
Equilibrium pH $= 10.4286 - 0.92328 - 1.62118 + 0.151 = 8.034$
If a pH of 7.75 was found in a water of this composition, the saturation index is
$I = \Delta pH = 7.75 - 8.03 = -0.28$

Tables 1–6, S. 98–100

The saturation index is negative, i.e., the water is aggressive

A sure opinion of the aggressive behaviour of a water should be voiced only when the data on the amounts of 1. excess carbon dioxide, 2. lime-aggressive carbon dioxide and 3. the saturation index all tend in the same direction. A deviation of one of the three indicators is generally due to errors of the analysis, especially the pH-measurement which is often needed, and to the determination of free carbon dioxide (*Zehender, Stumm* and *Fischer,* Mon. Bull. Schweiz. Ver. v. Gas- u. Wasserfachm. *36,* 269 [1956], *Hässelbarth,* GWF *104,* 89, 157 [1963], DIN Vornorm 50930).

Carbon dioxide preventing formation of an anti-rust protective layer; Carbonate balance
If a water contains just enough free carbon dioxide to maintain in solution the calcium hydrogen carbonate present, the desired protective layer in the pipeline network gradually forms. Too little carbon dioxide to keep the calcium hydrogen carbonate in solution means that the protective layer forms too quickly and undesirable incrustations are produced in the pipes.
Should, however, too much carbon dioxide be present – more than necessary to keep the calcium hydrogen carbonate in solution – no protective layer is formed, since the calcium hydrogen carbonate stays in solution. Further, the excess carbon dioxide attacks the iron pipes and rust lumps are formed. The total excess of free carbon dioxide, i.e., that over and above the "associated" carbon dioxide (p. 163), is therefore termed "carbon dioxide preventing formation of an anti-rust".
This protective layer is, in other words, the difference between the total free carbon dioxide and the "associated" CO_2. Values can thus be taken directly from Tillmans' table. It must be remembered, however, that very hard waters, especially those which have high permanent hardness, almost always have 2 to 5 mg/l of excess carbon dioxide and yet, as found in practice, do not attack the pipe system.
The excess carbon dioxide attacks metallic pipes, particularly those of iron, zinc and lead, dissolving up the metal. If an accurate determination of the metal is made at the beginning and at the end of a new pipe system, it is therefore possible to establish whether all the carbon dioxide present is "associated" (p. 163), or a part attacks the pipe as excess carbon dioxide and dissolves iron from the surface of the pipes.

Lead-attacking carbon dioxide: The oxygen content of the water is primarily responsible

Table 1. Constants of Tillmans' law

t [°C]	$K_{Tillmans}$ [m Val^{-2}]	t [°C]	$K_{Tillmans}$ [m Val^{-2}]
0	$9{,}371 \cdot 10^{-3}$	26	$2{,}387 \cdot 10^{-2}$
1	9,700	27	2,459
2	$1{,}004 \cdot 10^{-2}$	28	2,539
3	1,039	29	2,617
4	1,076	30	2,703
5	1,114	31	2,804
6	1,156	32	2,903
7	1,199	33	3,012
8	1,244	34	3,117
9	1,291	35	3,235
10	1,339	36	3,356
11	1,389	37	3,482
12	1,441	38	3,613
13	1,492	39	3,748
14	1,548	40	3,889
15	1,606	41	4,025
16	1,667	42	4,176
17	1,729	43	4,323
18	1,790	44	4,486
19	1,857	45	4,643
20	1,926	46	4,806
21	1,999	47	4,986
22	2,074	48	5,162
23	2,147	49	5,356
24	2,227	50	5,544
25	2,311		

Table 2. Constants of Tillmans' law

t [°C]	$K_{Tillmans}$ $\left[\left(CO_2 \frac{mg}{l}\right)^{-1} CaO°d^{-1}\right]$	t [°C]	$K_{Tillmans}$ $\left[\left(CO_2 \frac{mg}{l}\right)^{-1} CaO°d^{-1}\right]$
0	$1{,}5213 \cdot 10^{-4}$	26	$3{,}874 \cdot 10^{-4}$
1	1,575	27	3,992
2	1,630	28	4,123
3	1,687	29	4,248
4	1,747	30	4,387
5	1,808	31	4,552
6	1,876	32	4,712
7	1,947	33	4,889
8	2,020	34	5,060
9	2,095	35	5,251
10	2,173	36	5,448
11	2,255	37	5,653
12	2,340	38	5,864
13	2,422	39	6,084
14	2,513	40	6,313
15	2,607	41	6,535
16	2,706	42	6,780
17	2,807	43	7,018
18	2,905	44	7,282
19	3,014	45	7,538
20	3,127	46	7,802
21	3,245	47	8,095
22	3,367	48	8,379
23	3,485	49	8,694
24	3,615	50	8,999
25	3,752		

Table 3.

μ (total)	f (Tillmans)	μ (total)	f (Tillmans)
$0,27 \cdot 10^{-3}$	1,101	19,2	1,752
0,53	1,124	$19,6 \cdot 10^{-3}$	1,758
0,8	1,187	19,8	1,761
1,2	1,228	20,0	1,764
1,6	1,260	20,4	1,769
1,8	1,281	20,8	1,774
2,0	1,289	21,0	1,777
2,4	1,314	21,2	1,780
2,8	1,337	21,6	1,785
3,0	1,348	22,0	1,791
3,2	1,358	22,2	1,793
3,6	1,377	22,4	1,795
4,0	1,396	22,8	1,800
4,2	1,404	23,2	1,805
4,4	1,413	23,4	1,808
4,8	1,429	23,6	1,810
5,2	1,444	24,0	1,815
5,4	1,451	24,4	1,820
5,6	1,459	24,6	1,822
6,0	1,472	24,8	1,824
6,4	1,485	25,2	1,829
6,6	1,492	25,6	1,833
6,8	1,498	25,8	1,835
7,2	1,511	26,0	1,838
7,6	1,522	26,4	1,842
7,8	1,528	26,8	1,846
8,0	1,534	27,0	1,849
8,4	1,544	27,2	1,851
8,8	1,555	27,6	1,855
9,0	1,560	28,0	1,859
9,2	1,565	28,2	1,861
9,6	1,575	28,4	1,863
$10,0 \cdot 10^{-3}$	1,585	$28,8 \cdot 10^{-3}$	1,867
10,2	1,590	29,2	1,871
10,4	1,594	29,4	1,873
10,8	1,603	29,6	1,875
11,2	1,612	30,0	1,879
11,4	1,616	30,4	1,883
11,6	1,621	30,8	1,887
12,0	1,629	31,2	1,890
12,4	1.637	31,6	1,894
12,6	1,641	32,0	1,898
12,8	1,645	32,4	1,901
13,2	1,653	32,8	1,905
13,6	1,661	33,2	1,909
13,8	1,664	33,6	1,912
14,0	1,668	34,0	1,915
14,4	1,675	34,4	1,919
14,8	1,683	34,8	1,922
15,0	1,686	35,2	1,926
15,2	1,690	35,6	1,929
15,6	1,696	36,0	1,932
16,0	1,703	36,4	1,936
16,2	1,706	36,8	1,939
16,4	1,710	37,2	1,942
16,8	1,716	37,6	1,945
17,2	1,723	38,0	1,948
17,4	1,726	38,4	1,951
17,6	1,729	38,8	1,954
18,0	1,735	39,2	1,957
18,4	1,741	39,6	1,960
18,6	1,744	40,0	1,963
18,8	1,747		

Table 4. Constants of the Strohecker-Langelier equation. Concentration in mVal

t [°C]	*pK**	*t* [°C]	*pK**
0	8,901	16	8,492
1	8,878	17	8,468
2	8,851	18	8,445
3	8,825	19	8,422
4	8,798	20	8,400
5	8,771	21	8,376
6	8,745	22	8,354
7	8,718	23	8,333
8	8,692	24	8,311
9	8,665	25	8,288
10	8,639	30	8,196
11	8,614	35	8,100
12	8,589	40	7,996
13	8,565	45	7,923
14	8,540	50	7,844
15	8,515		

Table 5. Constants of the Strohecker-Langelier equation
Concentration: C_{CO_2} (combined) in mg/l (mol.wt. 22)
C_{CaO} in °d (mol.wt. 56)

t [°C]	*pK**	*t* [°C]	*pK**
0	10,6961	16	10,2812
1	10,6681	17	10,2575
2	10,6409	18	10,2351
3	10,6151	19	10,2116
4	10,5878	20	10,1901
5	10,5610	21	10,1660
6	10.5343	22	10,1439
7	10,5078	23	10,1231
8	10,4814	24	10,1011
9	10,4550	25	10,0781
10	10,4286	30	9,9858
11	10,4035	35	9,8893
12	10,3786	40	9,7857
13	10,3547	45	9,7129
14	10,3298	50	9,6341
15	10,3048		

Table 6.

μ (total)	lg *f* *(Langelier)*	μ (total)	lg *f* *(Langelier)*
$0{,}1 \cdot 10^{-3}$	0,024	$9{,}0 \cdot 10^{-3}$	0,153
0,2	0,033	10,0	0,158
0,4	0,045	12,0	0,166
0,6	0,054	14,0	0,173
0,8	0,061	16,0	0,179
1,0	0,067	18,0	0,185
1,3	0,074	20,0	0,190
1,6	0,081	22,0	0,194
1,9	0,087	24,0	0,198
2,2	0,092	26,0	0,201
2,5	0,097	28,0	0,205
2,8	0,102	32,0	0,210
3,2	0,107	36,0	0,215
3,6	0,112	40,0	0,219
4,0	0,117	45,0	0,224
4,5	0,122	50,0	0,228
5,0	0,126	60,0	0,233
6,0	0,134	70,0	0,238
7,0	0,141	80,0	0,241
8,0	0,147	100,0	0,245

for corrosion of lead. Waters poor in oxygen, e.g. ground waters from greater depth, do not attack lead at all. Lead oxide must first be formed on the lead pipe surface and this is then converted by the aggressive water into soluble lead salts.
Many waters which attack strongly lime, concrete, mortar and iron therefore cannot attack lead at all because of lack of oxygen; on the other hand, oxygen-rich, soft waters, even when they contain only little free carbon dioxide, often possess a powerful capacity for dissolving lead (e.g. rainwater) (*K. Höll,* Arch. f. Hygiene *113*, 283–295 [1935]). Harder waters attack lead more strongly only when the free carbon dioxide amount exceeds half of the amount of combined carbon dioxide. If the total free carbon dioxide is less than 20% of the combined, lead is never attacked (*K. Höll,* Ges. Ing. *58,* 323–328 [1935]).
There is always a fear of attack on lead by soft water, rich in oxygen, even when little aggressive carbon dioxide is present. The corrosion processes with newly laid lead pipes are particularly obscure. A lead corrosion experiment is thus often necessary.

10. Lead Corrosion Experiment

Lead strips or halved sections of lead pipes which have been defatted all round, cleaned with dilute nitric acid and well washed with water are placed in wide-necked cylinders, fitted with slantingly cut glass stoppers. The lead is cautiously covered with the water (which has not come into contact with the air) with the help of a small siphon tube. Water is passed through for a few minutes, the vessel then filled to overflowing and carefully stoppered.
After 12 hours, the lead pieces are withdrawn with a repeated back- and forward movement. The lead content of the water is then determined as on p. 131. In water mains, a new section of lead piping fitted with a tap at each end can be introduced into the system. The water is left for 10 hours under pressure and then withdrawn for testing. The first 50 to 100 ml are discarded.
The experiment should be repeated every three days, seven times in all, after the water has stood for a longer time in the pipe. A few ml of acetic acid are added to the samples, since the lead compounds are readily retained on the glass walls (*Jackson,* Water & Water Eng. *39,* [1937] and *K. Höll,* D. Apoth. Ztg. *50,* 126 [1935]).

11. Oxygen Determination

Approximate field determination: 2 ml of a manganese (II)-chloride solution[1] and 2 ml of sodium hydroxide[1] are added to a glass stoppered bottle filled to overflowing with the water (see below). The bottle is stoppered and shaken. The content of oxygen can be estimated from the colour of the precipitate, if necessary with the help of a colour scale according to *Hofer* (from the firm of Franz Bergmann, KG., Berlin-Zehlendorf and Hamburg). Oxy-

1 *Tablets, supplied by* Franz Bergmann KG., *Berlin-Zehlendorf, can also be used.*

gen-free water yields an almost white precipitate; water poor in oxygen, a light yellow precipitate; and water rich in oxygen, a brown one.
Gilreas' method is quite good for limnological field determination of oxygen (Z. Wasser u. Abwasser *33*, 464 [1935]). It uses 2 ml of 13% amidol solution and a preliminary addition of 1 ml calcium citrate solution (2 : 1) per 100 ml water. In the field, comparison can be made with standard coloured solutions from cobalt chloride and potassium bichromate.
Nowadays the oxygen measuring apparatus of *F. Tödt* is employed for control of waters (*F. Tödt* and *G. Petsch*, Ges. Ing. *76*, 104–106 [1955] and *F. Tödt*, Die elektrochemische Sauerstoffbestimmung, Berlin [1959]) (Apparatus from Franz Bergmann, KG., Berlin-Zehlendorf and Hamburg). One saves many single determinations of the oxygen by using this apparatus and discovers currents of waste water better in a river. Recognition of the thermocline in standing waters is also easy (see *W. Ohle*, Vom Wasser *19*, 99–103 [1953]). The apparatus in its newest form with gold amalgam electrode gives exact values also with hard waters. Further developments are the "Protech electrode" with silver oxide base and the "Protech-traverse" which protects the electrode and maintains a continuous stream in front of the membrane; it is coupled with a very precise thermometer (see also *F. Tödt*, Intermittierende elektrochemische Sauerstoffregistrierung in verunreinigten Gewässern, Dechema-Monographie 54).
G. Stracke's apparatus has proved useful to the Ruhrverband (Ruhr Association, W. Germany) for continuous determination of oxygen in aeration tanks (Vom Wasser *34*, [1967]). A new development is the physico-chemically functioning phase exchanger of the Chlorator GmbH of Grötzingen (see p. 214).
A continuous oxygen profile in waters can be measured by *Grashoff*'s "Sauerstoff und Temperatursonde" (Oxygen and Temperature Tester) (Firm of Hydrobios, Kiel, W. Germany).

a) Iodometric Oxygen Determination according to Winkler-Bruhns

Preparation: With the help of a plastic tube, the water to be examined is introduced into a 300 ml glass-stoppered flask (so-called oxygen bottle and accurate to 0.1 ml). The tube reaches right to the bottom of the bottle and the precautions given under "Sampling" (p. 11) are observed. By allowing the water to overflow for a longer time, the water in the flask will not have come into contact with air. The plastic tube is carefully withdrawn during the procedure or the oxygen bottle lowered, and the slanting or pointed[2] glass stopper fitted on. Care must be taken that no air bubbles enter the flask during filling. Transparent plastic tubes help in correct sampling.
A milkiness of very fine bubbles is often observed with underwater pumps. These are not air bubbles but are due to the swirl of the pump. This can be proved beyond doubt with oxygen-free water from a depth.
Stopper and flask fitment must be well greased. No air bubbles must be present. Gas bubbles which are formed some time after reagent addition are, however, harmless; they are nitrogen bubbles. The temperature must be noted at the same time and the barometric pressure in accurate investigations.

2 see *W. Ohle*, IVL-communication of the standardisation commission No. 3, 153.

Immediately after sampling, 2 ml of 40% iron-free manganese (II) chloride [3] and 2 ml of 33% sodium hydroxide are added by means of elongated pipettes. If nitrite is present, 20 drops of 5% sodium azide solution are added, regardless of whether overflowing occurs. After cautiously replacing the glass stopper, the solutions are mixed by inverting the flask several times, holding the stopper. The sample can be kept for a longer period in this condition, for accurate investigations under water. The sample flasks are wrapped in moist cloths on the way back to the laboratory *(W. Ohle)* and best transported in metal boxes *(R. Maucha)*.
The flasks containing the manganese precipitate must not be exposed to sunlight before the further examination.

After brief standing, 5 g of powdered potassium hydrogen carbonate can be added, or, better, a 20% solution of ammonium hydrogen carbonate, preventing access of atmospheric oxygen. The *Lübbert-Schneider* stopper is used for closing the flask.
100 ml bottles suffice for larger serial studies, e.g. for limnological purposes; only 1 ml of the manganese(II)- and NaOH solutions is then needed.

Titration: After allowing to stand for a longer time, the clear supernatant liquid is siphoned or sucked off until about 100 ml remain. To this are added 1 g potassium iodide and 5 ml 25% HCl or, especially in presence of ferric compounds, 85% phosphoric acid. The flask is closed and left for 10 minutes. It is not necessary to transfer to an erlenmeyer flask.
The KI is best added through a wide-mouthed funnel, so that crystals do not stick in the neck. 0,5 g KI and 2 ml HCl are enough with a 100 ml water sample.
After the 10 minutes, the liberated iodine is titrated with a 0.01N sodium thiosulphate solution, using the customary starch indicator at the end of the titration. Titration at this point must be especially slow to avoid overtitration. Any blue colour which appears later is of no importance.
The thiosulphate solution is prepared with boiled water or by diluting a freshly-prepared 0.1N thiosulphate solution which has been preserved by addition of 1% of isobutanol or amyl alcohol. The thiosulphate solution may be stabilised also through addition of 0.4 g NaOH or 1 g of salicylic or benzoic acid per litre.

The thiosulphate solution is best *standardised* with a potassium biiodate solution containing 0.325 g KIO_3/l. 10.0 ml of this solution are diluted in an erlenmeyer flask and 0.5 g KI and 5 ml of 10% HCl added. After standing, titration should be with 10 ml of 0.1N thiosulphate or a factor must be employed.
According to *R. Czensny,* standardisation should be undertaken with the same end volume as that of the normal titration, i.e., 100 ml.
The 1% starch solution, prepared by boiling, can be stabilised with 3% common salt.

Many *organic compounds* combine with part of the liberated iodine and thus *interfere* if present. According to *Winkler,* one treats 100 ml of water sample with 10 ml of 0.01N iodine, allows to stand for some minutes and then titrates with thiosulphate. The blank consumption of thiosulphate is added to the oxygen value.
With waters containing humins and those contaminated with many organic substances, it is best, as already mentioned, to add some potassium hydrogen carbonate, filter the manganese-containing precipitate after some time, and to wash it and transfer it to a flask with ground-glass stopper. It is then determined iodometrically in this flask after acidification. Still better is *W. Ohle's* procedure (see below).
Even a low nitrite content makes itself unpleasantly felt in the titration; the end-point

3 This must not yield any iodine colour in acidified KI-solution (add starch if necessary).

is no longer sharp as a result of return of the blue colour. The error is, in fact, small – 0.021 mg oxygen for 0.1 g N_2O_3. The addition of sodium azide, as given above, is advisable to the prevent this dragging of the end-point.
0.14 mg oxygen/l must be added for every mg of Fe^{2+}/l in waters rich in iron.
W. Ohle's procedure (see below) must be used if much organic matter and sulphites are present (as in receiving waters which have taken up sulphite waste liquors from wood pulp factories).

Calculation of the oxygen content: The exact content of the sample bottle must be borne in mind in the calculation. The oxygen content in mg O_2/l is given by:

$$\frac{\text{a ml 0.01 N sodium thiosulphate used} \times (0.08) \times 1000}{\text{volume of flask} - 4 \text{ ml}}$$

0.025 mg O_2/l can still be present even if the *Winkler* method gives no oxygen. A small amount is always introduced through the added reagents; it must be corrected for in accurate studies (0.1 mg O_2/ l on the average).

Conversions 1 mg O_2 = 0.70 ml O_2
1 ml O_2 = 1.429 mg O_2

Quotations of oxygen in ml/l are, however, inaccurate and misleading (*H. Schmassmann,* Schweiz. Z. f. Hydrologie [1949]). The oxygen value is thus quoted in mg O_2/l.

b) Oxygen Determination in Surface Waters and Contaminated Waters according to W. Ohle; Differential Iodine Method

Two oxygen bottles are filled with 100 ml as described on p. 11 and 1 ml of iodine solution is added to each with an elongated pipette.
This iodine solution is prepared by dissolving 5 g iodine and 100 g KI in 80 ml boiled water and diluting 10 ml of this solution with 100 ml of saturated (35%) common salt. 1 ml = 4 mg iodine and 80 mg KI.
Both oxygen bottles are carefully stoppered and shaken. To one is then added 0.5 ml manganese chloride solution and 0.5 ml of sodium hydroxide (100 g NaOH and 200 ml boiled water + 50 g KI). The mixture is shaken for 10 seconds after the stopper has been replaced; 3 ml of 50% sodium hydrogen sulphate are then added immediately. Only the sodium hydroxide solution is added to the second bottle but it receives 0.5 ml more of the hydrogen sulphate.
Titration must be carried out after 10 minutes and after the contents of the bottles have been transferred to erlenmeyer flasks; 0,01N sodium thiosulphate is used as in the standard procedure. The volume of thiosulphate required in the blank determination is subtracted from that in the determination with manganese chloride. As in the standard procedure, calculation takes account of the added reagent amounts but 0.12 mg O_2/l is subtracted from the final result to allow for oxygen introduced with the reagents.
Neither nitrite, sulphite, sulphide, hydrogen sulphide, iron nor manganese interferes.
The *Alsterberg* method has been superceded by *Ohle's* method.
The oxygen determination can also be conducted in the Hellige apparatus. The intensity of the yellow or brown iodine colour is measured in the 40 mm cuvette.

c) *Determination of Very Small Amounts of Oxygen*

An especially accurate method for determining oxygen in boiler feed water can be carried out in the Hellige apparatus with the help of the o-tolidine reagent. The ferroin method according to *Töller* has proved the best for very small amounts of oxygen (Veröff. der Verein. der Großkesselbesitzer [1954]). Small oxygen amounts in boiler water are determined by dissolving the manganese precipitate in 5 ml phosphoric acid, then adding 2 ml o-tolidine and measuring colorimetrically (*R. Freier,* Chem. Ztg. *76,* [1952]).
According to *E. Schumann,* oxygen in boiler feed water can be photometrically determined with precision as follows: 0.15 ml Ce(III)-chloride solution and 0.20 ml 40% sodium hydroxide are added to the contents of an oxygen bottle (ca. 100 ml), filled with sample as described on p. 11). The mixture is shaken up after stoppering the flask.
The Ce(III)-chloride reagent is prepared by dissolving 120 g of $CeCl_3 \cdot 7H_2O$ in 100 ml of boiled, distilled water, adding 3 ml hydrochloric acid* and boiling for a longer time. The volume is made up again with boiled water.
The sodium hydroxide also is made up with boiled water and boiled for a longer period.
4 ml of the tolidine reagent given below are then added to the contents of the flask and colorimetric or spectrophotometric evaluation is performed.

Tolidine reagent: 1 volume of sulphuric acid (1:1) is mixed with 3 volumes of a solution of 0.4 g o-tolidine in 100 ml of 0.1N HCl. The final solution can be kept for only 3 hours.

d) *Determination of the Oxygen Consumption of River Water and Wastewater*

The oxygen consumption or depletion is a measure of the amount of decomposable organic compounds contained in the water. It is thus determined in order to gain an idea whether much putrescible matter is present in a receiving water.
At the same time that the sample is taken for oxygen determination (p. 102), a second sample is taken with a glass stoppered bottle of known content (300 ml), as given on p. 102, for measurement of oxygen consumption. This second sample is kept for 48 hours at 22° C in absence of light and without adding any chemicals. The oxygen determination is then carried out as described above. The difference between the two values is the oxygen consumption (depletion).
Oxygen consumption must not be confused with oxygen deficit. Oxygen- or saturation deficit or deficiency signifies the difference between the actual amount and the saturation amount of oxygen in the water at the same temperature (see below).
Oxygen consumption can be quoted also as a percentage of the original oxygen content. Sometimes the values are given also as percentages of the saturation amount, when there is no deficit at the time of sampling.
The result is inaccurate if the residual oxygen, after determining oxygen depletion, is less than 2 mg/l. The biochemical oxygen demand (BOD) (BSB 5 in German) gives a more accurate idea of oxygen consumption (see p. 215).
The purely chemical oxygen depletion (CBS in German), due to H_2S, NO_2, phenols etc.,

can be obtained when the consumption is measured as above but after standing for only 1 hour (see *W. Leithe,* Wasser, Luft und Betrieb [1970]).

12. Oxygen Deficiency, Oxygen Saturation Deficiency

Oxygen deficiency is the amount of oxygen needed to bring the existing amount up to the theoretical saturation value at the same temperature and pressure. It is thus the difference between actual and saturation concentrations of oxygen.
Ground waters from a great depth and deep waters from eutrophic lakes have a high oxygen deficiency. Contaminated waters, whether ground or surface, also show a high deficiency, due to consumption of oxygen during decay processes. Theoretical values may be taken from the table below (p. 107).
A lower value represents oxygen deficiency; a higher value, supersaturation.
G. A. Truesdale in 1955 corrected *C. J. J. Fox*'s tables of 1909. *H. A. C. Montgomery* found in 1964 that *Truesdale*'s values are too low. The newly evaluated amounts lie between those of *Fox* and those of *Truesdale* (*H. A. C. Montgomery, N. S. Thom* and *A. Cockburn,* J. Appl. Chem. *14,* 280–296 [1964]).

13. Oxygen Saturation Index

The saturation index is the ratio, expressed as percentage, of the actual oxygen to the saturation oxygen concentrations.
Further information about evalution of the relative oxygen saturation can be found in the yearbook Vom Wasser (*22* [1955] *R. Burkhard*). The "Oxygen-calculator" is described there also.

Oxygen Saturation Value: By oxygen saturation value or oxygen saturation concentration is understood the actual oxygen concentration of a water which is in equilibrium with the atmospheric oxygen at the surface of the water and at the prevailing temperature.
Since the saturation values refer to normal atmospheric pressure of 760 mm, the altitude of higher situated waters must be allowed for, according to the formula:

$$\log B = \log 760 - \frac{h}{18400}$$

or, more accurately,

$$\log B = \log 760 - \frac{h}{18421\,(1 + 0.004\,t)}$$

where B = mean barometric pressure of the place, in mm mercury
h = altitude of the water level above sea level
t = mean air temperature.

New Table for Oxygen Saturation Concentration according to *Truesdale, Downing* and *Lowden* (1955) at a total pressure of 760 mm mercury (atmosphere saturated with water vapour)
Expressed as mg oxygen/l

t °C	0,0°	0,1°	0,2°	0,3°	0,4°	0,5°	0,6°	0,7°	0,8°	0,9°
0°	14,16	14,12	14,08	14,04	14,00	13,97	13,93	13,89	13,85	13,81
1°	13,77	13,74	13,70	13,66	13,63	13,59	13,55	13,51	13,48	13,44
2°	13,40	13,37	13,33	13,30	13,26	13,22	13,19	13,15	13,12	13,08
3°	13,05	13,01	12,98	12,94	12,91	12,87	12,84	12,81	12,77	12,74
4°	12,70	12,67	12,64	12,60	12,57	12,54	12,51	12,47	12,44	12,41
5°	12,37	12,34	12,31	12,28	12,25	12,22	12,18	12,15	12,12	12,09
6°	12,06	12,03	12,00	11,97	11,94	11,91	11,88	11,85	11,82	11,79
7°	11,76	11,73	11,70	11,67	11,64	11,61	11,58	11,55	11,52	11,50
8°	11,47	11,44	11,41	11,38	11,36	11,33	11,30	11,27	11,25	11,22
9°	11,19	11,16	11,14	11,11	11,08	11,06	11,03	11,00	10,98	10,95
10°	10,92	10.90	10,87	10,85	10,82	10,80	10,77	10,75	10,72	10,70
11°	10,67	10,65	10,62	10,60	10,57	10,55	10,53	10,50	10,48	10,45
12°	10,43	10,40	10,38	10,36	10,34	10,31	10,29	10,27	10,24	10,22
13°	10,20	10,17	10,15	10,13	10,11	10,09	10,06	10,04	10,02	10,00
14°	9,98	9,95	9,93	9,91	9,89	9,87	9,85	9,83	9,81	9,78
15°	9,76	9,74	9,72	9,70	9,68	9,66	9,64	9,62	9,60	9,58
16°	9,56	9,54	9,52	9,50	9,48	9,46	9,45	9,43	9,41	9,39
17°	9,37	9,35	9,33	9,31	9,30	9,28	9,26	9,24	9,22	9,20
18°	9,18	9,17	9,15	9,13	9,12	9,10	9,08	9,06	9,04	9,03
19°	9,01	8,99	8,98	8,96	8,94	8,93	8,91	8,89	8,88	8,86
20°	8,84	8,83	8,81	8,79	8,78	8,76	8,75	8,73	8,71	8,70
21°	8,68	8,67	8,65	8,64	8,62	8,61	8,59	8,58	8,56	8,55
22°	8,53	8,52	8,50	8,49	8,47	8,46	8,44	8,43	8,41	8,40
23°	8,38	8,37	8,36	8,34	8,33	8,32	8,30	8,29	8,27	8,26
24°	8,25	8,23	8,22	8,21	8,19	8,18	8,17	8,15	8,14	8,13
25°	8,11	8,10	8,09	8,07	8,06	8,05	8,04	8,02	8,01	8,00
26°	7,99	7,97	7,96	7,95	7,94	7,92	7,91	7,90	7,89	7,88
27°	7,86	7,85	7,84	7,83	7,82	7,81	7,79	7,78	7,77	7,76
28°	7,75	7,74	7,72	7,71	7,70	7,69	7,68	7,67	7,66	7,65
29°	7,64	7,62	7,61	7,60	7,59	7,58	7,57	7,56	7,55	7,54
30°	7,53	7,52	7,51	7,50	7,48	7,47	7,46	7,45	7,44	7,43
31°	7,42	7,41	7,40	7,39	7,38	7,37	7,36	7,35	7,34	7,33
32°	7,32	7,31	7,30	7,29	7,28	7,27	7,26	7,25	7,24	7,23
33°	7,22	7,21	7,20	7,20	7,19	7,18	7,17	7,16	7,15	7,14
34°	7,13	7,12	7,11	7,10	7,09	7,08	7,07	7,06	7,05	7,05
35°	7,04	7,03	7,02	7,01	7,00	6,99	6,98	6,97	6,96	6,95
36°	6,94	6,94	6,93	6,92	6,91	6,90	6,89	6,88	6,87	6,86
37°	6,86	6,85	6,84	6,83	6,82	6,81	6,80	6,79	6,78	6,77
38°	6,76	6,76	6,75	6,74	6,73	6,92	6,71	6,70	6,70	6,69
39°	6,68	6,67	6,66	6,65	6,64	6,63	6.63	6,62	6,61	6,60
40°	6,59	6,58	6,57	6,56	6,56	6,55	6,54	6,53	6,52	6,51

Further information may be obtained from articles by *H. Schmassmann* (Schweiz. Z. für Hydrologie *XI*, 430–462 [1949] and *XVIII* [1956]); *G. A. Truesdale, A. L. Downing* and *G. F. Lowden* (J. Appl. Chem. *5*, 53–62 [1955]); and *W. Ohle* (Mitt. der IVL-Methoden-Kom. No. 3 [1953]). *H. Schmassmann* has summarised the information and drawn attention to the necessity of considering the atmosphere, saturated with water vapour (Schweiz. Z. für Hydrologie, *XVIII* [1958]).
The oxygen saturation values depend on the salt content of the water. The value falls by from 0.006 to 0.017 mg O_2/l, depending on temperature, for each 100 mg Cl/l. The equation,

$$c = q .- \frac{2.65 \times Cl}{33.5 + t}$$

must be employed, where q is the value taken from the table giving oxygen saturation concentration (p. 107), Cl = concentration of Cl^- in mg/l and t = water temperature.

C. Technical-chemical Water Analysis

1. Determination of Iron

Occurrence: Ground water contains 0 to 10.0 mg Fe/l, rarely more and usually from 0.5 to 1–3. Surface water has appreciably less (up to 0.5 mg Fe/l) and alpine waters contain only traces. Higher contents are often found in the deeper regions of lakes and reservoirs; under certain circumstances, the iron is dissolved out of the sediments during the stagnation period and is carried up to the upper water layers during the circulation period.

Shaking test: Freshly sampled water in a sample flask shows the presence of iron in greater amounts through a separation on shaking vigorously for some minutes. Waters containing over 3 mg Fe/l yield a yellowish turbidity immediately, turning to a brown precipitate within a few hours. Waters having 1 to 3 mg Fe/l give the turbidity only after several hours and precipitation often takes 1 to 2 days. There are, however, some marked exceptions to this rule. For example, waters of lower pH precipitate the iron less easily and those of pH above 7 give a precipitate easily through aeration. In the former case, the iron content is easily estimated as too low; in the latter case, as too high.
The iron content of waters containing humins is high but in a form difficult to precipitate. In any case, the reaction to the shaking test is noted since it gives a hint about the method of iron removal (p. 175).

Test-tube estimation of iron (II) ion: 20 ml of water to be tested are mixed with a drop of sodium sulphide solution* (5 g Na_2S + 10 g water + 30 g glycerol) in a test tube. A brown-black coloration is yielded immediately with contents of over 3 mg Fe/l, brown with 2 mg/l, yellow with 1 mg/l and faintly yellowish with 0.3 mg Fe/l (viewed from above).

Many waters have a yellowish colour of their own, due to humic substances, and a reference standard of the untreated water should then be used in a second test tube.
Lead and copper in tap water can interfere. The colour due to iron is removed with a few drops of HCl, but that from copper remains.

Determination of iron (II) ions: 0.5 ml 0.1N H_2SO_4 and 1 ml of a 0.4% solution of α,α-bipyridyl in 0.1N H_2SO_4 are added to 100 ml of the water and mixed. A pink or red colour appears after standing for some time, if Fe(II)-ions are present. This reaction can be carried out quantitatively, e.g. by comparing with standard solutions of Fe(II) ammonium sulphate (Reagent No. 12) in Hehner tubes (for colour comparison) or at 522 nm spectrophotometrically after 1 to 2 hours reaction.
20 ml of bipyridyl solution in a 100 ml graduated flask can be treated on the sampling spot with water, the exact amount of which can be measured later in the laboratory. The red coloration which is yielded immediately is stable for days.
This method is recommended for limnological work and is preferable to the EV method with isonitrosoacetophenone. *W. Rodhe* had good results using this method, even with the low iron contents of waters from Swedish primary rock regions.
Test for Fe(III)-ions: 100 ml of the freshly sampled water are treated with 2 ml HCl and 2 ml of 10% potassium thiocyanate solution. The colour yielded is compared after 5 to 10 minutes in Hehner cylinders with those from reference solutions (from Reagent No. 12) made up at the same time, or with standard colours, e.g., *Meinck-Horn* (p. 23).
Total iron is yielded if the solution is previously oxidised with H_2O_2.

a) Colorimetric Determination of Total Iron

2 ml hydrochloric acid* and 0.5 ml of 3% hydrogen peroxide are added to a 100 ml water sample and the mixture is boiled for about 5 minutes or allowed to stand overnight. Heating is preferable with samples where part of the iron has already precipitated. The (cooled) sample is mixed with 2 ml of 10% potassium thiocyanate solution. After 10 to 20 minutes, the red colour is compared in Hehner cylinders, colorimeter or photometer with mixtures from Reagent No. 12, made up at the same time. The test in Hehner cylinders is sensitive to 0.05 mg Fe/l.
The comparison is very convenient with the help of the Meinck-Horn colorimeter (Firm of Franz Bergmann KG., Berlin-Zehlendorf and Hamburg). Photometric calibration curves can be used for comparison. One saves having to make up numerous reference solutions in both methods. The Hellige apparatus is less suitable for serial iron determinations than for single determinations in the field, since each sample has to be treated individually and must be estimated straight after addition of thiocyanate. The colour discs in the Hellige-comparator are unfortunately not set to maximum colour intensity.
The value of Fe(II)-ion is obtained from the difference between the values for total iron and for Fe(III)-ion (see preceding section).

Organic matter, especially humins, *interferes* through its own colour. In this case, i.e., when the $KMnO_4$-consumption exceeds 50 mg/l, 100 ml water are treated with 2 ml nitric acid and 2 drops of cold, saturated $KMnO_4$-solution. After allowing to stand some time, 2 ml potassium thiocyanate solution are added; the permanganate colour dissappears and the iron thiocyanate colour appears. The reference solutions must be treated in the same way or the colorations do not agree accurately.

Brown waters containing humins are best evaporated down with potassium persulphate and the residue is heated in a platinum dish or crucible until the sulphuric acid is expelled. The residue is taken up in 2 ml HCl and the above procedure followed after making up the volume.

Turbidities must not be removed by filtering since the filter paper adsorbs dissolved iron. The suspension must therefore be centrifuged but one should remember that the suspended material has adsorbed iron, contains it or consists of iron oxide. Determinations of iron in turbid waters are always problematic; this can be changed by the sampling procedure (p. 5).
The thiocyanate coloration is impaired in waters of very high salt content (e.g. brackish and mineral waters). The reference solutions must be made up with the corresponding salt concentrations or the iron must be determined titrimetrically on the spot (see p. 246).
If the water sample is more than 12 hours old, the walls and bottom of the vessel must be rubbed all over with a rubber policeman before the aliquot for iron determination is taken (p. 21). With a water sample older than 5 to 8 days, the whole contents of the bottle must be used after addition of acid.
1 mg Fe = 1.43 mg Fe_2O_3; 1 mg Fe_2O_3 = 0.7 mg Fe.

Phenanthroline Method

According to the new German standard procedures (EV) total iron is determined as Fe^{2+} with phenanthroline. 100 ml of water sample, acidified with a little H_2SO_4 at the time of sampling, are treated with 2ml of acetate mixture (40 g ammonium acetate + 50 ml acetic acid, made up to 100 ml with distilled water) and 1ml of 20% hydroxylammonium chloride; after mixing well, 2ml of 0.5% phenanthroline solution are then added and measurements made on the orange solution at 492 nm in the photometer or using filter Hg 492 in a filter colorimeter.
The acidified sample must be treated as described in the laboratory within 2 hours. The bipyridyl method is therefore sometimes more convenient.

2. Determination of Manganese

Occurrence: Only a few tenths of a mg/l in ground water, rarely over 1 mg Mn/l. Only traces in surface water and occasionally more in the deep water from lakes and reservoirs (p. 175).

Rough field determination: 250 ml of water in each of two erlenmeyer flasks are treated with 1 drop of methyl orange solution (1 : 5000). 1 ml of NaOH is then added to the first flask. After brief standing, 5 ml of 10% HCl are added to both flasks. The first solution is decolorised in the presence of manganese (down to 0.1 mg/l). This preliminary test cannot be made on chlorinated drinking water since the reaction is shown by chlorine.
According to *K. Schilling*, a manganese content of more than 0.165 mg/l can be recognised by adding 1 ml 0.01N $KMnO_4$ solution per litre water; it yields an immediate yellow-brown colour whereas water containing no manganese retains the violet permanganate colour (Vom Wasser *28* [1961]).

a) Colorimetric Determination of Manganese

To 100 ml water are added 10.0 ml nitric acid (d = 1.2)[1] and the amount of 0.02N silver nitrate solution which was needed for the chloride titration, plus an additional ml of this reagent. The mixture is brought to the boil and, if the original water was yellowish, boiling must be maintained 10 minutes.

Mineral waters, especially brines, containing much chloride are evaporated in a wide-necked erlenmeyer flask, after addition of a few ml of H_2SO_4, until white vapours are evolved. The residue is diluted and checked for absence of chloride by adding 1 ml of 0.02N silver nitrate solution.

10 ml of 10% ammonium persulphate solution are added to the boiling liquid and boiling continued for 5 minutes. The presence of manganese is shown by a pink to red-violet colour. The liquid is cooled and made up to 100 ml with purest distilled water (Reagent No. 10) which has been boiled with nitric acid. The colour yielded is compared in comparator tubes or Hehner cylinders with those from standard solutions. These tubes or cylinders must be previously rinsed with hot water containing nitric acid. The reference cylinders are filled with 100 ml of distilled water which has been boiled with 10 ml of nitric acid and cooled; 0.01N $KMnO_4$ solution is added until the colours are the same. Mixing must be with a glass rod or sphere and not by placing the hand over the open end (*K. Höll,* Z. Chem. Fabrik [1934]).

Amounts of manganese down to 0.03 mg/l can be determined. In accurate determinations, a larger amount of water must be evaporated for a repeat estimation should the result be negative.

Calculation: Under these conditions, 1 ml 0.01N $KMnO_4$ corresponds to 1.1 mg Mn/l. A solution of 0.2877 g $KMnO_4$/l may be used as standard solution, containing 0.1 mg Mn/l.

The colour comparison can be undertaken also with the drum colorimeter of *Meinck-Horn,* or by preparing calibration curves at 530 nm with the help of a spectrophotometer.

Interferences: Waters rich in salts or with a high permanganate consumption (over 60 mg/l) can be evaporated with sulphuric acid and some potassium persulphate and the residue ignited in a crucible until sulphuric acid vapours are evolved. The residue is dissolved up in water plus 10 ml nitric acid and treated further as above. Iron amounts exceeding 5 mg/l can also interfere; this can be prevented by addition of sodium phosphate.

The precipitated silver chloride turns violet under the action of light; this may give the impression that manganese is present. The manganese determination ought not to be conducted in sunlight for this reason.

The walls of the flask must be rubbed round with a rubber policeman if the samples are older (p. 21).

b) Manganese Determination in Waters rich in Chloride

The tiresome evaporating down to expel HCl with waters rich in chloride, mineral waters, brines and brackish water can be circumvented. This is done by adding to 100 ml sample, 10 ml of a formaldoxime solution (40 g hydroxylammonium chloride +

1 ca. 32%, prepared by diluting conc. HNO_3 (d = 1.4) with an equal volume of water.

8 g paraformaldehyde or 20 ml 37% (weight) formalin solution, made up to 1 l with distilled water) and of a ferrous ammonium sulphate solution (140 g $(NH_4)_2Fe(SO_4)_2$ + 1 ml conc. H_2SO_4, made up to 1 l); after 5 minutes, 5 ml of an EDTA solution (37.224 g/l of disodium-EDTA) are added. Photometric evaluation is made at 450–480 nm after 1 hour. Amounts down to 0.05 mg/l can be estimated without evaporating the sample.

3. Determination of Hardness

(See p. 178) for the concept of hardness)

Occurrence of hardness-formers: Waters from limestone and dolomite regions are hard. They have values of up to 50° total hardness and 25° and more of carbonate hardness; usual values are 10 to 20° total and 8 to 15° carbonate hardness.
Waters from primary rock regions are soft, possessing only 1 to 3° hardness. Surface waters are often softer than the ground waters from the same formation.

Rough determination of hardness: The "soap method" is still being used for boiler and similar water, since the works are tuned to this method; it is thus retained in this edition.

Soap method of Boudron and Boudet: 40 ml of water are taken in *Boudron* and *Boudet's* method (filled to the mark of a glass stoppered flask for this purpose). 0.1 ml phenolphthalein solution is added and neutralisation to this indicator carried out with 0.1N NaOH or HCl. A tap less-burette ("hydrotimeter"), fixed in a burette holder, is filled with soap solution to the top with the help of a pipette. The soap solution is allowed to flow slowly down a glass rod. The burette is taken between thumb and middle finger, with the index finger covering the burette opening at the top. The burette is held slantingly and solution is allowed to run out (if necessary, by raising the index finger) until the soap solution is at the zero mark. Soap solution is then added in a similar way to the water sample until the solution is at the burette graduation 1. The flask containing the sample is stoppered and shaken several times in a vertical direction. When no froth remains on the water surface, further amounts of soap solution are added from graduation to graduation. This is continued until froth of about 1 cm height remains for several minutes and no longer "rustles". If the opening of the flask at the beginning of the titration is held to the ear, this "rustling" ("Selter water sound") is heard each time that one shakes the mixture. This is no longer audible at the end-point of titration.
Froth formation is often hindered by the "cheesy" magnesium soap which is formed with waters containing relatively high amounts of magnesium. A false end-point, appearing too early, is caused by formation of large bubbles in the space above the water. The determination is then repeated with a suitably diluted water sample.
Waters requiring more than 15 to 20 graduations of soap solution must always be diluted with distilled, boiled water.
If less than 2 graduations of soap solution are needed, the determination is repeated according to *Splittgerber-Mohr* on a suitably larger (500 ml) volume in an appropriate vessel. It must be preceded by addition of some NaCl (ca. 0.2 g) and titrating with 0.1N NaOH to

phenolphthalein (pH 11 to 13). Free carbon dioxide consumes soap solution (see below). Feed waters which are strongly alkaline to phenolphthalein are neutralised with 0.1N HCl to a pale pink coloration. *Clark's* soap solution, diluted 5 times with 50% methanol or ethanol, is used to examine condenser waters; 1 ml of this solution (see below) is equivalent to 0.003° dH with a 500 ml sample of water (Richtlinien der Ver. der Großkesselbesitzer und *Ammer,* Vom Wasser [1939]).

Calculation: The burettes in use nowadays are graduated in German degress of hardness. Older burettes, graduated in French degrees may be used when a conversion factor is employed (multiplication by 0.56 to yield German degrees) (see p. 179).

Interferences: 7 mg/l of free CO_2 consume soap solution corresponding to one French degree of hardness; 12.5 mg/l, to one German degree.
The soap method fails in the presence of polyphosphates.

Determination with Clark's soap solution: A more dilute soap solution is used in *Clark's* method. It is therefore more accurate than *Boudron* and *Boudet's* method and is especially good for very soft waters and for determining residual hardnesses of softened waters. This residual hardness is due principally to magnesium compounds.
At least 100 ml water are used. According to *Splittgerber-Mohr,* as much as 500 ml is employed for residual hardness. 1 ml of diluted *Clark* soap solution then corresponds to 0.003° German hardness (see above).
The various soap solutions are commercially available. Water samples for determination of residual hardness must be kept only in plastic bottles, paraffined bottles or, if need be, in Jena glass bottles.
The firm of Heyl, Hildesheim, W. Germany, is now offering a reagent, "Durognost", for rapid determination of hardness of water and boiler feed water; it is suitable also for water controllers and in steam boiler establishments. "Duroval" is also available for determining total hardness and "Duroval C" for determining the carbonate hardness. The firm of Merck, Darmstadt, W. Germany, is shortly to bring out a rapid checking apparatus on the market.

a) Accurate Determination of the Hydrogen Carbonate Ion (and carbonate hardness according to Lunge)

100 ml of water to be examined are placed in an erlenmeyer flask of 200 to 250 ml capacity. Precisely 0.1 ml of methyl orange solution (0.1% aqueous solution-DIN 8106) is added and the solution titrated with 0.1N HCl to the just recognisable colour change (yellow to orange-yellow). The titration must be performed in bright daylight, but not in direct sunlight, and against a white background. An untitrated sample of water + indicator can serve as comparison colour at the side. If the titration exceeds 8 ml of the 0.1N HCl, the determination must be repeated with a smaller size of sample, or the liberated CO_2 must be expelled by blowing in air and titration continued if the yellow colour then returns (see below).
If the orange colour and not the yellow appears on adding the indicator, the water has no carbonate hardness.
If higher concentrations of iron and manganese are present, 0.1° hardness per mg Fe or Mn/l must be subtracted from the total result. A very high sulphate content brings about rather too high values for carbonate hardness.

Interferences: The colour of moorland waters interferes; this can be prevented by filtration through active charcoal. Free carbon dioxide reduces accuracy with very soft water, according to *Biskei* (Z. Anal. Chem. [1937]). This applies equally to mineral waters, rich in carbon dioxide and also to very hard ground waters. The free CO_2 must be removed by warming and occasionally the CO_2 liberated during titration must be expelled. *Mortimer's* mixed indicator is good for peaty water; it consists of 20 mg methyl red + 100 mg bromocresol green in 100 ml alcohol (Merck).
Hard waters should not be left standing for days in half-filled bottles before examination; calcium carbonate then precipitates and adheres to the walls. This leads to too low values for carbonate and total hardness and to too high values for the "associated" carbon dioxide (p. 163).

Calculation: The ml of 0.1N HCl required for 100 ml sample give the carbonate hardness in mval/l; multiplied by 2.8, this yields the carbonate hardness as German degrees.
See p. 179 concerning "apparent hardness".

b) Accurate Measurement of Total Hardness according to Blacher

The water sample which has been titrated with 0.1N HCl to methyl orange is used in the determination of total hardness. The CO_2 liberated by the acidification is expelled by briefly boiling or aspirating air through for some time (5 minutes with the water pump). The colour change is then much sharper.
The water sample, cooled from boiling if necessary, is mixed with some drops of 1% phenolphthalein solution and titrated dropwise with 0.1N sodium hydroxide to a weakly pink colour. This colour is removed by adding a drop of 0.1N HCl.
The neutralised sample is titrated immediately with 0.1N potassium palmitate solution (Reagent No. 14) until the phenolphthalein red colour appears strongly.
The red colour is again removed with 0.3 ml of 0.1N HCl. Should less than 0.3 ml of the 0.1N HCl be required, the titration to the intense red is repeated. Should more than 0.3 ml be required, the extra amount beyond 0.3 ml must be subtracted from the palmitate consumption.

Very hard waters give an unclear colour change. If the titration requires altogether more than 8 ml potassium palmitate solution, it must be repeated after suitable dilution with distilled water and boiling and cooling.

Calculation: 1 ml of the palmitate solution consumed corresponds to 2.8° German hardness = 28.0 mg CaO/l, when a 100 ml water sample is used. The result is accurate to about 0.3° dH.
Water hardness has been given recently in mval/l. It then suffices to quote the ml consumed in the titration as mval/l.
0.1° dH is subtracted from the result for every mg Fe or Mn/l in waters rich in iron – but only when these two are still in solution. The values for very soft waters (below 1.0° dH) are about one-third too high (CO_2 interference etc.). Consequently N/28 potassium palmitate solution is best used with these very soft waters; 1 ml of it is equivalent to 1.0° dH for a 100 ml water sample.
Interference by humic substances (brown-coloured peaty waters) can be eliminated by evaporating 200 ml water with some potassium chlorate and HCl. The dry residue is taken up with hot, distilled water. After cooling and adding 2 drops of methyl orange and 1 drop

of phenolphthalein, the solution must be accurately neutralised. The titration with potassium palmitate is carried out on this neutralised solution.

c) Calculation of Total Hardness from Calcium and Magnesium Hardnesses

The number of mg/l of CaO found for the water is divided by 10 to give the calcium hardness. The magnesium hardness is converted to calcium hardness by dividing the number of mg MgO/l by 7.19; or by dividing this number of mg MgO/l by 10 and multiplying by 1.399. Calcium hardness plus magnesium hardness converted to calcium hardness figures gives total hardness.

d) Rapid Method for Determining Total Hardness with EDTA (Titriplex)

Merck of Darmstadt, W. Germany, has brought out standard solutions of ethylenediaminetetraacetic acid (Complexone), based on the work of *Schwarzenbach* (Helv. Chim. Acta *29,* 811 [1946]). These have the names Titriplex A (for hard waters) and Titriplex B (for soft waters). Both contain a zinc compound, so that accurate results are obtained also with waters containing no or little magnesium and a sharp colour change is always obtained.

Procedure: An indicator tablet is added to a 100 ml sample (of hard water). The tablet contains a mixed indicator with Eriochrome Black and dissolves at once on swirling round. 1 ml of concentrated ammonia (d = 0.910) is added and the solution titrated with Titriplex A until the red colour changes to pure green.

Calculation: The number of ml of Titriplex A required (for 100 ml sample) multiplied by 5.6, gives the degree of total hardness.
100 ml of *soft water* (up to 3° hardness) are warmed to 40° C, an indicator tablet and then 1 ml of the ammonia added, and the solution shaken as above. The warmed solution is slowly titrated with Titriplex B until the green colour comes. The consumption of Titriplex B in ml gives directly the degree of (total) hardness.
Useful results are obtained for soft waters, using Titriplex B. The author's experience shows, however, that they are regularly somewhat lower than those obtained with the standard method according to *Blacher* using potassium palmitate. Values about 5% too low are obtained for hard waters when using Titriplex A.
The method fails with iron-containing waters; there is no clear end-point. Added Na_2S, as recommended by Merck, is also without effect. Experiments of the author show that quite tolerable results can, however, be obtained by addition, before titration, of a few drops of a 5% sodium phosphate solution, or, more recently, 1–2 ml of triethanolamine.
Titration is possible also with a 0.1M Titriplex solution, 1 ml of which is equivalent to 4.008 mg Ca and 2.431 mg Mg.
Work with concentrated ammonia is unsatisfactory in a drinking water laboratory, even when performed in the fume cupboard.

e) Titrimetric Ca- and Mg-determination with Complexone; Method based on the New EV (German Standard Methods)

100 ml of water under examination are titrated with 0.1N HCl until the colour of the mixed indicator changes from green to red. A further 0.5 ml of the acid is added and the liberated CO_2 expelled by boiling.

Mixed indicator: 100 ml of a 0.03% alcoholic solution of methyl red are mixed with 15 ml of a 0.1% aqueous solution of methylene blue.
The pH of the sample after cooling to 30°–40°, and of a blank is adjusted to 12–13 by adding 0.4 ml of a 15% NaOH solution. Murexide indicator is immediately added and titration carried out with Complexone to the colour change from red to blue ("green" is given in error in the EV). The new indicator Calconcarbonic acid (3-hydroxy-4- [2-hydroxy-4-sulphonaphthyl-1-naphthylazo]-2-naphthoic-acid) (Merck 4594) has proved very good. (6.65 g Complexone III in 1 l distilled water). Titriplex III of the firm of Merck and Idranal III of Riedel de Haën, Seelze, W. Germany, can be used as well as the Complexone of the firm of A. G. Siegfried, Zopfingen (Switzerland). Absolutely copper-free distilled water must be used to dissolve these titration substances. All three are the disodium salt of ethylenediaminetetraacetic acid.
100 ml of distilled water, treated under the same conditions, not titrated but containing 0.2 to 0.5 ml of Complexone, serve for colour comparison. The main determinations must always be carried out singly, otherwise calcium and above all, magnesium could become insoluble. Even duplicate determinations should not be made simultaneously.

1 ml of the above Complexone solution corresponds to 1 mg CaO = 0.7147 mg Ca.

Magnesium determination: The above solution is acidified with 0.5 ml 25% HCl for the magnesium determination and heated for a short while at 60 to 80° C to destory the murexide indicator. As soon as the light red colour of the mixed indicator is seen, 5 ml of ammonium buffer (54 g ammonium chloride + 350 ml of 25% ammonia, made up to 1 l with distilled water), plus 1 ml of 25% ammonia are added to the still warm sample, bringing the pH of the solution to about 10.
0.1 ml of Eriochrome Black solution (0.2 g Eriochrome Black T and 0.5 ml of 25% ammonia, made up to 100 ml with distilled water) is then added and the solution titrated at 40 to 50° with Mg-Complexone III solution (prepared by dissolving 6.65 g Complexone III and 5.0 g magnesium complex in distilled water and making up to 1 litre) until the colour changes from wine red to inky blue.
The determination of calcium and magnesium with Complexone is very accurate and convenient with soft waters. The end-point is not so sharply recognisable and the accuracy lower with hard waters, especially those containing iron and when they are rich in magnesium and in calcium sulphate. Waters containing much magnesium become turbid when the murexide is destroyed; this can be prevented by adding more HCl.

1 ml Complexone III solution is equivalent to 0.719 mg MgO
1 mg MgO = 0.603 mg Mg

Interferences: Results are inaccurate with waters containing iron and if these amounts ex-

ceed 2 mg Fe/l, the method fails altogether. The colour change is, however, more easily recognised if some sodium phosphate is added.
There should be no turbidity as a result of separation of calcium when sodium hydroxide is added; hence a sample which has been cooled too far before titration must not be rewarmed after the addition of alkali.

Non-carbonate hardness
Calculation of the non-carbonate hardness: If the carbonate hardness is subtracted from the total hardness, the non-carbonate hardness is obtained; this was formerly termed the permanent hardness.

4. Determination of Calcium and Magnesium

Preparation: 400 ml of soft water, 200 of hard are taken. To iron-containing waters, 0.5 ml of 3% hydrogen peroxide and 2 ml of hydrochloric acid* are added in a beaker and the volume reduced to about 100 ml by evaporating down.
Excess ammonia is added to the boiling liquid and the mixture heated a few minutes more over a small flame to ensure complete precipitation of the iron. The precipitated iron hydroxide is filtered off and washed briefly with hot distilled water.
0.5 g of ammonium persulphate is added to waters containing humins before evaporating down.

a) Gravimetric Calcium-determination

The filtrate, freed from iron and, if necessary, from organic matter (see above), is acidified in a beaker with acetic acid and brought to the boil. 10 to 20 ml of a hot 4% ammonium oxalate solution are added dropwise. The mixture is heated for a further hour in the covered beaker, on the water-bath or with a small flame.
After standing for a longer time, when the precipitate has fully settled, the solution is filtered (Schleicher and Schüll No 589[2]) and the calcium oxalate precipitate washed with hot ammonium oxalate-containing water and finally briefly with distilled water. The filtrate is ready for the magnesium determination. The funnel with filter is placed in another beaker. As a precaution the tube of the funnel is washed with distilled water, on the outside also. The filter is then pierced with a pointed glass rod. The precipitate is dissolved from the filter with hot dilute HCl and washed into the beaker with hot, acidified (HCl) distilled water. This is especially to be recommended with waters which are rich in magnesium. The solution is treated with excess ammonia and acidified again with acetic acid. The precipitation of calcium oxalate is then repeated in the acetic acid-containing solution, using a few ml of ammonium oxalate solution. A magnesium-free precipitate of calcium oxalate is thus obtained. It is again washed with a little water containing ammonium oxalate and twice with the smallest possible amounts of pure distilled water.
The precipitate is introduced into an accurately weighed porcelain filtering crucible which has stood for an hour in an oven at 105° C and has cooled in a desiccator. The precipitate is washed with water containing ammonium oxalate and with a little distilled water. The crucible is dried for 2 to 3 hours at 105° C in the drying oven and weighed after cooling in a desiccator.

Interferences: Waters rich in calcium sulphate are rendered ammoniacal before evaporating down. They would otherwise bump fiercely, despite addition of HCl, and also give a precipitate of calcium sulphate which adheres strongly to the walls.

Calculation: mg calcium oxalate = 0.2743 mg Ca or 0.3838 mg CaO
mg Ca = mg CaO × 0.7147; mg CaO = mg Ca × 1.399.

b) Titrimetric Calcium-determination

The calcium oxalate precipitate can be determined volumetrically instead of gravimetrically. The second (Mg-free) calcium oxalate precipitate is collected on a smooth, compact filter and washed with hot water, containing acetic acid, until the oxalate reaction is no longer shown.

The filter is placed in a 300 ml erlenmeyer flask and the still moist precipitate is washed into the flask with hot dilute sulphuric acid (about 20 ml, 1 + 5) after piercing the filter with a pointed glass rod. Care must be taken that no filter fibres enter the flask; these seriously interfere with the result (saccharification of the cellulose).

The liberated oxalic acid is titrated in hot solution with 0.05N $KMnO_4$ solution until the first drop giving a permanent faint pink colour. After adding the first 10 to 20 drops, one must wait a little until the reaction gets going. According to *Kuisel,* the end-point is better recognisable if a few drops of ferroin solution (e.g., Redoxindikator Merck) are added; this gives a sharp change from red to blue.

1 ml 0.05 N $KMnO_4$ corresponds to 5 mg Ca/l (with a 200 ml sample).
1 ml 0.05N $KMnO_4$ = 1.4 mg CaO or 1.0 mg Ca.

c) Titrimetric Magnesium-determination

For the magnesium determination, the filtrate from the precipitation of calcium oxalate (see above) is evaporated down to about 30 ml, best after having added 3 ml HNO_3. Phenolphthalein is added to the boiling solution and then ammonia until a red colour is present. 10 ml of 7% diammonium hydrogen phosphate solution and 25 ml of 25% ammonia are then added. The covered beaker is left standing till the next day to ensure complete precipitation. The precipitate of magnesium ammonium phosphate is then filtered and washed with 2% ammonia until the filtrate no longer gives a chloride reaction i.e., the washings must give no turbidity when acidified with HNO_3 and treated with silver nitrate solution. The filter with the magnesium precipitate is dried in the oven at 50 to 70° C.

The dried filter is placed in a 100 ml beaker and covered with 20 ml of 0.1N HCl. After allowing to stand for some time, with occasional swirling of the contents, 0.1 ml of 0.1% alcoholic Dimethyl Yellow solution is added and the solution titrated with 0.1N NaOH to the colour change from red to yellow. The Mg-content is calculated from the difference.

Calculation: For a 200 ml sample of water, 1 ml of 0.1N HCl corresponds to 6.08 mg Mg/l.
1 ml 0.1 HCl = 2.016 mg MgO = 1.216 mg Mg
1 mg Mg = 1.658 mg MgO
1 mg MgO = 0.603 mg Mg

d) Colorimetric Magnesium-determination

Small amounts of magnesium (less than 10 mg Mg/l) are determined colorimetrically with *Titan Yellow.* To 50 ml of the water are added 1 ml sulphuric acid (1:35), 8 ml of 1% starch solution, 10 ml of saturated calcium sulphate solution, 1 ml of 0.05% Titan Yellow solution and finally 5 ml of 2N NaOH, in this order, shaking after each addition. The volume is then made up to 100 ml with distilled water and, after 5 minutes, the solution compared with reference solutions made up at the same time. The standard solution used for this purpose contains 2.5341 g of $MgSO_4 \cdot 7H_2O$ in 500 ml, i.e., 1 ml contains 0.5 mg Mg.

e) Determination of Magnesium and Calcium Chlorides

Potash mine wastes carry $CaCl_2$ and $MgCl_2$ into the receiving waters and ground water. Drinking water must not contain over 170 mg/l of $MgCl_2$.
300 ml of the water are evaporated to dryness in a porcelain basin. The residue is fully dried out by heating in a drying oven at 110° C. After it has cooled, the residue is powdered with a pestle and stirred up with 20 ml of absolute alcohol. The supernatant alcohol is decanted through a compact filter after the residue has settled out. This extraction is repeated 6 times, using a maximum of 200 ml of alcohol.
The combined filtrates are evaporated to dryness and the residue is taken up with water. Calcium and magnesium are determined in this solution, as described in the preceding sections; the results are expressed as $CaCl_2$ and $MgCl_2$.

5. Hydrogen Sulphide and Sulphides

Rough field determination: The test through smell (p. 16) suffices in most cases. If a chemical test for H_2S is desired, a 100 ml flask is three-quarters filled with the water and a strip of paper, moistened with lead acetate solution, is attached to the inside of the stopper. The presence of H_2S is shown by a yellow to brown colour of the test paper.
Combined hydrogen sulphide, i.e., sulphides, can be detected if the water is acidified.

a) Colorimetric Determination of small H_2S-Amounts

A glass stoppered bottle (so-called oxygen bottle, of 300 ml) is, as described on p. 11, is filled to the top of the neck by allowing the water to flow through for a longer time; the water sample thus does not come into contact with the air. 5 ml of alkaline lead solution (see below) are then added with an elongated pipette. The slanting stopper is replaced and the contents mixed by inverting the flask. Any colour is compared immediately with that from standard solutions after carefully transferring to Hehner cylinders; or compared in the Hellige comparator with colour discs; or evaluated photometrically at 380 nm.
Sulphide ion can be quantitatively determined after expelling the hydrogen sulphide with nitrogen.

Standard sulphide solution: 100 mg of purest, dry sodium sulphide ($Na_2S \cdot 9H_20$) and 50 g of sodium nitrate plus a little glycerol are dissolved in water and the volume is made up to 100 ml. 1 ml = 0.1419 mg H_2S.

Alkaline lead solution: 25 g of Rochelle salt, 5 g of NaOH and 1.0 g of lead acetate are dissolved in 100 ml of distilled water.
The reference solutions are best made up with oxygen-free water (boiled and cooled, or distilled water prepared free of oxygen by means of nitrogen).

Interferences: Any interference through the colour of the water itself (humic substances) can be eliminated by using for the reference solutions the same water under examination but after having expelled H_2S from it by aeration; alternatively, the colour of the untreated water can be determined with standard sulphide solutions.

b) Iodometric H_2S-determination

If the presence of larger amounts of H_2S can be inferred from the smell, these are combined chemically at the time and place of sampling. Sampling is as above, in a litre flask with stopper or an oxygen bottle. A few crystals of cadmium acetate are added, or 3 ml of a 5% solution of cadmium acetate containing acetic acid, according to the German standard procedure (EV), or 2 ml of 20% zinc acetate solution, according to *K. E. Quentin.*
After the precipitate has settled, the greater part of the supernatant is drawn off and the precipitate brought on to a filter. Precipitate + filter are treated with 0.01N iodine solution in a glass stoppered flask and acidified with 10 ml hydrochloric acid*. Back titration with 0.01N thiosulphate solution is performed after 15 minutes. 1 ml of consumed iodine solution is equivalent to 0.1704 mg H_2S or 0.1603 mg S^{2-}.
The 1% starch solution is made up with 3% common salt solution to render it more stable. In the calculation, the amount of added reagent must be subtracted from the amount of water used.
If cadmium sulphide does not separate, direct titration can be carried out, as described below.

c) Direct iodometric titration of H_2S

If only small amounts of H_2S are present, the water can be sampled in a litre flask as described above and treated directly with 10 ml of 0.01N iodine solution; after a brief wait, the solution is back titrated with 0.01N sodium thiosulphate solution.

Interferences: Organic matter interferes through its colour and iodine consumption. This can be allowed for by measuring the iodine consumption of a water sample freed from H_2S by aeration. To determine sulphide, free H_2S is first expelled by passing CO_2 or nitrogen through a measured volume of the water until the smell of H_2S has vanished.

6. Residue on Evaporation and Ignition; Loss on Ignition

200 ml of the water are taken for determination of the residue on evaporation. The sample is evaporated to dryness in an ignited and accurately weighed platinum dish on the waterbath. The residue is heated in an oven at 110° C for 1 to 2 hours to constant weight. The dish

is allowed to cool in a desiccator and then quickly weighed. The increase in weight of the dish, multiplied by 5, gives the residue on evaporation in mg/l. Should the residue be less than 50 mg/l, the determination is to be repeated with a larger sample.

According to the new EV (Einheitsverfahren = Standard procedures), the drying must be carried out at 180° C, as is already customary in mineral water analysis and for boiler waters.

If large amounts of undissolved material are present, the *total residue* of unfiltered water is determined and then the *residue on evaporation* of a filtered sample (Schleicher and Schüll paper No. 589[2]) giving total dissolved material.

The standard flask with the water sample must not be warmed on the water-bath, as is otherwise habitual in quantitative analysis, because carbonates would separate on the walls, yielding too low values for the residue.

Determination of the residue on ignition. The residue from evaporation in the platinum dish is heated to dull redness for 15 minutes by means of a burner at low setting or in a muffle furnace at 600 to 650° C. Any resultant smell is noted. Contaminated waters, which usually contain proteins, give rise to an odour like burnt hair. One can recognise a larger amount of organic matter through the darkening of the residue just before ignition sets in.

Decomposed carbonates are reformed by evaporating down again after addition of ammonium carbonate solution and then igniting quite gently; or also by moistening with a little ammonium nitrate solution and igniting. The residue from ignition should be a clean white, although this is not so with waters rich in iron. Ignition is then carried out after adding ammonium nitrate. The residue from ignition is weighed after cooling in a desiccator. The increase in weight of the empty dish, multiplied by 5, gives the ignition residue in mg/l.

Loss on ignition: The difference between the residues on evaporation and on ignition gives the loss on ignition. It may not be considered as organic material; some inorganic compounds also lose weight, e.g. nitrate, bicarbonate, carbonate etc.

The firm of W. C. Heraeus, Hanau, W. Germany, supplies an oven for this purpose, designed according to instructions of K. Höll.

7. Electrolytic Conductivity

The electrolytic conductivity of the water gives a clue to the amount of salts or dissociated material present. This plays a part in the control of interceptors and river courses, for example, which take variable amounts of effluent; and of ground water which can be influenced by surface water. A conductivity determination is likewise useful for control of demineralisation plants (p. 74).

The conductivity determination enables a rapid judgment to be made of the salt content at various depths in the waters of coastal and tropical regions. It is often necessary there to ascertain to what depth the water of low salt concentration floating above a layer of salty water extends. Even in fresh water regions, conductivity data can provide information about varying concentrations of inorganic material (meromictic lakes).

Electrolytic conductivity is expressed through the reciprocal of the electrical (specific) resist-

ance in Ω (ohms), referred to a cube of water of side 1 cm at 20° C. The unit is the mho (Siemens in Germany).
The firm of WTW of Weilheim, Bavaria, provides a handy apparatus for determining conductivity.
Conductivity and pH-measurements can be made with the Labor-Lyo-Ionostat of the firm of Lautenschläger of Munich. Continuous measurements can be made with the Dephimeter according to *Klutke* (Firm of Pancontrol, Frankfurt/Main) and the apparatus of the firm of Wösthoff of Bochum, W Germany.
Good distilled water should have conductivity values of less than 0.3×10^{-6} mho.cm^{-1} ($< 5\ \mu$ mho · cm^{-1}). The conductivity determination is best done at 18° C. Converted data are inaccurate.

8. Interferometer Value

Using an instrument such as the Zeiss interferometer (firm of Carl Zeiss of Jena, E. Germany), the light diffraction of the untreated water can be compared with that treated with barium sulphate. This gives information about the presence of colloidal matter in the water. The colloids are adsorbed by the barium sulphate and this yields differences in the interferometer values[2].

9. Density (Specific Gravity)

Specific gravity and density are measured at 20° C. Values are given to the fourth decimal place for medicinal springs. The determination is made with calibrated pycnometers of 100 or 250 ml content. They are left for an hour in a thermostat at 20° C. Three measurements to the fourth decimal place are carried out with mineral water, best in pycnometers with ground-glass fitting thermometers.
The pycnometer must be cleaned from time to time with alcohol, acid and distilled water.
Calculation: According to the formula:

$$d_{20^\circ} = \frac{(W_1 - L) \times (0.9982)}{W_2 - L}$$

where W_1 = weight of the pycnometer containing the water under investigation, W_2 = weight of it when containing distilled water, and L = the weight of the empty pycnometer. 0.99823 is the density of distilled water at 20° C.
Reference to water at 4° C has been abandoned.
The density of soft waters, low in mineral content and containing carbon dioxide, can be below unity.

2 *W. Prée,* Anwendungsmöglichkeiten des Zeiss'chen Flüssigkeit-Interferometers bei Trinkwasseruntersuchungen, Thesis, Dresden 1931 (Applications of the Zeiss Liquid-Interferometer in Studies of Drinking Water).

The expression "specific gravity" is no longer used. It was the ratio of the weights of equal volumes of a solution and of water.
E.g., d_{20}°(pyc.) = 1.010 g/ml

10. Radioactivity of Drinking Water and its Measurement

The radioactivity of materials, including water, is an emission of radiation which arises during the decay of radioactive substances.
Distinction must be made between the natural and the artificial radioactivity of the water. Natural radioactivity is never harmful to man and animal; as known, it is utilised for medicinal purposes (radioactive medicinal waters). Artificial radioactivity is harmful or can have a harmful action beyond a certain toleration limit in so far as it is due to "contamination" (radioactive contamination through wastewaters or air). The artificial radioactivity of the relatively innocuous ^{24}Na is an exception.
Nowadays radioactive isotopes are known of most elements, not only of the radium-thorium group but also of the "harmless" elements, such as potassium, hydrogen and oxygen. Natural radioactivity comes principally from radioactive potassium, ^{40}K, which occurs everywhere in nature[3]; and also from both the components of the water, hydrogen and oxygen, e.g., from the radioactive hydrogen isotope, tritium, ^{3}H.
Natural radioactivity is harmless because the isotopes creating it have only short lives; their half-life is a matter of only hours or days. Artifical radioactivity is due on the other hand to the long-life isotopes of strontium, radium, krypton and thorium with half-lives of several years, or of ^{238}U, with a half-life of 4.5×10^9 years.
The strontium isotope ^{90}Sr, one of the most dangerous radioactive isotopes, has nothing to do with the ordinary strontium occurring in mineral waters and medicinal springs. Such questions are always being asked of the spring specialist nowadays by laymen. ^{90}Sr and ^{85}Kr play a big part in the contamination of receiving water with effluents of atomic energy plants and in decontamination of these wastewaters. The limiting amount of ^{90}Sr in drinking water, reckoned for the total population, is 8×10^{-8} μ C/ml.
The average consumption of 2.2 litres water per day over a period of 70 years is taken for calculating the maximum permissible concentration.
The radioactive emission consists of α-, β- and γ-radiation, both corpuscular (α- and β-radiation) and electromagnetic (γ-radiation). The β-radiation plays the principal part in natural radioactivity and γ-radiation in artificial radioactivity. γ Radiation is the more dangerous since it has far greater penetrative power and it is harder to gain protection from it. Most of the artificial radionuclides are β-emitters and therefore total β-activity is measured in supervising waters (*S. Müller*, Dechema Monographie, Vol. 30 and *K. Haberer*, Radionuklide im Wasser, Thiemig-Taschenbuch 17, Munich 1969).
According to the report of the "Sonderausschuß Radioaktivität Bundesrepublik Deutschland" (Special Commission on Radioactivity in the Federal German Republic),

3 In nature there is 93.08% ^{39}K, 0.0119% ^{40}K and 6.91% ^{41}K. ^{40}K is a β-emitter and has a half-life of 3.8 days.

Stuttgart [1958], publisher G. Thieme, the mean specific strontium activity of milk increased by a factor of 3 between 1955 and 1957. This commission is in the process of establishing the normal values for foodstuffs and water and to follow subsequent development as a result of atomic explosion tests and the peaceful uses of atomic energy. *G. Gad* and *T. Kempf* (Ges. Ind. *79,* 84–87 [1958]) have established the radioactivity level of ground water in the Berlin area; and *H. Unger* (Öffentl. Gesundheitsdienst *21,* 16–25 [1959]) for Bavaria. *Wieser* (Lit. Ber. No. 4, p. 236 [1959]) found in the Lake of Constance that the radioactivity had increased in 40 metres depth from 0.02 to 0.22 $\times 10^{-10}$ C/l and that it had a total value of 0.43 to 0.60 $\times 10^{-10}$ C/l on the surface.

a) Measurement of the Radioactivity

Nowadays one must reckon with a large number of radionuclides in water. Not all of these can detected with chemical methods but can be on the basis of their radiation. Uranium is an exception, which can be detected in amounts down to 0.012 ppm via uranium fluoride.

Radioactivity is measured by counting the number of nuclear disintegrations per minute per unit volume, e.g. in 1 litre of drinking water. If artificial radioactivity is suspected, e.g. through pollution by radioactively contaminated waste water, the number of disintegrations must not exceed 22 per minute; that represents 0.1 $\times 10^{-7} \mu$ C/ml (microcurie/ml). The natural β-radioactivity, coming from ^{40}K, of the ground water and thus of ordinary drinking water, is only 1 $\times 10^{-8}$ to 1 $\times 10^{-10} \mu$ C/ml. These extraordinarily small amounts in drinking water can usually be determined only after concentrating the acidified water, or in the residue from evaporation or ignition. It is done after a procedure in which silicates are separated by evaporating down with HCl, alkaline earths by precipitation as sulphates, and potassium by precipitation as potassium aluminium sulphate (*E. Wosahlo,* Gas- u. Wasserfach *99,* No. 14 [1958] and Handbuch der Lebensmittelchemie (Handbook of Food Chemistry), Vol. *8,* p. 771 [1970]; *K. Haberer* also gives additional information in the series of publications of the W. German ministry for atomic energy and water economy, Nr. 6, 1958). Hard waters, rich in salts, make measurement difficult; only small amounts of the evaporation residue can be introduced into the counting pans. Losses of radioactive material occur during evaporation down; these have to be estimated by parallel treatment of standard preparations A compensation is that evaporation removes radon, a source of error leading to too high values with an immersion β-counting tube.

The radiation of usually 300 mg of a residue from evaporation is measured in a counting pan and compared with the value from standard preparations.

K. Aurand (Z. Röntgenpraxis *19,* 266, 276 [1966]) reports the apparatus aspect of determining radioactive matter in water and effluents.

α- and β-radiation are measured with the Geiger-Müller counting tube, the end window counter and the methane gas-flow counter. Scintillation counters are used to determine γ-radiation. The number of impulses is registered in a measuring cabinet. An idea of the nature of the emitter is obtained from the way in which the impulse decays.

Firms in German-speaking areas which provide measuring instruments are: Elektro Spezial, Hamburg (Philips); Frieseke & Hoepfner, Erlangen. W. Germany; Kirem, GmbH., Frankfurt/Main (Hartmann and Braun System); Landis & Gyr, Zug, Switzerland.

The portable apparatus according to *D. E. Schmitt* (firm of Franz Bergmann KG., Berlin-Zehlendorf) serves for preparation of radioactive samples of water.
Kirem GmbH. of Frankfurt/Main supplies a complete "Drinking water control plant" of the firm of Landis & Gyr.

Determination of Radon
Measurement of the natural radioactivity of medicinal springs, the content of radon, is performed with the Fontaktoscope (supplied by Kirem GmbH of Frankfurt/Main).
The electrostatic Fontaktoscope has been ousted recently by the vibrating reed electrometer.
K. E. Quentin and *G. Schretzenmayr* (Z. Leb. Unt. u. Forschung *139,* 87–96 [1969]) report on the up-to-date state of radon determination in mineral and medicinal waters.

b) Decontamination

Decontamination is the removal of radioactive elements from polluted objects and water.
According to *W. H. Frank*'s experiments at the Dortmund Municipal Works (W. Germany), 75–95% of the radioactive fission products can be removed in artificial concentration of ground water if pre-filtration is used. 99% of the retained radionuclides were held in the upper 5 cm of the sand filter beds. The two-stage sand filters were more effective than iron hydroxide and aluminium hydroxide adsorption, followed by a rapid filter (GWF *105,* No. 50 [1964]). *G. von Hagel* confirmed these results (Veröff. Inst. für Siedlungswasserwirtschaft, Hannover, 1964, No. 16).
W. Herrmann (Series of Reports (Schriftenreihe) of the Deutscher Arbeitskreis Wasserforschung (German Working Group for Water Research, No. 11) reported the removal of radionuclides in the Krefeld (W. Germany) waterworks. Up to 65% of the strontium was taken out after addition of iron sulphate, with the help of the "Krefeld blanket filter". The result was even better when several such filters were used in series. Aluminium sulphate proved less suitable there.
Many and detailed experiments have demonstrated the varied effectiveness with different radionuclides of the normal procedures for conditioning, e.g. precipitation with lime, lime-soda process, flocculation process with iron and aluminium salts – even though 30% decontamination is always attained. *H. Bernhardt* (Deutsche Beiträge Wasser für den Frieden, 76–103 [1967]) describes very interesting experiments at the Wahnbach reservoir, W. Germany.
Cation exchanger filters will be used on a large scale in the event of a catastrophe. Sawdust has also proved useful in decontamination, especially where it was treated with sodium silicate and aluminium sulphate solution. Bentonite is a useful material too.
Swiss experience shows peat to possess a high exchange capacity. 99% of ^{90}Sr (1000 pC/l) was retained from cistern water. It is best to mix the peat with sawdust, so that the filter runs better (*M. M. Bezzegk,* Inform. Sheet 10 of the Fed. Europ. Water Protection).
K. Haberer treats exhaustively the various decontamination processes in the monograph "Radionuklide im Wasser".
The application of hydrosilicates (e.g. DEK–1) by the North German Chemical Works, Hamburg (20–100 mg/l water) is new.

The firms of Krupp of Essen, Berkefeld of Celle and Seitzwerke of Bad Kreuznach (all W. Germany) offer small equipments for deactivation.
S. Krawczynski and *B. Kamellakopulos* (GWF *102*, 601–603 [1961]) report on decontamination of radioactive effluents in small purification plants.
The International Commission for Radiation Protection (ICRP 1959) has fixed 100 pC/l (100 μC/ml) as the maximum permissible limiting concentration for drinking water.
The simplest method of decontamination is leaving the water in tanks. Many nuclides decay rapidly, e.g. ^{131}I in 7.7 days and ^{103}Ru in 40 days. Only the particularly dangerous ^{90}Sr has an especially long half-life (see above).
The firm of Berkefeld, Celle (W. Germany) furnishes a mobile decontaminating plant.

D. Artificial Water Additives

(to be carried out immediately after sampling)

1. Free Chlorine and Combined Available Chlorine

a) Free Available Chlorine

In chlorinated water, chlorine can be in the form of free, elementary chlorine or of hypochlorite; this is termed "free available chlorine" (formerly "free chlorine").

b) Combined Available Chlorine

"Combined available chlorine" consists of chlorine compounds, such as inorganic and organic chloramines, monochloramine, dichloramine and chlorinated organic amines; these act as oxidising and sterilising agents.

c) Total Chlorine (Available Chlorine)

Total chlorine is the sum of free chlorine and all chlorine substitution products of oxidising activity; this is now termed "available chlorine" also.

2. Determination of Free Available Chlorine

a) Colorimetric Procedures

Tolidine method
The colorimetric method of determination with o-tolidine has been given up because it has been shown to be too insensitive, inaccurate and unspecific (limit of detection >

0.3 mg/l, error ± 0.275 mg/l chlorine according to *U. Hässelbarth,* Z. Anal. Chem. *234,* 22–37 [1968]). The method with diethyl-p-phenylenediamine (DPD), due to *Palin,* has now replaced it.

The Colorimetric DPD-Method

The colour of water samples (containing 0.1 to 4 mg Cl/l) to which DPD reagent has been added can be easily and sufficiently accurately compared with colour discs in the Lovibond comparator (German suppliers, firm of Akdolit, Hans Börner, Düsseldorf) or the Hellige comparator (firm of Hellige, Freiburg i. Breisgau). The reagent, diethyl-p-phenylenediamine, yields a red coloration which can be adapted to colorimetric determination in accordance with Beer's law. The DPD tablet A is crushed with a glass rod in a 13 mm cuvette with a few drops of the water under investigation. The cuvette is filled up with the water and the contents swirled round. The value in the comparator is read off immediately.

b) Titrimetric Determination of Free Chlorine with DPD

To 5 ml of DPD reagent (Reagent No. 10a) and 5 ml of phosphate buffer (Reagent No. 24) are added 100 ml water sample if it contains a larger amount of chlorine. After 5 minutes, the solution is titrated with ferrous ammonium sulphate solution (Reagent No. 12a) from red or reddish to colourless (1 ml = 0.1 mg Cl). Work is best with microburettes and a magnetic stirrer.

3. Titrimetric Determination of Total Chlorine

Total chlorine (i.e. free available chlorine + combined available chlorine) is determined in the same way as just mentioned, except that 1 g of potassium iodide is added as well as the 5 ml amounts of DPD and of phosphate, and that the 100 ml water are added after mixing the reagents.

Colorimetric determination of total chlorine

After the value for free available chlorine has been read in the Lovibond or Hellige comparator, a reagent tablet B is added to the cuvette and dissolved by mixing. After 2 minutes, the colour intensity is evaluated again. Combined available chlorine leads to an intensification of the existing red colour.

4. Determination of Combined Available Chlorine

The value for combined available chlorine is the difference between total chlorine and free available chlorine.

c) *Determination of Chlorine dioxide*

According to *M. A. Post* and *W. A. Moore,* chlorine dioxide can be colorimetrically determined in acetate-buffered solution (pH = 4.1 to 4.4) and in presence of iron (III) chloride, using 1-amino-8-naphthol-3,6-disulphonic acid. The bluish-red coloration attains its maximum intensity after 20 minutes (Z. Anal. Chem. *176,* 295–296 [1960]).

5. Chlorine Demand (Chlorine Combination Capacity, Breakpoint)

Almost all waters have the property, depending on their chemical nature, of combining with small amounts of chlorine. This was formerly termed chlorine capacity and is now spoken of as chlorine demand, chlorine combination capacity or break point. It is due to iron (II), manganese (II), sulphide and nitrite ions.

For the determination of *chlorine demand during a short period and with increasing reagent addition,* increasing amounts of 0.1% chlorine water (0.05 up to 3 ml) are added to several one litre samples of water. After mixing well, the mixture is left 10 min, 1 g of KI then added and, after a further 10 min, plenty of starch solution is added and the iodine from the unused chlorine back titrated in a large beaker, using 0.01N thiosulphate solution (1 ml = 0.355 mg Cl/l). The amount of chlorine titrated is subtracted from the amount added to the water sample. The titration value taken is of that litre sample showing a chlorine excess equivalent to less than 1 ml of 0.01N thiosulphate solution.

Chlorine demand (organic) (Chlorzehrung in German): The amount of chlorine which is consumed by organic matter over a longer period may be thus termed. Excess chlorine is added to the water and the unused amount is determined after standing. One-litre water samples are treated with the amount of chlorine corresponding to the chlorine demand, plus 3 extra ml of chlorine water. After 30 minutes at 22° C, if necessary in an incubator, unused chlorine is estimated as in the previous section. The amount of chlorine consumed during this time is the chlorine demand (organic).

6. Ozone

Ozone can be determined in a 200 ml water sample by adding 0.5 g KI and 0.5 ml H_2SO_4 and titrating with 0.01N thiosulphate solution. 1 ml of 0.01N thiosulphate solution is equivalent to 0.24 mg O_3/l. Free chlorine reacts similarly.

In *F. Zehender* and *W. Stumm*'s method for determining small amounts of ozone, 100 ml of water sample are treated with 10 ml of a 0.5% manganese sulphate solution which contains 0.5 ml sulphuric acid. One minute later, 5 ml of a 0.1% o-tolidine solution[1] is added and well mixed. After a further minute, the intensity of the yellow colour formed is measured colorimetrically or photometrically at 440 nm (e.g., with the Elko II instrument). As standards,

1 0.1 g tolidine is dissolved in 50 ml water and the solution poured into 50 ml of dilute hydrochloric acid.

ozone solutions are used, the concentrations of which have been measured iodometrically. (*F. Zehender* and *W. Stumm*, Mitt. Lebensmittelhygiene *44*, 206–213 [1953]).
All vessels must be scrupulously clean and the temperature watched accurately. Chlorine, Fe^{2+} and NO_2^- ions interfere.
Ozone determination according to *Hettche:* Small amounts of ozone can be determined with o-tolidine in the same way as is chlorine. 5 ml of tolidine reagent (Reagent No. 32) are added to 100 ml of water and colour comparison is undertaken in the Hellige comparator. The value read off must be multiplied by 0.67. This method has been termed the easiest and best by *F. Sulzer* (Schweiz. Z. für Hydrologie *20*, 16–29 [1958]). This applies, however, only to the normal ozone concentrations of up to 0.4 mg O_3/l. Dilution should be avoided. A colorimetric ozone determination is possible also with N, N-dimethyl-p-phenylenediammonium dichloride.

7. Hydrogen Peroxide

In recent times hydrogen peroxide has been finding use for sterilising water, e.g. in USA and USSR. For example, 100.000 typhoid bacteria and paratyphoid bacteria per litre are killed off with 3 mg H_2O_2/l. The viruses of poliomyelitis are killed within 30 minutes by 6 to 7 mg H_2O_2/l (Chem. Abstracts *66*, 98393r [1967] and Lit. Ber. 1968, p. 85).
1 ml of a titanium-containing solution is added to a 100 ml water sample and the resulting yellow colour compared with those from H_2O_2-standards treated in the same way at the same time. The titanium solution is prepared by melting together 1 g of titanium dioxide and 20 g of potassium pyrosulphate, dissolving the cooled mass in 20 ml of sulphuric acid (1:3) and making up to 100 ml. A reference solution is prepared by diluting an accurately standardised 3% solution of hydrogen peroxide.

8. Silver

(Katadyn process, p. 66)

1 litre of the water plus 1 ml of nitric acid are evaporated down to 50 to 100 ml. 2 ml of 20% Rochelle salt solution are added and ammonia added until the solution reacts alkaline to litmus. 1 ml of a 1% solution of gum arabic is then added and finally 0.5 ml of the reagent p-dimethylaminobenzylidenerhodamine (0.03 g in 100 ml of acetone, according to *Feigl*). Silver is shown by a reddish colour. The amount can be colorimetrically determined with the help of reference solutions (reaction time of 5 minutes). Standard solution: 0.1445 g silver sulphate and 5 ml of dilute sulphuric acid are diluted to 1 l; 1 ml = 0.1 mg Ag.
Water sterilised with silver should contain about 0.1 mg Ag/l.
According to *O. Hettche* (Ges. Ing. *74*, 164 [1953]), 2μg of Ag in 0.5 ml of water can be detected with dithizone without evaporating down (see p. 251).
The Deutsche Katadyn GmbH of Munich is introducing a patented apparatus for quantitative silver determination.

9. Determination of Aluminium

Rough field determination: 3 drops of acetic acid and 1 ml of morin solution (0.005 g in 100 ml of 96% ethanol) are added to a 20 ml water sample. The resulting fluorescence is compared with those from standard solutions treated in the same way (see below).

a) Colorimetric Aluminium Determination

If the water under investigation contains iron, 100 ml of it are first treated with a large excess of sodium hydroxide and the solution is boiled. The iron thereby separates and is filtered. The filtrate is neutralised with acetic acid and an extra 1 ml of acetic acid added.
To this solution are added 1 ml of 1% sodium thiosulphate and 1 ml of 0.1% sodium alizarinsulphonate solutions. After mixing, 25% ammonia is added until the colour changes. After 10 minutes, the solution is again acidified with acetic acid.
The colorimetric comparison is carried out in Hehner cylinders or colorimeter glasses or in a spectrophotometer, using reference aluminium solutions which are made up at the same time. The standard solution for this purpose contains 0.168 g ammonium alum ($Al(NH_4)(SO_4)_2 \cdot 12H_2O$) in 1 l of water. 1 ml = 0.01 mg Al.

b) Aluminium Determination according to Gad

100 ml of the water are titrated in an erlenmeyer flask with 0.1N HCl to the colour change of methyl orange indicator and the free carbon dioxide is then expelled by blowing through air. 1.0 ml of gum arabic solution (10% solution, extra white, Merck) is added and then 0.5 ml of alizarin solution (0.1 g alizarin in 100 ml of absolute alcohol) dropwise, swirling the contents round continuously. Finally, 0.5 ml of 10% sodium acetate solution is added and the mixture filled into a colorimeter cylinder. Comparison is performed as above. See above also for the standard solutions.

Lovibond procedure
Using the Lovibond Comparator of the firm of Tintometer, Salisbury, England, amounts of aluminium from 0.1 to 0.8 mg/l can be determined.
Another photometric aluminium determination, due to *G. Giebler,* uses aluminon solution at pH 5.3. The red coloration which develops is measured after 15 mins, using the Elko II instrument and filter S 53. Iron can be removed beforehand by converting to a complex with bathophenanthroline and extracting this with chloroform.

10. Determination of Polyphosphates

The phosphate polymers are precipitated at room temperature with 2% barium chloride solution from a 250 to 500 ml water sample which has been acidified with HCl. The precipitate is filtered, well washed and dissolved in dilute nitric acid. The insoluble barium sulphate is filtered off. 15 ml of HNO_3 are added to the filtrate and the solution is boiled for 15 minutes. This converts the polyphosphates into orthophosphate which is determined.

For identifying metaphosphates, the tests with m-phenylenediamine (5% aqueous solution of the hydrochloride) and with o-tolidine (0.5 g o-tolidine in 6 ml glacial acetic acid, made up to 50 ml) are recommended.
The determination may be simplified by boiling the 100 ml water sample for 1 hour with 1 ml hydrochloric acid under reflux and then estimating the increase in phosphate content as on p. 44.

11. Determination of Sulphite

The water, boiler feed water or effluent to be investigated is filled into an oxygen bottle (p. 102) without coming into contact with air, using the rubber sampling tube (p. 11) or the doubly pierced stopper (p. 13). The bottles is filled to overflowing. 3 ml of 0.1N or 0.01N iodine solution and of 25% phosphoric acid are introduced into the sample with an elongated pipette. The flask is closed without air bubbles' gaining entry, and shaken. After 10 min. standing, the excess iodine is back titraded with 0.1N or 0.01N sodium thiosulphate solution, making the usual addition of starch solution towards the end of titration.
Direct titration with iodine solution in the presence of starch to the blue colour can also be performed but must be done slowly.
1 ml 0.01N iodine solution is equivalent to 0.4 mg SO_3.

12. Determination of Hydrazine

The determination of hydrazine with the Bayer reagent is described on p. 232.

E. Determination of the Metal Ions, etc.

1. Determination of Lead

Occurrence of lead: Seldom in natural waters, e.g., 0.1 to 0.5 mg/l in lead mine regions. Up to and above 5 mg/l in aggressive waters in lead pipes, but usually 1 to 2 mg Pb/l.

a) Sampling and Preparation of the Sample for Lead Investigation

Correct sampling is a condition for lead investigation. One must first make sure that the water has really stood for some hours (9 to 12, p. 154) in the lead pipe. Statements of the house resident are notoriously unreliable. It is thus best to seal the tap beforehand.
A 500 ml sample is taken in a sudden burst, after having allowed 100 to 300 ml to flow away slowly first. Since lead can be deposited on the walls of the sampling vessel while standing

and therefore evade study, 2 or 3 ml of acetic acid are added to the sample as soon as it is taken.
After allowing the water to run for a longer time, the samples for other determinations (carbon dioxide, hardness, etc.) are taken. Samples which are sent in for analysis are almost entirely poured into a beaker and 5 ml of dilute acetic acid are added to the sample bottle. This is shaken vigorously and the poured-out amount is then returned from the beaker to the acetic acid in the sampling bottle. Any separated flakes of iron must be brought into solution with acetic acid (if necessary by warming) since these have usually taken up part of the lead (*K. Höll,* Deut. Apotheker-Z. [1935]).

b) Colorimetric Determination of Lead

100 ml of the prepared water sample (see above) and of the reference solutions are placed in viewing tubes or Hehner cylinders, 10 drops of dilute acetic acid are added to each and the mixtures shaken. 10 ml of 20% ammonium chloride solution, containing 2% acetic acid, are then added, the mixtures shaken and allowed to stand briefly. 2 drops of sodium sulphide solution* are finally added, followed by cautious mixing to prevent formation of turbidity through introduction of air[1].
The resulting yellow, yellow-brown or brown colour is compared with those of the reference solutions[2], made up at the same time; this must be done within 1 to 2 minutes. Subsequent addition of standard solution to the sulphide mixture is to be avoided since turbidities then occur which falsely imply higher lead content. The depths of layer should differ by only little.

Sensitivity: 0.1 to 0.2 mg Pb/l can be detected in this way. Waters containing more than 3 mg Pb/l must be diluted. Amounts of below 0.1 mg Pb/l can be concentrated by shaking a larger amount of water with some calcium carbonate, dissolving this filtered product in dilute acetic acid and proceeding as above. See p. 250 for the detection of traces of lead.

Interferences: 3 ml of dilute sulphuric acid are added instead of the 10 drops of acetic acid when copper is present at the same time. A yellow coloration can then come only from the copper (see below).
One must work rapidly *if much nitrite is present,* or else the sulphur which separates as a turbidity falsifies the results; alternatively, a mixture of 2 ml of Rochelle salt solution, 2 ml of NaOH and 3 drops of 10% potassium cyanide solutions are added instead of the acetic acid, and then the sodium sulphide.
Iron(II)-compounds do not interfere. Iron(III)-compounds yield turbidity which can be prevented, however, by addition of 0.5 g sodium pyrophosphate or 1 ml of Rochelle salt solution before the sulphide is added.
The detection of lead is upset by the *colour of peaty waters.* Such waters are rendered weakly alkaline and shaken with pure crystallised calcium carbonate. The carbonate adsorbs practically all the lead. It is filtered and treated on the filter with hot dilute acetic acid. The carbonate and the basic lead salts go into solution. The filter is washed through and the filtrate diluted and lead determined according to the above method (cf *K. Höll,* Deut. Apotheker-Z. [1935]).

1 *Best mixed with a large "mixing bulb", p. 25.*
2 *Reagent No. 6.*

2. Determination of Copper

Occurrence: Copper does not occur in significant amounts in natural waters. It plays a part as a trace element in mineral waters. Tapwater with aggressive properties may attack copper parts of the pipe system and dissolve amounts of several mg Cu/l.

Rough determination: 20 ml of water in a test tube are treated with 3 drops of sodium sulphide solution*. A yellow-brown colour in transmitted light is yielded if copper is present in large amounts; the liquid appears yellow if only little copper is in the water (on the assumption that the water itself is not coloured) (see below). Any colour from copper disappears when 3 drops of 10% potassium cyanide are added.

a) *Colorimetric Determination of Copper according to Winkler*

5 to 10 drops of 10% Rochelle salt solution are added to a 100 ml sample of water, the mixture is shaken and then 3 drops of ammonia are added. After 2 minutes, 2 drops of 5% potassium ferrocyanide solution are added, yielding a reddish coloration. The intensity is compared with those from reference solutions made up at the same time. These standard solutions contain 0.393 g crystallised copper sulphate in 1 l water (1 ml = 0.1 mg Cu).

b) *Copper and Lead together*

In order to detect lead in the presence of copper, the above liquid is treated with 2 drops of 10% potassium cyanide solution. The colour changes to greenish yellow. The intensity of colour in both cylinders must be the same. To both solutions are added 10 ml of a 20% ammonium chloride solution containing 5% ammonia, and 3 drops of sodium sulphide solution. The copper reference solution having no lead is fully decolorised; if lead is present as well as copper, a yellow colour appears which can be compared with that from standards (see under "Lead").
Interference by high iron contents can be circumvented by adding 2 ml of 50% Rochelle salt solution before the above reagents (*J. Müller,* Ges. Ing. *70,* 407 to 410 [1949]).
The determination with sodium diethyldithiocarbamate is specific for copper. 10 ml of 0.1% aqueous reagent solution are added to 100 ml of the water plus ammonia. Traces of copper yield a golden yellow colour which can be colorimetrically or photometrically determined; or the coloured complex can be extracted with chloroform and the extract evaluated at 432 nm in a spectrophotometer such as Elko II.
Rochelle salt or sodium hexametaphosphate are added to hard waters and mineral waters, or the water is brought to pH 9 with ammonium citrate.

3. Determination of Zinc

Occurrence: Amounts of up to 5 mg Zn/l can be found in aggressive waters which have stood in zinc pipes; usually the content is 1 to 2 mg/l. Several mg/l of zinc are found in the waters from zinc mines. Variable amounts occur in industrial waters (e.g. pickling liquors and from munition works) and their receiving waters.

Preliminary tests: One first tests with hydrogen sulphide whether other metals are present as well as zinc. The H_2S is passed through successive water samples which have been acidified with mineral acid, acidified with acetic acid, and rendered alkaline with ammonia. If a white turbidity is observed only in the solution containing acetic acid, zinc alone is present.

Zinc in presence of Cu and Sn: If a dark brown turbidity is noted in the water acidified with mineral acid, copper (and perhaps tin) is present. A larger sample of the water is then acidified with nitric acid and some copper sulphate is added (see *Höll,* Deut. Apotheker-Z. [1935]). H_2S is passed through for a longer period and the solution is left overnight in a closed vessel. After filtering, the filtrate is boiled to expel H_2S. 3 ml more of dilute nitric acid are added and boiling is continued. Any iron, manganese or aluminium present are precipitated from the boiling solution with ammonium hydroxide in excess, and filtered off. The filtrate is tested for zinc (see below).

Zinc in absence of Cu, Fe and Sn: 100 ml of water or the above filtrate which contains zinc as the only metal, are cooled and treated with 25 ml of 25% sodium acetate solution, 3 ml of acetic acid, 2 ml of 10% gum arabic solution and 2 ml of potassium ferrocyanide solution (5%) or 0.1 ml of sodium sulphide solution*, swirling after each addition.

In both cases, the whitish turbidity yielded after longer standing is compared with those given by standard solutions. These last-named are prepared by dissolving 0.100 g analytical grade zinc metal in dilute HCl and diluting to 1 litre. 1 ml of this solution contains 0.1 mg Zn.

Interferences: Organic matter can interfere. Yellow or brown waters containing humins are boiled with bromine water and excess bromine expelled by further boiling. The solution is filtered and made up to the original volume.

Zinc in presence of iron: W. Hermanowicz and *E. Sikorowka* (Chem. Zentralblatt *124,* 4585 [1953]) proposed a colorimetric method for determining zinc without interference from iron. 1 ml of 5% potassium fluoride solution is added to a 25 ml sample of the water and the solution filtered. The filtrate and reference solutions are treated with 1 ml of 25% HCl, 1 ml of 0.03% crystal violet solution and 1 ml of 20% potassium thiocyanate solution and colorimetric evaluation carried out without delay.

Larger amounts of zinc can be precipitated and gravimetrically determined as quinaldinate salt with 3% quinaldinic acid from the heated water sample at pH 5 (sodium acetate/acetic acid buffer).

According to *A. Bilikova,* Zincon is most suitable for microdetermination of zinc; it yields a stable blue complex at pH 9.

4. Determination of Potassium[3]

From 1 to 10 litres of the water are treated with a little acetic acid and concentrated to ca 50 ml. This solution is filtered and warmed to 60 to 70° C. One crystal of aluminium chloride is

3 Worked out by the author, based on the work of *P. Raff* and *W. Brotz* (Z. Anal. Chem. *133,* 241–248 [1951]).

added to the weakly acid (acetic acid) solution, followed by 0.1M (3.4%) sodium tetraphenylborate solution[4] in small portions until the clear solution above the settled precipitate no longer yields a turbidity or precipitate. 1 ml of reagent corresponds to about 3.5 mg K. The mixture is allowed to stand a little longer at the reaction temperature and is then filtered through a filter which has been dried as 110° C and weighed. The precipitate on the filter is well washed with water containing acetic acid; filter + precipitate are then dried in an oven at 110° C and weighed after cooling in a desiccator.

The filter can be conveniently replaced by a porcelain filtering crucible A 2; it can be cleaned with acetone after use.

The limit of detection is about 50 μg K/l.

To avoid transporting large amounts of water and the time-consuming evaporating down, the water can be passed on the spot through suitable filtering pipes containing the ion exchanger I of Merck, Darmstadt, W. Germany. The liberated potassium in the filtrate after elution with dilute HCl is determined quantitatively.[5]

Calculation: The gravimetrically measured amount of potassium tetraphenylborate is multiplied by 0.1091 to yield potassium.

Limit of error: ± 0.3% when only little potassium is present in the concentrate.

Interferences: Alkaline earths interfere only with waters of considerable hardness. It is then advisable to precipitate calcium and magnesium along with the sulphates by using barium hydroxide and to remove excess barium hydroxide by passing CO_2 into the hot, concentrated filtrate from the precipitate. After acidifying the filtrate with acetic acid, the potassium can be precipitated.

The influence of ferric salts can be eliminated by complex formation with NaF.

Interference by ammonium salts can be prevented by adding some formalin to the solution after rendering strongly alkaline (Chem. Ztg. No. 1 [1954]).

a) Titrimetric Determination of Potassium

According to *P. Vogler,* the precipitate of potassium tetraphenylborate can be dissolved in 50 ml of acetone and, after adding 10 ml of potassium bromide solution and a drop of mixed indicator, is titrated with 0.01N silver nitrate solution to the colour change from green to red-violet. (1 ml 0.01N $AgNO_3$ = 0.3909 mg K)

The mixed indicator is prepared by dissolving 300 mg of eosin and 5 mg of dimethylaminoazobenzene in 150 ml of acetone and mixing with 50 ml of an aqueous solution of 350 mg methyl green (light green) (Fortschritte der Wasserchemie, No. 1 [1964]).

H. Flaschka's iodometric determination of potassium would thus appear to be outdated.

Flame-photometric determination of potassium

The flame-photometric determination of potassium is the most convenient and accurate,

4 The substance can be obtained from the firm of Heyl & Co. Berlin, under the name of "Kalignost". If it does not give a clear solution, 0.5 g of purest aluminium hydroxide is added.

5 *E. Abrahamczik,* Mikrochemie *25* [1938] (Experiments with neopermutite) and *Folke Nydahl,* Verh. der IVL, Vol. XI, Stuttgart [1951] (Experiments with cationic ion exchangers).

especially when a caesium-aluminium radiation buffer is used (Merck). Measurement is at the potassium line of 768.2 nm.
100 ml of potassium stock solution (190.7 mg analytical grade KCl in 1 litre distilled water solution) and 100 ml of radiation buffer (50 g of analytical grade caesium chloride and 250 g aluminium nitrate in 1 litre distilled water solution) are made up to 1 litre with distilled water for the potassium reference solution.
Flame photometers are offered in W. Germany by Netheler u. Hinz of Hamburg, Zeiss-Opton of Oberkochen and Kirem of Frankfurt.

5. Determination of Sodium

Sodium occurs or ought to occur in normal drinking water in amounts of only a few mg/l. The American Health Association has fixed 20 mg Na/l as limit (Chem. Abstracts *66*, No. 10 [1967]). Sodium plays a decisive part in the water retention of the body.
If in rare cases a direct determination of sodium is necessary, this is done most quickly and accurately by flame photometry, as for potassium. 0.0003 mg of Na per ml water can be detected in this way (Standard-Werk der "Flammenphotometrie", R. Herrmann, Berlin and Göttingen [1956]). The firms of Dr. Lange, Berlin-Zehlendorf, Zeiss-Opton, Oberkochen and Kirem, Frankfurt/Main supply flame photometers in W. Germany.
The stock sodium solution contains 0.2542 g NaCl + 0.1907 g KCl + 0.6109 g LiCl, made up to 1 litre with distilled water. 1 ml = 0.1 mg of Na, K and Li. The sodium line 589.3 nm is used for evaluation. Blanks and replicate measurements are needed.
Sodium is usually determined from the difference between the total amount of alkali chlorides and the amount of potassium. To this end, 1 to 2 l of the water is concentrated and treated with 5% barium hydroxide solution until sulphates have been completely precipitated. The mixture is heated and filtered and CO_2 passed through the boiling filtrate. After filtering again, the filtrate is treated with a little HCl and evaporated to dryness. The residue is dried in an oven at 110° C. The residue, consisting of alkali chlorides, is weighed and the potassium amount subtracted.
The gravimetric uranyl acetate method can be employed if larger amounts of sodium are expected.

6. Determination of Chromium

5 ml of 1N silver nitrate solution, 10 ml of 10% ammonium persulphate solution and 6 ml of 1N HCl are added to a 100 ml sample of the water. This is filtered and 2 ml of 1:1 sulphuric acid and 0.4 ml of a solution of diphenylcarbazide in acetone are added to the filtrate. The red-violet colour resulting is evaluated colorimetrically or photometrically at 540 nm within 5 to 10 min.
If small chromium amounts are present, the water sample is treated with about 10 mg of Fe (III) and the iron is precipitated with ammonia; the chromium is adsorbed in its entirety on the ferric hydroxide.

To determine total chromium, the water sample is treated with alkali and bromine water. After boiling off unused bromine and cooling, the above procedure is used.

7. Determination of Arsenic

Occurrence: Arsenic occurs very sporadically in ground waters, e.g. through washing-out of ash- and rubble slopes and earth from cemeteries (up to 0.2 mg/l). It is found in receiving waters supplied by effluents. In rare cases, it occurs from geological origins in spring waters. One encounters arsenic more often in medicinal springs (e.g. Dürkheim Max-Spring, W. Germany).

Approximate field determination: 5,0 ml of conc. H_2SO_4 are added to 100 ml of the water in a round-bottomed 150 ml flask. Interfering gases escape from the warm solution. After cooling, a few copper-coated pieces of zinc are added; hydrogen is evolved and an Allihn tube is fitted on to the flask mouth. A cap of filter paper is held across the opening of the Allihn tube with a rubber band and a drop of 66% silver nitrate solution is placed on to the filter paper. Observation is maintained for a longer time. A yellow coloration of the reagent spot indicates arsenic.

Interferences: Larger amounts of organic matter retard the reaction. Wastewaters and waters containing humic material must therefore be evaporated down with nitric and sulphuric acids and the residue submitted to wet combustion.
Evaporation of the last amounts of hard waters with sulphuric acid is beset with difficulties; the following procedure is thus recommended:
More accurate determination of arsenic: To 100 to 1000 ml of clean, colourless water is added 0.5 g of ferric alum and the solution is warmed; ammonia is then added. The ferric hydroxide which separates carries practically all the arsenic with it. It is filtered off, washed and dissolved in 20 ml of arsenic-free dilute sulphuric acid. The solution is filled into a round-bottomed flask of 50 ml capacity and some pieces of copper-coated zinc[6] are added. The flask is closed with a stopper carrying an Allihn tube and a tube of 5 mm diameter. The wider part of the Allihn tube contains a small wad of cotton wool which has been saturated with 5% lead acetate and dried. In the glass tube is a strip of mercuric bromide paper. This reagent paper is prepared by saturating the especially firm reagent paper No. 575 (or, according to *Lockemann,* No. 589 Blauband or No. 606 of the firm of Schleicher & Schüll) with 5% alcoholic mercuric bromide and allowing to dry. The contents of the flask are swirled round from time to time.
Arsenic is shown by a yellow to yellow-brown colour of the mercuric bromide strip within 12 hours. Amounts as low as 0.005 mg arsenic (As) can be determined by comparison with the colorations from known quantities of arsenic. The reaction time can be appreciably shortened through warming. According to *J. Bodnar, E. Szep* and *W. Cielsky*, the determination with tin and hydrochloric acid gives better results (Z. Anal. Chem. *115,* 412 [1929]).

6 The zinc pieces must be well washed after coating with copper since traces of copper impair the detection of arsenic.

The mercuric bromide strip can be replaced by an absorption tube containing a solution of silver diethyldithiocarbamate (5% in pyridine) which is attached to the flask in which hydrogen is developed. The red coloration is evaluated photometrically at 546 nm.

Reference arsenic solution: 0.132 g of arsenic acid plus 10 ml of 1N sulphuric acid are diluted to 1 litre with distilled water. 1 ml of this solution = 0.1 mg As. If it is diluted a hundredfold, 1 ml = 1 μ g As.
1 mg As = 1.85 mg AsO_4; 0.7 mg As = 1.3 mg $HAsO_4$.

8. Determination of Selenium according to Quentin

K. E. Quentin and *L. Feiler* (Vom Wasser *34* [1967]) were the first to report the importance and determination of selenium. According to their experimental work, the best method of determining selenium is spectrophotometrically with 2,3-diaminonaphthalene, which yields the yellow piazaselenol. Amounts from 0.001 to 0.25 mg Se/l can be determined.
The limit of 50 μ g Se/l is recommended by the International Standards for Drinking Water and by the European Standards.

9. Determination of Tin

Occurrence: Only in traces in tap water. To determine it, 2 to 3 litres of water are evaporated down with HCl several times and finally with nitric acid. Silica is removed by heating with 1 ml of concentrated H_2SO_4 and 5 ml of HF and fusing the residue with soda and sulphur. The extract of the melt is added to very dilute sodium sulphide solution. Tin is shown by a yellow colour (compare with standard solutions).
If larger amounts of tin are present, tin sulphide is precipitated in the reaction with sodium sulphide; it can be gravimetrically determined in this case.

F. Determination of Fluoride Ion (according to Sanchis)

Occurrence: Normal ground waters contain 0.05 to 0.5 mg F/l[1]. Much more is found in some medicinal springs, especially in alkaline sulphur springs (*K. E. Quentin,* Ges. Ing. *78,* 1–4 [1957]).
K. Höll found huge amounts of fluoride ion as a characteristic component of geyser waters and hot springs (Heilbad and Kurort *16* [1964] and *20* [1969]).

1 According to investigations of the Institute for Water, Soil and Air Hygiene, Berlin and of *A. Kaess,* Vom Wasser *23,* 70–94 [1956].

Exactly 5 ml of the zirconium-alizarin reagent (No. 34) are added to 100 ml of sample and of standard solution. After mixing, the solutions are left 1 hour at room temperature. The colours are compared in Hehner cylinders, Nessler tubes, colorimeter or photometer. Both the water under investigation and the reference solution must have exactly the same temperature from the beginning. Checking with a thermometer is essential in this case. The higher the fluoride concentration is, the paler is the solution; the colours go from reddish through orange to yellow. Amounts from 0.01 to 0.15 mg F per 100 ml can be determined. If fluoride contents exceed 0.15 mg, the solutions are suitably diluted. A solution of 0.221 g NaF in 1 litre is used as standard; if diluted 10 times, the final solution contains 0.01 mg F per ml.

Interferences: Interference is caused by amounts of aluminium of over 0.5 mg/l; of iron of over 2 mg/l; and of manganese of over 0.2 mg/l. Chlorides and sulphates interfere when their amounts exceed 500 and 200 mg/l, respectively. High phosphate concentrations cause too high values for fluoride to be obtained.
In these cases, equivalent amounts of the interfering compounds must be added to the reference solutions also, or fluoride must be removed by distillation after adding 0.2 g of silica powder and 25 ml of perchloric acid (chloride is bound with silver perchlorate). According to experience of ORCA[2], this steam distillation is to be recommended for every water. Interference by manganese is eliminated by adding 1 drop H_2O_2 to every 100 ml of water and removing excess of the H_2O_2 with 1 drop of 3% KI solution and titrating the iodine formed with 0.1N $Na_2S_2O_3$. Free chlorine also impairs the zirconium lake formation somewhat; this can be prevented by addition of sodium thiosulphate (*Jacobson,* J. Amer. Waterworks Ass. *44,* 1169–1173 [1952]).
K. E. Quentin, J. Indinger and *S. W. Souci* (Z. Lebensmittel-Unters. u. Forsch. *109,* 213–218 [1959]) worked out highly sensitive methods for detecting fluoride.

1. New Working Procedure

K. E. Quentin and *J. Indinger* (Z. Lebensmittel-Unters. u. Forsch. *110,* 249–260 [1959]) studied in detail the above methods for determining fluorine; their work merits attention. Fluoride is best separated from interfering materials and enriched by distillation at 135° C from a perchloric acid solution containing 53.6% acid by weight (2 vol. 70% $HClO_4$ + 1 vol, distilled water). Halogens must be completely held back from distillation by adding silver perchlorate.
Quentin and *Indinger* describe precisely the colorimetric determination of fluoride in the distillates using a modified zirconium-alizarin and zirconium-eriochrome-cyanine method. The original work (1959, see above) must be consulted for the exact procedural details, since it is comprehensive. This is unavoidable in analyses of medicinal waters.

Serial investigations of fluoride ion
The new method with ion-specific electrodes of the Orion Research Incorporated of Cam-

2 *ORCA = European Working Commission for Fluorine Research.*

bridge, USA, represents a great saving of time, especially in serial investigations and control of fluoride additions to drinking water (see also p. 22).

2. Significance of Fluoride Ion for Health

According to *Rieder* (J. Amer. Waterworks Ass. [1935]), the lower limit of danger to health is 0.7 mg F/l. Regular consumption of water containing more than 2 mg F/l gives rise to symptoms of drinking water fluorosis; the teeth show "mottled enamel". An excessive fluorine content can be reduced through magnesite filters *(O. Hettche)*. Waters rich in fluoride are treated in the USA with Defluorite, an activated aluminium or tricalcium phosphate (bone charcoal) (Ind. Eng. Chem. *29,* 424 [1937], see also *K. Haller,* Städtehyg. *3,* 73/74 [1952]). Waters occur in the USA which have 10 mg F/l and more (*F. Sierp,* GWF [1956]). Waters containing 4.3 mg F/l are claimed to be drunk in Isparta (Turkey) regularly without harm (*S. Velicangil* and *S. Eser,* Z. f. prophylak. Med. No. 2 [1957]). In geyser waters and other hot springs in the Rocky Mountains and in Iceland, *K. Höll* found very high fluorine amounts (up to 26 mg F/l) (Heilbad und Kurort *16* [1964]). A limit of 1.5 mg F/l has been set by the European bureau of the WHO on account of the damage due to fluorosis (see *H. Kruse* [54]).
The firm of Lurgi, Frankfurt am Main, has developed a process for total removal of fluoride using a specific ion exchanger, for Kaolack in Senegal (*W. D. Schmidt,* German contribution to "Wasser für den Frieden" [Water for peaceful purposes], Bonn 1968).
Active carbon serves also for removing fluoride ion (see, e.g. *H. S. Horowitz* and *F. J. Maier,* Publ. Health Rep. *82,* 965–973 [1967]).
If drinking water is to be treated with fluoride, aluminium sulphate must not be used at the same time in conditioning the water; the complexes formed would be ineffective.
If the water has less than 0.5 mg F/l, damage due to fluorine shortage arises (increase of dental decay). The fluoride level of water in the USA is augmented nowadays, e.g. with "Flural", to ca 1 mg F/l (*W. E. White,* J. Amer. Waterworks Ass. *44* [1952]) or with fluorosilicic acid (Water and Sewage Works *101* [1954]). Already one fifth of the population of the USA obtains artificially fluoridated water from waterworks plants. In recent times, sodium silicofluoride has been added, and sodium fluoride in the smaller plants. Exhaustive information about the fluorine problem is to be found in *H. J. Schmidt's* "Karies-Prophylaxe", published in 1951 by the Hüthig-Verlag in Heidelberg. Short reviews are given by *H. Müller* in Deut. Leb. Rundschau *47,* 254–260 [1951] and *H. Kruse,* GWF *94* [1953]. *H. Hugelmann* (GWF *94* [1953] has reported on the fluoridation of drinking water in Kassel, W. Germany. According to *S. Gericke,* 1 mg F is to be regarded as the daily human demand (Z. f. prophylakt. Med. No. 1 [1957]). *C. Parma* (Dt. Zahnärztebl. No. 2 [1956]) states that an optimum fluoride amount in food has a favourable effect not only on the teeth but also on the whole skeleton (see also *H. Hornung, F. Boettcher* and *Harmsen,* Städtehygiene *11,* 243 [1970]).
The DVGW rejects fluoridation of drinking water (*N. Deliva,* Zeitschrift No. 8 [1955] and again in the conference in 1966). It is also declined in Switzerland (*E. Bosshard,*

Mon. Bull. Vers. Gas- und Wasserfachm. *35*, 236–237 [1955])[1]. As a result of the clause in the German Foodstuff Law, fluoridation of drinking water will not be generally allowed in Germany (strict demand for natural purity of foodstuffs, see also p. 148 et seq.).
The Federal German Ministry of Health has nevertheless instituted experiments for a longer time on the effects of fluoride, at the instigation of the "Ausschuss für Zahngesundheit" (Commission for Dental Health); one must await the results.
The WHO recommends strongly fluoridation of drinking water, based on twenty years of experience without disadvantageous effect (Städtehygiene 1970, p. 165–167).
The fluorine balance of the body during the first ten years of life is of importance for the whole life.
The DVGW has recently again uttered a warning in an "Erklärung zur Trinkwasser-Fluoridierung" (Declaration on fluoridation of drinking water) (Korrespondenz Abwasser No. 7 [1966]). Drinking water should not be a vehicle for forced medical treatment! It is highly problematic to consume daily the necessary amount of fluoride (1.0 mg) in a litre of water and then to avoid food, rich in fluorine, such as certain fish, meat and vegetable sorts so as not to exceed the maximum amount of 1.5 mg F per day. *K. E. Quentin* has treated definitively the question of fluoridation of drinking water in the Deut. Zahnärztlich. Z. *23*, 141–146 [1968].

G. Determination of Iodide Ion (according to the new EV)

Occurrence: Stooff states that the iodine content of ground water lies between 0.01 and 20 μg/l.
Potassium carbonate solution is added to 2 to 10 litres of water until it has an alkaline reaction to phenolphthalein. The solution is concentrated to about 100 ml by evaporation. It is then filtered and further potassium carbonate added (1 to 2 drops per litre of water used). The solution is evaporated to dryness and heated in a crucible but without igniting. Cautious combustion is necessary if organic matter is present. The dry residue is powdered and extracted several times with warm 96% alcohol.
The alcoholic extract is taken to dryness and the residue dissolved in 30 ml of water. 1 ml of phosphate mixture[2] and then 5 ml of 0.1N bromine water are added. This is mixed and 2 ml of 2N formic acid are immediately added. The solution is shaken for 2 minutes, during which it becomes decolorised. 1 ml of 1N KI solution is introduced and, after 1 minute, the mixture is titrated with 0.005N sodium thiosulphate solution, using starch solution as indicator. A microburette is advisable. 1 ml of 0.005N thiosulphate solution corresponds to 0.1058 mg iodine.

1 Basel, however, introduced fluoridation of drinking water for 250.000 inhabitants in 1962 (sodium silicofluoride).
2 The mixture contains 20 g monosodium phosphate, disodium phosphate and sodium pyrophosphate, dissolved in water and made up to 250 ml.

Interferences: Nitrites interfere and must be removed. Nitrates interfere since they may yield nitrites during the ignition of the organic matter.
See p. 248 for more accurate methods.

H. Determination of Silica

Occurrence: There are a few mg/l in almost every water, usually 5 to 8 mg/l. Occasionally up to 50 mg/l may be found, especially in soft waters and in silicate-mineral waters. Lakes have only 1 to 2 mg/l and often only traces in the spring.

Significance of the silica: Silica plays a part in formation of the protective layer in pipes and is nearly always detectable in the scale (fur) of the pipe.
Silica must not be present in boiler feed water for high pressure boilers (p. 231).

1. Colorimetric Determination according to Winkler

5 ml of freshly prepared 10% ammonium molybdate solution and then, immediately, 5 ml of 10% HCl are added to 100 ml of the water, swirling the contents after each addition. After 3 minutes, 1 ml of 10% oxalic acid solution is added in order to eliminate interference by any phosphate.
The yellow colour produced is compared with that of reference solutions in Hehner cylinders or a colorimeter after 10 to 30 min reaction time, or evaluated photometrically at 720 nm. A 0.5% borax solution[1] serves for comparison, to which the potassium chromate standard (0.53 g of K_2CrO_4 in water, made up to 1 l) is added. 1 ml of this silica standard, diluted to 100 ml with borax solution, corresponds to 1 mg SiO_2/l.
Waters containing over 10 mg SiO_2/l must be appropriately diluted before the determination; those having less than 1 mg/l must be concentrated by evaporation or the metol method must be employed (see below).
The colour comparison can be made also in the Hellige comparator (Colour disc No. 3060/90).

Interferences: If the determination cannot be carried out immediately on samples contained in glass vessels, the results are inaccurate; this is due to separation of silica on to the surface, especially noticeable with ordinary glass. Special samples are thus taken in plastic bottles. Plastics do, however, take up silica from the water. Too little silica is thus found in samples which have stood for a longer time in plastic bottles, this applying particularly to mineral waters rich in silica.
Hydrogen sulphide interferes in this method and must be removed beforehand by boiling or shaking; interference by phosphate is avoided by oxalic acid (see above).

1 This solution must be made up with boiled, CO_2-free distilled water for accurate determinations.

High contents of salts interfere. Brackish waters and brines must consequently be diluted before the determination. Interference by iron (by higher iron contents of 3 to 5 mg/l) can be eliminated by adding HCl and extracting the iron complex with acetylacetone, using chloroform (acetylacetone-chloroform, 1 + 1). Sometimes it is enough to oxidise Fe^{2+} to Fe^{3+} with 0.01N $KMnO_4$.

One must always check that the water used for dilution is free of silica.

Colloidal silica is not determined by the above method. To accomplish this, 100 ml of water are warmed with 0.4 g of sodium carbonate for an hour on the water-bath. After cooling, the solution is neutralised with dilute HCl and one procedes as above. Merck's analytical grade sodium bicarbonate can be used for this; it yields a blank of 0.4 mg SiO_2/l in the determination, which must be subtracted from the result.

Determination of very small amounts of silica

If the silica content is very low, the above standard procedure is used. After the addition of oxalic acid, 4 ml of metol solution are added (2 g of metol + 2 g of potassium hydrogen sulphite are dissolved in 100 ml of water). The blue colour yielded is evaluated spectrophotometrically at 720 nm.

I. Detection of Ground Water Contamination through Mineral Oil

Proportions of 1:100.000 to 1:1.000.000 of mineral oil in drinking water can be detected through their influence on taste, especially with benzine (*Müller,* GWF *93* [1952]). Kerosine and other petroleum products impart an unpleasant taste to drinking water already in 1 part per 5 million (*W. Zimmermann,* Städtehygiene [1955]).

Contamination through mineral oil thus can be easily detected by taste. According to *J. Holluta,* the threshold odour number for benzine is 0.05 mg/m³.

1. Chemical Investigation

Analytical detection of such small amounts of mineral oils is difficult, however. The first valuable work on this problem is due to *B. Nietsch* (Z. Angew. Chem. *66,* 571–572 [1954], Vom Wasser *21* [1954] and Mikrochimica Acta 171–178 [1956]).

The contaminated water can be extracted with petrol ether and the solvent evaporated off in a dish. The residue is examined for mineral oil in the quartz lamp; its presence is shown by a yellowish-white to bluish-white fluorescence.

Mineral oil can be concentrated by shaking 1 l of water sample with about 0.5 g of magnesium oxide which adsorbs it. The filtered oxide can then be examined in the quartz lamp. The wet filter containing the oxide is laid under the lamp and a positive test is the appearance of spots of bluish-white fluorescence-visible still at a dilution of 1 to 10^9.

Fluorescence is observed without concentration when the contamination is greater. The water is introduced in a dish into the rays of the UV-lamp. High pressure lamps of 365 to 366 nm are the best for this observation of fluorescence (Firm of Heraeus, Hanau, W. Germany). Amounts of 0.02 mg/l of mineral oil can still be detected on the thin-layer chromatogram after extraction with CCl_4 and isothermal distillation (*G. Giebler, P. Koppe* and *H. T. Kempf*, GWF *105*, 1039–1042 [1964]). The Rf-values permit distinction of fatty materials from mineral oil, if necessary with the help of parallel runs with water to which the same mineral oil or oil is added.

P. Ladendorf has reported other methods and their sources of error in Vom Wasser *29* [1962]:

J. G. Sherratt's method consists of warming the water sample in a two-necked flask and collecting the volatile mineral oils in tubes with active charcoal, from which they can be eluted with acetone. The extinction at 490 nm with acidified Teepol solution (Shell Co.) is measured, using, e.g., the Elko II (Analyst *87*, 595 [1962]).

According to *C. Rübelt, R. Schweissfurt* and *W. Zimmermann*, the water can be extracted with CCl_4 and very small amounts of mineral oil in the extract determined through infrared absorption in quartz cuvettes at 3.2 to 3.6 μ (GWF *108*, 893–900 [1967]).

Similar experiences have been reported by *H. Hellmann* (Deut. Gewässerkundl. Mitt. *13*, 19–24 [1969]) and *L. R. Baynon* etc. from the Dutch Stichting Concave, The Hague [1968].

Various values have been quoted for the solubilities of mineral oils in water. *H. Eilers'* values (Wasser, Boden, Luft *50*, 58 [1960]) are.

n-heptane	45	mg/l
n-hexane	86	mg/l
o-xylene	136	mg/l
n-pentane	350	mg/l
toluene	470	mg/l
benzene	800	mg/l
benzine	31	mg/l

together with mean solubilities of:

petroleum	10	mg/l
diesel oil	22.5	mg/l
automobile gasoline (car petrol)	149	mg/l

P. Ladendorf (loc. cit.) states that the light mineral oils and motor fuels are not quantitatively determined by the extraction methods. They are obtained, however, when the isothermal distillation of easily volatile extraction agents, such as petrol ether, is used.

Mineral oil residues adhere easily to the walls of the sample bottles and have to be extracted also from them to be determined (private communication of *Dr. K. Knie*).

P. Ladendorf has developed a new pycnometric method which depends on differential measurement on pure tetrabromoethane and the tetrabromoethane extract of the water sample at pH 5 (limit of determination: 0.3 mg/l) (Vom Wasser *29*, 119–141 [1962]).

2. Removal of Mineral Oil from the Water

The mineral oils not only impart a most unpleasant taste to the water but also introduce carcinogenic materials into it (p. 78); their removal from water is thus necessary. This is best done with the help of active carbon or active carbon filters (*K. Fricke*, Ges. Ing. *74*, 394–396 [1955] and *Daur*, GWF *96*, 153–155 [1955]). Water contaminated with mineral oil can be purified on the small scale also for domestic consumption, e.g. with the handy filter tubes containing active charcoal fillings of the firm Berkefeld of Celle, W. Germany.

3. Seepage of Mineral Oil

Several comprehensive expert opinions about the seepage of mineral oil in various types of soil have been compiled, in part as a result of the instigation of the Federal German Ministry of Health. Examples are: *E. Holstein* on "Möglichkeiten des Schutzes des Grundwassers bei der Lagerung von Heizöl" (Ways of Protecting Ground Water in Storage of Heating Oil) (1963): Professors *E. Becksmann, H. Billib, W. von Engelhardt* and *W. Zimmermann* on "Verhalten von Erdölprodukten im Boden" (Behaviour of Petroleum Products in the Ground). Professors *Michels, Udluft* and *Zimmermann* and Herr Dipl.-Ing. *Nabert* have reported on drinking water from contaminated ground water; the LAWA (Länderarbeitsgemeinschaft) (Working Group of the German Länder) appointed a specialist commission to report on "Lagern von wassergefährdenden Flüssigkeiten" (Storage of Liquids potentially dangerous to Water) (this expert report is set down in DIN 6608-Directives for Storage Containers of Steel): *W. Zimmermann* carried out extensive experiments on mineral oil seepage in the laboratory and outside (GWF *105*, 1033–1038 and 1089–1092 [1964]). *H. Billib* carried out detailed work on the behaviour of El-heating (fuel) oil and of benzine in loess soils (Schriftenr. d. Inst. f. Wasserwirtschaft der TH Hannover, No. 5 and GWF *107*, 138–142 [1966]) on the problem of movement of mineral oil products in the soil. Another publication is by *W. Zimmermann* in the Schriftenr. Wasser u. Abwasser No. 13, 1962.

Mineral oil is more rapidly decomposed in oxygen-containing ground water than in water which is free of oxygen. The danger is thus greater in waters of the latter sort (loam and clay soils). The consumption of oxygen in the former leads, however, to oxygen-free zones in the ground water, with all their consequences (*H. Sontheimer* and *W. Kölle*, Erdöl u. Kohle *20*, 648–655 [1967]).

K. H. Wallhäuser found that certain bacteria, such as *Desulfovibrio desulfuricans* and *Hermodendron* species decompose mineral oils in soil but need oxygen for this. The breakdown is very slow (Helgol. Wiss. Meeresunters. *16*, 328–355 [1967]). This is roughly confirmed by the results of *H. Krieger, R. Schweissfurth, R. Mertens* and *K. E. Heyl* (GWF *105*, 1033–1038 [1964]). According to *H. Sontheimer* and *W. Kölle*, the microbial decomposition in surface waters takes place faster at higher water temperatures (22° C) than at lower (5° C) (Erdöl u. Kohle *20*, 648–655 [1967]).

Mineral oil seepage is thus accompanied by considerable dangers for the ground water since the oils

1. spread out widely and uncontrollably in the soil. Mineral oils behave differently in different soil types. Practical experience of the author has shown that even soils considered to be impermeable to water (e.g., loess loam) are permeated in a short time by mineral oil (e.g., a metre in 20 to 30 min.). Benzine penetrates every soil faster than does water.

2. remain for a long time unchanged in the ground or spread on the ground water and little by little dissolve in the ground water (table on p. 144).
3. are noticeable in million-fold dilution in drinking water.

There are over 25.000 petrol (gasoline) stations and 12.000 petrol lorries (gasoline trucks) in the Federal German Republic; in addition there are millions of users of fuel oil containers. Prof. *Jaag* has established that 30% of the fuel oil containers in Switzerland were damaged.

Water regimen laws published in W. Germany on 27/7/57 and 19/2/60 forbid storage of all materials which could cause a detrimental pollution of the ground water or other disadvantageous changes in it; this applies generally and not only in the vicinity of waterworks and drinking water protective areas. In a subsequent extension of the law (of 6/8/64), the vicinity of pipelines is included.

According to a expert report of *Nabert, mineral oil containers* must not be stored in the nearer protective zone of a waterworks or spring, and in the wider zone of 2 kilometer radius, only under definite conditions. Double-walled steel tanks with inner and outer envelope of plastic (surer than in a concrete bath) are the safest according to *E. Holstein's* report mentioned above. *H. Kruse* considers that fuel oil tanks should be constructed only at least 300 metres from water supply plants. Only safety concrete baths (coated with bitumen) should be used in the 300 to 500 m range (Schriftenr. a. d. Gebiet d. öff. Gesundheitswesens, No. 8).

The DVGW has published working sheets W 345 and W 801 on the protection of drinking water plants and on measures to be taken in the waterworks and piped system in the event of a catastrophe. It has published also directives for legal sanction of storage containers for oil and fuels.

Optical and acoustical alarm equipment (e.g. FG 1-Sonde of the firm of Felten and Guillaume) constitutes and additional safety measure in the event of a leak in an especially susceptible place (e.g. medicinal springs).

Mineral Oils and Carcinogenic Substances

Never knowledge, won in particular by *J. Borneff* and co-workers, indicates that *contamination of ground water by mineral oils* not only upsets the palatability but introduces carcinogenic (cancer-producing) materials into it (Arch f. Hygiene 1959–1964, see also our p. 78). Certain herbicides, being used more and more for selective weed destruction in agriculture, augment the danger of mineral oil contamination of ground water; this applies likewise to the chlorinated hydrocarbon pesticides.

Nitrates disappear from ground water during seepage of motor fuels. The mineral oil adhering to the soil particles evidently impedes soil aeration, so that reductions occur. It is a fact the nitrates reappear in the ground water some time after the mineral oils have been washed out by water seepage. According to *K. Stundl* (Zentralblatt f. Bakt. II, 35–41 [1959]), the soil bacteria are variously affected by motor fuels. *J. Müller* (GWF *93*, 98 [1952]) has reported notable results concerning contamination of ground water with petrol (gasoline). *A. Janke* has shown that fluorescence-bacteria and mycobacteria slowly decompose the hydrocarbons.

Oil binding in oil-contaminated waters: Surface water can be freed from floating mineral oil when an "oil binder" is strewn on it. Useful materials are the recently marketed modified

vulcanisation products *Ekoperl* (firm of *E. Michels*, Essen) and *Perlit*. The oil binders possess a high adsorption capacity for the floating oil and, after adsorption, remain on the surface. They can then be taken off. The success of this treatment is greater, the sooner it is applied, since part of the mineral oils slowly dissolves in the water (see above). For this reason, these oil binders are now kept readily available by those concerned in their use in case of catastrophe. (DVGW working sheet W 806).
"*Carexet 7664*" of the Esso company is a new spray agent against oil contamination.
The Bundesanstalt für Gewässerkunde (Federal German Institution for Water Matters) in Koblenz has enquired into the efficacy of various oil binders. Of those based on wood, peat, plastic and minerals, the last-named proved the best (Wasser und Abwasser No. 1 [1965]).
With smaller streams, planks can be laid at right angles to the direction of flow, so as to dam the oil surface, with or without using oil binders.
Mineral oil which has run out on to the surface of the ground can similarly be treated with the mentioned oil binders before it soaks into the ground.
When *motor fuels seep* into the ground, they carry the lead tetraethyl into the ground water. Car petrol contains 0.6‰ lead and aviation petrol, 1.5‰, as anti-knock agent.
The law (German) of 1/1/69 about removal of used oil is to encourage recuperation of this oil. This prevents, directly or indirectly, its entry into waters.

Contamination of the Water through Chemicals

Not only mineral oils endanger the ground water through seepage from fuel oil tanks, oil pipelines, petrol lorries (gasoline trucks) etc.; solvents of all types, used in the preparation of agricultural pest control agents, etc., and the agents themselves, e.g. herbicides, fungicides, are contaminants.
Chemical warfare agents are a threat to drinking water in times of crisis. Arsenic-containing compounds of this sort can be determined according to the procedures for arsenic estimation.
Fluorine-containing carboxylic acids can be estimated easily through their fluorine content; this applies equally to the poisonous alkyl phosphates DPF, and other highly toxic fluorine compounds.
This is more difficult with LSD (lysergic acid diethylamide); 1 mg amounts of it have a strongly toxic action.
All these materials can be removed from drinking water with highly active A-carbon. Active carbon has proved to be a panacea in connection with water. If a chlorination or ozonisation plant is also available, purification or detoxification of the drinking water is technically possible.
(see also *O. R. Klimmer*, Katastrophenmedizin *5*, 19–24 [1969]).
Most effective against organic phosphate esters of the Sarin type (so-called esterase blockers) is treatment with hypochlorite solution in combination with active carbon.
Very large amounts of active carbon are generally necessary for detoxification (*K. F. H. Mehls*, GWF *102*, 1–5 [1961]).
Radioactively contaminated water can be decontaminated with cationic ion exchangers (see *Mehls* as above). Ion exchangers have a limited time of action, however (ca. 1 year storage time).

VI. Assessment of the Drinking Water and Service Water for Central Water Supply

A. Treatment and Conditioning of the Water

1. General

Even though only a small fraction of the water is used for drinking purposes and preparation of food, separation of drinking and service water is not domestically feasible for reasons of public health (as already said at the beginning).

a) Basic Demands on Tap Water

The water supplied to a central water system must be above reproach hygienically (p. 51) and also usable for household purposes, such as preparation of food and drink and for washing, bathing and cleaning. In addition, it should not attack the pipes, not take up iron, lead, zinc, copper and other metals. It should not dissolve lime and mortar from the tanks. Attacks on pipes by aggressive waters have not infrequently been the cause of foul water penetrating into the system.

Next to the hygienic demands, a principal requirement of tap water is that the water is able to form a protective layer in the pipe system (see below).

Along with the sanitary question of dissolving of lead from lead pipes, economic considerations play a part here; destruction and clogging of pipes in the municipal water supply system have an expensive sequel which is out of all proportion to the cost of processing and additives.

b) Drinking Water Conditioning Regulation (foreign matter decree) of 19/12/59 (W. Germany)

(Regulation concerning the addition of foreign material in conditioning drinking water, enacted by the Federal (German) Ministry of the Interior (Home Office) on 19/12/59. Federal Law Gazette, Part I, No. 52, p. 762 [1959], modified on 27/6/60, Federal Law Gazette 1960, I, p. 479).

During the printing of the 3rd edition of this book, the above decree was published, based on §§ 3 and 5a of the foodstuff statute of 17/1/36 and the supplementary law of 21/12/58. Fairly clear directives are thereby given to the waterworks for the technique of conditioning drinking water and for the maximum permitted amounts of additives.

The following additives are permitted for drinking water after complete conditioning:

1. Chlorine
Sodium hypochlorite
Calcium hypochlorite
Bleaching powder
Magnesium hypochlorite

in amounts of 0.3 mg/l at the most, calculated as available chlorine. In the decree is written that the chlorine content of the drinking water can be raised to 0.6 mg/l when this is temporarily necessary for adequate sterilisation of the drinking water.

Ammonia
Ammonium salts

The maximum concentration of ammonium ion, including the natural content of ammonium ion in the treated water, shall be 0.6 mg/l at the most.

2. Ozone
3. Potassium-, sodium- and calcium salts of the mono- and polyphosphoric acids

in drinking water in amounts not exceeding 5 mg/l, calculated as phosphorus pentoxide.[1]

4. Silica and its sodium compounds

in an amount of 40 mg/l in drinking water (calculated as SiO_2).

5. Silver
Silver chloride
Sodium-silver chloride complex
Silver sulphate

in drinking water in amounts not exceeding 0.1 mg Ag/l.

Permitted additives to drinking water to combine with free carbon dioxide are:

Calcium carbonate
Magnesium carbonate
Partially-ignited dolomite
Calcium oxide
Magnesium oxide
Calcium hydroxide
Sodium carbonate
Sodium hydroxide

The pH-value of water which has been treated with these materials must not exceed 7.5. This requirement has been modified in the meantime: The pH limit is now 8.5 and, with soft waters (up to 5° dH), pH 9.5 (Federal Law Gazette I, p. 479 [1960]).

Permitted additives to reduce an increased alkalinity or to establish a definite pH-value of drinking water are:

Sulphuric acid
Acidic salts of sulphuric acid
Hydrochloric acid

in amounts up to those which do not upset the carbonate equilibrium.

1 According to an expert opinion of the Institute for Water, Soil and Air Hygiene of the Federal (German) Health Office, additions of up to about 7 mg P_2O_5/l are to be regarded as fully harmless for the human organism (see also *P. Höfer*, Ges. Ing. 77 [1966]). 5 mg P_2O_5/l = 6,7 mg PO_4/l.

After conclusion of water conditioning, the following processing chemicals may be present as residues:

Ferric chloride
Ferrous sulphate
Ferric sulphate
Chlorinated copperas

in amounts not exceeding 0.2 mg Fe/l (including the natural iron content of the water).

Potassium permanganate

in amounts not exceeding 0.1 mg Mn/l (including the natural manganese content of the water).

Aluminium sulphate
Aluminium chloride
Sodium aluminate

in amounts not exceeding 0.2 mg Al/l (excluding the natural aluminium content of the water).

Sulphur dioxide
Sodium sulphite
Calcium sulphite

in amounts not exceeding 5 mg SO_3/l.

Sodium thiosulphate

in amounts not above 0.5 mg S_2O_3/l.

Clays
Active charcoal

in amounts of 0.5 mg/l at the most.

Every new addition of any of these materials is to be published in the daily press in future by the waterworks.

The permitted residual amounts of iron and manganese appear to be rather too high, considering the directives for conditioning drinking water and the DIN 2000 [17]; that for silver seems too low for effective sterilisation of certain waters.

Coagulant Aids

Coagulant aids are "technical aids" in the sense of the foodstuff statute, which do not enter in the foodstuff, water; they do not fall under the above mentioned regulations for conditioning of drinking water, since they are no longer present in the pure, processed water (e.g. polyacrylamides such as Sedipur). The coagulation aids must nevertheless be unobjectionable from a sanitary point of view. (*U. Hässelbarth,* Veröff. Wasserchemie, Karlsruhe No. 3 [1969]).

Industrial (Process, Service) Water

The new conception of *"service water for foodstuff concerns"* will be introduced in the coming drinking water statute. The provision of industrial water will be equivalent to that for communal purposes. The bacteriological and chemical concentration limits will be the same.

The nature of the water is highly important in *dairies* and has sometimes led to considerable annoyance (Hagen, W. Germany). Dairy water must be sterile, free of iron and manganese and not be too hard nor contain too many salts.

In *sugar manufacture,* soft water containing no microorganisms and little nitrate and mineral matter is used in the diffusers.
Brewing requires also water as free as possible from microorganisms and containing neither iron, manganese nor ammonium ion. Water containing little sodium chloride is needed in the *malt house. Starch* and similar factories need sterilised water, not too hard, and with no iron, manganese, and, in particular, no nitrite and ammonia.
Other works need water, free of iron and manganese, e.g., laundries; factories for making artificial silk and other textiles; canneries; paper mills; power stations; distilleries; bleaching plants; margarine factories.
Canneries can employ only soft water, of low carbonate content on account of the possible danger to vitamins. Clear ice manufacturers and producers of spirits use only soft water containing no iron.
Laundries still require soft water, since the synthetic detergents have not rendered the soaps wholly superfluous.

2. Corrosion

Corrosion is the attack on metals, begun and maintained by electrochemical processes, which gradually leads to destruction of the particular material.
Corrosion may be of purely electrochemical nature, e.g. due to stray currents, which bring about anodic dissolution of the metal. These cases will not be discussed here, chiefly because they have lost significance through the predominance of alternating current supply.
In the discussion below, the formation of galvanic local elements as a result of chemical reactions between metal and water plays the principal part.

a) Corrosive Power of the Water (cold water)

Hard waters, containing oxygen but no aggressive carbon dioxide (p. 162) form on the inner surfaces of the pipes a protective layer of calcium and magnesium compounds with variable amounts of iron and silica. All waters which are not in the carbonate equilibrium and hence have low pH-value (7.0 and less), i.e., hard and soft waters containing aggressive CO_2, attack the pipes and the material of the collecting tanks. They dissolve out the metal of the pipes, iron, lead, zinc and other metals from the walls of the pipes and also the calcium carbonate from the concrete and mortar of the tanks. Oxygen is important with iron and lead, since waters containing little or no oxygen always attack iron but never attack lead. More information is to be found in the relevant chapters.
The corrosive power of the water is increased when it contains less salt and is softer. This applies to attack on both metals and concrete. In the latter case, however, a higher content of salts, especially sulphate, plays a large part; concrete containers are strongly attacked by waters containing much sulphate, and sometimes become leaky. Chemical changes lead to formation of sulphoaluminate crystals, referred to in practice as "Candlot's salt" or, in German, "Zementbacillus"; this causes expansion, ultimately destroying the texture of the concrete (p. 209).

Very soft waters containing much aggressive CO_2 and with pH-values below 7 always act especially destructively on all metals. The attack on the pipes leads to solution of more or less large amounts of the particular metal. This process is of sanitary importance with lead and zinc, since both metals can be harmful to health.
Along with the free carbon dioxide, the carbonate hardness is decisive; the formation of a protective layer depends on its magnitude.
The product of corrosion of iron sometimes becomes deposited also at the point of origin, forming lumps of rust. The fact that a tapwater is free of iron does not therefore always mean that no corrosion has occurred.
Standing water effects greater corrosion in the pipes than flowing water. An unsuitable type of water therefore has an especially evil influence in the terminal parts of the pipe system. It can be said in general that the slower the rate of flow, the greater is the corrosion. The reduction of oxygen content becomes particularly noticeable when the rate of flow is small. Pipe corrosion and yellow-brown turbid water as a result of shortage of oxygen is thus encountered often in end pipe sections. A distinction is hence made in the DIN-Norm 50930 between rates of flow above and those below 0.5 m/sec.

b) Increase of Aggressivity on Mixing Two Waters

When two waters, in themselves not aggressive, are mixed, a product is yielded which often possesses aggressive properties. This is always the case when the two waters have differing carbonate hardnesses, even when both are completely in equilibrium (see below). Even a soft water which has been fully deacidified and contains absolutely no free carbon dioxide yields an aggressive product on mixing with a hard water in the carbonate equilibrium. In central water supply, waters must thus be mixed first and then deacidification performed after. *L. W. Haase* ([37], p. 84) quotes figures.
Two aggressive waters do not lose their aggressivity on mixing. Even when only one water is aggressive, there is no loss, so that there is no "dilution" of aggressivity.

Water conditioning must always aim at provision of a protective layer through the action of the water, i.e., establishment of the carbonate balance and adequate oxygen in the water.

According to *K. E. Oehler,* two waters cannot be completely homogeneously mixed in the pipe system but only in tanks and then the inlets must be suitably positioned (GWF *110,* 860–865 [1969]). We have, however, accomplished in Hamelin complete mixing of two waters in the feed water of the local system, by bringing the conducting pipes together; we wish to publish these results.

Control of mixing
Our experience shows that the completeness of mixing of two waters can be checked by plotting on squared paper the data (e.g. chloride content or degree of hardness) which differ most for the two waters; the value found for the mixed water is plotted too.
The diagram gives an example of a mixture of 80% of the water from waterworks II with 20% of that from waterworks I. Water I has 183.0 mg Cl/l and water II, 38.0 mg Cl/l. The two values are joined with a straight line. The mixed water had 67.0 mg Cl/l. This point is marked on the connecting line and a vertical line, parallel to the ordinate (y) axis, intersects the abscissa (x-axis) at the sought mixing proportion (20 : 80).

The more the analytical results for the two waters differ, the more exactly can the mixing proportion be evaluated. If the diagram is appropriately arranged, the oxygen and CO_2 values can also be considered; these play the most important parts in problems of mixed waters. When it is necessary to "blend" a too hard or too salty water, the above diagram helps in preparing a mixed water of desired degrees of hardness, salt content and CO_2 content.

Diagram

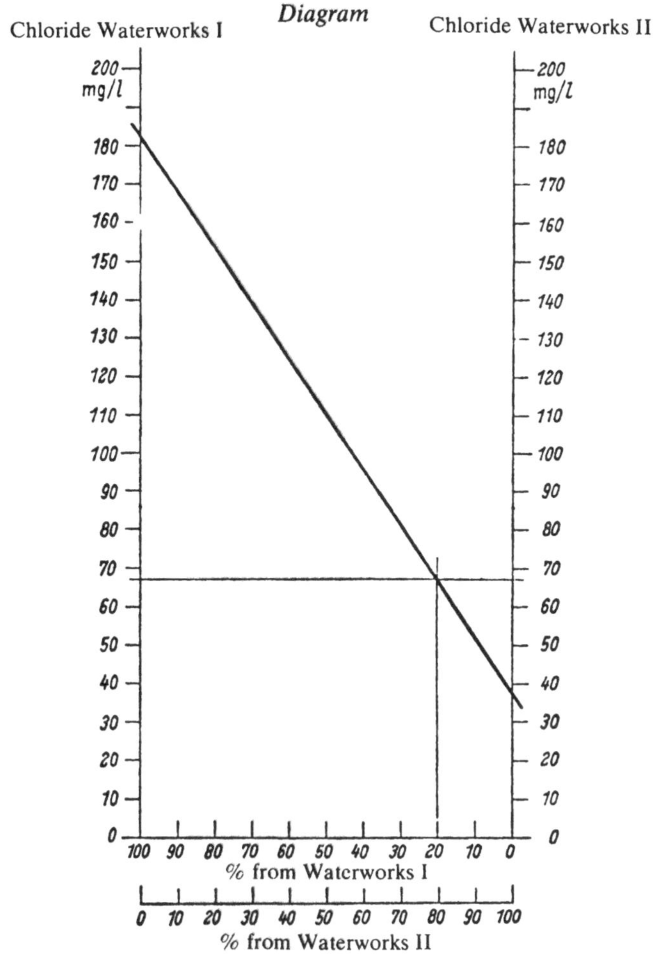

c) Pipe Corrosion and Water Hygiene

The corrosion of water pipes has a technical-economical aspect, related to the cost of keeping the costly system in order; and also a hygienic aspect in so far as corrosion to a hole could result in foul and waste water being sucked into the pipe system. This is quite possible when the speed of flow of the tap water and its pressure vary. This has an extremely unpleasant sequel for those responsible, as the case of Altötting (Bavaria, W. Germany) showed. Protection of the water pipe system is therefore in every connection one of the most important duties of the superintendent of the waterworks. An intact system is vital for the waterworks concern on account of the losses of water; and for the public as a result of danger of infection through penetration of micro-organisms into the system via corrosion sites, external (p. 156) and internal, and via leaky places at pipe junctions. The "Sander apparatus" (firm of Sewerin of Gütersloh, W. Germany), for example, serves for searching for leaks. The DIN-Norm leaflet 1988 gives precise directives for pipe joints, armatures, etc. The DVGW-working paper W 345 regulates the technique of connections for drains, sink fitments and rubber joints.

Reference may be made to the "Judo insulation pieces" with special gasket fittings (firm of Jul. Dopslaff, Stuttgart).

H. Hugelmann reported on mechanical stress and strain on pipes at the DVGW-conference on damage to water pipe systems in 1961.

d) Behaviour of Metal Materials towards Water

Lead: Lead is attacked by waters containing carbon dioxide; conditions are that the water contains oxygen and that its content of free CO_2 exceeds 20% of the combined CO_2 (see *K. Höll,* Ges.-Ing. *58,* 323–326 [1935]).

Considerable solution of lead always occurs when the ratio of free to combined CO_2 is 1 : 2. All waters of pH-values below 6.8 have at least some solubility power for lead. A protective layer of basic lead carbonate is formed by hard waters containing no aggressive CO_2. Lead is in no case attacked if oxygen is absent (converse of iron, see below). Cases have recently come to the notice of the author, however, where strongly alkaline water, namely very soft, conditioned reservoir water of pH-values 9.0 and 9.5, brought about marked corrosion of lead. The lead concentrations of 2 to 5 mg/l thereby yielded in the water soon led to lead poisoning of the consumers (Deister region near Hannover, W. Germany). These experiences from practice do not conflict with the experimental results of *J. Müller* (Vom Wasser, *18,* [1950/51]), since the pH-values were a little below 9.0 in his experiments.

The new regulations for conditioning drinking water (p. 148) will prevent too intensive alkalisation in future (see above).

Unfortunately the general literature on drinking water contains no references to the dangers of corrosion of lead pipes through strongly alkaline waters. Only *L. W. Haase* [*37*] mentions the inferior formation of protective layers by surface waters with low carbonate and carbon dioxide contents, and the formation of plumbite in lead pipes (see above). According to him, chemical deacidification plants should leave no residual excess of alkali in the water after reaction is complete.

High content of silica and of phosphate prevents attack on lead, even by aggressive waters. Silicates and phosphates are thus added to aggressive waters as a protection for lead pipes (p. 172).

H. Haupt pointed out the corrosion-inhibiting action of naturally occurring small amounts of phosphate and silicate in the water. *J. Müller* was able to bring about formation of a protective layer in lead pipes by means of small silicate additions to the very soft reservoir water of the Dresden waterworks, Coschütz (Vom Wasser *18,* 226–253 [1951]).

K. Höll ascertained that the pipe system in Brünnighausen (W. Germany) was intact, al-

though the water was soft and contained considerable excess carbon dioxide. This can be due only to the high natural content of silica (over 20 mg SiO_2/l).
Large amounts of salts (> 500 mg Cl/l) increase solubility of lead. Brackish and sea water have an external corrosive influence on lead pipes.
A high nitrate content augments the uptake of lead (East Friesland and East Westphalia, Germany, p. 69). Lead is attacked also by water containing much ammonium ion and by water which has been treated by hypochlorite.
Some lead is nearly always dissolved from new lead pipes, even when the water is not aggressive (*G. Gad* and *K. Naumann,* Ges.-Ing. *74,* 85 [1953]).

e) Water Pipes

Iron: Iron is attacked and dissolved by waters containing carbon dioxide which prevents formation of a protective layer against rust (p. 164). In the absence of oxygen, iron enters solution as ferrous hydrogen carbonate; oxygen access leads to precipitation as ferric hydroxide. If the oxygen content is excessive, especially when it is not in genuinely dissolved form (milky turbidity of the water), iron is likewise attacked, wholly independently of the remaining components in the water. Pitting by corrosion is very dangerous.
A higher but not excessive oxygen content in the water is hence desirable for transmission in iron pipes (6 to 10 mg O_2/l, depending on the water temperature); this amount counteracts attack on the iron.
Every water attacks the iron pipes when the oxygen content is insufficient, even when the other corrosion factors do not favour attack. The situation is thus diametrically opposed to that with lead. The oxygen concentration should never be allowed to fall below 4.0 mg/l in the system and below 2 mg/l in the terminal reaches.
A high pH-value of the tap water is good for iron pipes, as shown by long-standing experience with the Söse (Harz, W. Germany) piped water system (*Wiendel,* Neue Deliwa No. 1 [1958]). Iron pipes are also protected from attack by carbon dioxide when the water has a high natural content of phosphate; this is shown by experience. Accordingly, larger amounts of phosphate are added artificially to aggressive waters with no natural phosphate content (p. 172). Large amounts of chloride are said to be bad for iron pipes. Brackish water attacks pipes strongly from the outside. The loss through corrosion is doubled when the salt content is 5‰ (*N. A. Malischewski,* Wass.-Vers. u. san. Technik, No. 10 [1956] and *H. E. Hönig,* Metall u. Wasser, Essen [1965]). Some of the iron pipes in Eschershausen (W. Germany) have been completely corroded in 30 years by the chloride-containing water (ca. 1000 mg Cl/l, p. 74).
A protective layer is formed more easily in cast iron pipes as a result of their rougher surface; they are thus preferred to steel pipes. Hence steel pipes are now being provided also with cement mortar lining for use with slightly aggressive waters. They have long been in use in the USA. They cause an initial increase of pH.
Waters containing much sodium hydrogen carbonate instead of, or as well as, calcium hydrogen carbonate attack iron pipes (such "exchange waters" are encountered in coastal areas).
Unprotected iron pipes are attacked by hydrogen sulphide and sulphides (*Steinrath,*

DVGW-conference on damage to water pipe systems, in 1961). Hydrogen sulphide is frequently formed biogenically from sulphate in moorland soils.

It can happen in closed systems for removing iron, that all the water oxygen is consumed. The water, then with no or little oxygen, yet containing some free carbon dioxide, thus becomes aggressive to iron. The phenomenon of *referrisation* is then met. Air must thus be added to closed iron removal plants, with the help of an air compressor, until about 6 to 8 mg O_2/l are present in the pure water.

A protective layer against rust is formed when the water does not contain the CO_2 which militates against this; oxygen is needed also for this to occur.

In rare cases, even strongly aggressive waters do not attack the pipes. Investigation then usually shows that the water contains much phosphate or silicate.

External attack on the pipes

Iron pipes are also readily attacked from the outside by ground water, especially in acidic peaty soils, ground containing slag, digested sludge, loam and clay soils with little calcium and also by salty ground water. Steel pipes are more susceptible to attack than cast iron pipes. Cast iron pipes have some other advantages concerning corrosion. They are more resistant than steel pipes against soft water of high oxygen content and aggressive properties *(L. W. Haase)*. High-alloyed austenitic cast iron pipes are appreciably more resistant to corrosion than normal cast iron pipes. Spheroidal graphite tubes are, however, not superior to ordinary cast iron in this sense.

The opinion is often voiced that iron pipes are not attacked in peaty soils because there is no oxygen present. This is certainly not true for normal acidic boggy ground, from the viewpoint of both the carbon dioxide and the biology (see above).

Sulphate contained in water-saturated clay soils can be reduced to sulphide ion by certain bacteria; this can give rise to considerable corrosion. The biogenic change may go further in some soils, to sulphuric acid (*A. Kaffka,* Arch. Hyg. 1967 and *W. Kirsch,* Korrosion im Boden, Stuttgart, 1968).

Cast iron pipes have a relatively high bending strength, especially the "Duktil-Rohre".

Iron pipes which have been specially treated with asphalt or bitumen must be used in unfavourable soil. They must be embedded in a layer of 30 to 50 cm (12 to 20") of sand or gravel (*I. Müller,* Wass.-Wirtsch. u. Techn. *2,* 275–281 [1952]). Best is a special protective coating according to *Schlumberger.* Attack on the metal is favoured by chloride concentrations in excess of 200 mg Cl/l, particularly with soft waters (*H. Klut,* Trink- und Brauchwasser, Berlin [1924]). Wrought iron pipes ("Habitrohre") coated well with bitumen have proved satisfactory in border-line cases. Iron pipes are strongly attacked in regions of brackish and sea water (Emden Waterworks, W. Germany). The iron pipes have to be protected there by melted bitumen or coal-tar pitch, so-called "synoplasts".

Pipes of asbestos cement and plastic are better in such areas (see below). As has already been said, standing water causes greater attack, particularly on iron pipes, than flowing water. An unsuitable water composition thus has an especially bad effect in the dead ends of the pipes.

Corrosion damage can be avoided by "cathode protection" also (see *M. E. Parker,* Kathodenschutz, 2nd Edition, Essen [1965]).

Steel pipes with a plastic covering coating have been used recently where soil conditions were unfavourable.

Nature of the water and the pipe system

The limits below are valid for galvanised and ungalvanised iron pipes (saturation index = 0):

Galvanised pipes −0.5

Carbon dioxide preventing formation of protective layer against rust = 0

pH-value equal to or just beton the equilibrium value;

Carbonate hardness:	flow speed up to 0.5 m/sec	= 6.0° dH
	flow speed above 0.5 m/sec	= 2.0° dH
Oxygen content:	flow speed up to 0.5 m/sec	= 6.0 mg/l
	flow speed above 0.5 m/sec	= 2.0 mg/l
Chloride content:	150 mg Cl/l	

Copper: Copper is little attacked by the water. Water, free of oxygen, attacks copper as little as lead. Very soft waters, containing much oxygen, attack copper only initially. Copper suffers no attack from the natural salts and acids of the water. These also do not attack the cuprous oxide layer which soon forms on new copper pipes.

A high salt content acts corrosively on copper (*L. W. Haase* and *O. Ulsamer,* Die Umschau, No. 4 [1934]). Brines are known not to attack copper pipes, provided they are free of oxygen, which is generally the case. The attack of oxygen-containing, aggressive waters on brass is preferentially on zinc (*L. W. Haase,* Ges.-Ing. No. 49, [1953]).

Special bronze alloys and phosphor-bronze can be used for the tappings of brine springs.

The harder the water, the better it is for copper tubes. Warm water attacks copper less than cold water.

Copper is strongly corroded by waters containing H_2S. This is particularly important for the tappings of medicinal springs. Phosphor bronze or plastic are better materials for pipes in these cases (see below).

Cell-formation (Local Elements) with Copper Pipes

Errors of installation are always being made by placing copper pipes before iron pipes. The former may be installed only after the steel pipes, in the direction of flow; non-observance of this leads to galvanic cell formation (local elements). The situatuon is reversed with copper and zinc pipes. Then, a galvanic cell is formed if the copper pipes precede the zinc; zinc enters solution, thereby impairing the consumers' health.

Copper pipes, in themselves very resistant to corrosion, have one disadvantage; the surfaces of copper pipes become pitted where iron or manganese separates. The water must therefore be free from iron and manganese and must not undergo referrisation in preceding iron pipes.

Copper pipes were excavated in Herculanaeum which had been laid there 3000 years earlier.

Brass pipes containing a certain amount of arsenic are stable to salt water.

Zinc: Zinc is dissolved from the walls of the pipes by air-free waters containing carbon dioxide; this occurs especially with soft waters. According to *Baylis,* all waters of pH below 6.5 dissolve zinc from the pipes. Galvanised zinc pipes should thus be laid only for hard

water possessing no aggressive properties. Alkaline waters, i.e. those of pH-value greater than 9.0, also attack zinc pipes, dissolving the metal as zincate. Chloride and sulphate aggravate the corrosion of zinc, likewise larger amounts of nitrate (over 20 mg NO_3/l).
Aluminium: This metal turns brown or black in warm, hard water (blackening of aluminium utensils in well water) (*D. Altenpohl,* Metalloberfläche *9,* p. 118–121 [1955]). Aluminium is stable in sea water and also in sulphur waters (*K. E. Quentin u. Pachmeyer,* 1964).
Tin is attacked only little by water. Tin free of lead must be used for plating iron pipes, or else the lead can be dissolved out.
Lead is attacked above all by all soft waters and also hard waters containing excess carbon dioxide (p. 148).
Plastic pipes
Pipes of plastic are being used more and more with aggressive waters and soils (*W. Reinhold,* Kommun. Wirtschaft 7, [1956] and *M. Mosler,* ibid. *8,* 231–237 [1957]).
Plastic pipes have the advantage of not being attacked by any water. They suffer no damage from carbon dioxide, humic acids, sulphates and chlorides in tap water or soil in any amount. Thus plastic pipes have proved very useful in medicinal baths and for transport of mineral water (*Wevelmeyer,* Heilbad und Kurort [1958]). There are, nevertheless, products which are unsuited to aggressive waters. According to *W. Reinholdt,* pipes of plastic based on PVC (polyvinyl chloride) are more suitable than those based on polyethylene (Kommun. Wirtschaft 7, 316–318 [1956]). *J. Böing* reached the same conclusion another way (Z. f. Bakt. Org. *168,* 324–328 [1957]).
Some plastic materials, notably poor *polyethylene* pipes, serve as nutrient for bacteria. Before purchase, one should test if the powdered pipe material is taken up by aggressive water. *F. Scheminsky* showed through fluorescence analysis that some types of plastic release organic compounds into the water (Fundamenta balneologica *1,* [1958]). This may explain why, in controls of drinking water, multiplication of the bacterial number has been found when the water lay stagnant in plastic pipes. *W. Ahrens* and *C. Siegert,* however, state that this multiplication is found only temporarily, at the beginning of use of new plastic pipes (GWF *98,* 345–351 and 661–663 [1957]). *W. Zimmermann* found more favourable bacterial numbers with PVC pipes than with those of polyethylene (Städtehygiene 7, 266–268 [1956]). *W. Schwartz* and *A. Müller* studied these problems thoroughly and discovered that plastic pipes, like all others, lead to initial bacterial multiplication when water stagnates in them (Kunststoffe *47,* 583–588 [1957]). No bacterial multiplication was found by *H. Unger* in plastic pipes from polyvinyl chloride (Wasser u. Boden *12,* 144–148 [1960]). The DVGW has brought out a test mark, the DVGW-sign for drinking water pipes from plastic. According to the DVGW pamphlet W 325, plastic pipes for this purpose must not favour growth of microorganisms and formation of algae; they must also impart no smell or taste to the water and not show the peroxide reaction (from benzoyl peroxide, used from the preparation). No regenerated material may be used in this preparation. The DVGW working leaflets W 320 to 323 regulate the installation, testing under pressure and pipe connections for plastic pipes (see also DIN 19532).

If one respects the old rule of sampling, i.e., that the water is run for 15 minutes beforehand, plastic pipes will show no disadvantage vis à vis metal pipes, from a bacteriological point of view.

One point must be mentioned with plastic pipes, namely, that many plastics are prepared with stabilisers which contain lead, cadmium and barium. Water investigations in Belgium and the Netherlands have shown the presence of small amounts of lead in water in experimental pipes from various plastics; these concentrations were initially as high as 1 mg/l, falling later to 0.3 mg/l and less. The Netherlands Kiwa Institute objected to one of eight sorts, because it regularly gave more than 0.3 mg Pb/l (*R. Buydens*, Bull. Centr. Belg. d'Etude des Eaux 84–88 [1958] and *W. D. Tiedemann*, J. Am. Water Works Ass. *46*, 775–785 [1954]).

In the meantime, the preparation of plastic pipes is being adapted more and more to the interests of water supply. The introduction of the DVGW test mark for plastic pipes has contributed to this (see above). *H. Sontheimer* and *J. Wagner* thus obtained better results in their experiments; only at the beginning was it possible to detect release of lead into 0.01N and 0.001N sodium hydrogen carbonate solutions (GWF *110*, 487–492 [1968]).

Plastic pipes have served well for carrying hydrogen sulphide waters from medicinal springs also. Plastic riser and main pipes should thus be preferred for conveying H_2S-containing ground water to the conditioning site in the waterworks.

Plastic pipes have proved satisfactory also with acidulous waters and for the tappings of very deep wells (*H. Herterich* and *F. Schwille*, Heilbad u. Kurort *10*, 56–60 [1958]).

The plastic era can be said to have begun in water transport technique in 1959, when prices became more reasonable. Progress was thus made technically in laying the pipes and in corrosion resistance to every type of water and ground; and in water hygiene also (p. 148). The Federal (German) Ministry for Atomic Energy and Water Economy has thus recommended the "Vestolen pipes" of the Hüls chemical works, which are supplied in lengths up to 1000 m (see Bund.Gesundh.Bl. 5, 242 [1962]).

Water pipes of plastic, reinforced with polyester, are a new departure. They are especially resistant to pressure and heat.

Plastic pipes have smooth walls and hence show little friction and no incrustations.

Asbestos cement pipes

Asbestos cement pipes, such as *Eternit*, *Fulgurit* and *Toschi*, have been finding extensive application recently for water pipe lines. They show high internal and external resistance to concentrated salt solutions. Even high sulphate content does not damage them. According to *Carière* (GWF No. 2 [1956], extensive experiments in the Netherlands have demonstrated that Toschi pipes are not corroded by waters containing 2000 mg $CaSO_4$/l and 5000 mg Na_2SO_4/l and $MgSO_4$. *P. H. Perkins* (J.Am.Water Works Ass. *50*, 257–262 [1958]) arrived at the same conclusion. Asbestos cement pipes are also resistant to electric currents.

Toschi pipes have proved very useful for tapping and conveyance of the strongly aggressive thermal, acidulous brines of high hydrogen carbonate content in Bad Oeynhausen.

Humic acids damage the asbestos cement pipes as little as does hydrogen sulphide. Only larger amounts of aggressive carbon dioxide occurring with low carbonate hardness causes damage. Linings of bitumen within and covering with coal-tar pitch externally are necessary with these waters. A few mg/l of aggressive carbon dioxide are unimportant, however. Pressure piping of Eternit is suitable for water conveyance across country, even in peaty and salty soils. It has been proving satisfactory in the Süderditmarschen (N. Germany).

Asbestos cement pipes have been finding considerable use in the water sphere since the Oettel coupling has enabled them to be joined better.
Hagusta pipes and Hagusta filter pipes are insensitive to any corrosion.

Drinking water Containers

Behaviour of the water in containers: Most waterworks are equipped with tanks to be able to cope with spasmodic water consumption and the rush demands at certain times of the day; and as a reserve for fire extinguishing. Sometimes the water first comes into contact with the air in these. Oxygen-free waters then take up oxygen, so that even a smaller amount of iron makes itself unpleasantly noticeable by colouring the water or yielding a deposit. Any manganese in the water is likewise deposited. Should, thus, on rare occasions, the purified water contain small amounts of iron and manganese, unnecessary contact of the water with air should be avoided. This can be accomplished, for example, by introducing the water into the containers from below and preventing the tanks from emptying too often.
If there is neither iron nor manganese and too little oxygen in the water (p. 193), it is allowed to fall into the tank from an inlet which is set up as high as possible. Oxygen is then taken up and the excess carbon dioxide is slightly reduced. In this way it may be possible to save having an aeration plant. This cannot be done, however, with harder waters which are just in carbonate equilibrium (p. 163) and with those which contain too little "associated" carbon dioxide (p. 163); separation of calcium would then occur. Such waters must enter the tanks from below and all cascades should be avoided.
Special attention must be devoted in future to achieving a good, uniform flow through the drinking water tanks (*Fr. Dosch* [33a] and *J. Reitinger,* Gas, Wasser, Wärme *23,* 33–39 [1969]). The same applies to the coating of the tanks and plastic top coatings (*H. Barth,* GWF *110,* 483–486 [1969]).

3. Hydrogen Ion Concentration, pH

The hydrogen ion concentration gives the existing, actual concentration of hydrogen ions. In contrast, acidity and alkalinity give the potential amount of hydrogen and hydroxyl ions, as consumption of normal alkali or acid, respectively.
Pure water is only very slightly dissociated into its ions, hydrogen and hydroxyl. Since it is neutral, it contains equal numbers of hydrogen and hydroxyl ions. One litre of chemically pure water contains a ten millionth of a gram of hydrogen ions, 10^{-7}. A higher concentration, e.g., a millionth of a gram per litre, 10^{-6}, yields an acidic water. If the concentration is lower, e.g. 10^{-8}, the water is referred to as alkaline. This applies to all aqueous solutions which contain electrolytes – acids, bases or salts. The values can lie between 10^{-1} and 10^{-14}.
The actual reaction is given by the pH-value instead of the hydrogen ion concentration. pH (pondus Hydrogenii) is the negative logarithm of the hydrogen ion concentration. E.g., [H] $= 10^{-7}$, pH = 7.

pH below 7 = acidic
(pH 1 to 7.0)
pH = 7 = neutral
pH above 7 = alkaline
(pH 7.0 to 14.0)

The pH value is important in water technology. Almost all waters of pH below 7.0 have the property of attacking materials (excepting very hard waters, for which the carbonate balance is at pH 6.9). They generally contain free excess carbon dioxide (p. 164) which brings about corrosion of pipes and destruction of concrete and masonry of the collecting tanks. All waters of pH values exceeding 7.0 (7.0 to 7.5) can also be aggressive when their calcium salt content is low (p. 164). A difference of a few tenths of a pH unit is significant, since it is a logarithmic quantity.

There are often ground waters and surface waters containing soda and of pH above 9.0 in warm countries (e.g., the Hungarian "Puszta", Neusiedler See in Austria and in Jugoslavia). Such high pH values are significant for corrosion of metal pipes (except iron) (p. 154). Standing waters have pH values of 9 and more during daytime in summer, as a result of the carbon dioxide assimilation of the plankton and of submersed flora. This occurs less frequently in reservoirs, which usually have soft water, since they contain little plankton.

Ground waters normally have pH values between 6.0 and 8.0, mostly from 7.0 to 7.5. Values of below 6.0 also occur frequently. These last-named, generally rich in carbon dioxide, are highly aggressive and must be deacidified before being fed into the pipe system. Tap water should have pH values of over 7.0, from 7.3 to 7.5 as far as possible; they should be in the carbonate equilibrium (p. 163). Newly bored wells in calcareous soils often have water of pH above 8.0. Similarly high pH values are often shown also by waters from freshly bricked dug wells or water tanks. The pH value falls after a few weeks.

4. Carbon Dioxide

Occurrence: Free and combined carbon dioxide is found in every natural water, at least in amounts of a few mg/l. Free carbon dioxide is largely dissolved in the gaseous form as CO_2 and is present as hydrated carbon dioxide, carbonic acid H_2CO_3, in only small amounts (0.7%); the acid is dissociated into hydrogen and bicarbonate ions.

The amount of free carbon dioxide in ground water is usually 10 to 20 mg CO_2/l but waters with 30 to 80 mg/l and more are frequently found. Naturally there are transitions to the mineral waters which contain 1000 mg CO_2/l and above, e.g. acidulous waters which must have at least 1000 mg CO_2/l (p. 237).

Percolating water or seepage, i.e., the water which seeps through into the ground water, becomes charged with the carbon dioxide formed biogenically in the uppermost soil layers; it is hence rich in carbon dioxide and aggressive (see *K. Höll,* Vom Wasser *30,* [1963]). The soil air possesses about 30 to 50 times as much carbon dioxide as the atmosphere.

Surface waters generally contain much less free carbon dioxide than ground waters; typical amounts are a few mg CO_2/l, and, at the most, 10 to 15 mg/l (in winter). In summer, often for a long period, all the free carbon dioxide and sometimes also part of the bicarbonate carbon dioxide is consumed in the surface layers of many lakes, rich in nutrient, by the microorganisms of the water (plankton material); the bicarbonate carbon dioxide is consumed as well by submersed aquatic plants. The water becomes alkaline to phenolphthalein, with a pH value exceeding 8.0. This occurs frequently in the waters of warm countries.

In the deeper zones of the lakes, below the thermocline, the CO_2-conditions during the stag-

nation period can be quite different from those in the surface layers of the pelagial. Equilibration of the layers occurs in the circulation periods of spring and autumn.
The supply of carbon dioxide in the surface layers is replenished from time to time through fermentation in the sludge region and upward movement of the gas.
The combined carbon dioxide is usually as ionised hydrogen carbonates of calcium and magnesium. This determines the carbonate hardness of the water. In warm countries, one finds also calcium and magnesium carbonates and sometimes sodium hydrogen carbonate and carbonate, especially in surface waters, free of carbon dioxide (so-called soda lakes, such as the Neusiedler See in Austria). Ground waters with small amounts of sodium bicarbonate are found also in German coastal regions and, in rare cases, even inland, e.g. near Bad Nenndorf. The bicarbonates, or hydrogen carbonates, are dissociated in water into Ca (or Na) ions and HCO_3 – hydrogen carbonate – ions.
The carbon dioxide content has scarcely any *hygienic significance*. Free carbon dioxide is very occasionally a decomposition product of decay processes. The content of combined carbon dioxide is then increased also, as a result of solubility processes in the soil (seepage, see above).
A certain content of carbon dioxide gives the water a pleasant, refreshing taste.

5. Free Carbon Dioxide

a) *Technical Significance of the Carbon Dioxide Content of the Water*

The content of free carbon dioxide of a water plays a very large part in water technology. The free carbon dioxide, particularly the excess carbon dioxide (see below) is able to attack and dissolve the metallic material of the pipes and also the calcium carbonate in mortar and concrete. Water for central water supplies must therefore be freed from its excess CO_2 before feeding into the mains; this is termed deacidification in water conditioning technique. A small excess of 1 to 3 mg CO_2/l is tolerable only with very hard water. Small amounts of excess carbon dioxide have a far more unfavourable effect in soft and medium hard waters than in hard waters.
The corrosion effect is the sum of that due to the aggressive carbon dioxide and that due to lack of oxygen.
The corrosion through free carbon dioxide (surface corrosion) differs from that due to oxygen (pitting).
If deacidification in the first conduit sections between the gang of wells and the water works is not possible or is not practicable in small plants, steel pipes, well treated with bitumen, are used, e.g. wrought iron pipes ("Habitrohre") (Röhrenverband GmbH, Düsseldorf) or asbestos cement pipes (e.g. Eternit). Recently, plastic pipes or pipes of V2A steel have been used in special cases. The width of the pipes should then be as small as possible, to permit rapid flow. Only pure (phenol-free) petroleum bitumen may be employed for the inside coat of water pipes, or else chlorinated water would yield the unpleasantly tasting chlorophenols (see also p. 78 in connection with carcinogenic substances). The material Inertol is more suitable for internal coating.

In all other cases, waters with aggressive properties must be deacidified before being fed into the network (see p. 166) or, in border-line cases, treaded with polyphosphates (p. 172). We thus distinguish:

Water

a) aggressive with excess CO_2 =	b) with deficit of associated CO_2 =	c) equilibrium water in which carbonate equilibrium holds
calcium dissolving and metal-attacking	*calcium-separating*	*forming protective layer*
pH value is below the equilibrium pH	pH value is above the equilibrium pH	pH value is the equilibrium pH

b) "Associated" carbon dioxide

Not all the free carbon dioxide takes part in the attack on lime but only that which is not regarded as "associated" carbon dioxide. The associated CO_2 is the free carbon dioxide necessary to maintain the calcium hydrogen carbonate in solution. The carbon dioxide over and above it is the excess carbon dioxide. Small amounts of calcium hydrogen carbonate, corresponding to a hardness of less than 2.0° dH, do not require associated CO_2; the total free carbon dioxide of quite soft waters is thus aggressive (see Tillmans' table).
The associated CO_2 increases with increase in the content of calcium hydrogen carbonate; extra calcium hydrogen carbonate therefore requires extra associated CO_2 (see under lime-aggressive carbon dioxide, p. 165).
Opinions still vary about the behaviour of the magnesium hydrogen carbonate in relation to associated CO_2. Formerly, it was believed that magnesium hydrogen carbonate required no associated carbon dioxide but nowadays one assumes that it is needed. *L. W. Haase* [37] even says that more carbon dioxide is needed to maintain the magnesium-CO_2 equilibrium than the calcium-CO_2 equilibrium. Elsewhere in his standard work on corrosion questions [37] *L. W. Haase* says that larger amounts of free excess CO_2 are of little harm in the presence of magnesium compounds, especially magnesium hydroxide. In practice, an intermediate situation will be found. Larger amounts of magnesium hydrogen carbonate will certainly need some associated CO_2 to maintain the magnesium-carbon dioxide equilibrium.

c) Carbonate Balance and Anti-rust (Protective) Layer

The carbonate equilibrium (balance) of a water represents a state in which the water shows neither a deposit of calcium nor aggressive properties. A water is in the carbonate equilibrium when it possesses exactly the right amount of associated carbon dioxide. A water of this sort gradually forms in the iron pipes a protective layer of calcium carbonate crystals with inclusions of magnesium and iron (III) compounds – the *anti-rust layer*. This layer is formed, however, only when sufficient oxygen is present in the water and in waters of carbonate hardness of at least 2.0° dH.

At least 3 mg O_2/l must be present in the moving water of the mains; but 6 mg O_2/l needs to be present in the more static water of the terminal pipe sections.

The *anti-rust layer* in the water pipes is absolutely vital for a regular water supply. Polyphosphates are added nowadays to the water in newly constructed pipe systems so as to accelerate formation of this protective layer; this is done even where the water has no aggressive properties. Disodium phosphate is used also, sometimes.

Water deficient in associated CO_2 scarcely occurs in nature but is frequently encountered after conditioning. This leads to undesirably large lime incrustations in the pipes, to formation of lime flakes and ultimately to obstruction of the pipe interior.

Very soft waters never form a protective layer, even when all the carbon dioxide has been removed (e.g. deacidified reservoir water). Phosphate must be added to such waters before they are fed into the pipe system (p. 172).

Water with excess carbon dioxide, i.e., over and above the associated CO_2, forms no protective layer, attacking the pipe material instead (see below).

The carbonate balance is highly dependent on temperature. This sometimes plays a considerable part in the water works. If the temperature of the water rises in the pipe network, as for example in the height of summer with pipes laid near the surface, then the water has too little associated carbon dioxide, corresponding to this higher temperature, and there is a danger that incrustations will be formed in the pipes.

If the water becomes cooled at the water works or in the conduit, there is a risk of its attacking the pipes, even when the conducted water is not aggressive. *W. Ohle* occasionally found in lakes, water which was supersaturated with calcium carbonate, although none separated; he referred to "metastability".

Too high pH values prevent formation of an anti-rust layer. Since the necessary amount of calcium is lacking in soft reservoir waters, e.g. in the conditioned Sösetal barrage water, no real protective layer is formed despite the addition of soda (pH of over 9.0). Excessive alkalisation of the water will not be permitted in future according to the new drinking water conditioning statute (p. 148).

d) Excess Carbon Dioxide

pH value in equilibrium waters

Waters in the carbonate equilibrium state normally possess a pH value of ca 7.0. Hard waters containing much combined carbon dioxide and much associated free CO_2 often have low pH values of 6.9, even 6.8, despite being exactly in the carbonate equilibrium state. In contrast, waters of pH 7.5 can sometimes have not insignificant amounts of excess carbon dioxide. Hence the pH value alone of a water is not enough to permit a judgment of its aggressivity. The carbonate equilibrium is established at pH ca 7.9 for a carbonate hardness of 1.0° dH; and at pH 6.9 by a carbonate hardness of 20.0° dH.

Carbon dioxide for the anti-rust layer

The desired protective layer against rust is gradually formed on the walls of iron pipes when a water is just in the carbonate equilibrium condition, i.e., when just enough free CO_2 is in the water to keep the calcium hydrogen carbonate present in solution.

If too little free CO_2 is present, i.e., less than needed to maintain the calcium hydrogen carbonate in solution, the protective layer is formed too rapidly and the precipitation leads to unwanted incrustations on the pipes.
If, however, too much free CO_2 is present, i.e., more than required for maintaining in solution the calcium hydrogen carbonate, no anti-rust layer is yielded since the calcium compound remains dissolved.
Unaffected ground waters are found in Lower Saxony which, on calculation, contain too little free CO_2. Yet, when their high magnesium content is taken into consideration, it is found that they are in the carbonate equilibrium state. This supports the opinion that magnesium compounds require no associated carbon dioxide.
After taking the magnesium compounds into consideration, one finds that some of these hard waters still contain excess carbon dioxide. It may be assumed with *Heckmann,* that this excess CO_2 is associated with the noncarbonate hardness, which is high.
U. Hässelbarth demonstrated the significance of the noncarbonate hardness, especially the magnesium hardness. According to his fundamental research, a water of carbonate hardness exceeding 12° has more aggressive carbon dioxide than that calculated from the *Tillmans* table. This applies especially to higher magnesium hardnesses.

The free available carbon dioxide, over and above the associated, i.e., the real excess carbon dioxide, is that which prevents formation of the anti-rust layer. It goes beyond this prevention activity in that it even attacks the iron and lead pipes of water mains system.
If a water is not in the carbonate equilibrium state, the total excess CO_2 is to be regarded as aggressive to metals. The situation differs in attack on concrete and mortar (see below).
It is a good idea to build control sections of pipe into a system when it is newly laid by the water works. This enables the formation of the protective layer or attack on the pipes to be checked. Special control sections should be introduced later if any unusual observations are made, to be able to watch corrosion or deposition.

Lime-aggressive carbon dioxide

A water which contains more than the associated CO_2 amount attacks lime also. The lime-aggressive CO_2 is a part of the excess CO_2, namely that which is able to dissolve calcium carbonate as hydrogen carbonate and maintain it in solution. This solubilisation process yields more calcium hydrogen carbonate which, in its turn, requires more associated free CO_2 to stay in solution. In the conception of attack on lime, this associated carbon dioxide is to be subtracted from the excess CO_2, that hindering formation of the anti-rust layer. The remainder is the lime-aggressive CO_2.

The lime-aggressive carbon dioxide is thus not identical with that which prevents formation of the anti-rust layer.

Lead-aggressive carbon dioxide

The processes involved in the attack of water on lead are different again. This depends not only on the carbon dioxide but, as already stated above, primarily on the oxygen content (see also p. 97).
Lead is attacked only when oxygen is present and this is so, even when the water contains a great deal of aggressive carbon dioxide.
One must always reckon with attack on lead by very soft waters, even when there are only traces of free CO_2 present, as, especially, with rain water. Corrosion by hard waters is to

be feared, however, only if the free CO_2 amounts to more than 20% of the combined–provided, as just said, that oxygen is there. Should this proportion be less than 20%, practically no uptake of lead occurs in hard waters (*K. Höll,* Ges.-Ing. *58,* 323–324 [1935]).
Solution of lead in the water is retarded or completely inhibited if a high content of silica is present (over 10 mg SiO_2/l). (*J. C. Threshe* and *J. F. Beale,* The examination of waters and water supplies, London [1925]).
Phosphate in the waters also counteracts solution of lead (*B. Haupt,* Vom Wasser *14,* [1939]). Hence three factors may check the action of lead-attacking carbon dioxide: shortage of oxygen, silicates and phosphates.

Summary
Carbon dioxide

Carbon dioxide in water

Free carbon dioxide
- Associated CO_2 (harmless to pipes, concrete)
- Excess CO_2 (aggressive)
  - Total excess: Prevents formation of anti-rust layer
  - Partial excess: Lime-aggressive: attacks concrete

Combined carbon dioxide
- Fully combined CO_2 e.g., $CaCO_3$ in carbonates
- Half combined CO_2 e.g. $Ca(HCO_3)_2$ in hydrogen carbonates
- (both:) Not aggressive

6. Deacidification of the Water

Various procedures are used to remove carbon dioxide from the water. Considerable specialist knowledge is necessary to ascertain the correct deacidification process, which depends on whether the water is soft or hard, or contains much or little CO_2, or much or little iron.

a) *Mechanical Deacidification*

For hard to medium hard waters (more than 6–8° dH)

a) The water is almost in the carbonate equilibrium state and lacks only oxygen. Closed aeration is then chosen, since no loss of carbon dioxide must take place.
b) The water contains excess carbon dioxide and requires an increase in oxygen concentration. Open aeration is best here.

Summary of aeration processes

1. Hard water with little excess carbon dioxide: Trickling, aeration in cascades
2. Soft water with only slight excess carbon dioxide: Spraying or chemical deacidification
3. Quite soft water with marked excess carbon dioxide: Atomising, perhaps, doubly, and chemical deacidification

Trickling

Trickling over cascades or aeration on hurdles is a method for hard waters containing small amounts of aggressive CO_2 and at least 5 mg/l of associated carbon dioxide. In the first process, the water is allowed to trickle over stone steps; in the second, over piles of twigs or wooden grids. Plastic grids have been used recently. During the trickling process, the water becomes charged with oxygen and loses most of its excess CO_2; iron is deposited as ferric hydroxide at the same time. A certain residual carbon dioxide remains which is the reason why this procedure is utterly unsuitable for soft waters. It is good for hard waters for this very reason – that the associated CO_2 is not removed. Incrustations would otherwise be formed in the pipes, especially if the trickling plants have not the correct dimensions.
These types of aeration plant are scarcely built any more. This is because of the danger of contamination of the water through sludge accumulation. They are used only as a pre-treatment in removal of iron.
Installation costs are medium; working costs, very low.

Spraying

Both hard and soft waters of higher CO_2 content can be treated by spraying through sieves, shower roses or perforated sheets. Spraying from 2 to 3 metres height is naturally more effective than trickling. The effect can be increased further by using so-called baffle plates. One must ensure that the necessary amount of associated carbon dioxide remains in the water to prevent separation of calcium.
Installation costs are high; working costs, medium.

Atomising

Atomising through jets is still more effective than spraying. Jets of various types are used, including the Amsterdam jets, Schlick or Körting spray nozzles and, nowadays, high-power centrifugal jets of, e.g. the firm of Lechler, Stuttgart. The water is atomised from below upwards in closed chambers, the air being introduced laterally through shutters. At the time of construction, the jets are regulated to suit a definite pumping performance; if this performance fluctuates according to water consumption, it is absolutely essential to switch on or off an appropriate number of jets; otherwise, the rise and hence the effect is no longer attained and there is too much excess carbon dioxide in the resulting water.
Atomisation can be considered for waters containing much aggressive carbon dioxide and iron. Double atomisation may have to be performed if the carbon dioxide content is very high; this is not appreciably dearer. The very fine division of the water droplets from atomisation is reduced as the content of organic matter in the water rises; it is thus a less suitable method for water containing a lot of organic matter.
Installation costs are high; working costs, rather high.
J. Grim developed a streaming apparatus for deacidification of water, in which large amounts of air are mixed into the water, like in water pumps for reducing pressure. It makes use of the water pressure of the entering water and is thus economical (lectures at Ludwigsburg, W. Germany, 1968).

Intensive aeration

A new procedure for intensive aeration has been introduced in water works practice during

the last few years, for waters which are difficult to condition (e.g., soft water containing much CO_2 as well as H_2S). It comprises atomisation and introduction of air from below from the base of the filter; the air is withdrawn with an exhaustor (see also p. 194).
Installation costs are high, but less than above; working costs, rather high.

Vacuum sprinkling and degassing
For waters containing very much carbon dioxide and too much oxygen, vacuum sprinkling is to be considered. Since most of the oxygen is removed, the procedure is not so good for central water supply but it has been used with conspicuous success for processing boiler feed water. Degassing under reduced pressure at 101–105° C is a new, still better process for this.
Installation and working costs very high.

Assessment of the aeration processes
There is a danger in the aeration procedures that carbon dioxide removal will go too far; calcium deposition then occurs if the associated CO_2 is lacking. There is also a danger of inadequate removal, so that aggressive properties remains.
Disadvantages of the aeration processes: The processes always leave 6 to 8 mg/l of free CO_2 in the water. This is aggressive CO_2 in conditioned pure soft waters and has to be removed by chemical means.
Aeration procedures are hence suitable only for waters containing 5 to 18 mg/l associated CO_2 and a carbonate hardness ranging from 7° (minimum) to 12° (maximum) dH. Chemical supervision is necessary, at least at greater intervals.

Advantages of the aeration processes: The advantage of aeration is that it does not need too much attendance and that an often highly desirable enrichment of oxygen is achieved at the same time as the deacidification. Deep waters, rich in carbon dioxide, often lack the oxygen required for forming the protective layer. The loss in pressure is thus tolerated in the water works. *L. W. Haase* states that shortage of oxygen is much worse than excess oxygen for the water pipe system.
Aeration in closed systems must not yield a milky turbidity in the purified water (p. 194).

Soft waters, rich in carbon dioxide

Soft waters, i.e. those of about 2 to 3° dH, which contain much free CO_2 and for which any unnecessary increase in hardness is to be avoided, are best treated first by an aeration process (e.g. atomisation) and then deacidified using one of the chemical methods given below.

b) Chemical Deacidification

Soft waters, containing no iron and a small to medium amount of aggressive carbon dioxide

Marble filtering: The water is passed through filter vessels or open basins which contain fine-grained marble. This yields a harder product, corresponding to the amount of lime-aggressive carbon dioxide. The increase in carbonate hardness is twice that occurring in the other processes given below, with calcium hydroxide and magnesium oxide; this is often desirable with very soft waters however.

The marble filters must be cleaned by backwashing from time to time.

The marble process has proved satisfactory only with waters of hardness less than 6–7° but more than 2° dH.

The rate of filtration should not exceed 1 to 2 m/h. The water should spend at least 40 minutes in the filter. The dimensions of the filter are calculated according to this requirement.
10 mg/l of free CO_2 to be removed yields an increase in hardness of 1.27° dH. 1 g of CO_2 combines with 2.3 g of marble.

The marble process is totally unsuitable for waters which contain iron. Iron compounds are deposited on the marble grains and render them ineffective. The raw water should best contain no trace of iron. The same applies to manganese-containing waters.

The filter performance is reduced by turbidities and by colloidal humic substances in the water. The deacidification effect is reduced by very low temperature. Thus, the action of the marble filters at 6° C is only one third of that at 20° C.
Deacidification through marble filtration has the advantage that it requires little attendance and no dosage equipment.
Installation costs are medium; working costs are small; considerable space is needed.

Soft waters, with or without iron, containing medium amounts of carbon dioxide

Deacidification with magnesite according to Tillmans: Before the second world war, filter material of ignited, granulated magnesite was used in Germany with success for deacidifying water; Germany had its own magnesite deposits at that time. They are much more economical than marble filters and filter material of ignited dolomite. The upper limit of hardness is not so low as for the marble filtering but water of hardness exceeding 15° dH should not be subjected to magnesite filtration.
Magnesite filtration removes iron at the same time. Any changes in carbon dioxide and iron contents play a smaller part in magnesite filtration; this facilitates the work of the water works organisation. An iron concentration of 3 mg/l should nevertheless not be exceeded. Although the magnesite filters are about 5 times as effective as marble filters, the increase in hardness in only one-half. 10 mg CO_2/l corresponds to a rise in hardness of 0.65° dH. The rate of filtration is 8 to 20 m/h.
Installation and working costs are medium; little space is required.

Deacidification with dolomite filter masses

Magno- or Akdolit-filtration: Filtration through dolomite filter masses is the ideal method of deacidifying water. Partially ignited, granulated dolomite – Magno and Akdolit – are used in open filter beds or closed filter vessels, with a layer thickness of 1–2 m. The speed of filtration is 5 to 10 m/h in open and 15 to 20 m/h in closed filters, depending on the hardness of the water. The plants require only a minimum of supervision and are not at all sensitive to interference. Backwashing must be carried out at certain times, usually once or twice per week, depending on the purity and iron content of the water. The backwashing must be regularly and carefully performed. The filter mass is filled up at definite intervals (about twice a year), depending on the composition of the water. These deacidification procedures are second to none as far as the small demands of attendance and control are concerned.

J. Holluta and *U. H. Armbruster* (Vom Wasser, *22*) have given precise data about the sizes of dolomite filters and the rates of filtration; these data have been contested by those practically concerned (*F. Brummel,* Ges. Ing. *83,* 1–5 [1962]).

By means of special equipment, the Magno-works has greatly reduced the filter capacity necessary for deacidification (German Patent, accepted 1965).

The filtration material is obtained by calcining dolomite, a naturally occurring double salt of calcium and magnesium carbonates, at ca 600–700° C. At this temperature, the magnesium carbonate is converted to magnesium oxide whereas the calcium carbonate remains practically unchanged. The calcined product is used in granular form for the filter filling. It is supplied in several grain sizes. The granulation should be as uniform as possible on account of the backwashing. The smaller the granules, the larger is the effective surface of the filter. The filter dimensions can be made smaller if fine granules are employed. A larger grain size is advisable for harder waters, however ("Akdolit-split" and "Akdolit-gran").

The ignited and granulated dolomite filter masses, the Magno- and Akdolit masses, contain as active components, from 22 to 27% magnesium oxide and 60 to 70% calcium carbonate. According to *J. Holluta* and *H. Armbruster* (Vom Wasser *22,* [1955]), it is principally the calcium carbonate which takes part in deacidification of soft waters; and principally magnesium oxide with harder waters. The author has obtained many analytical results showing that it is overwhelmingly increase in magnesium hardness which accounts for rise in hardness of all waters during filtration. This can fluctuate a little, depending on the running time of the filter filling.

Magno- and Akdolit filters have proved especially useful with soft waters of 2 to 8° carbonate hardness. The dolomite filter masses are serviceable also with waters of medium hardness (carbonate hardness up to 12°). *J. Holluta* and *H. Armbruster* state that they can be used with carbonate hardnesses exceeding this value of 12° dH with a special filter disposition (Vom Wasser, *22*). The dolomite filters are ideal for quite soft waters when they contain at least 10 mg/l of free carbon dioxide. Reaction of this free CO_2 leads to a desirable increase in carbonate hardness which then helps formation of a protective layer in the pipes. If the quite soft waters contain only a few mg/l of free CO_2, the ultimate carbonate hardness of the filtered water is often insufficient for formation of this protective layer. This frequently occurs in the deacidification of reservoir waters. Addition of phosphate is then advisable (p. 172).

The dolomite process cannot be used if there is much sulphate hardness present in the water (over 150 mg SO_4/l). The new filter materials "Magno-Neutra" and "Akdolit Special" are then suitable. These are, respectively, a synthetic, granular, magnesium hydroxide preparation and an ignited, specially treated magnesite.

If clarification of the water is necessary at the same time, it is best to use the aluminium sulphate process in combination with addition of soda. When the Magno- or Akdolit filtration process is subsequently used, reaction of the CO_2 liberated from the sodium carbonate yields a small increase in hardness.

For economic reasons, an aeration procedure can be applied beforehand to waters containing very much carbon dioxide (p. 160). With soft waters which have a high content of CO_2, the aeration process is always followed by a dolomite filtration; not all the excess CO_2 is removed entirely in the aeration and the treated water would remain aggressive. Small

amounts of iron have no influence on the filtration through dolomite filter masses. The iron flocculates out in the upper parts of the Magno- and Akdolit filters. Two working procedures are thus united in the dolomite filtration of waters containing iron and aggressive carbon dioxide. The iron content of the water to be purified should not be too high, though; the upper limit is about 2 to 3 mg Fe/l. The dolomite process is thus far superior to the marble process for waters containing iron. This applies to both the rate of filtering and the increase in hardness. To react with 1 g of CO_2, 1.3 to 1.4 g of dolomite filter mass is needed (including loss through washing) and 2.3 g of marble. The increases in hardness per 10 mg/l of CO_2 which reacts, are 1.0° (dolomite) and 1.27° dH (marble).
The author has observed that filtration through dolomite is often accompanied by a slight increase or decrease in the silica content of the water. This depends evidently on the chemical composition of the filter mass and partly on the alkalinity of the raw water introduced.
Magno-Syn is a synthetic dolomite filter mass. This has already shown its practical value where used for deacidification of only slightly contaminated waters of high CO_2 content. Although the Magno-Syn filter mass comes rather dearer, this is compensated by the smaller consumption of filter mass and smaller initial costs.
When several dolomite filter masses are used in parallel filters, care must be taken to load all the filters equally. The water becomes too alkaline from the filters through which the flow is less than that calculated. Apart from this undesirable rise in pH, the lower parts of the filter mass tend to cake. Particular value must thus be laid on the correct sizes of the deacidification filters when they are planned with dolomite masses.
Better results can be obtained with a coarser grained filter when the pH of the water is too high. Since the purified water is always rather strongly alkaline after each new filter filling is put in, it is best to work with coarser dolomite first.
Filtration through dolomite filter masses is especially advantageous for smaller towns and communities, where the water works are far away, if only because of the smaller attendance requirements.
Decarbolith process: Decarbolith is being used in the DDR at the present time, as a substitute for the above alkaline filter masses. It is an ignited dolomite from the Gera vicinity.

Dolomite filters can be used as open or closed filters. The latter have the advantage that the loss of pressure is not entire and double pumping is avoided. They are also better from a hygienic standpoint. As in the marble procedure, fluctuations of rate of filtration should be avoided.

Hard and soft aggressive waters, with and without iron

Calcium hydroxide process: The older calcium hydroxide process also can be successfully used for hard and soft waters containing much free CO_2 and iron. It consists of dosage with lime water (with ca. 1.3 g CaO/l) or with milk of lime (a suspension of calcium oxide with water).
In the original calcium hydroxide process, the *Bücher* process, powdered calcium hydroxide used to be added to soften the water. It is scarcely used now, since such a plant demands very careful surveillance. If the operating personnel do not fulfil this, there is a danger that calcium will separate in the pipes. The firm of Wallace and Tiermann, Günzburg, W. Germany, supplies plants for dosage with dry calcium hydroxide.
Lime water and milk of lime also must be added with great care or else here, too, pipe incrus-

tations will occur. Accurate, continuous control of the pH of the purified water is thus necessary. The pH control cannot, however, be carried out immediately after the apparatus for dosage by calcium hydroxide since the neutralisation does not occur instantaneously; a reaction tank is interposed. pH values of 9.0 and more are found straight after the lime has been added and the desired value, 8.0 to 8.5, is attained only after several hours. The end value should not exceed pH = 8.0 for hard waters and should be 8.3 to 8.5 for soft waters. More accurately expressed, it should be adjusted to the carbonate equilibrium. A too high alkalinity causes the water to take up lead from lead pipes (p. 154).
The carbonate hardness increases by 0.064° dH per mg of free CO_2 which reacts.
Calcium dosage is suitable only for larger water works with appropriately trained personnel.
Installation costs are fairly high; working costs, medium; considerable space is required.

Soft waters containing small or medium amounts of aggressive carbon dioxide

For such waters, e.g. in reservoir water supply, dosage with sodium carbonate or sodium hydroxide would be considered in some cases. Very accurate supervision is needed of the dosage plants (continuous measurements of pH) and of the nature of the raw water to be treated (fluctuations of the CO_2 content). Formation of a protective layer is, however, not achieved. See *Wette,* GWF *85,* 551 [1942], for information on deacidification with soda. *W. Hopf* reported the deacidification of bank-filtered Rhine water (GWF *109,* 193–197 [1968]).
Installation costs are small: working costs, high; little space is required.

Continuous control of the water deacidification

The conditions for establishing the carbonate equilibrium must be permanently maintained to preserve the pipe system. If not, the formation of the protective layer is repeatedly upset. Even waters which yield calcium deposits do not always provide adequate formation of an anti-rust layer, since the forced deposits are highly porous.
It is therefore best to control permanently the quality of the purified water. Two equipments have become known for this during recent years. One apparatus depends on estimating the capacity of a water to dissolve marble, using differential conductivity measurements (*Häselbarth,* GWF *105,* 1937 [1964]) (firm of Dr. Thiedig & Co., Berlin); the other measures pH changes on filtering through calcium carbonate (*G. Axt,* Vom Wasser, *31,* 311 [1965]) (firm of Kratzenstein & Henke, Wuppertal).

c) Phosphate Dosage for Protection of the Pipes against Corrosion

Addition of polyphosphates is preferable to treatment in deacidification plants for borderline cases of aggressive CO_2, i.e., small (2 to 6 mg/l) amounts of excess CO_2, for waters of medium or greater hardness and also for some which are soft (see below). Phosphate dosage is much more economical.
Polyphosphates and condensed phosphates possess the property as cathodic inhibitors of preventing attack on the pipe material by aggressive water. This is effected by the formation of a thin protective layer of ($FeCa(PO_4)_2$) on the inner walls of the pipes. As anodic inhibitors they prevent also the formation of calcite crystals in the water and hence of scale in pipes and tanks (*Shyamala, Venn* and *Subramanyan,* Corrosion *11,* [1964]).

The visible success of polyphosphate dosage is first that there are no more brown colorations and deposits of iron in the terminal pipe sections. The oxygen content in these end parts of the system does not fall any more.
The polyphosphate treatment has been in use for a long time in the USA, termed the "threshold treatment" (US patent Rosenstein process). In Germany it has been introduced since the end of the war on a considerable scale thanks to the pioneer work of *F. Schönaich* (Die Wärme *63* [1940] and Gas, Wasser, Wärme *12* [1957]).
Protection against corrosion is accomplished also in galvanised (zinc) and lead pipes.
According to the newest information, treatment with zinc-containing polyphosphate solutions is especially good for protection against corrosion in service and cooling water plants and in air-conditioning plants. This composition of the solution for dosage prevents or retards the change of polyphosphate to orthophosphate due to the unavoidable warming. This is especially noticeable economically in circulation of cooling water.

Chiefly sodium tripolyphosphate, which has an alkaline reaction, is used to prevent corrosion by the above-mentioned waters.

Disodium phosphate is used at first if one wants to bring about rapid formation of a protective layer in newly laid pipes; this applies particularly to galvanised pipes and lead mains.
It must be emphasised, however, that a *protective layer can be formed only when the tap water contains at least 6.0 mg* O_2/l, just as with any other protective layer formation also. Otherwise, corrosion is prevented but without formation of a protective layer. Water of low oxygen content must consequently be aerated beforehand.
Phosphate dosage is carried out by introducing 1% polyphosphate solution directly into the pressure main (Venturi principle) or, in larger plants, by using proportioning pumps. "Phosphate sluices" are practical and economical for very small plants.
At the beginning, 5–7 g amounts of P_2O_5 are added per cubic meter (250 gallons), later about 4–5 g and finally, i.e., after some months, 2–3 g. More phosphate is needed if the speed of flow in the pipes is low; and less phosphate, if it is high.
The amount of phosphate falls in the pipe system, especially at first, when the protective layer is being formed. It should be ensured by chemical tests that the pipe extremities really do contain at least 0.5–1.0 mg P_2O_5/l. Formation of a protective layer is impeded if too much polyphosphate is added, since it combines with the calcium complex.

Disavantages of the phosphate dosage

The polyphosphate addition acts unfavourably on water via waste water unless the clarification procedure of this waste water includes the "third purification stage" (see p. 227). Phosphate dosage is thus permitted in Switzerland only in processing plants for warm water (Circular of the EGA, 1966).
According to more recent investigations, water rich in phosphate gives rise to bacterial multiplication in the pipes; this causes loss of oxygen and imparts an odour to the water.
At the moment, the cost of the chemicals comes to about 0.5 pf/cubic metre (ca. $^1/_2$ U.S. cent or less than $^1/_4$ British new penny per 1000 gallons).
Holluta states that phosphate dosage must be performed before any necessary chlorination.

It has been shown in practice that rust lumps and layers in the pipe system are gradually dissolved by the phosphate-treated water and that any obstructed parts are freed.

The polyphosphate solutions must not be made up at the water works by warming or heating, or else the polyphosphates lose their effectiveness.

Metaphosphate is unsuitable for formation of the protective layer (yet for complexformation of hardness, see p. 187).

d) Protection against Corrosion by means of Treatment with Silicates

Experience has shown that a high content of silica inhibits attack on the pipes. For 25 years in the USA, silicates have therefore been added to many waters which contain little. Waterglass solutions are usually employed. The New York water works also does this. Amounts of 8 to 30 mg SiO_2/l are added.

Plants are constructed in W. Germany by the firm of Henkel & Cie., Düsseldorf. See *Böttcher* (GWF *93,* [1953]) for the *Ferrosil process.*

The silicate protective layer is formed relatively fast but adheres less well than that from the polyphosphate, as shown in practice. The silicate protective layer is said to suffer damage at a high rate of flow.

The Silicophos process (firm of Giulini Brothers, Ludwigshafen, W. Germany) produces a better protective layer, for which the phosphate components are more effective (expert opinion of the Inst. f. Wasser-, Boden- und Lufthygiene of 10/8/54).

7. Iron and Manganese

Occurrence: Iron and manganese often occur together in water, both generally as hydrogen carbonates. Many ground waters of the N. German plain contain appreciable quantities of iron, namely, 1 to 3 mg/l. Amounts up to nearly 10 mg Fe/l are found also and, further, all transition intermediates to the mineral waters which contain 10 to 50 mg/l (e.g., Pyrmont, W. Germany, 20 mg Fe/l). This limiting value for "steel springs" is 10 mg Fe/l (p. 239).

Iron occurs organically combined as humate in moor waters. It is met as iron sulphate in the ground water from brown coal regions, occasionally along with free sulphuric acid.

Running waters usually have little iron (up to 0.3 mg/l) unless they come from moorland areas or are fed by wastewater tributaries.

The content of manganese is lower and rarely exceeds 1 to 2 mg Mn/l. Manganese shows up unpleasantly in piped water sometimes even in amounts of only a few tenths of a milligram per litre.

Iron and manganese in household water: Waters rich in iron are unsuitable for household purposes since they yield spots on washing (rust stains). Manganese yields dark brown stains which are very difficult to remove.

a) Iron and Manganese in Tap Water

Water for a municipal piped supply system should be practically free of iron and manganese. Many enterprises need iron-free water, e.g. dairies, breweries, laundries, dyeworks, paper mills and many others.

The presence of iron and manganese compounds in water, especially in hard water, leads to encrustations and sludge in the pipes and also deposits in water meters. The deposits may be so large that the inner pipe diameter is significantly reduced, especially when iron and manganese bacteria contribute.
There are namely, bacteria which, through slime secretion, are capable of thread-like growth together in iron-containing water. They include principally *Leptothrix ochracea, L. echinata, L. trichogenes, Gallionella ferruginea, G. minor, Crenothrix, Clonothrix fusca* and *Cl. polyspora* as iron and manganese bacteria; and *Bacillus manganicus, Lept. echinata* and *L. lopholea* as manganese bacteria (see p. 367 for further information). The growth of these bacteria is inhibited when the iron content is below 0.1 mg Fe/l. With higher amounts of iron, they develop in the pipe system and a spontaneous mass expansion of the iron bacteria can take place; this can have disagreeable consequences for the water works and the pipe system.
Manganese often plays a big part in reservoirs, especially in the hypolimnion; *H. Bernhardt* has frequently reported on this (see also *G. Clasen*, Städtehygiene 7, 1969).
Consequently, iron-containing water should be relieved of its iron before introduction into the pipe system. According to the "Hygienische Leitsätze für die Trinkwasserversorgung" (Hygienic principles for drinking water supply) [*14*], the iron content must not exceed 0.1 mg/l. The concentration of manganese may not exceed 0.05 mg Mn/l. Up to 1 mg Fe/l and 0.3 mg Mn/l are permissible with individual wells (hand pumps). 0.2 mg Fe/l and 0.1 mg Mn/l are tolerable with small water works. Dairies need water free of iron and manganese, especially for butter making.

b) Removal of Iron

Open iron removal: Iron can be removed from many waters by aeration. The water is allowed to trickle over wooden frames or, more recently, frames of plastic (formerly coke filters). The ferrous hydrogen carbonate of the water is oxidised by atmospheric oxygen to ferric hydroxide which separates.
If the carbon dioxide content is high, the rainfall *(Prof. Lehmann)*, atomisation or baffle plates process (p. 167) as pact plate procedures are employed (p. 167).
A settling tank for the separating ferric hydroxide and a gravel filter must be attached to the open iron removal plant, especially if there is a high content of iron ("Flocculation tanks").
The carbonate balance must not be displaced during the open aeration or else there may occur pipe incrustations and even caking of the filters through deposition of calcium carbonate on the filter grains, leading to failure of the filters. If this once occurs, the filter can be regenerated only by energetic backwashing at at least 25 m/h.
Closed iron removal: Contact of the water with the outside air is not hygienically desirable. Closed iron removal plants are therefore being built more and more, especially when no deacidification is necessary at the same time. Closed iron removal plants consist of a filter vessel in which the water is forced by compressed air through gravel layers of varying granular composition (1–2–3 mm). The iron separates in the upper filter layers. The filter can be intermittently rinsed by backwashing using compressed air.
In older plants, the complete filter sand had to be taken out every 2 or 3 years and washed (except with the Bollmann filters), the so-called *"filter washing"*.

Recently most firms have been providing filter bottoms where the filter material can be cleaned perfectly by a combined water-air rinsing; removal of the filters is then superfluous. Excess air must be removed with a self-operating de-aerator or else milky turbidities are observed by the consumer when he draws water; these are regarded askance. Closed iron removal plants must have a *running-in time:* enough ferric hydroxide must first separate on the filter to accelerate the iron removal as adsorbent.
Atmospheric oxygen must be supplied continuously in closed iron removal, namely 0,5 to 1 l of air per g Fe (theoretically 0.143 g oxygen; 1 m^3 of air = 300 g O_2. This is accomplished with the help of a compressor or through air intake (a sniffler). If the air supply is inadequate, pure water may be obtained which contains little or no oxygen; this takes up iron again in the pipe system. Equipment for controlling the oxygen content of the pure water is thus advisable in larger plants. The Chlorator-Gesellschaft, Grötzingen, W. Germany, for example, supplies such apparatus.

Iron removal in open and closed plants is easy at pH-values above 7 and medium carbonate hardness. With very hard waters, the equilibrium pH-value of which is ca. 6.9–7.0, and at low carbonate hardness it is difficult. The iron removal plant must precede the chlorination plant to prevent any possible interference; an exception is where HClO is used which assists iron removal.
The closed iron removal plant must be preceded by an open iron removal plant (see above) with reaction tanks for waters containing more than 4 mg Fe/l; or, in the case of waters of low CO_2-content, by a closed iron-removal plant with reaction tanks or double filtering with two gravel filters in series.
The rate of filtering should be about 15 m/h at the most.

The pure water from closed iron removal in sand and gravel filters has a slightly increased free CO_2-content; this comes from the iron hydrogen carbonate. If the iron amounts are high, the carbonate equilibrium of a water can thus be displaced easily. Some of the CO_2 is, however, taken up by the precipitated ferric hydroxide.
Closed iron removal is made more difficult if large amounts of chloride are present.
Iron removal by aeration is not possible with *waters containing humic substances,* which almost always contain iron; the filter grains become too smeared. Methods of chemical treatment (see below) must then be applied. Closed iron removal from waters containing sodium hydrogen carbonate is also more difficult.

Chemical iron removal: Iron can be taken out also by chemical means, during which the water likewise does not come into contact with the outside atmosphere. *Closed filters containing granular, ignited magnesite or dolomite* (Magno or Akdolit) are used; the water is passed through at about 15 m/h. The iron flocculates out in the uppermost filter layer and carbon dioxide is released at the same time (p. 170). The separated iron is taken off the filter at intervals by backwashing. These filters otherwise scarcely require attendance. The raw water should, however, not contain too much iron (not more than 3–5 mg/l). Reaction tanks must otherwise be introduced beforehand.

Iron removal from peaty waters: Iron cannot be removed from peaty waters since it is combined as humate. It is therefere important when planning iron removal plants to know whether organically combined iron will be present. This is recognisable from the high $KMnO_4$ consumption. In such cases, the organic matter must be precipitated with aluminium

sulphate or sodium aluminate, or removed by permanganate oxidation or super-chlorination according to the ADM-procedure (p. 64).
According to recent investigations, hypochlorite salts can be used successfully for precipitating organic matter and iron. The firm of Neue Continentale Bau-Ges., Hannover, W. Germany, undertakes this.
One can give no general rules for the most expedient and economical technique for iron removal. It is often most suitable to ascertain this through experiment. The Accelator – or Reactivator-procedures (p. 226) achieve complete iron removal in difficult cases in a very economical way.
There are no hygienic misgivings about high iron and manganese content. For individual water supplies a certain iron or manganese content does not thus have to be judged so strictly (see below), especially when the iron does not separate immediately. Iron becomes a nuisance when it separates from water which is standing, yielding a precipitate or turbidity. This applies not only to individual water supplies. Iron concentrations exceeding 2 to 3 mg/l influence taste markedly, imparting an astringent, almost ink-like taste. Water with much iron is unsuitable for preparing tea and coffee and also foods.

Complex formation of iron with polyphosphates: If small amounts of iron are present in drinking water (0.1 to 0.3 mg Fe/l), the brown colour or any precipitation in the pipe system or at the consumer's can be prevented by addition of tetrasodium pyrophosphate; this forms complex salts with iron and manganese.
The "dose" is four times the amount of iron, as phosphate (*Holluta*). This addition of phosphate is particularly advantageous with waters which require no other conditioning. It enables financially poorer communities to save installing expensive plants for iron removal.
Waters free of oxygen ought really to be aerated but installation of an aeration plant can be avoided if sodium tripolyphosphate is added as well as the tetrasodium pyrophosphate; this impedes further uptake of iron by the water in the pipe system.

Iron removal on the small scale: Iron removal is often necessary on the small scale also, for the reasons given above. For domestic purposes, this can be done most simply by aeration or boiling and leaving the water to stand in wooden or earthenware containers. Practically iron-free water can be siphoned off after any deposit has settled.

c) Removal of Manganese

In some cases manganese separates along with the iron during iron removal; e.g. in treatment with lime and frequently, too, when dolomite filter materials are used, such as Magno and Akdolit. In other cases, removal of manganese is rendered appreciably more difficult and not only when the manganese occurs, though rarely, as sulphate. In this last-mentioned instance, only the calcium hydroxide procedure is successful; otherwise "Mangankies", a filter sand containing manganese dioxide, accomplishes a good removal of manganese. Ordinary gravel filters require longer "running-in" before manganese removal proceeds to full satisfaction; MnO_2 has first to be deposited on the filter grains. This can, however, be accelerated by adding $KMnO_4$ to the raw water; it yields a precipitate of MnO_2 on the filter

aggregate. 1.91 mg $KMnO_4$/l are needed for 1 mg Mn/l. Removal of manganese is greatly facilitated by intensive aeration. The Neue Continentale Bau-Ges. of Hannover-Linden, W. Germany, has special experience in this sphere.
Even when iron and manganese can be easily taken out by closed iron removal in a gravel filter, so that the iron separates in the upper and the manganese in the lower part of the filter, experience shows that it is nevertheless advisable to connect two filter units in series. The filters for iron removal are cleaned by backwashing at a much higher speed and intensity than the manganese filters, which require an especially slow and gentle rinsing flow.
For removal of manganese, the pH-value must be as high as possible, at all events higher than for iron removal, i.e., above 8.5. This value is always exceeded when the above chemical additions are made. The lime-addition plant cannot, however, be considered for removing manganese in central water supplies.
In many cases good service has been obtained from the ready prepared filter masses such as: Mangan-Permutit (firm of Permutit AG., Duisburg, W. Germany); Akdolit-Mangan (firm of Akdolit GmbH., Düsseldorf) and Magno-Entmanganungsmasse (firm of Magno-Werk, Duisburg, see *K. Schilling* Vom Wasser *28,* 185–202 [1961]).
Manganese can be removed from waters containing much organic matter, especially humins, only after precipitation with aluminium sulphate.
It is a general rule that iron must be removed from iron-containing water *before* manganese removal; it may have to be deacidified beforehand also.
Waters containing iron can be more easily freed from manganese by aeration than can iron-free waters.
High ammonia and also chloride contents interfere with manganese removal.
Acidic waters cannot be freed from manganese by aeration, not even the most energetic aeration. It is possible then only with the more powerful oxidising agents, such as chlorine and $KMnO_4$.
Manganese can be stabilised in water by treatment with phosphate; *Illig* uses 3–5 g of sodium hexametaphosphate (glassy phosphate) per cubic metre of water (J. Amer. Water Works Assoc. *52,* 867–874 [1960]).
For many years, manganese has been removed in Dresden with the help of sand filters, on the grains of which manganese bacteria had been cultivated to form a blanket layer. Filtration effects total removal of manganese. See also *H. Brantner* and *H. Börner,* Österr. Wass. Wirtsch. *18,* No. 11 [1966]).

8. Hardness of the Water

a) Concept of Hardness

The hardness of a water is dictated by its content of calcium and magnesium salts. Water containing much calcium and magnesium is termed hard; that containing little, soft. This is expressed numerically through the degree of hardness. One (German) degree of hardness corresponds to a concentration of 1 part of lime (CaO) in 100,000 parts of water. Unfortunately there is so far no international unit. In France, 1 degree of hardness is a concentration of 1 part of $CaCO_3$ in 100,000 parts of water.

1 German degree of hardness, d H°	= 10 mg CaO/l = 7.14 mg Ca/l = 17.9 mg $Ca(HCO_3)_2$/l
1 French degree of hardness, French H°	= 10 mg $CaCO_3$/l
1 English degree of hardness, Engl. H°	= 10 mg $CaCO_3$/0.7 l
1 Engl. degree of hardness	= 0.8° dH
1° dH	= 1.25° Engl. H°
1 German degree of hardness	= 1.79 French degrees of hardness
1 French degree of hardness	= 0.56 German degree of hardness
1 French degree of hardness	= 0.7 English degree of hardness
1 American degree of hardness	= 1 mg $CaCO_3$/l
1 American degree of hardness	= 0.056° dH

Degree of hardness should now be given in mval. 1 mval means that 1 litre contains 1 milliequivalent of the material concerned.

Degrees (German) of hardness are converted to mval by dividing by 2.8.

Total hardness

In natural waters, calcium and magnesium are largely combined with carbon dioxide, namely as hydrogen carbonate. Usually a small amount is combined also as sulphate, chloride, nitrate, silicate and phosphate. Waters are found in some gypsum regions, in which the calcium occurs only as calcium sulphate.

The sum total of all these calcium and magnesium compounds yields the total hardness, expressed as mg CaO/l, 1° of total hardness = 10 mg CaO/l or 7.14 mg MgO/l. To calculate total hardness, the weights of magnesium compounds as MgO are converted to the equivalent amounts of CaO by multiplication with 1.3999 and the total is divided by 10. Since 10 mg CaO are equivalent to 7.14 mg MgO, 1 degree of magnesium hardness = 7.14 mg MgO/l.

Carbonate hardness

When water is boiled for a longer period, the calcium and magnesium combined with carbon dioxide are almost entirely precipitated as insoluble carbonates. One refers thus to a temporary or transient hardness, now generally spoken of as carbonate hardness. The carbonate hardness is thus due to the hydrogen carbonates of the calcium and magnesium. This is expressed as the amount of CaO equivalent to the total alkaline earth bicarbonates. Carbonate hardness is usually less than total hardness. If it is greater, one speaks of "apparent carbonate hardness"; the alkali bicarbonates are calculated from the difference between total and carbonate hardnesses. A better term is "exchange hardness", since it is due to natural ion exchange in the soil. A good example is given by the well waters in the "Seewinkel" on the Neusiedler See (Austria), in the experience of the author.

The expression "carbonate hardness" is not really quite correct, since the alkaline earths, as already stated, exist in the water as hydrogen carbonates (with very few exceptions). But the concept of carbonate hardness is so firmly established that it should be retained. This has been done in the new standard procedures of the German authorities for water investigation.

Non-carbonate hardness

The fraction of the calcium and magnesium remaining in solution as sulphate, chloride and nitrate (silicate and phosphate) after boiling constitutes the residual hardness, formerly also referred to as permanent or mineral acid hardness. This is now spoken of more accurately as non-carbonate hardness (not to be confused with the "residual hardness", p. 184 and 231).

One must thus distinguish:

Total hardness: total of alkaline earths combined with carbonic, sulphuric, hydrochloric, nitric and phosphoric acids, and expressed as mg CaO/l (1° dH = 10 mg CaO/l).

Carbonate hardness: (also carbonic acid, temporary or transitory hardnesses) = that part of the calcium and magnesium which is combined with carbon dioxide; expressed as mg CaO/l.

Non-carbonate hardness: (formerly permanent, mineral acid or sulphate hardnesses) = that part of the calcium and magnesium which is combined with sulphuric, hydrochloric, nitric or phosphoric acids, likewise expressed as mg CaO/l.

Subdivision of the waters according to degrees of hardness
The approximate terms used are in the table:

Degree of hardness	Term
0– 4° dH	very soft
4– 8° dH	soft
8–12° dH	medium hard
12–18° dH	fairly hard
18–30° dH	hard
30° dH and over	very hard

b) Occurrence of the Hardness-constituents

Waters from limestone, gypsum and dolomite regions are hard. Ground waters with total hardness of 30° dH and more are often encountered there. In rare cases, waters with 100 and higher degrees of hardness are found (water works Artern up to 140° dH; Schwäbisch Hall, occasionally up to 105° dH; and in the area of Bad Nenndorf). Very hard waters come also from the Keuper gypsums, zechstein and Jura formations, e.g., the Münder marl. Waters from primary rock regions, basalt and other volcanic rocks usually have only 1 to 2° dH (Alpine regions, Black Forest, Thüringen, Norway). A low degree of hardness (ca. 10° dH) is found also for waters from new red sandstone regions, e.g., in the GDR. Higher non-carbonate hardness is found in water from upper red sandstone regions.

c) Hardening of Ground Water through Refuse Dumps

Decomposition of the vegetable part of refuse, especially summer refuse, liberates large amounts of carbon dioxide. If refuse tips are extracted by seeping rainwater, this CO_2 enters the ground water. It brings about uptake of calcium from calcium-containing soils and consequently an increase in hardness. Fluctuations in the carbonate hardness and calcium content of a ground water can therefore help confirm suspicions that the ground water has been affected by refuse dumps. As the author has said in various places, the uptake of arsenic from the ashes in the refuse is of great significance (see also p. 191).

Refuse dumps can also harden the ground water directly (*E. Lang*, Z. für Städtehygiene [1932]; *G. Haupt*, GWF [1935]; *W. Langer*, Abh. a. d. Arbeitsgeb. d. Inst. f. Wa-, Bo-, Lu-Hygiene No. 19, 113–130 [1961]). *G. Haupt* reports, for instance, an increase in hardness of a ground water by 1.6° to 20.8° dH as a result of a rubbish dump 700 yards away. According to *E. Lang*, increased hardness of ground water can be traced to colliery and slag heaps.

B. Rössler reported considerable contamination of ground water through refuse tips in the Rhineland (Vom Wasser *18*, 1950). A detailed account of the influence on ground water of refuse tips has been given by *E. Nöring, A. Golver* and *G. Matthess* in the report of the Hydrological Congress IAH, 1965.

Increase in sulphate of ground water often occurs further away from the deposition of refuse, i.e., in the aerobic zone in which the H_2S, yielded by decay, is oxidised to H_2SO_4.

Ammonia, nitrite, nitrate and sulphate likewise occasionally indicate the influence of rubbish dumps.

d) Sanitary Significance of Water Hardness

Even though high water hardness has no direct deleterious effect on health, very hard water is all the same not good as drinking water on account of its laxative action. Water of over 25° dH is not acceptable for drinking purposes. On the other hand, a certain degree of hardness is favourable for tooth and bone formation. The inhabitants of regions of very soft water suffer from bad teeth, e.g., in the whole of Norway. Dental decay due to soft drinking water is frequent in the Vosges also (*F. Bentz*, GWF [1942]). Similarly decayed teeth are encountered everywhere in the Azores, Iceland and the Canary Islands (*K. Höll*).

A. I. Bokina has demonstrated that high water hardness favours gallstone maladies and that radioactive calcium, ^{45}Ca, is directly concerned with the stone formation. He gives 10–20° as the upper limit for hardness (Hygiene and Health (Russian) *30*, 3–7 [1965]).

Medium hardness is thus the most suitable, from both health and technical (service water) reasons.

According to investigations in the USA (*Schroeder*), Japan (*Kobayashi*), England (*Morris*) and Sweden (see Bundesgesundheitsbl. *8*, 351 [1965]) (Cause of death 420 and 422) there is a certain dependance of various cardiovascular illnesses on the calcium content of the water, based on the mortality statistics. A calcium content of from 30 to 80 mg CaO/l is the best for human health, according to these findings. Statistus show that drinking waters of very low, especially zero, calcium content, such as occur in the Canary Islands for example, yield higher mortality figures from heart and vascular diseases. The higher the calcium content of a drinking water, the fewer the cases of heart disease encountered in the population.

Water hardness and drinking water conduits

A definite hardness is necessary to transport drinking water in pipes. Soft waters do not lead to formation of a protective layer and, through the ever-present aggessive CO_2, attack the material (p. 163). A higher degree of hardness of the tap water is necessary, especially with lead pipes; the lead entering solution would then lead to lead poisoning.

Waters containing less than 30 mg/l of calcium hydrogen carbonate form no protective layer and must therefore be hardened (p. 172). Extensive experience is needed for this, however (see *H. Wette,* GWF *90,* 121–126 and 154–158 [1949]; GWF *91,* 49–55 [1950]).
On the other hand, considerable hardness is also technically unsuitable, especially when the water is just in the carbonate equilibrium or when there is even a deficit of free CO_2.
Apart from this, hard waters can lead to undesirable calcium deposits when air access causes increase in oxygen concentration (e.g. after collective tanks) (see *Klut,* Trink- u. Brauchwasser, p. 68) or the temperature changes (p. 164).
The associated and aggressive carbon dioxide must for all these reasons be determined with great care and this should be left only to those who are really experienced and qualified.

e) Significance of Water Hardness for Service Water and Technical Purposes

Just as "fur" or "scale" forms in domestic kettles from hard water, so deposits of boiler scale are formed in warm water supply plants and heating plants with water of higher degrees of hardness. The deposit has a low heat conductivity which often has a highly unfavourable influence on the technical process. In recent times, the polymeric phosphates and metaphosphates have provided a way and means of preventing this formation of deposit in the heating technique. The poly-metaphosphates have the property of combining with the hardness-provoking components of the water to give stable products and thereby to prevent their separation in the warmth (see also p. 187).
The "magnesium rod" has proved valuable in heating technique for impeding deposit of scale on the walls of boilers.
The poly- and metaphosphates have been of great service in laundering technique since they prevent formation of calcium soaps and permit formation of lather immediately, even with hard waters. They prevent at the same time the harmful deposition of calcium soap on the fabric (see below).
The hardness of the water is of great importance for numerous industrial purposes. Harder water is detrimental in the textile industry, in laundries, bleaching and dyeing works. Water of low hardness is necessary for paper and chemical pulp manufacture, for tanning and in the preparation of spirits and liqueurs. The quality of a beer is largely dependent on the degree of hardness of the water used for brewing (p. 234). Soft water is required for preparing canned vegetables.

Hardness of water for domestic purposes
The hardness of water is noticed also during its household uses. For example, peas do not become soft when cooked in hard water. The aroma of tea and coffee is impaired by the water hardness. The coffee distributors accommodate their coffee sorts to the water hardness in the consumer regions. Cocoa drinks are less palatable with hard than with soft water: they are also less "smooth" and are lumpy. The tannins from tea yield deposits or turbidities with the calcium salts of hard water.
Human skin becomes rough from washing in hard water, and the pores become blocked with the deposited calcium and magnesium soaps. This can cause inflammation of sensitive skins.

Hardness of water and laundries

Hard water is very disadvantageous in laundering. The calcium and magnesium salts of the water yield insoluble calcium and magnesium soaps with the fatty acids of the soap, so that a part of the soap employed is consumed. The harder the water, the more soap is used up to precipitate these hardness-producers, before foam is formed and the washing action begins. The hardness of the water thus plays an important part in large laundries; they soften their water when no soft water is directly available, since this is more economical than the increased soap consumption. 1 cubic metre of water (ca. 250 gallons) requires ca. 150 g (ca. 5 ounces) soap per degree of hardness; i.e., a water of 20° dH needs 3 kg (ca. 7 lbs) of soap per cubic metre.

Nowadays the hexatrens, e.g. hexatren C5 of the firm of Giulini, Ludwigshafen, W. Germany, are used in order to prevent interference in textile establishments through the softened water being taken up in the piped water system of the factory.

The increased soap consumption is noticeable also in domestic washing, apart from the fact that the fibres of the material are damaged in the course of time by the calcium and magnesium soaps which adhere firmly to the fabric. This reduces appreciably the length of life of the washing. It is thus customary to use rainwater for washing in regions of hard water, or to make use of water-softening agents.

Water hardness has become a less serious matter in *domestic washing* since the introduction of the synthetic washing agents (detergents).

f) "Fur" or "Scale" Deposit

When hard waters are warmed or boiled, calcium, magnesium and also iron and manganese salts separate out and form the so-called boiler scale. Pipes for warm water can thereby become gradually blocked. As mentioned above, this can be prevented by addition of polyphosphates. Boiler scale can be dissolved with acid solutions containing additives, such as certain resins. *Coatings of scale in boilers* cause appreciable heat losses: a thickness of 3 mm means a 20% increase in coal consumption.

Carbonate hardness yields a softer deposit than does non-carbonate hardness. Scale containing gypsum is especially unfavourable and dangerous in steam boilers (p. 231).

g) Softening of Service Waters

Lime dosage = decarbonisation of waters with chiefly carbonate hardness: The lime treatment procedure is the simplest and oldest method of softening, due originally to *C. Bücher* ("Bücher-process"), who employed dry lime. Nowadays, one uses milk of lime or calcium hydroxide solution to precipitate the hardness formers and also to deacidify the water. The amounts to be added must be accurately calculated stoichiometrically in each case and the plants must be continuously controlled, best with automatic pH-measuring apparatus (Balke-Riwag decarbonisation process).

This process enables water to be softened to 2 to 4 degress of hardness, depending on the magnitude of the non-carbonate hardness. To attain this effect with waters of high magnesium bicarbonate hardness, lime in excess (ca 25 mg/l) must be added. The excess must be removed before the water enters the supply system, e.g. by passing CO_2 into the softened water.

Decarbonisation with calcium hydroxide is not easy if much chloride is present, however, since the solubility of the calcium carbonate, produced through addition of lime, is increased.

The Wirbos process of the Permutit AG firm uses special contact material (calcite crystals) and a vortex to separate the calcium carbonate quickly. Other rapid reactor processes function in this way, e.g. the "Circulan process" of the firm of H. Koppers of Essen.

These procedures are used at higher temperatures also (60° C and above) with considerable success, e.g., in boiler operation.

This contact process suffers interference and fails at very low temperatures (below 6° C), however.

Addition of acids: Decarbonisation can be carried out also by addition of acid to the water (process of the firm of Balke, Bochum, W. Germany). The process must naturally be superintended very closely and is thus applicable only in special cases. The pure water contains the CO_2 set free and is thus highly aggressive.

The lime-soda process: Lime and soda are added simultaneously to waters which have both high carbonate and non-carbonate hardness; the latter is then precipitated along with the former. It thus finds the most use, even for central water supplies in the USA. The softening extends down to 1 to 2° dH. The use of so-called pre-heaters enables the precipitation to be shortened markedly in time and to be more complete.

The cost of installation is rather high and the working costs are moderate.

The caustic soda process is less customary; it has proved useful with waters of low carbonate hardness, especially when the carbonate hardness equals the permanent hardness; and with waters of relatively high magnesium hardness.

Installation costs are less than with the lime-soda process but working costs are somewhat higher.

The soda-regenerative process is applied when hardness is chiefly non-carbonate, for boiler feed in conjunction with boiler water return feed ("Neckar" process of the firm of P. Müller, Stuttgart). The raw water must contain only little CO_2 and the boiler pressure must not exceed 20 atmospheres.

The trisodium phosphate process (Budenheim process): All hardness constituents are precipitated by addition of trisodium phosphate in this procedure. A flocculent sludge is formed which does not burn on, even in the boiler. Hardness can be reduced to 0.1° residual value.

The procedure is, however, too expensive in many cases, on account of the highly priced chemicals. It is therefore used often only as an extra process for removing residual hardness after the lime-soda process. Installation costs are low. The trisodium phosphate process has been introduced everywhere as a result of its success and safety in service. *The "Röhrenverfahren«*, of the firm of Budenheim may be specially mentioned, in which the water is heated, degassed and partly softened in pipes before reaching the reactor with trisodium phosphate; this saves much room and iron.

Trisodium phosphate is also added to completely softened boiler feed water and distillate from high-pressure boilers for reasons of safety.

Firms which carry out construction in W. Germany are the Chem. Fabrik Budenheim, Mainz-Budenheim and those concerned with water purification.

h) Ion Exchange Procedures

Softening of service water through cation exchange

Permutit and Lewatit processes: In contrast to the previously mentioned processes, the Permutit and Lewatit-resin processes do not depend on any addition to the water; the water is filtered through a granular exchange mass in a closed system. Any fluctuations in the water hardness are without significance in these processes. They are particularly good for soft waters which are to be prepared for boiler feed purposes. Since the pure waters suffer no loss of palatability, these processes in themselves would be suitable for central drinking water supplies; the filter materials are, however, expensive. Only at first is there an impaired taste.

The Federal German Health Office (Bundesgesundheitsamt) has published recommendations about ion exchange treatment of service water for food product factories (Bundesgesundheitsblatt, Vol. 6, 1963).

Waters containing organic matter and iron are not suited to treatment with basic ion exchangers, or special arrangements must be made.

Permutit process: In cation exchange, the water to be softened is passed through a filter filled with granular material. Permutits are sodium aluminium silicates, the sodium of which is exchanged for the calcium and magnesium of the water. From time to time, the Permutit filter is regenerated by passing 5–10% sodium chloride solution through it. Salt which has been denatured with Crystal Ponceau 6Rb is used (40–70 g NaCl per °dH/m^3). The hardness can be reduced to 0.02° dH. Some waters take up a lot of silica from the Permutit filter.

Neopermutit, a granular polystyrene resin which is employed nowadays, functions much more quickly than Permutit, both in softening and regeneration. Permutit RS, a sulphonated styrene resin, possesses an especially high exchange capacity and, at the same time, a particularly low salt consumption. The cost of the salt for regeneration is often not considered in planning softening plants; it can be quite large in larger plants, however.

Regeneration can be carried out with an acid instead of with common salt. This yields the H-form of the cation exchanger. This can be exchanged against the calcium and magnesium cations but gives an acidic filtrate.

The washing losses are less with the green-black Neopermutit than with Permutit, since it has a larger density and is therefore not so easily stirred up; and since the mechanical friction is less with the smooth spheres. No release of silica occurs as with Permutit.

The firm of Permutit AG, Berlin and Duisburg, and the relevant firms concerned with water preparation, undertake equipping work.

Lewatit- and Wofatit processes: Lewatits (Bayer, Leverkusen am Rhein) and Wofatits (Farbfabriken Wolfen near Bitterfeld) are granular ion exchanger masses, based on phenol and polystyrene resins. They yield a completely softened water, of practically zero degree of hardness. They are insensitive to high temperature and acids. Wofatit and Lewatit filters are regenerated like Permutit by using common salt solution. The Duolites are similar.

These basic ion exchangers are uninfluenced by fluctuating degrees of hardness of the water. Waters containing organic matter cause the filters to become less efficient exchangers. High iron content interferes with the filter function of many exchanger masses.

The present-day high pressure boiler establishments and atomic power stations would be inconceivable without ion exchangers.

Supervision of exchange filters: The volume of the exchanger mass is considerably greater in the moist state than when dry. Accordingly, the mass must not be filled in the dry state into the filter vessel. The filters must also not dry out, e.g. before regeneration, but must run continuously as submerged filters. The exchanger mass must be protected from contamination. The raw water has therefore to be completely clear and to yield no deposit. Further, it must be free of oil, which reduces the exchange capacity, and should contain no H_2S.
The loss of exchanger mass should not exceed 3% per year if its quality is good.

Regeneration
When the residual hardness after using the cationic exchange filter rises by more than 0.1°, regeneration should be performed. This is carried out with 5–10% solutions of common salt. The pure water is controlled with soap solutions (p. 113) or with Titan Yellow indicator (0.01% solution in NaOH), which must not yield a red coloration. The completeness of salt removal is checked by conductivity measurement, bearing in mind the effect of any CO_2 present which falsely implies higher salt content.
In the *regeneration of cation exchange filters,* the most concentrated salt solution is added to the filter at the end. Natural brines can be used at the beginning or a type of pre-regeneration carried out, even when they contain considerable amounts of hardness-producing agents. There are precise requirements in the USA about the nature of the NaCl used for regeneration (J. Amer. Water Works Ass. *42,* [1950]). Duty-free salt, denatured with Crystal Ponceau (see above) of suitable grading and purity, is used in Germany.
The artificial resin ion exchange filters require only half the salt for regeneration needed for the zeolites (Eng. News Rec. *44,* 42–44 [1950]) which anyhow have a much smaller exchange capacity and are consequently no longer used.
70 g common salt are necessary for regeneration per cubic metre of water flowing through and per degree of (total) hardness.
The filter function is markedly reduced at very low water temperatures.
After regeneration, the filter is washed with softened water until the emerging water is free of salt (see p. 41 for test).
Anionic exchangers are regenerated with 10% sodium carbonate or 4% sodium hydroxide solution. 1% NaOH is used in the stage for removing silica.
The filter containers must be well protected against acid. The eluents from basic ion exchangers are highly aggressive since all their free carbon dioxide is present as aggressive CO_2. All the pipes in the plant are thus endangered, except those made of acid-resistant steel or of plastics such as Vinidur or Mipolam etc. The containers for hydrogen ion exchange are thus rubber-proofed by the particular manufacturers, and the containers for other ion exchangers are provided with stoving varnish. The effluents can be neutralised by using Akdolit-special filters or Akdolit hydrate addition (Akdolit GmbH, Düsseldorf).

Complete demineralisation
Synthetic resin masses are available also for anion exchange. If cation- and anion exchange filters are connected in series, complete softening and demineralisation is accomplished. Certain exchangers also remove silica from the water, so that an ion- and silica-free water is obtained (aqua deionisata). The total removal of salts can be carried out also in the "mixed bed process". This consists of a mixture of granular cationic exchangers in the H-form and anionic exchangers in the OH-form, in the filter vessel. In a single working stage this provides water completely free of ions. The DVGW-circular No. 49, *K. Schilling,* contains further information about cationic exchanger processes. Mixed bed filters are more difficult to regenerate than separate ion exchanger filters.
Salt removal with synthetic resin exchangers is markedly cheaper than distillation for water of low salt content. With hard waters, it is best to carry out preliminary softening with calcium hydroxide before cationic exchange, or else the sodium content in the exchange will be too high.

Softening on the small scale: Permutit filters (Permutit AG, Duisburg) and Lewatit filters (Berke-Filter-Ges., Celle and Karl Klein and Sohn, Mannheim) are primarily those considered for softening service water on the small scale. Various water softening filters, e.g. the Puritfilter of the firm of Mayer KG of Wiesbaden, W. Germany, are used in the spirits industry.
Equipments for "magnetic water treatment", e.g., with the Cepi apparatus, are without effect. Unanimity was shown about this at the conference in Göttingen, W. Germany, on economics of water (see also *K. E. Oehler,* Vom Wasser, 58–76 [1967] and *G. Greiner,* ebenda p. 77–81).

j) Addition of Polyphosphates

In more recent times, the polymeric phosphates have furnished ways and means of preventing formation of scale in water heating technique. The polymeric ortho- and metaphosphates, especially sodium hexametaphosphate (glassy phosphate) and sodium tripolyphosphate, have the property of converting the hardness constituents of water into stable complexes and thereby of preventing their separation. Phosphate addition is, however, ineffective with waters containing only calcium sulphate.
Addition of condensed or polymeric phosphates can prevent excessive deposition of hardness-formers in the pipe system and separation of calcium salts during heating in boilers, heating coils and warm water units etc. These methods have proved very useful in Germany and they are, moreover, highly economical.
As mentioned in the discussion on corrosion prevention (p. 172), treatment with phosphate has been in use in America already for thirty years, originally for preventing scale formation. Polyphosphates have, namely, the property as cathodic inhibitors, of preventing attack on the pipe material by aggressive water; as anodic inhibitors, they prevent also the formation of calcite crystals in the water and consequently of a solid deposit.
Best for *preventing scale deposit* and excessive separation of calcium in the pipe system is *the complex sodium metaphosphate,* from which the weakly acid monosodium phosphate is formed by hydrolysis.

The phosphate dosage thus provides:
1. prevention of corrosion
2. avoidance of scale deposit and of pipe corrosion
3. stabilisation of iron and manganese in the water, prevention of iron deposition in the pipe system and of brown coloration of the tap water.
4. gradual breakdown of old pipe incrustations

"Petzoldt 600" possesses very high complex-forming capacity for the hardness-formers (firm of F. Petzold, Münich); it can be used also for clearing up encrusted systems.
The polyphosphates should not be dissolved in hot solution before addition or they lose their effectiveness as a result of depolymerisation.
The polymeric phosphates are added in amounts of 1 to 2 g/cubic metre of water; it is done with proportioning pumps or "phosphate sluices" (these are filter vessels built into the pipe system).

Preparation of warm water
It must be borne in mind in the preparation of warm water that stabilisation of the hardness at above 65° C is incomplete. At 65° C, the efficiency is still 95%; at 80° C, however, only 60%; and at still higher temperatures, phosphate addition has no effect.

Phosphate dosage on the small scale
The *"Microphos-process"* of the firm of Benkiser, Ludwigshafen, W. Germany, has gained popularity in small scale plants and has ousted the older "Chromium-process" in warm water supply. It comprises automatic addition of polyphosphate through a phosphate sluice (see above).
Phosphate treatment in small apparatus has led to markedly excessive dosage in some cases, an offence against the regulations governing content of foreign matter, Excessive addition leads to digestive disturbances *(H. Haak,* Der Sanitär-Install. *9,* 80–85 [1966]). Installations exist nowadays, however, to prevent over-addition (see *K. Marquardt,* Der Sanitär-Install. *10,* 310–315 [1967]).

Sanitary Significance
Although metaphosphates are substances which *precipitate* protein and are thus not without importance for the human organism, a sanitary commission in the USA has concluded after years of study that the metaphosphate amounts added to drinking water are unobjectionable. Up to 10 mg/l are claimed to have no influence on health (see also *P. Höfer,* Ges.-Ing. *77* [1956]). According to an expert opinion of the Institute for Water, Soil and Air Hygiene of the Federal German Health Office (IIIB-A 687 of 10/8/1954), there is no objection to adding up to 7 mg P_2O_5/l to drinking water in the form of ortho- and polyphosphates. According to the new drinking water *regulations* of 1959, up to 5 mg P_2O/l are permitted (p. 148).
Pharmacological feeding experiments have shown that food with 1% added hexametaphosphate is tolerated by animals without harm (*B. Behrens* and *K. Seelkopf,* Arch. f. exper. Path. u. Pharmakol. *169,* 241 [1956]).
Phosphate addition sometimes acts unfavourably in the pipe system as a result of marked multiplication of bacteria. The phosphate nutrient occasionally increases strongly a low

bacterial content of the tap water. Such water must thus be completely free of bacteria when phosphates are to be added, and should have a low permanganate consumption.
Addition of phosphate to piped water has now been forbidden in Switzerland, also because of the effect on surface waters which results from the phosphates gaining access to the lakes in wastewater and providing an initial nutrient for algae development.

Phosphate dosage of cooling and boiler feed water
Polyphosphates have proved useful technically also in the *preparation of cooling water.* The temperature in the cooling process must not exceed 75° C, however, or calcium separates on the cooling surfaces despite the phosphate addition. Hardness formers are especially important in *steam boiler plants.* Small degrees of hardness can be tolerated in *low pressure boilers* if they are complexed with trisodium phosphate and the calcium-containing sludge is removed.
Complete removal of hardness is necessary for the use of *high pressure boilers* (p. 230 and 232). This is for reasons of works technique and heat exchange (heat is lost through the scale which is otherwise formed). Only ion exchange processes come into consideration, therefore (p. 185 for further information).

9. Metals, etc.

a) Lead

Occurrence: Lead is found in rare cases in natural ground and surface waters which have contact with lead ore deposits.
Of sanitary importance are especially cases where aggressive waters dissolve lead from lead pipes. Soft waters, rich in CO_2–all waters in fact with a pH-value of under 7–tend to dissolve lead. The amounts of lead which are encountered range from 0.1 to 10 mg/l, generally lying between 1 and 3 mg/l. Waters which contain much oxygen also attack lead, whether they have excess carbon dioxide or not.
The uptake of lead takes place with most waters gradually during standing in the pipe; the lead content increases for a period of about 10 hours, then remaining constant or, more rarely, decreasing. Some highly aggressive waters can take up appreciable amounts of lead in one hour in the pipe and show decreased lead content after 10 hours.

Sanitary importance of the lead content: Continuous uptake of even very small amounts of lead can lead to chronic lead poisoning. Frequent control of soft waters in the central supply is thus absolutely necessary, especially on account of possible fluctuations in the carbon dioxide content. It has been mentioned already on p. 154 that soft alkaline waters can dissolve lead also.
The use of lead for piping water has been forbidden for a long time in the German Lands of Württemberg, Oldenburg and Hessen.
Traces of lead are taken up by every water and lead is for technical reasons virtually indispensable; a small lead content has thus been tolerated for a long time. According to the hygienic directives for drinking water supply [*14*], "perhaps still permissible is a content of 0.3 mg Pb/l after standing for 9 hours – roughly overnight – in the pipe".

The World Health Organisation originally proposed 0.1 mg Pb/l as the maximum value for lead-containing water *(K. Kruse, 21)*. The value of 0.05 mg Pb/l was suggested in 1963 in the International Standards for Drinking Water of the WHO, however.
It was mentioned above that some aggressive waters take up lead in a shorter period and that a decrease is to be expected after a longer time. According to *J. Harmand* (Nancy, France, 1941), autochthonous lead content should not exceed 0.1 mg Pb/l.
Further information about the large number of cases of poisoning in Leipzig in 1930 is to be found in "Die Bleivergiftung durch Leitungswasser" (Lead poisoning though piped water) by *Fuchss, Bruns* and *Haupt;* and in "Bleivergiftungen auf Helgoland" (lead poisoning in Heligoland) by *K. Höll,* Arch. f. Hygiene [1934]
It was mentioned already on p. 158 that lead is yielded even by plastic pipes which have been stabilised with lead. One should thus look for the DVGW-test mark when purchasing plastic pipes.
Lead can be dissolved from lead pipes by water as a result of stray electric currents in the ground; this leads to lead poisoning. The DVGW and the association of electricity works in W. Germany have worked out together directives for earthing of water conduits (GWF *96,* 332 [1955]). All earthing of or connecting with a source of current to water installations is forbidden in the USA (GWF *95,* 334 [1954]).
Appreciable amounts of lead get into the soil from the anti-knock material, tetraethyllead, in the exhaust from the intense automobile traffic today (2–3 g of Pb per 60 miles of road). Vegetables and grass take it up from the seepage (Literaturber. *14,* 135 [1966]). The cultivation of potatoes and vegetables is now forbidden in the USA in the vicinity of motorways with considerable traffic (see also *H. O. Leh,* "Verbraucherdienst" 1966, Issue B 11).

Removal of lead: According to experiments of the author, lead is best removed from water by shaking with active carbon and then filtering. Fine-grained, ignited magnesite is also very good. It renders the water free of lead immediately by filtration and goes a long way *(K. Höll,* Arch. f. Hygiene *113* [1935] and Deutsch. tierärztl. Wschr. *43,* 339 [1935]). Dolomite filters can be used also (firm of Akdolitwerke Hans Börner, Düsseldorf and Magno-Werk, Duisburg).

b) Copper

Occurrence: In Germany copper is practically never found in natural water. Copper (up to 3 mg/l) is attacked at first and dissolved by strongly aggressive waters, rich in oxygen. The dissolved copper gradually separates from hard waters as a basic salt, however. Copper sulphate is added to swimming bath water to combat algae and for sterilisation (amounts of 1 to 10 mg/l; see also p. 205).

Sanitary importance of the copper content: A chronic copper poisoning is unknown (see *H. Oettel,* Ullmanns Encyklopädie, Vol. 11, 3rd Edition 1960, p. 213 and 258). Acute copper poisoning has also occurred only rarely. Copper salts, even verdigris, do not possess the poisonous character which has been ascribed to them. The concentrations occurring in drinking and baths water could never cause poisoning; moreover, a copper-containing water is so unpalatable that nobody would drink it (see below). Human beings consume 4 to

5 mg copper daily in normal food. The DAB V (German Pharmacopoeia V) quotes 1 g of copper sulphate as maximum dose.
Copper is present in all foodstuffs. For example, flour has up to 8 mg/kg, beans up to 11 mg/kg, lentils up to 6.8 mg/kg and kidneys up to 8 mg/kg. The average value for foodstuffs is over 4 mg Cu/kg. *Splittgerber* quotes the case of a schoolboy who consumed 5 to 30 mg of copper daily for 50 days without suffering any harm.
Very large amounts of copper were tolerated without damage by animals in animal experiments in the veterinary college at Hannover; the author established no great accumulation of copper in organs.
Copper salts impart a metallic, bitter "after-taste" to water. According to *Klut,* 2 mg of copper sulphate per litre are noticeable in the taste. Copper may be removed with Magno filters or Akdolit filters (*K. Höll,* Arch. f. Hygiene *113,* 113 [1935]).
The European office of the WHO has set the limiting value as 0.05 mg Cu/l (see *H. Kruse, 21*). It is included among the "interfering" but not the "harmful" materials. Consequently, up to 3 mg/kg is permitted in the pipe system with newly fitted copper pipes. In 1963, the WHO recommended 1 mg Cu/l as the limiting value for drinking water.

c) Arsenic

Occurrence: Arsenic occurs only seldom in ground water. Some soils contain arsenic. Mineral waters often contain large amounts of arsenic, up to 20 mg As_2O_3/l (Dürkheimer Max Spring). The limit for arsenic-containing medicinal waters is 0.7 mg As/kg. Ordinary ground water contains 0.1 mg As/l at the most. In the region of Bad Dürkheim (W. Germany), ground waters with arsenic contents up to 0.25 mg As_2O_3/l occur. Occasionally, river waters also have arsenic in 0.05 mg As/l amounts and more, coming from waste waters. Ground water, too, may acquire arsenic from the drainage from dumps and from the effluents of tanneries, chemical and smelting works.
Risky concentrations of arsenic may be reached in well waters in the neighbourhood of irregularly deposited rubbish dumps. It is a fact that the ash from brown coal contains appreciable amounts of arsenic which is dissolved out by rainwater or ground water. According to the author's experience, an influence on the ground water can be more easily demonstrated through the highly sensitive arsenic detection than through the increase in hardness (p. 180).

Hygienic assessment of the arsenic content: Since chronic arsenic poisoning exists, the maximum permissible arsenic content has been established as 0.2 mg As/l (Standard demands of the WHO, Schrift. Reihe Wass., Boden- und Lufthygiene, *14a* [1960]).
The International Standards for Drinking Water of 1963 take 0.05 mg "Arsenic"/l as the limit, giving no further details.

Removal of arsenic: H. Stooff and *L. W. Haase* (Vom Wasser, Vol. *12*) state that arsenic can be removed by treating with ferric chloride or sulphate and then filtering over Magno mass. This applies only to inorganic arsenic, however. Arsenic is much more active in the form of the arsines (e.g. in poison gases) than in inorganic form.
Arsenic is removed from drinking water in Taiwan by catalytically-functioning iron sulphide filters, followed by a sand filter (Conference on Waste Waters, Tokio [1965], Lit. Ber. *13,* 2 [1965/66]).

d) Zinc

Occurrence: Zinc occurs in ground water of narrowly restricted zinc mining areas *(Stolberg).* Zinc is found also in drinking water where it has stood in zinc galvanised pipes; amounts up to 5 mg Zn/l are then found, but seldom more. Soft waters containing carbon dioxide and waters with high chloride and sulphate content dissolve zinc easily from galvanised pipes and gradually destroy them (p. 158). Water rich in nitrate likewise easily dissolves zinc from zinc pipes (see below).

Sanitary importance of the zinc content: Damage to health from water containing much zinc has not yet been proved with certainty. Generally speaking, zinc is considered to be one of the less poisonous metals. An adult consumes 12 mg of zinc daily in his food. One kg of meat contains 50 mg of zinc on an average. According to *Lewin* (Gifte und Vergiftungen, Berlin [1929]), damage to health ensues only after longer consumption of drinking water containing more than 8 mg Zn/l. Up to 5 mg Zn/l is permitted in water according to American directives. A well water in Tullendorf containing 7 mg Zn/l has been drunk for 100 years without harmful effects. 35% of the drinking waters in Finland contain over 8 mg Zn/l and some have 50–60 mg/l *(E. Halme,* "Vitalstoffe" [1961]).
E. Halme regards a too high content of zinc in food and drinking water as the most important cause of cancer frequency, since he has established that cancer patients accumulate zinc. Experiments with animals have confirmed his clinical-chemical discovery.
Zinc affects taste only when 5–10 mg/l are present. Water containing zinc tends to be cloudy and turbid water is not gladly drunk. Amounts as low as 2 mg/l yield opalescence, depending on the hardness. If nitrate-containing water stands in galvanised pipes, part of the nitrate may be reduced to nitrite, especially with aggressive waters; this accentuates the attack on the zinc.
The limiting concentration has been fixed internationally at 5 mg Zn/l, on the recommendation of the WHO. Work of *U. Rüdt* and *C. L. Kruse* shows that no more than 2 mg Zn/l should be present for reasons of palatability (Ges. Ing. *89,* 338–341 [1968]).
According to *H. V. Herbst,* amounts of zinc exceeding 1 mg/l have an unfavourable effect on water fauna. Gammarids which are fed on by trout, are destroyed when the concentration of zinc exceeds this amount (Gewässer u. Abwässer, No. 44, 1957).

e) Tin

Damage to health through water which contains tin is so far unknown. Tin is less poisonous and is also much more resistant to aggressive water than is zinc. It is, however, attacked by strongly alkaline water.

f) Aluminium

Aluminium has no influence on health in the amounts in which it occurs in normal water (ca 1 to 2 mg Al/l). When water is treated with aluminium sulphate, the pure water product must not contain more aluminium than the raw water ($<$ 0.2 mg Al/l); otherwise the limit is 0.5 mg/l (Lit. Ber. Vol. *14,* 5 [1966]) (p. 150).

g) *Chromium, Cadmium, Antimony, etc.*

Chromium occurs in small amounts in ground water in America and is stated to be harmless in amounts up to 0.05 mg Cr/l (*C. Muehlberger,* J. Amer. Water Works Ass. *42* [1950]). The WHO has quoted this value of 0.05 mg Cr/l as the maximum.
R. U. Byerrum (Eng. Bull. *45,* 1–8 [1961], Lit. Ber. p. 273 [1961]) reported studies on the chronic toxicity of cadmium and hexavalent chromium in drinking water.
The limit for cadmium of 0.05 mg Cd/l has been adopted in the WHO-table on suggestions from Denmark; cadmium can be dissolved by water from certain plastic pipes.
According to *J. R. Ball* (Water Research *1,* 805–806 [1967]), cadmium is a powerful fish poison. Amounts as low as 0.01 mg Cd/l are harmful to trout.
The Public Health Service demands a maximum content of 0.01 mg Se/l and 1 mg Ba/l.
An upper limit of 0.05 mg Sb/l is to be considered for antimony according to *e. V. Arzamestev* (Chem. Abstracts *62* [1965]).
K. P. Selyankina and *E. V. Nekrasova* (Chem. Abstracts *69,* [1968]) quote 0.1 mg/l for titanium in waters.

10. Oxygen

Occurrence: In variable amounts up to saturation in ground and surface water (saturation value at 0.1° C is 14.5 mg O_2/l). Oxygen supersaturation is often observed in surface waters with marked plant production. Waters from deep wells are free of oxygen, like water from the depths of some lakes. All intermediate amounts are found between these extremes.
Genuine ground water, i.e., one that has not been influenced by surface water, normally contains no oxygen; at 150 feet depth, oxygen is never found.
If, therefore, water containing oxygen is obtained from over 60 feet depth in a pumping trial or in working wells, there is ground for suspicion.
It must be remembered in investigations on lakes, that littoral and profundal, i.e., epilimnion and hypolimnion, vary greatly in relation to oxygen and carbon dioxide contents. This is always the case with eutrophic lakes during the summer and winter stagnation periods. Oxygen supersaturation often occurs in the epilimnion as a result of the plankton activity; yet in the hypolimnion there is sometimes no oxygen at all just below the thermocline and even hydrogen sulphide may be present in the profundal, without the lake being contaminated.
Importance of the oxygen content: Oxygen is necessary in surface water for animal and plant life. The oxygen content thus plays an important part in assessing deliveries of waste-water into interceptors (p. 213).
The oxygen content of water *in iron pipes* is important in so far as all waters of low content (below 2 to 3 mg O_2/l) attack the iron and dissolve it out. *L. W. Haase* says that tap water takes up iron again even when the oxygen content is below 4 mg O_2/l, if the water is not very hard. This can occur not only with waters which are naturally poor in oxygen, but also in those cases where water in dead ends loses much of its oxygen through chemical or biogenic[2] processes or where the oxygen is used up in closed iron removal processes.

2 When organic compounds separate out in the mains, organisms colonise and consume oxygen.

Sufficient oxygen must be present in the water to enable the protective layer to be formed (p. 163). If the oxygen content is less than 2 mg/l, aeration is always advisable[3]. Recent information shows that water devoid of oxygen can never form a protective layer, even when the other factors are favourable. Addition of phosphate to water containing little oxygen is generally useless. A minimum content of 6 mg O_2/l is a condition for the successful application of this technique. Aeration is performed with the help of an air compressor or with the Venturi tube, 2–4 mg O_2/l suffice with waters of high silica and nitrate contents.
A closed, intensive aeration is achieved with the apparatus of the Neue Continentale Bau-Ges., Hannover, W. Germany; this has proved successful even where difficult manganese removal and simultaneous removal of H_2S were performed.
Aeration in the closed system must not be pressed so far that the emerging tap-water is cloudy (see below). If necessary, one must introduce a pressure reduction unit after the compressor.
Water which is supersaturated with oxygen induces formation of lumps of rust in iron pipes; this applies expecially to water from which gas bubbles separate or that to which air has gained access as a result of leaky valves, piston pumps and sockets.
Cloudy water through air should be avoided since it is unpopular with the consumer. It can arise domestically when a very cool water, saturated with oxygen (e.g. spring water) warms up, e.g. through neighbouring warm water pipes. For instance, if tap water at 10° C (11.4 mg O_2/l) is warmed to 20° C (9.2 mg O_2/l), 2.2 mg O_2/l must separate as fine bubbles, especially when the tap is opened and the pressure thereby reduced.

A minimum amount of oxygen is thus necessary in cold piped water; yet the smallest amount of oxygen in warm or hot water is detrimental.

A typical example of this is the hot water supply of Reykjavik in Iceland, from the naturally occurring hot water of the geysers. If traces of oxygen gained access to the piped water in the container or the pipes, e.g. when running light or by air suction, considerable corrosion resulted.
The amount of oxygen plays an important part in boiler feed water (p. 232). Water containing more than 0.3 mg O_2/l must be freed from the oxygen for use in high pressure boilers. This may be done with sodium sulphite (deoxygenation process), or, recently, with sodium dithionite; or by heating with iron borings, also nowadays with hydrazine or by thermal degassing.

11. Hydrogen Sulphide

The occurrence of hydrogen sulphide may be traced to putrefaction processes in the soil; this is highly unsatisfactory from a hygienic point of view. It scarely arises in deep water (well depths of 60 feet or more) but such waters often contain hydrogen sulphide (and no oxygen) which is formed by reaction of iron sulphide with carbon dioxide. Hydrogen sulphide is formed in soils also through the action of sulphur bacteria. Traces of H_2S are frequently found in peaty waters.

3 Cf also *L. W. Haase,* Werkstoff u. Korrosion *1,* 4–9 [1950].

Hydrogen sulphide in water is harmful to lead pipes, since the lead sulphide formed is soluble and can lead to lead poisoning. The H_2S can be removed by aeration over coke filters or cascades, and, in concentrations of 1 to 2 mg/l, by atomisation; the necessary amount of oxygen is added at the same time. Copper pipes also are attacked by sulphur-containing water (see also p. 157).

The aeration aggregate of the Neue Continentale Bau-Gesellschaft of Hanover, W. Germany, enables a considerable increase in oxygen concentration to be attained and thereby H_2S is expelled to the last trace, even when present in higher concentrations; it is led away through exit ventilation tubes.

This closed process of sulphur removal, especially when done under increased pressure (which permits uptake of more oxygen) is vastly superior to the open aeration (see also *S. W. Wells,* J. Amer. Water Works Ass. *46,* 160–170 [1954]). Total removal of sulphur is possible, too, according to the Schiezelt-patent, through perforated ring sprayers (firm of H. Reisert & Co., Celle, W. Germany). H_2S is more easily removed from acidic than from alkaline waters. With the latter, lowering of the pH facilitates the procedure of simple aeration. To prevent the pure water from being aggressive, it is then subsequently necessary to raise the pH-value again. Chlorination after aeration renders removal of sulphur more sure.

B. Examples of Analyses of Tap Water and its Assessment

1. Survey of Analysis Values found

a) Examination of the Water from the Iberg Spring (W. Germany)

General:

Appearance when sampling : completely clear and colourless in a 1.5 metre viewing tube
Turbidity and colour values not measurable
Appearance after standing 24 hours : unchanged
Odour and taste : no special features; good
Temperature of the water : 8.9° C; of the air : 19.0° C.

Hygienic-chemical analysis:

Ammonium ion	: 0 mg NH_4/l (zero value means less than 0,01 mg/l)
Nitrite ion	: 0 mg NO_2/l (zero value means less than 0.01 mg/l)
Nitrate ion	: 9.0 mg NO_3/l
Phosphate ion	: 0 mg PO_4/l (zero value means less than 0.005 mg/l)
Chloride ion	: 13.5 mg Cl/l
Sulphate ion	: 23.3 mg SO_4/l

$KMnO_4$-consumption	: 1.8 mg $KMnO_4$/l
Urochrome	: 0 mg/l (below 0.1 mg/l)

Technical-chemical analysis:

Total iron	: 0 mg Fe/l (zero value means less than 0.01 mg/l)
Iron in filtrate after 24 hours standing	: 0 mg Fe/l
Manganese	: 0 mg Mn/l (zero value means less than 0.01 mg/l)
Total hardness	: 14.56° dH (German degrees)
Carbonate hardness (temporary)	: 10.92° dH (German degrees)
Non-carbonate hardness (permanent)	: 3.64° dH (German degrees)
pH-value	: 7.33
Total free CO_2	: 17.0 mg CO_2/l
Of this, associated CO_2	: 15.0 mg CO_2/l

Lime-aggressive CO_2:

a) calculated theoretically	: 2.0 mg CO_2/l
b) experimentally through the Heyer marble-solution test	: 0 mg CO_2/l
Iron-aggressive CO_2 (preventing formation of the protective layer against rust)	: 2.0 mg CO_2/l
Lead-aggressive property	: absent
Alkalinity	: 3.9 mval/l
Combined CO_2	: 85.8 mg CO_2/l
Oxygen	: 8.1 mg O_2/l
Hydrogen sulphide	: 0 mg H_2S/l
Calcium ion	: 59.3 mg Ca/l
Magnesium ion	: 27.1 mg Mg/l
Residue on evaporation	: 290.0 mg/l
Residue on ignition	: 238.0 mg/l
Loss on ignition	: 52.0 mg/l
Silica	: 8.5 mg SiO_2/l

Assessment

The water from the Iberg spring possesses a high degree of purity. It contains no trace of nitrite, ammonium and phosphate ions. Nitrate, chloride and sulphate are present in only very small amounts. The permanganate consumption–an index of the content of organic matter – is extremely low.

On the basis of the chemical analytical data, the water of the Iberg spring can be termed very pure at the time and, in conjunction with the satisfactory bacteriological findings, hygienically flawless.

Iron and manganese cannot be detected in the water.

The water has a relatively favourable degree of hardness for the region; it is practically in carbonate equilibrium and is thus not corrosive to the pipes and containers of the system. In this connection, the pH-value also is especially favourable.

Taking all in all, the Iberg spring provides a good drinking water.

b) Waterworks L.

General:

Appearance when sampling	: completely clear and colourless in a 1.5 metre viewing tube
Smell and taste	: no special features
Temperature of the water	: 9.8° C; of the air : 18.5° C.

Hygienic-chemical analysis:

Ammonium ion	: 0 mg NH_4/l (zero value means less than 0.01 mg/l)
Nitrite ion	: 0 mg NO_2/l (zero value means less than 0.01 mg/l)
Nitrate ion	: 15.0 mg NO_3/l
Phosphate ion	: 0.005 mg PO_4/l (zero value means less than 0.005 mg)
Chloride ion	: 10.6 mg Cl/l
Sulphate ion	: 16.0 mg SO_4/l
$KMnO_4$-consumption	: 4,6 mg $KMnO_4$/l
Urochrome	: 0 mg/l (below 0.1 mg/l)

Technical-chemical analysis:

Total iron	: 0.8 mg Fe/l (zero value means less than 0.01 mg/l)
Manganese	: 0.11 mg Mn/l (zero value means less than 0.01 mg/l)
Total hardness	: 3.92° dH (German degrees)
Carbonate hardness (temporary)	: 0.42° dH (German degrees)
Non-carbonate hardness (permanent)	3.50° dH (German degrees)
pH-value	: 5.65
Total free CO_2	: 48.0 mg CO_2/l
Of this, associated CO_2	: 0 mg CO_2/l
Lime-aggressive CO_2:	
a) calculated theoretically	: 48.0 mg CO_2/l
b) experimentally through the Heyer marble-solution test	: 44.0 mg CO_2/l
Iron-aggressive CO_2 (preventing formation of the protective layer against rust)	: 48.0 mg CO_2/l
Lead-aggressive property	: strongly positive
Alkalinity	: 0.15 mval/l
Combined CO_2	: 3.3 mg CO_2/l
Oxygen	: 5.6 mg O_2/l
Hydrogen sulphide	: 0 mg H_2S/l
Calcium ion	: 18.94 mg Ca/l
Magnesium ion	: 5.43 mg Mg/l
Residue on evaporation	: 175.0 mg/l
Residue on ignition	: 155.0 mg/l
Loss on ignition	: 20.0 mg/l
Silica	: 7.0 mg SiO_2/l

Assessment

The above results show that the water is of pure quality. Since it strongly attacks pipes and containers, it must be deacidified as soon as possible. Moreover, the iron and manganese contents are too high for municipal piped water. These shortcomings could be eliminated by filtering through alkaline filter masses such as Akdolit or Magno, yielding a fully acceptable water.

c) Water from a Deep Well in C.

General:

Appearance when sampling	: clear, almost colourless
Turbidity and colour values	: colour value of 10 mg Pt/l
Appearance after standing 24 hours	: yellowish, cloudy with brownish deposit
Smell and taste	: tastes strongly of iron
Temperature of the water	: 10.2° C; of the air : 18.5° C.

Hygienic-chemical analysis:

Ammonium ion	: 0.4 mg NH_4/l (zero value means below 0.01 mgl)
Nitrite ion	: 0 mg NO_2/l (zero value means below 0.01 mg/l)
Nitrate ion	: 0.5 mg NO_3/l
Phosphate ion	: 0.01 mg PO_4/l (zero value means below 0.005 mg/l)
Chloride ion	: 25.0 mg Cl/l
Sulphate ion	: 32.0 mg SO_4/l
$KMnO_4$-consumption	: 4.7 mg $KMnO_4$/l
Urochrome	: 0 mg/l (below 0.1 mg/l)

Technical-chemical analysis:

Total iron	: 3.5 mg Fe/l
Iron in filtrate after 24 hours standing and shaking	: 0.5 mg Fe/l
Manganese	: 0.45 mg Mn/l
Total hardness	: 6.72° dH (German degrees)
Carbonate hardness (temporary)	: 2.52° dH (German degrees)
Non-carbonate hardness (permanent)	: 4.20° dH (German degrees)
pH-value	: 6.44
Total free CO_2	: 58.0 mg CO_2/l
Of this, associated CO_2	: 0.5 mg CO_2/l

Lime-aggressive CO_2:

a) calculated theoretically	: 57.5 mg CO_2/l
b) experimentally through the Heyer marble solution test	: 48.4 mg CO_2/l
Iron-aggressive CO_2 (preventing formation of the protective layer against rust)	: 57.5 mg CO_2/l
Lead-aggressive property	: strongly positive
Alkalinity	: 0.9 mval/l
Combined CO_2	: 19.8 mg CO_2/l
Oxygen	: 0 mg O_2/l
Hydrogen sulphide	: 0 mg H_2S/l
Calcium ion	: 31.8 mg Ca/l
Magnesium ion	: 9.77 mg Mg/l
Residue on evaporation	: 194.0 mg/l
Residue on ignition	: 178.0 mg/l
Loss on ignition	: 16.0 mg/l
Silica	: 6.0 mg SiO_2/l

Assessment
The results of the chemical investigation show in agreement with the bacteriological findings and with the inspection of the site, that the water at the time was pure and hygienically acceptable[1].
Since the water is highly aggressive, it cannot be used as tap water without preparation. Iron, lead and zinc would be otherwise attacked in the pipes and dissolved out. The solution of the last-named two would be harmful to health (lead poisoning) and the entry of iron into solution would create technical problems (corrosion of pipes, brown coloration of the water) and impair taste. The corrosion of the pipes could lead to dangers of epidemics from seepage of foul and waste water into the system. The water could attack even concrete and mortar in the waterworks containers, rendering them leaky.
Mixing with the rather harder water of the old wells would increase the aggressivity still further, in contrast to the opinion held at the place.
The mixed water from all wells must thus be deacidified. Further, the iron content is very high. Removal of acid and iron is thus urgently necessary. This is best done by an aeration process, e.g. atomising through jets, with a settling basin and filter to follow. This also will not suffice to remove the last aggressive CO_2 and manganese. For this, subsequent passage through alkaline filter masses, such as Akdolit (firm of Börner & Co., Düsseldorf) or Magno (firm of Magno-Werk, Duisburg) is necessary.

d) Water from the Deep Well of the Firm of B.

General:

Appearance when sampling	: clear and colourless
Appearance after standing 24 hours	: faintly opalescent, no sediment
Smell and taste	: no special features
Temperature of the water	: 12.0° C.

Hygienic-chemical analysis:

Ammonium ion	: 0.05 mg NH_4/l
Nitrite ion	: 0 mg NO_2/l (zero value means below 0.01 mg/l)
Nitrate ion	: 18.0 mg NO_3/l
Phosphate ion	: 0.06 mg PO_4/l
Chloride ion	: 90.0 mg Cl/l
Sulphate ion	: 610.0 mg SO_4/l
$KMnO_4$-consumption	: 4.5 mg $KMnO_4$
Urochrome	: 0 mg/l (below 0.1 mg/l)

Technical-chemical analysis:

Total iron	: 0.84 mg Fe/l
Iron in filtrate after standing 24 hours	: 0.03 mg Fe/l
Manganese	: 0.35 mg Mn/l
Total hardness	: 58.80° dH (German degrees)
Carbonate hardness (temporary)	: 15.80° dH (German degrees)

1 *The small ammonium ion content is due to reduction processes in the oxygen-free ground water and not to contamination.*

Non-carbonate hardness (permanent)	: 43.00° dH (German degrees)
pH-value	: 6.88
Total free CO_2	: 95.0 mg CO_2/l
Of this, associated CO_2	: 75.0 mg CO_2/l
Lime-aggressive CO_2:	
a) calculated theoretically	: 20.0 mg CO_2/l
b) experimentally through the Heyer marble-solution test	: 19.8 mg CO_2/l
Iron-aggressive CO_2	: 20.0 mg CO_2/l
Lead-aggressive property	: weak
Alkalinity	
Combined CO_2	: 124.3 mg CO_2/l
Oxygen	: 2.8 mg O_2/l
Hydrogen sulphide	: 0 mg H_2S/l
Calcium ion	: 237.0 mg Ca/l
Magnesium ion	: 131.7 mg Mg/l
Residue on evaporation	: 1112.0 mg/l
Residue on ignition	: 996.0 mg/l
Loss on ignition	: 116.0 mg/l

The water is pure. It can be regarded as hygienically satisfactory from the chemical analysis data and the bacteriological information[2].

The iron content, the high degree of hardness and the excess CO_2 together with the low pH-value make it unfavourable from technical viewpoints. Iron bacteria gaining access to the pipe system cause the water to become brown, with red-brown deposits in the tanks. The red-brown deposits consist of growths of the iron bacteria, *Gallionella ferruginea,* and of manganese bacteria.

The growth of bacteria and its sequels would cease forthwith if the iron were removed from the water. Existing iron and manganese bacteria could be killed off by one or two chlorinations with bleaching powder paste; they would then be removed by vigorous rinsing of the pipe system.

Removal of iron from this water is not very easy because of the considerable hardness. Since the fairly large manganese amounts must be removed at the same time, only a vigorous aeration process could be considered. Extra addition of chemicals (potassium permanganate or chlorine) is unavoidable to remove the final traces of manganese.

The high permanent hardness of the water is to be attributed to its content of calcium sulphate. This is not favourable in drinking water but is very bad for boiler feedwater. For this latter purpose, the relevant part of the water must be softened; this is probably best done by using the lime-soda process for todays' boilers.

2 *See footnote on p. 199.*

e) Favourable and Unfavourable Values for Tap Water (under the conditions given on p. 68–78 and 151–200

	Favourable or tolerable	Limiting value	Unfavourable	
Ammonium ion mg NH_4/l	0 — 0.1	0.2	0.3 — 0.4 — 0.5	and more
Nitrite ion mg NO_2/l	0	0.05	0.1 — 0.15— 0.2	and more
Nitrate ion mg NO_3/l	0 — 10.0—20.0	30.0	40.0 — 50.0 — 60.0	and more
Chloride ion mg Cl/l	0 — 10.0—20.0	30.0	40.0 — 50.0 — 60.0	and more
Organic matter – mg/l of $KMnO_4$ consumption	0 — 6.0	12.0	20.0 — 30.0 — 40.0	and more
Urochrome mg/l	0	0.1	1 — 50	and more
Phosphate ion mg PO_4/l	0 — 0.01	0.02	0.1 — 0.15— 0.2	and more
Sulphate ion mg SO_4/l	0 — 25.0	50.0	75.0 — 100.0 — 125.0	and more
Iron mg Fe/l	0 — 0.05	0.1	0.15— 0.3 — 0.5	and more
Manganese mg Mn/l	0 — 0.01	0.03	0.1 — 0.15— 0.2	and more
Total hardness °dH	3.5°— 5.0°	10.0°	15.0°— 20.0°— 30.0°	and more
			1.0°— 2.0°— 3.0°	
Carbonate hardness °dH	2.0°— 5.0°	8.0°	15.0°— 20.0°— 25.0°	and more
			1.0°— 1.5°	
Non carbonate hardness °dH	0° — 5.0°	10.0°	15.0°— 20.0°— 25.0°	and more
Calcium, mg CaO/l	20.0 — 50.0	100.0	150.0 — 200.0 — 250.0	and more
			107° — 140° — 178.8°	and more
or mg Ca/l	14.0 — 36.0	70.0	10.0 — 15.0 — 20.0	
			7.0 — 11.0 — 14.0	
Magnesium, mg MgO/l	10.0 — 50.0	100.0	50.0 — 200.0 — 250.0	and more
or mg Mg/l	6 — 30.0	60.0	90.0 — 120.0 — 150.0	and more
pH-value, alkaline	7.1 — 7.5	8.0	8.5 — 9.0 — 10.0	and more
acid			3.0 — 5.0 — 7.0	
Excess CO_2 (aggressive CO_2) by				
a) soft waters		1.0	3.0 — 5.0 — 10.0	and more
b) hard waters	1.0 — 2.0	3.0	10.0 — 20.0 — 30.0	and more
Hydrogen sulphide, mg H_2S/l		0	1.0 — 2.0 — 3.0	and more
Oxygen mg O_2/l	6.0 — 8.0	10.0	11.0 — 12.0 — 15.0	and more
Residue on evaporation, mg/l	100 —500.0	1000.0	1500.0 —2000.0 —3000.0	and more

VII. Examination and Assessment of Swimming Bath Water, Water for Building Purposes, Surface (Receiving) Water and Wastewater

A. Swimming Bath Water

Swimming bath hygiene has made gratifying progress in the past five years. Swimming bath water must basically now have the *quality of drinking water,* i.e. the bacterial number should be below 100 over the whole bath and no *B. coli* should be detectable in 100 ml of water. The points listed below must be respected in order to accomplish this:

1. *A circulation plant* must be available which functions well and can be adapted at any time to the number of visitors. Complete circulation must take place at least 4 to 6 times during the day.
2. *The hydraulics of the swimming bath* must be in order. This means that no dead volume or rotation zones may be present in the water, since chlorine or other added sterilisation agents are excessively consumed at these places; this leads to high bacterial numbers. This is more important when the baths water is warmed, as is being done more and more. A suitable colour test can show whether the *water is flowing uniformly through the pool.*
3. The amount of water used must be worked out from the relation: 0.5 persons at the most per cubic metre/day.
4. *The swimming bath water must be permanently sterilised,* since every bathing visitor introduces a vast number of micro-organisms into the water. At all times and at all places in the pool, including the exits to the circulating plant, at least 0.3 mg/l of active chlorine (see p. 126) must therefore be present. The baths attendant must check this by chlorine determinations using the Lovibond-comparator (see p. 127), performed several times each day at various places and at various depths.
5. *At least 5% of the contents of the pool must be replaced daily by fresh water.* This latter must also have been sterilised.
6. The "Wiesbadener Rinne" (Wiesbaden gutter) is now outdated. It is now recognised that the swimmers are disturbed by the chlorine atmosphere which collects above the water surface and by the carbon dioxide and water-vapour breathed out by the bathers and set free from the baths water. Consequently only *the high overflow channel along the outside of the baths* is now built and water is filled to the upper edge of the pool (*H. Willach,* Archiv. d. Badewesens 22, 145–147 [1969] and *K. Riedle,* ibid. No. 5).
7. The high situation of the water level in modern swimming baths enables the uppermost layer, which contains relatively many microorganisms as a result of disappearance of

the sterilisation agent, to be taken off easily, especially during the daily "cleaning" of the surface. In open-air baths there are also dust, pollen, soot, leaves, cosmetics, oils for protection against sunburn and for massage, to be reckoned with. These contaminate the upper water layer but can likewise be easily removed, thanks to the high water level.

8. The high situation of the water level eliminates the *dirty sides,* found with low water levels, which were a permanent source of infection.
9. Modern *swimming baths are rectangular.* They are more uniformly swept through in the *direction across the pool* than, as formerly, along the length from the shallow to the deeper end. The "more daring" types of pool of some architects are outdated from the hygienic point of view, as a result of their inferior hydraulic aspect. The water introduced forms a layer, the position of which varies according to the temperatures of added and pool water (as with reservoirs, see p. 59). This must be borne in mind particularly with warmed baths, now being increasingly used. Vertical water movement is therefore now coming in (*S. Langer,* Arch. d. Badewesens, 357 [1964]).
10. *The baths water must always be completely clear and colourless.* Visibility right to the bottom is a principal requirement to be fulfilled by the bathing attendants. Hence, the circulation water must always be treated with aluminium sulphate before filtering (3–4 g/m^3). At the same time, the pH of the water, which should lie between 6.8 and 7.2, must be continuously controlled. The pH can often rise rapidly in summer because there are always some algae which take up CO_2 from the water. Matter flocculates out in the baths at pH values exceeding 7.2. Sometimes it is enough to remove the film of oil from the surface. A small amount of aluminium sulphate (1 mg/l) is good for this.
    If aluminium sulphate is added all the time, the swimming bath water will soon become more or less aggressive, depending on the added volume of fresh water and on the carbonate hardness. This can be rectified by using dolomite filter masses in the circulation plant if the pH should fall below 6.5. According to the author's experience, extra CO_2 enters the water from the flame where the circulated water is directly heated, e.g., with natural gas. A dolomite filter layer is then usually necessary in the filter room.
11. Surprisingly, *no slime is deposited on the bottom of the pool* when the water flows through completely uniformly. Insoluble organic matter, which is an excellent bacterial carrier, is then wholly removed in the circulation plant. Less chlorine is thus needed in order to keep the limiting amount of 0.3 mg Cl/l. The hydraulic situation is governed by the disposition of the jets.
12. The redox potential must be maintained at ca. 700 mV for pH 7 together with the control of the chlorine concentration and pH value, as said. This is important for the bactericidal action of the chlorine or hypochlorous acid. Work of *U. Hässelbarth* has shown this (Bundesgesundheitsblatt *12,* 251–256 [1969]). See p. 86 for the redox measurement.
13. The number of visitors should be restricted in all baths and restricted in each individual hour. If the limit is set for a whole day, the suitable number can be exceeded in a particular hour.
14. Chlorine is the principal sterilising agent in W. Germany. At the values given above for pH and redox-medium, the chlorination is really treatment with hypochlorous acid

(HClO). At least at these values, the proportion of hypochlorous acid is a maximum and it is the real active entity in the chlorination. The Elektrochlor process of the firm of Elektrochlor GmbH., Stuttgart-Hohenheim is thus equipped with a pH-regulator. If pH value, redox medium, potassium permanganate consumption and the hydraulic aspect are all in order, less chlorine is consumed and the swimmers are less disturbed. The bactericidal action of chlorination falls as the pH increases, i.e., when less hypochlorous acid is then present.

15. As far as possible, the content of ammonium ions should always be below 0.05 mg/l. Ammonium ions are not only indicators of purity but, further, they yield chloramines with the chlorine used for disinfection. These are less bactericidal than the chlorine and make their presence felt through their strong odour and as eye irritants (*O. Hettche).* This latter effect applies particularly to trichloramine.
16. The permanganate consumption serves, together with the ammonium and phosphate contents, for judging the purification effect. Its value lies below 12 mg/l and the ammonium ion content is less than 0.05 mg/l (quality of drinking water) if the circulation and the hydraulics are in order.
17. *Chlorine dioxide* alone has not proved satisfactory for sterilisation of swimming bath water. It is suitable only together with chlorine for reducing odour, effecting stronger oxidation and sterilising. It is, however, then difficult to maintain the limit of 0.3 mg Cl/l in the exit water, even when only 7% ClO_2 is added to the chlorine dosage. Re-formation of chlorite is prevented by addition of chlorine. *Ozone* alone also is not adequate. Although ozone sterilisation is so favourable with drinking water, since no outside materials thereby enter the water and because ozone has an especially powerful action against bacteria and viruses, yet ozone disinfection in baths is controversial. *On no account may ozone be present in the swimming pool.* It may be added only before the filter plant and all residual ozone must be taken out by means of active charcoal filters after action.

The swimming pool water must contain no ozone because it easily escapes into the air and irritates the bathers and, in particular, is detrimental to the health of the attendants. Existing ozone plants of this type have had to be modified after a short time. The maximum allowable concentration of ozone in air is about 0.1 ppm (see *K. Höll,* Heilbad u. Kurort *21,* No. 7 [1969]).

Ozonisation of the circulating water effects powerful oxidation of the organic substances[1] and the chlorine consumption is markedly reduced. One must remember, however, that toxic materials can be formed via certain intermediate ozonides which escape control. Many highly toxic compounds are known today which have toxicity limits in the nano- and picogram region. (*E. Habermann,* Katastrophenmedizin *5,* 35–45 [1969]).

The ozone process is expensive and there is also the price of the necessary additional chlorination of the water. Ozone is accorded a good viricidal action but this is of little value in a baths establishment; in the bath itself, where the viruses first turn up, it can have no action.

Hydrochloric acid accumulates if the same water is frequently chlorinated. It then attacks concrete and destroys the walls of the bath.

1 *except chlorinated urea.*

This reduces the carbonate hardness and displaces the carbonate balance (p. 163). Continual use of aluminium sulphate to improve the circulated water brings about a much greater reduction of carbonate hardness and displacement of the carbonate balance.
The use of bleaching powder solution for sterilisation is unsatisfactory since it raises the pH value too much and thereby greatly reduces the bactericidal activity. It gives rise to a particularly strong odour.
Solid chlorination preparations have proved to be highly satisfactory, especially for small and very small baths. Parakaporit of the firm of Bayer, Leverkusen, is rather dearer but is more active and goes further. Another is HTH[2] of the firm of Hoelzle & Chelius, Frankfurt/Main. Both consist of calcium hypochlorite. The HTH has a somewhat stronger smell.
In *Höll's copper sterilisation procedure,* only a single treatment with 3 to 6 mg/l of cupric sulphate or chloride is needed, depending on the quality of the water. A complete sterilisation is accomplished by this procedure and it prevents in addition any formation of algae in the pool and on its walls. This enables the prescribed depth visibility from the diving stages to be achieved. The water is moreover freed from colloidal material by flocculation and acquires an attractive bluish colour. The copper content of the baths water can be controlled with supplied by the firm of Franz Bergmann KG, Berlin-Zehlendorf.
The disadvantages of chlorination, such as eye irritation, the odour of chlorine and formation of chlorophenols, are not encountered with this process. Likewise avoided is the tiresome emptying of chlorine cylinders at the close of the season.
the method described on p. 133, or with the help of a small apparatus for bath attendants, If desired, the copper can be removed again afterwards, so as not to burden the interceptors when the water is let out. This is done with dolomite filter masses.
In recent years, *Fekta,* a chlorine-free preparation of the firm of Goldschmidt AG, Essen, has had profitable application for preventing formation of algae in swimming pools. A single application suffices for a few weeks. It formerly led to some froth formation but this is scarcely the case now. Micropur has been a good material for small swimming baths (100 g Micropur per 10 cubic metres capacity). It is supplied by the Deutsche Katadyn-Gesellschaft, Munich.
The Bayer works of Leverkusen are introducing Dimanin A on to the market for prevention of algae. It contains a quaternary ammonium base. The *Petunia process* depends on removal of the CO_2 which prevents algae from growing in the baths water. Supplying firm is the Chemische Fabrik Petunia, Grötzingen, W. Germany.

1. Chemical Examination of the Swimming Bath Water

Determinations to be carried out

pH value (p. 84)	Urea (see below)
Redox-Potential (p. 86)	Indican (see below)
Chlorine (p. 127)	Carbonate hardness (p. 114)

2 High Test Hypochlorite.

Ammonium (p. 27)
Protein ammonia (p. 30)
Nitrite (p. 31)
Nitrate (p. 33)
Urochrome (p. 46)
Oxygen content (p. 101)
Oxygen after standing for 48 hours (p. 105)

2. Hygien-chemical Assessment of the Swimming Bath Water

The content of nitrogen compounds should not be significantly higher in the baths water than in the fresh water. The ammonium and nitrate ion concentrations are raised through pollution with urine. *Mahla* determined the extent of urine contamination of the baths water on the basis of the nitrate content (Arch. Hygiene *231* [1933]). Change of water is needed if the concentration rises by more than 10 mg NO_3/l.
The regulations in Lower Saxony forbid a concentration higher than 0.1 mg/l of ammonium ion in baths water. 5 mg O_2/l should still be present after 24 hours consumption. Other parts (Länder) of W. Germany now have the same or similar regulations.
Protein ammonia must be present in baths water only in traces. A few mg/l of protein ammonia are found in strongly polluted swimming bath waters. Such an evil state of affairs must be remedied immediately. Entry of 1 ml of urine into 100 l of water ultimately yields a concentration of 0.1 mg/l of protein ammonia.
The determination of amino acids also can be utilised to enable a judgment to be made (p. 207).
If the baths water contains greater amounts of settleable matter, the circulation period must be shortened. This must be allowed for in planning when the raw water is not completely pure. Nitrite must not be detectable in swimming pool water.
According to *W. Jung,* contamination through urine can be accurately established by a phosphate determination (GWF 1961).
U. Hässelbarth has given useful tips for surveillance of swimming baths (Bundesgesundheitsbl. *8,* 353–357 [1965]).

3. Detection of Urine Components in Swimming Bath Water

Older procedures: Detection of urea with xanthydrol. The procedure of *Fosse* (C. R. Acid. Sci. [Paris] *158,* 1076 and 1588 [1914] and *159,* 253 [1914]) was checked by *I. Wagner* (Z. U. Lebm. *84,* [1942]). In the procedure, 5 l of water under investigation are treated with 2 ml of nitric acid* and 1 g of sodium acetate, and the mixture is evaporated to dryness on the water bath. The residue is taken up in a few ml of water and washed into a beaker. The total volume should be 10 ml. 15 ml of glacial acetic acid are added, the mixture is filtered and the residue on the filter washed with 3 to 5 ml of glacial acetic acid.
5 ml of a 10% solution of xanthydrol in methanol are added to the filtrate in 1 ml portions every 10 minutes, swirling round after each addition. After standing for a longer time, the precipitate is collected in a weighed porcelain filtering crucible and washed with methanol. After drying briefly in a drying oven and cooling in a desiccator, the crucible is reweighed.

The increase in weight (dixanthylurea), multiplied by 142.9, yields the amount of urea in 1 l of water.
1 litre of urine in 100,000 litres of water can still be detected.
Heavy metals, halides and hydrogen peroxide interfere.

Indican test: Contamination of the baths water through urine can be demonstrated also with the help of the indican determination. According to *Jolles'* procedure (Ber. Deut. Pharm. Ges. *30,* 421–442 [1920]), 3 to 5 l of water are evaporated down to 250 ml. If nitrites are present, 1 to 3 g of Mohr's salt (ferrous ammonium sulphate) are added to this concentrate. The resulting mixture is concentrated further to 10 ml and filtered. To the filtrate, 1 ml of 5% alcoholic thymol solution or α-naphthol solution is added. After mixing well, 10 ml of fuming hydrochloric acid containing 0.5% ferric chloride are added. The mixture is left for 10 minutes, stirring frequently, and then extracted several times with 5 ml portions of chloroform.
Should more than 0.3 mg indican/l be present in the original water sample, the chloroform layer is reddish violet. Comparison can be made with the colours from standard solutions of urinary indican (Merck, Darmstadt), using a stock solution of 10 mg/l.
The test can be carried out also with success on samples of ground water; the indican is not completely retained by the soil particles.
The reaction product of indican with ferrous ammonium sulphate decomposes rapidly in water. The xanthydrol method is thus better.

Griess reaction: Some sodium hydroxide solution and a few crystals of p-diazobenzenesulphonic acid are added to about 100 ml of swimming pool water in an especially well cleaned flask fitted with a glass stopper. Contaminated water is recognised through the yellow colour yielded on shaking vigorously.
The reaction is not specific for human faecal matter, however. It is shown also by phenols, cresols and other compounds. A yellow coloration is in any case a bad sign, since it does not occur with pure water.

Amino acids according to Folin-Ciocalteu: The *Folin-Ciocalteu* reagent is prepared by dissolving 100 g of sodium tungstate and 25 g of sodium molybdate in 700 ml of water and adding 50 ml of phosphoric acid (d = 1.7) and 100 ml of fuming nitric acid (d = 1.19). The mixture is boiled for 10 hours under reflux in a 2-litre flask. 150 g of crystalline lithium sulphate, 50 ml of distilled water and 5 drops of bromine are then added. After boiling for an additional 15 minutes, the solution is cooled. The colour then changes from green to yellow. The volume is made up to 1 litre. The reagent can be kept in dark bottles.
It detects tyrosine, cystine and tryptophan, but neither urea, alanine nor glycine.
The *Folin-Ciocalteu* reagent is now supplied by chemical firms, e.g. Merck of Darmstadt.
Procedure: 250 ml of the filtered water sample are neutralised with 0.1N HCl and evaporated almost to dryness in vacuo. The residue is taken up in water.
10 ml of a sodium carbonate solution (15 g of calcined soda in 100 ml of distilled water) and 10 ml of the vacuum concentrate are placed in a test tube. A reference amino acid solution[3]

3 110 mg of tryptophan, 140 mg of tyrosine and 50 mg of cystine are dissolved in 300 ml of 1N H_2SO_4. This stock solution is diluted by 5 before use. 1 ml then corresponds to 0.2 mg of amino acid.

is prepared in a second test tube from 1 ml of standard amino acid, 4 ml of distilled water (Reagent No. 10), 1 ml of calcium chloride solution (32.7 g crystalline $CaCl_2$ in 50 ml of water, plus 5.1 mg crystalline $MgCl_2$, made up to 100 ml with distilled water) and 10 ml of the above sodium carbonate solution. Regardless of the turbidity which forms in this reference solution, 1 ml of *Folin-Ciocalteu* reagent is added to both solutions. They are shaken up and warmed for 10 minutes on the water bath at 37° C. After allowing to stand for 20 minutes, the clear supernatants are evaluated colorimetrically.
The new procedure for detection of the urinary colouring matter, urochrome, due to *O. Hettche,* yields approximate quantitative data and hence provides a good index for contamination of the swimming bath water (p. 46).
Apart from the chemical and bacteriological controls of the swimming pool water, continuous biological control should be undertaken of fungus germs, worm eggs, Cercaria etc. (e.g., using the nannoplankton net of *Höll* p. 350).

B. Water for Building Purposes

The ground water must be accurately tested as well as the ground itself when larger concrete buildings or extensive concrete pipe lengths are to be built or laid. If not, there is a danger of a large silo collapsing or a whole sewage pipe system corroding completely and ceasing to function; both these have occurred in W. Germany.
Concrete and mortar are destroyed by waters possessing aggressive properties. The quality of the ground water at the site of concrete buildings and foundations is therefore of great importance. The investigation should comprise the following:

Free carbon dioxide (p. 86)
Lime-aggressive CO_2 (p. 87)
pH, hydrogen ion concentration (p. 83)
Hydrogen sulphide and sulphide ion (p. 119)
Free mineral acids (H_2SO_4, p. 83)
Sulphate ion (p. 47)
Chloride ion (p. 40)
Organic matter (p. 38)
Carbonate hardness (p. 112)
Total hardness (p. 112)
Permanent hardness (p. 117)
Magnesium sulphate and chloride, calcium sulphate (p. 118)
Ammonium ion in wastewaters (p. 27)

Sampling from bore holes requires considerable skill. Best for this are a hand pump without valve, the water drawer according to *Gad* (both supplied by Franz Bergmann KG, Berlin) or the drawing equipment of *Friedinger* (made by Feinmechan. Werkstätten, now Hydrobios of Kiel, W. Germany). Pumping away is still the best of all, however.

1. Assessment of the Harmfulness to Concrete

Water which comes into contact with concrete structures must be colourless. Yellowish or brownish water contains humins and humic acids which are highly detrimental to concrete.

This applies particularly to brown moor waters which often contain free sulphuric acid as well as the humic acids (practical experience in the Hamburg area).
The potassium permanganate consumption should thus not exceed 25 mg/l or else the concrete must be specially protected (see below).
Waters of pH values below 7.0 must be regarded as harmful to concrete, and extremely harmful if the pH is 6.5 and below. Waters of such acidity generally have low carbonate hardness and contain aggressive carbon dioxide. This aggressive CO_2 dissolves the calcium salts from concrete and mortar and gradually destroys these building materials. Flowing water with these properties performs this very rapidly. The different types of cement yield concretes of varying resistance to this destructive influence. Resistance increases in the order; Portland cement, blast furnace cement, high alumina cement.
Organic acids usually act less destructively on concrete. Sometimes they even yield difficultly-soluble compounds with the cement calcium, which then make the concrete water-tight (e.g. oxalic or tartaric acids).
Waters of pH exceeding 7.0 can also cause damage to concrete and cement if the sulphate content is high. Calcium and magnesium sulphates and, to a small extent also the corresponding chlorides, destroy concrete with "drifting" phenomena ("Zementbacillus" = Candlot's salt). Mainly the compound $Al_2O_3 \cdot 3CaO \cdot 3CaSO_4 \cdot 32H_2O$ (Ettringite) is formed as fine white needles.
Harmful to concrete are also waters containing hydrogen sulphide and sulphides and larger amounts of ammonium salts. The concrete must thus be specially protected from *sulphur waters* and from wastewaters (as a result of the H_2S formed during putretaction); this applies especially to the zone above the water level, where H_2S escapes, e.g., in sewers.
Further concrete is attacked by waters containing sodium hydrogen carbonate, as encountered in coastal areas and in medicinal springs.
Blast furnace slag in the soil is very harmful to concrete buildings and pipes since it yields sulphate (*R. Lautrich,* Der Abw. Kanal., 2nd Edition, Hamburg 1966).
Concrete can be adequately protected against weakly aggressive waters by certain additions, e.g. Puzzolan or Rajasil (Marthahütte, Marktredwitz, W. Germany) (see under "Sulphate-Resistant Cements").

2. Concentration Limits for Harmfulness to Concrete

The potassium permanganate consumption should be less than 25 mg/l, and less than 50 mg/l for wastewaters. The pH value must be above 7.0 and under no circumstances below 6.5. The carbonate hardness must not be less than 2.0° dH and no aggressive carbon dioxide should be present in the water.
The sulphate content is to be below 250 mg SO_4/l for static water and not above 100 mg/l for flowing water, provided it is hard water; 50 mg/l is the upper limit for soft water. Experience at the Achensee power station (Austria) has shown that only 70 mg SO_4/l in ground water flow caused damage to concrete (*F. Gutberlet,* Das Baugewerbe No. 14, 1958). The water of the Achensee is soft but contains no aggressive carbon dioxide.
Water from the R. Weser (W. Germany) containing 100 mg SO_4/l caused damage to concrete at the Weser lock at Dörverden.

The ammonia content should not exceed 1.0 mg/l and hydrogen sulphide and sulphides should not be detectable.
Brackish and sea water attack concrete strongly, particularly porous concrete. Hence aluminous cement is used for making concrete at the sea-shore (*A. Hummel* and *K. Wesche,* Beton im Seewasser, Berlin [1956]).
The draft of DIN 4030b gives directives concerning harm to concrete but adheres too much to the concentration limits of the concrete industry (see under 7.).
All the practical experience just mentioned is inadequately considered in the tables of the cement industry. According to these tables, water containing 15–30 mg/l of lime-dissolving carbon dioxide is only "weakly corrosive"; no distinction is made even as to whether the water is hard or soft, standing or flowing.
Modern concrete manufacturing technique aims at the lowest possible water-cement values (below 5–6), optimum grain composition of the aggregates and completely pounded fresh concrete; concrete is thus more resistant nowadays. Notwithstanding, it is certain that normal concrete is endangered by all waters which turn litmus red and rosolic acid yellow and those which contain over 300 mg SO_4/l.
Prestressed steel in reinforced concrete is attacked by waters rich in chloride, which can penetrate through the finest cracks.
Wastewaters of all types attack concrete and mortar. This is due partly to hydrogen sulphide present, partly to ammonia and acids.
Water components such as free CO_2, sulphate, chloride etc. reinforce each other in the attack on concrete. If the concentrations of several of these constituents approach the threshold values, the water is branded as concrete-aggressive. This is considered sensibly in the technical directives of the GDR.

3. Sulphate-resistant Cements

The Dyckerhoff Cement Works (Lengerich in Westphalia) recently brought out a sulphate-resistant cement, termed *Sulfadur-Cement,* from which the aluminium-calcium-containing product, mentioned above, cannot be formed. High sulphate contents of ground and river waters have thus no influence on structures built with this cement. It is rather dearer than the normal Portland cement but neither clinker construction nor mixing of Portland cement with additives is needed, saving money and time (Quality group Z 225 according to DIN 1164).
The Heidelberg Portland Cement Works recently introduced the Dur-Atherm Cement as a highly sulphate-resistant product. It is stated also to yield concrete of high resistance to aggressive CO_2, since it is prepared with blast furnace slag (in contrast to the Sulfadur-Cement). Both these two products have in addition a very low setting heat (Quality class of the Dur-Atherm HOZ 275). Anti-Sulfat Cement of the Portland Cement Works is resistant only to sulphate (PZ 375 C_3A-free).
Concrete pillars must be surrounded with clinker if the ground water is aggressive. The trench carrying iron pipes must be surrounded with 20″ of neutral sand or the ground water led away through drainage pipes.

In border-line cases, as firm as possible a concrete (i.e. compact, smoothed concrete from cement with a minimum amount of calcium (blast furnace cement or high alumina cement) must be used, preferably after treatment with bitumen (DIN 4030b Deutsche Normen). Pounded loam around the foot of the pillars gives additional protection to the concrete.
Concrete tanks for water storage are always provided with protective coatings such as of Inertol (firm of Lechler-Chemie, Stuttgart). Aggressive waters nevertheless gradually attack the concrete. They gain access to it through very small cracks in the coating. Ultimately, the extended action of the water dislodges the coating. The only sure protection against stored aggressive waters is a layer of tiles.

4. Water for Concrete Mixing

Water containing humic acids must not be used for concrete mixing. The potassium permanganate consumption of the water must not exceed 20 mg/l and the sulphate content, 250 mg/l.
Neither water containing salt in concentrations exceeding 0.15% NaCl, nor that containing oil is suitable for concrete mixing.
High alumina cement is especially unsuitable for concrete preparation with salt-containing water which must therefore not be used.

5. Concentration Limits for Water in Contact with Iron and Steel

Ground water with more than 250 mg SO_4/l destroys laid iron pipes also from the outside. Further they suffer external attack in loam and clay soils containing little calcium and especially in peaty soils and those containing calcium sulphate. The same applies to zinc pipes. When zinc pipes are laid, the sand lime mortar should be used and not gypsum sand mortar.
Concrete and iron are to be provided with water-repellent coatings in all cases of aggressive ground water; they are then pointed up with asphalt and covered with tar board, or given a coating of adhesive with a putty-knife.
Iron and steel are attacked by soft waters of low carbonate hardness and with aggressive CO_2 and pH values below 7.0.
Soft waters with high nitrate content (over 20–50 mg/l) act corrosively on iron, steel and zinc parts, even when there is no aggressive CO_2 in the soft water.
A chloride content exceeding 100 mg/l in soft ground waters has a damaging effect on iron and steel, especially if magnesium chloride is present, e.g. in water from the R. Weser. This is important with reinforced concrete since the reinforced steel is attacked by waters rich in chloride (concrete bridge construction).
Sea water and brackish water attack iron and steel since they contain so much magnesium chloride.
If the ground water has a pH of 6.5 or less, iron and steel tanks should have cathodic protection. This applies with especial force to mineral oil tanks on account of the danger of oil

seepage from leaks due to corrosion. Cathodic protection in ground waters of pH below 6.5 has been prescribed recently.

C. Surface and Receiving Waters

In the preceding sections, hints are found in numerous places about peculiarities in the examination of surface water. In the general investigation of suitability for drinking water purposes, the removal of turbidity from samples of surface water creates difficulty. Some ways of accomplishing this are found in the section on "General Principles of Chemical Work" (p. 20).
In other respects, the methods of examining surface water are the same as those for ground water and spring water, for example:

Ammonium ion	Acid combining capacity, SBV.
Nitrite ion	Alkalinity, now acid consumption
Nitrate ion	Oxygen
Potassium permanganate consumption	Oxygen consumption
Chlorine number	Iron
Chloride ion	Manganese
Phosphate ion	Combined carbon dioxide
Sulphate ion	Carbonate hardness
pH-value	Total hardness
Free carbon dioxide	Residue on evaporation
Acidity, now base consumption	Residue on ignition

The biochemical oxygen demand (BSB_5 in German) is measured also.

1. Contamination of Waters

Investigations by *K. Höll* over many years (Vom Wasser *31*, 26–42 [1964]) showed that uninfluenced waters have a constant, pure composition, even in times of heavy precipitation, independent of the season (high summer and low winter temperatures).
Contamination of our waters has increased enormously in recent times. Continuous chemical and biological control of the watercourses is therefore necessary. This is entrusted to the authorities for investigation of river water. The call for protection of waters, from the international society for applied and theoretical limnology, illuminates these problems in the journal "Korrespondenz Abwasser", No. 3 [1966].
The "European Water Charta" was proclaimed in Strasbourg on 6/5/68 by the consultative group of the European Council in presence of the European Foreign Ministers. It contains the necessary water-protective measures.
Unfortunately little has changed since the entering into force of the "Water Economy Law" of the W. German Federation (of 27/7/57, Federal Law Gazette I, p. 1110 [1957] and modifications of 19/2/59, Federal Law Gazette I, p. 37 [1959] and of 6/8/64, Federal Law

Gazette I, p. 611 [1964]) as a covering federal law, and the corresponding laws of the individual Länder of 1960 and 1961. The law of 1/10/60 regulates the maintenance in a clean state of the federal waterways (Federal Law Gazette II [1960]). The Federal German Institution for Water Matters in Koblenz supervises this. It is regrettable that the Federal Constitutional Court declared this law of maintenance in a clean state to be void in its decision of 30/10/62. For the moment, further pollution of the W. German rivers must thus be reckoned with. The Land Hessen has had a very good "stream law" since 1887.
The second "Water Economy Law" of 6/8/64 is aimed at protecting waters against mineral oil contamination. Stricter legal regulations concerning the composition of diverted wastewaters have existed in Switzerland since 1966.
Schmidt-Berger, Heilbad u. Kurort *16,* 72–76 [1964] and *R. Sievers,* Wasserrecht Vol. VI,1 in *H. Uhle,* Verwaltungsgesetze des Bundes und der Länder (Administrative Laws of the Federation and the Länder), Cologne 1964, give information about the legal consequences of a contamination of waters and ground water.

Assessment: The oxygen consumption, like the oxygen content, plays an important part in judging a water contamination. Contamination through wastewater usually lowers the oxygen content of the surface water considerably; the fish stock is thereby imperilled, and often even destroyed.
Determination of the settleable and suspended matter is important in investigations of receiving waters, since, in quiet parts of these interceptors (flow speeds below 0.3 m/sec), it can deposit and form sludge banks. These can in turn lead to secondary river water pollution with oxygen depletion of the water.
Samples in studies of interceptors are taken just above the point of entry of the wastewater (from 5 to 50 yards away, depending on the rate of flow) and also at about 50 to 100 yards below this point of entry. The contaminants are more or less rapidly decomposed, depending on the biological self-purifying power of the receiving water. Further still beyond the entry point of the wastewater one thus finds normal values again in the chemical examination. The sampling spot must thus be chosen carefully. The current in the interceptor is important in this since "strands" of waste water can form. Water samples are consequently taken from the thread of the current and at both banks. If samples are taken at several sites of the stream or river course, an idea is gained about the degree of loading through the wastewater and about the self-purifying capacity of the receiving water. Even considerable pollutions disappear in about 2 km (1¼ miles) stream course, excepting industrial wastewaters, e.g. from sugar and paper works, and naturally salt tributaries. The rate of flow is important for the assessment.
The pH value of the water is significant for the fish stock. The pH of fish waters must lie between 5.0 and 9.0 (*Schäperclaus*)[1].
Assessment of contamination of a receiving water: The assessment of contamination of a receiving water must be based always on analysis values of samples taken from above and below the entry point of the wastewater. Distinct increases in the amounts of nitrogen compounds, organic matter and settleable matter show that the inflow of wastewater is not toler-

1 Cf. also *W. Ohle,* Z. für Fischerei *36,* 185–191 [1938].

able for the receiving water, or that the wastewater is not being sufficiently purified. The Prussian water regulations, along with the new Länder laws, still determine the introduction of wastewater into water courses of first to third order in N. Germany.

The chlorine number is a good measure of the influence on the interceptor. If it exceeds the permanganate number, the contamination is too excessive for the interceptor.

Measurement of oxygen content of the interceptor is of great importance. The temperature must be accurately determined for this, so as to be able to work out the oxygen saturation value. As said already above, the oxygen consumption plays a significant part, yielding information about the eventual pre-loading. The oxygen saturation value of 40% can be regarded as the lowest limit for loading a receiving water with wastewater. About one third of the load comes from insoluble and about two thirds from dissolved material. See *K. Imhoff* (Ges.-Ing. *71*, 269–271 [1959]) concerning the "wastewater load" of interceptors and plans about it. When the water is highly contaminated, the oxygen consumption is often so large that all the oxygen disappears within a few hours.

Flowing water must have at least 2.5 mg O_2/l to keep alive the aquatic organisms on which fish feed *(Ambühl).*

H. Nold and *H. Schmassmann* (Erg. d. Wiss. Untersuch. d. Schweiz. Nationalparks, Vol. IV) carried out exhaustive studies of oxygen saturation in the streams of high mountain areas. Atmospheric pressure and temperature exerted considerable influence there.

The oxygen content of standing waters depends greatly on the plankton. During the day, assimilation of the plankton yields oxygen but at night it disappears through the dissimilation. This occasionally leads to death of fish in the early morning.

Determination of the oxygen content

Oxygen content is determined electrochemically in running waters, in the whole river channel. *F. Tödt's* oxygen measuring apparatus is now used only with membrane-covered electrodes and with automatic cleansing and rinsing through. A new development for continuous oxygen determination is the phase exchanger of the form of Wallace & Tiermann, Chlorator GmbH, Grötzingen.

Further information is to be found from *K. Wuhrmann, F. Zehender* and *K. Woker,* Vierteljahresschr. naturf. Ges. Zürich *92,* 198 [1947] and Z. f. Hydrologie *11,* 20 [1948].

H. Liebmann established 0.2 to 2.0 mg NH_4/l as the limits of harmfulness to individual fish species.

If wastewater carries nutrient into standing water, "water bloom" due to algae may arise. Metabolic products of a blue algae "water bloom" harm the fish and, if drinking water is derived from the lake as source, can cause health damage (*G. Vogler,* Arch. Hygiene No. 1, 1967).

Floating masses of sewage fungi in streams into which wastewaters have been run, indicate extensive pollution of the water. The nets and fish baskets are clogged by the fungus so that fishing is impaired. When fungus clumps settle in quiet parts of the stream, secondary decay occurs, accompanied by further oxygen consumption and additional damage to the fish situation.

Kolkwitz's "Clump catcher"[2] enables a quantitative estimate to be made above and below a contamination spot in a part of the river.

2 Supplied by Franz Bergmann KG, Berlin-Zehlendorf.

The presence of dead fish may be related to metal poisons. Many sorts of fish can tolerate only a few mg/l of copper, lead and zinc, for example.
In the investigation of receiving waters it is important to inspect closely the bed of the river or stream, in combination with a biological examination of it (*R. Weimann,* Techn. Berichte des Wirtsch. u. Verk. Min. Nordrhein-Westfalen No. 8 [1951]). Larger amounts of fallen leaves can lead to considerable loss of oxygen in the deep parts of standing water (*E. S. Chase,* J. New Engl. Water Works *71,* 307–312 [1957]).
So-called *"thermal contamination" of streams,* e.g. from the cooling water of power stations and, in future, of atomic power stations, accelerates biological processes. *Knöpp* states that the rate of decomposition in water at 30° C is five times that at 5° C. This depends, however, on massive multiplication of microorganisms, naturally including pathogenic bacteria too. Degassing takes place also as a result of the rise in temperature. Loss of oxygen is the consequence, perhaps eliminating the living basis of the fish (with certainty if they are sensitive). Plankton growth is hindered through the loss of CO_2 from the water, and this in turn impedes biogenic oxygen production.
The determination of phenols in interceptors and river courses is important. These often gain access to river water via industrial wastewaters. This causes damage to the fishery and leads also to the unpleasantly tasting chlorophenols when the water is conditioned as drinking water by chlorination. The discharge of phenol-containing wastewaters into rivers is strictly supervised by many authorities. The contamination of the Ruhr in W. Germany may be given as an example (GWF 1929).
A continuous check must be maintained also on salt addition to rivers through the effluents of the potash and other industries; determinations include those of the alkali and alkaline earth metals as well as of chloride.
Waters strongly polluted with organic matter show zero oxygen content after 24 hours at the latest, and 100% oxygen consumption. The biochemical oxygen demand was introduced to achieve a numerical differentiation of the more extensive degrees of pollution. *Biochemical oxygen demand* (BOD; BSB_5 in German) gives the degree of loading of a receiving water (see below).
F. Grabbe describes a "warning apparatus" for oil (Industrieabwässer 45–48 [1968]). If the river water contains mineral oil, the current between two electrodes in the apparatus is reduced and the uptake of water can then be halted.

2. Determination of the Biochemical Oxygen Demand

Dilution method: The biochemical oxygen demand is the amount of oxygen consumed in 5 days at 20° C through the biological self-purification of the water, i.e., during bacterial degradation of the organic compounds present (expressed in mg/l or g/m³).
The receiving water under investigation, or wastewater, is diluted with specially prepared water (saturated but not supersaturated with oxygen) immediately after sampling. This dilution water is prepared in 5-litre containers from pure tap water or distilled water by frequent aeration and standing. About 2–4 weeks are needed to attain equilibrium. A few mg of $NaHCO_3$ are added per litre to the distilled water. After the 2–4 weeks, no further oxygen

consumption is shown and the containers are closed and stored in the dark. The oxygen content is measured in each case. It must not exceed the saturation value.
The drawn water for investigation is mixed with the dilution water in a 2-litre vessel. The extent of dilution is governed by the potassium permanganate consumption of the water under investigation and ranges from 5:1 to 20:1. The biochemical oxygen demand of normally loaded receiving waters and, discharges from percolation filters is 1–2 times the oxygen consumption determined by the $KMnO_4$-method (p. 40); for wastewaters clarified by sedimentation, the ratio is 3–5.
The water mixture is siphoned into 300 ml glass-stoppered bottles (oxygen bottles) up to the mouth of the bottle. The stopper is replaced and the bottle left at 20° C in the dark. The oxygen content is determined as described on p. 102, after 1 hour and after 5 days. The former measurement gives the purely chemical oxygen demand. From 3 to 5 mg O_2/l should still be present after 5 days.

If:
the dilution (ml of studied water in 1 litre of mixture) = m
original O_2 content of the dilution water = a
O_2 content of the dilution water after 5 days = d
O_2 content of the mixture after 5 days = b
O_2 content of the mixture at the moment of preparation = c
then the BSB_5 (BOD) in mg/l = $a - (a - c)\,m + m\,(d - b) - d$

Higher concentrations of poisons such as phenols, copper, free chlorine etc., or strongly acid or alkaline solutions *interfere*.
Concerning the evalution of the BSB_5, see also *R. Weimann*, Schriftenreihe d. Ausschusses f. Kulturbauwesen in Westdeutschland, No. 2 and *W. Niemitz*, Arch. f. Hygiene *148*, 288–298 [1964]).
H. E. Klotter and *E. Hantge* reported in detail on the evaluation of biological investigations on waters and their relation to the biochemical oxygen demand (BSB_5) (Wasserwirtschaft *56*, 21–26 [1966]). Values expressing the "*biological condition*" of a stream were calculated (see also Lit. Ber. XIV, p. 49 [1966]).
Continuous measurements of the BSB_5 (BOD) can be carried out with the Sapromat of the firm of Voith, Heidenheim/Brenz, W. Germany (*H. Liebmann* and *K. Offhaus* [1966] and the less expensive apparatus of the firm of Hoelzle and Chelius, Frankfurt am Main).

3. Biomass Titration for Determination of the Biologically active Components of a Water

According to *G. Bringmann*, the "biomass titration" is the amount of organic substance in definite unicellular test organisms which can be formed in a water sample under certain conditions. The biomass titration of test organisms is determined as a function of non-mineralised nitrogen compounds of the water, through cultures of *Escherichia coli;* and as a func-

tion of mineralised nutrient through cultures of *Scenedesmus*. The biomass titration enables the saprobic load of a water through organic nitrogen compounds to be determined. At the same time it provides a means of comparing acute saprobic load and general eutrophication in a water (further details: *G. Brinkmann*, Ges.-Ing. *77*, 374–381 [1956]; *G. Brinkmann* and *R. Kühn*, Ges.-Ing. *79*, 329–333 [1958]; and an exhaustive treatment of this physiological procedure in biological water analysis in the Handbuch der Lebensmittelchemie Vol. VIII 1200–1228 [1970].
The directions for the investigation are voluminous and unfortunately cannot be included within the scope of this book (see foreword).

4. Determination of the Phenol Content

Phenols in river water (Hinden-Splittgerber procedure):
The reagent is a solution of 1.38 g of p-nitroaniline in 310 ml of N hydrochloric acid, made up to 2 litres. 20 ml of this reagent solution are cooled in ice and decolorised with sodium nitrite. The water under investigation is treated with sodium hydroxide solution immediately after sampling. After solid has settled, the clear solution is siphoned off and 20 ml of it, together with 30 ml of N sodium carbonate solution, are added to the 20 ml of the above reagent solution immediately after decolorisation mentioned above. Samples of grossly contaminated river water and of wastewater are distilled.
The red colour yielded in 10 minutes is compared with that from a reference solution which is treated in the same way. This solution contains 0.5 g of m-cresol and 0.5 g of phenol in 1 litre of water. 1 ml contains 1 mg of phenols. It is appropriately diluted.
Colour comparison is possible in the range 0.1 to 2.0 mg of phenol per litre. Solutions of higher concentration are diluted, those of lower concentration are enriched by distillation. The chief phenols reacting are xylenol, resorcinol, α-naphthol, cresol and phenol. Hydroquinone, pyrocatechol and chlorophenols do not react. A photometric determination can be performed at 530 nm.
According to *H. Könen*, the phenol determination suffers interference by sulphite waste liquor (Z. Anal. Chem. *128*, 127 [1948]).

Colorimetric phenol determination according to Martin: This is a specific and highly sensitive method *(R. W. Martin*, Anal. Chem. *21*, 1419 [1948]). 100 ml of the water sample, neutralised if necessary, are treated with 0.6 ml of a 2% solution of 1-phenyl-2,3-dimethyl-4-aminopyrazolone (analytical grade, e.g. Merck), 2 ml of 5% ammonia and 2 ml of 2% potassium ferricyanide solution (shaking after each addition). The red colour formed is compared with those from the standard phenol solutions made up at the same time, using a colorimeter or photometer (460 nm). Ferrous ions interfere.

5. Quantitative Determination of Detergents (Tensides)

Detergents, also termed tensides, are surface-active, synthetic washing agents. They are replacing ordinary soap more and more for washing textiles. Distinction is made between anion-active, cation-active and non-ionic agents.

Up till recently, detergents made their presence most disagreeably felt in receiving waters (reached via waste waters) through marked foaming. Mountainous froth was sometimes formed, especially at dams.
There was no way of removing detergents from wastewaters until a few years ago. The situation has now improved since the *detergent decree* of 1/10/64 entered into force. Only those detergents which can be easily broken down biologically are now allowed. When the old stocks are used up, this contamination through detergents will cease. The easily decomposable, "soft" detergents are based on cetylbenzenesulphonate which is 80% broken down biologically. The residual 20% are giving cause for new concern as a result of accumulation through wider use.

New standard method for determination of detergents
The special commission for detergents proposed in 1960 the method of *J. Longwell* and *W. D. Maniece* (Analyst *80,* 167 [1955] as standard procedure for determining anionic detergents. It is as follows:
To 100 ml of the water under investigation (wastewater after it has settled) in a separating funnel are added 10 ml of alkali phosphate solution, 5 ml of neutral methylene blue solution and 15 ml of chloroform (analytical grade). The mixture is shaken after each addition. The amount of sample is chosen according to the content of detergent. It should have between 20 and 150 μg. This must be established through preliminary experiments. Smaller samples are taken where the water contains more detergent, and are made up to the 100 ml.
The mixture is shaken gently for 1 minute, uniformly with two movements per second. The chloroform phase is run off into a second separating funnel which contains 110 ml of distilled water and 5 ml of acidic methylene blue solution. Two further ml of pure chloroform are then added to the first separating funnel and the mixture is shaken once more. The chloroform from this is run into the second funnel.
Any emulsion can be broken with a glass rod.
The second separating funnel is shaken just like the first. The total chloroform phase is filtered into a 50 ml standard flask through a small cottonwool plug, saturated with chloroform. The plug is washed through with 2 ml of chloroform.
The contents of the first funnel are extracted by shaking with two more 10 ml portions of chloroform. The chloroform phases are filtered into the standard flask and the volume is made up to 50 ml.
Comparison is made at 650 nm with standard solutions of Manoxol (disodium dioctylsulphosuccinate of the firm of Chem. Werke, Hüls; the German standardisation compound is tetrapropylenebenzene-sulphonate, "TPS" of Merck, Darmstadt).
W. K. Fischer (Münchener Beiträge, Vol. 9, *24,* 46 [1967]) reports full details about sources of error, improvements and abbreviations of this official procedure.
Filtrates from membrane filters must not be used for determinations of detergents since the filters contain detergents. (*W. K. Fischer,* Z. Anal. Chem. *227,* 116–120 [1967] and Henkel-Referate 4 [1969]).
Reagents. Alkali phosphate solution: 10 g of Na_2HPO_4 are dissolved in about 200 ml of distilled water and the pH value of the solution is adjusted to exactly 10 by adding sodium hydroxide solution. The volume is then made up to 1 litre with distilled water.

Neutral methylene blue solution: 0.35 g of methylene blue (Bayer DAB 6, firm of Chroma Ges., Stuttgart-Untertürkheim) is dissolved in 1 litre of distilled water.
Acidic methylene blue solution: 0.35 g of methylene blue is dissolved in ca. 500 ml of distilled water, 6.5 ml of conc. H_2SO_4 are added and the total volume ist made up to 1 litre with distilled water.
(see *H. Wachtel,* "Aquarienhygiene", Stuttgart 1963 for information about the *water of aquaria*)
Pesticides
Waters may be contaminated also by the pesticides from wastewaters and from tillage of the land. An example is Toxaphene dust (*H. Althaus* and *B. de Jong* GWF *106,* 559–564 [1965]).
Pesticides are very difficult to detect in waters and this requires considerable outlay of apparatus. It is therefore confined at the moment to a few specialist institutes. The pesticides are extracted with active carbon from larger amounts of water, and the extracts treated chromatographically. Quantitative determination is by gas chromatography using a microcoulometer or electron capture detector. *K. E. Quentin* gives further details in the Handbuch der Lebensmittelchemie Vol. VIII, p. 633 et seq (1970).

D. Wastewater (Sewage)

1. Sampling

What has been said about correct sampling of drinking water applies with special force to sampling of wastewater. Of primary importance is the right choice of sampling site (especially for sampling in sewers) and time (e.g., of peak occurrence). It is usually necessary to sample both clarified and unclarified waste water at various times of the day and to prepare mixed samples from various individual samples.
Only the specialist should take samples of wastewater. Samples which are sent are useless for investigation.
The concentration of wastewater fluctuates markedly during the day. Information must thus first be collected from the attendant at the clarifying plant about the peaks and concentration situations. If this proves impossible, samples must be taken every 15 or 30 minutes (day and night, if necessary) and mixed samples prepared from these. Sampling should not be extended over 6 hours with unclarified water containing putrefiable matter, or else part of the samples must be conserved immediately (see below). When sampling over many hours is not possible, one must choose the times of maximum discharge and maximum contamination.
To be able to investigate the action of a clarifying plant, care must be taken that *corresponding samples* are taken from one and the same wastewater surge in the different purification stages. The rate of flow through the single stages (clarification basins, percolation filters, secondary settling tanks) must be established for this (e.g., through experiments with common salt, chloride determinations, tests with the dye uranine (sodium fluorescein), or with radioactive isotopes).

The clarified wastewater must be sampled as near as possible to the outflow place. The *Dinse*-beaker is especially suitable for sampling wastewater. It is submerged in the water flow with a supporting holder (firm of Franz Bergmann KG., Berlin-Zehlendorf and Hamburg). Two oxygen bottles also can be fitted into the *Dinse*-beaker. *Prütz's* "Stockklammer" ("stick clamp"), on which all vessels can be dipped into open waters and tanks, is very practical too (supplied by Otto Otto of Hildesheim, W. Germany).

Amount of sample: 2 litres are needed for a normal investigation of wastewater. Glass stoppered bottles of Jena glass are the most suitable. Samples are conveniently sent in aluminium bottles with an inner protective lacquer coating, and plastic bottles. Conserved samples are needed in addition.

Conservation of samples: All domestic and municipal wastewater samples undergo rapid changes as a result of bacterial processes. Conservation is thus advisable, especially when immediate investigation is not possible. It is carried out with 3 ml H_2SO_4 $(1+1)$ per litre for subsequent determination of the $KMnO_4$-consumption, of ammonium ion and of organic nitrogen compounds; with sodium hydroxide (3g/l) for the phenol determination; and with 2 ml chloroform per litre for the remaining determinations. An untreated sample is taken for chloride and pH determinations. Hydrogen sulphide is fixed with cadmium acetate. For the BSB_5 (BOD) determination also, an untreated sample is taken (see p. 215).

Investigation on the spot: As with drinking water (p. 14), the temperatures of wastewaters and the air at the site are measured. Parallel also are records of the external nature, colour, odour, transparence, depth of visibility and reaction (p. 16 et seq). The odour demands special attention; it may be faecal, putrefying, acidic or foul. The behaviour of the sample during several following days is observed. Other determinations etc. on the spot are: the methylene blue test; determinations of oxygen, BOD, phenols and hydrogen sulphide; approximate field determinations of nitrogen compounds (ammonia, ammonium-, nitrite- and nitrate ions) and of pH. Samples to be conserved are taken finally.
The determination of the degree of turbidity is performed as on p. 18.

2. Some Special Tests of Wastewater

Determination of "settleable matter": The value of the "settleable matter" gives an approximate measure of the amount of fresh sludge to be expected.
It is carried out on wastewater in special conical settling vessels of 250 to 1000 ml (according to *Imhoff, Spillner* or *Schulze-Forster* and provided by the firms of Franz Bergmann KG of Berlin and Hamburg, and of Feddeler of Essen) which have graduations in ml at the bottom (apex). The measured amount of water is allowed to stand for 2 hours in them. After one hour, the glass is spun several times on its axis to free matter adhering to the walls, so that it can settle also to the bottom. This freeing movement is repeated just before the second hour elapses. The ml of settled matter is calculated for 1000 ml sample.
This determination must be carried out if possible straight after sampling and at the latest after 2 hours. If not, flocculation may occur, falsifying the results. The effluent from good functioning primary clarification plants should not contain more than 0.3 ml/l. According

to the "Normal Values for Wastewater Purification Procedures of the LAWA" (1970), this value is attainable for domestic wastewaters, even when only primary purification is employed. It is always practicable for industrial wastewaters also.

Determination of suspended matter: Suspended matter is determined in a porcelain filtering crucible or, better, in an asbestos-filled Gooch crucible. 200 to 500 ml of wastewater are placed in a measuring cylinder. After leaving to settle for a short period, the sample is transferred in portions to the filtering crucible which has been ignited and weighed. It is then sucked through with a water pump. The bulk of the suspended matter is brought on to the filter at the end. Measuring cylinder and crucible are rinsed with distilled water.

The crucible is then heated for an hour in a drying oven at 105° C, cooled in a desiccator and weighed. The result is quoted as mg of suspended matter per litre. The total residue and the loss of weight on ignition of the suspended matter can be given also.

The residue on ignition of the suspended matter is determined by placing the crucible in a crucible support and heating at 600° C with a Bunsen flame or in a muffle furnace. After cooling in a desiccator, the crucible is reweighed. The weight of residue is expressed as mg/l.

The loss on ignition is the difference between total residue of suspended matter and residue on ignition, all in mg/l. If the ignition temperature is kept approximately at 600° C, the loss on ignition is attributed only to organic substances.

The calculation is as follows:

Weight of the crucible	a
Weight of crucible + total residue	b
Weight of suspended matter	b–a
Weight of suspended matter after drying at 105° C	c
Weight of crucible + residue on ignition	d
Weight of residue on ignition	d–a
Weight of organic suspended matter	c–a

The HCl-soluble part of the residue on ignition is determined by treating this residue in the crucible several times with 10% HCl, washing with distilled water and applying suction. The crucible is redried in the oven and reweighed. The loss in weight gives the HCl-soluble part.

G. Giebler and *T. Kempf* found fibre-glass filters to be good for determining suspended matter (Z. Anal. Chem. *199*, 23–24 [1964]).

Biochemical oxygen demand: see p. 215 for determination.

The determination of oxygen and of oxygen consumption is done on unfiltered waste water, as given on p. 102–106.

Methylene blue test (test for putrescibility and stability): The test for the putrescibility of a wastewater gives information about the sulphur-containing organic compounds and the degree of loading for the receiving water. The wastewater is run into a 50 ml glass stoppered flask containing 0.3 ml of 0.05% methylene blue solution, up to the top of the neck. The stopper is fitted so that no air bubble remains in the flask. As soon as possible after sampling, the flask is placed in an incubator at 37° C for 1 to 6 hours and is observed frequently during this time. If no decoloration takes place during the 6 hours, the water is

considered no longer putrefiable. Alternatively, the mixture is kept for 48 hours at 22° C and watched for decoloration. The test can be performed on filtered water also.
The other determinations are carried out on filtered waste water.
Filters of crêpe paper are used for filtration: the usual filter papers retain colloidal parts.
Testing for hydrogen sulphide is initiated on the spot. A strip of filter paper saturated with lead acetate solution is suspended in its moist state above the liquid in the sample bottle which is about four fifths full and at 20 to 22° C. The vessel is occasionally gently moved to and fro. Hydrogen sulphide is shown by a brown coloration. It is often not recognisable through its odour when the wastewater has a putrefying smell. Gentle warming is necessary if the content of hydrogen sulphide is low.
For a more accurate determination of hydrogen sulphide on the sampling spot, 10 ml of a 5% cadmium acetate solution are added to a 250 ml sample of water. The precipitate is filtered in the laboratory after it has settled, using an asbestos-filled Gooch crucible. The bulk of the cadmium sulphide precipitate is brought on to the filter with the last 10 to 20 ml of solution. The asbestos plus precipitate is returned to the sample flask and treated with 10 ml of 0.1N iodine solution and 10 ml of hydrochloric acid (d = 1.12). After leaving for 2 hours, the solution is titrated with 0.1N thiosulphate solution (starch solution as indicator; 1 ml of 0.1N iodine solution = 1.7 mg H_2S).
Further determinations to be carried out on the filtered or decanted samples in every investigation of wastewater are:
Permanganate consumption; chlorine number; chloride content[1]; ammonia; nitrite; nitrate; phosphate; organic nitrogen (total nitrogen less inorganic nitrogen and albuminoid nitrogen); residue on evaporation; residue on ignition; pH-value. All these determinations are carried out as described under drinking water but after appropiate dilution and clarification or decoloration. The total nitrogen is determined in the usual way according to Kjeldahl. If necessary, the phenol determination as described on p. 217 is performed.
The Magnos apparatus of the firm of Hartmann & Braun, Frankfurt/Main and the apparatus of the firm of Chlorator GmbH., Grötzingen, W. Germany are suitable for *continuous oxygen determinations* in plants for sludge activation. The former is based on paramagnetic measurement of the oxygen which is expelled with nitrogen; the latter, on phase exchange, *G. Stracke's* apparatus determines the oxygen electrochemically; it fails with fouled sewage.
Another development is the "BSB-Wächter" ("BOD-sentinel") of the Passavant Works of Michelbach, W. Germany. It is based on manometric measurement of BOD according to *Warburg*.
The Dephimeter of the firm of Krohne, Duisburg, is used for continuous determination of salt content.
Cyanides: See p. 229 in the section on trade wastewaters for cyanide determination; the A–Z test of *H. Knöpp* is found there also.

1 For this purpose, turbid wastewaters containing iron are clarified by shaking with chloride-free zinc oxide and filtering. Organic matter can be removed by boiling 100 ml wastewater with $KMnO_4$-solution and then with added alcohol to remove excess $KMnO_4$. After filtering into a 200 ml flask and making up to the mark, the titration is carried out with 100 ml (= 50 ml of sample).

3. Assessment of the Wastewater

The assessment of a purification plant for wastewater demands highly specialised experience. Only a brief survey can thus be made in this book (furhter details in *W. Husmann,* Praxis der Abwasserreinigung-Purification of Wastewater in Practice, [1969]).
Rapid removal of undissolved material and its separation from the liquid wastewater phase is important for the total purification process.
The effectiveness of the primary (sedimentation) treatment, i.e., the difference between the content of suspended matter in the unclarified and the clarified wastewater, should be 60 to 70%, expressed as percent of the unclarified water. When the sedimented wastewater leaves the plant (settling basins, Emscher-, Kremer- or OMS tanks), it should contain less than 100 mg total suspended matter per litre, and not more than 0.3 ml of suspended matter should separate from 1 litre in 2 hours.
Water leaving biological purification plants must be non-putrescible and should contain no more than 0.3 ml of suspended matter per litre; that leaving primary (sedimentation) plants, likewise not more than 0.3 ml/l. The percolating filters would be overloaded otherwise.
Very important indications of the biological purification are increase in nitrate together with decrease in $KMnO_4$-consumption and, above all, in the BSB_5 (BOD).
The effectiveness of the biological purification can be gauged from the reduction in $KMnO_4$-consumption which should be at least 50 to 60% with respect to the unpurified wastewater. The ratio of ammonia to organic nitrogen is highly significant. The former is much the higher in purified, non-putrescible wastewaters; the reverse applies to putrescible, raw wastewaters, i.e., the organic nitrogen value is higher.
The ratio of permanganate number to chlorine number (p. 72) can also be utilised.
The $KMnO_4$-consumption should fall by 75 to 90% after irrigation field treatment.
The methylene blue test is important for the assessment of the putrescibility (see above). A putrefying wastewater is much more difficult to purify biologically.
Permanganate consumption and chloride content yield clues to the concentration of wastewater in domestic and municipal effluents (see below).
Most domestic wastewaters have a pH of ca. 7.4, municipal wastewaters of ca. 8.0. Acid fermentation occurs at pH below 7.0.
The pH value of industrial wastewaters plays a decisive part in the corrosion of slide valves and destruction of the concrete of sewer pipes.
The biochemical oxygen demand of the clarified wastewater and of the receiving water must be evaluated above and below the position of inflow into interceptors of low flow. It often comes about that even purified wastewaters load the receiving water markedly and bring about a powerful consumption of oxygen.
The outflows from biological purification plants should have a BOD of below 30 mg/l for percolation filters and of ca. 30 mg/l for activated sludge plants.
The latter procedure is benefited by high iron content of the wastewater, since the sludge settles better and phosphate is largely combined. This is an advantage for the receiving water (p. 213). High iron content is, on the other hand, unfavourable for trickling filters.
Thumm gives the following values as normal for different wastewaters:

Normal values	Undissolved matter	Evaporation residue	$KMnO_4$-consumption	NH_3	Cl
1. Dilute wastewater	up to 300 mg/l	up to 500 mg/l	up to 200 mg/l	up to 30 mg/l	up to 100 mg/l
2. Medium wastewater	up to 500 mg/l	up to 1000 mg/l	up to 300 mg/l	up to 50 mg/l	up to 150 mg/l
3. Conc. wastewater	over 500 mg/l	over 1000 mg/l	over 300 mg/l	over 50 mg/l	over 150 mg/l

4. Sewage Clarification

Distinction is made between mechanical clarification and biological purification of the wastewater. The former separates undissolved sediment; the latter involves biogenic degradation of dissolved and colloidal organic matter. The first stage of mechanical clarification is mostly a screening or sieve plant to retain the coarsest sediment (coarse screens to retain clogging constituents, and fine screens). A newer departure is replacement by a comminutor. This is succeeded by a grit chamber, especially with the mixed system (sewage and street gulleys), and often also a grease separator. The most important part of every clarification plant is the settling basin, where the sedimentation takes place. The settled sediment is putrefied in these digestion chambers or those straight below it, within hours or days. After the digestion process, the still fluid sludge is let on to sludge drying beds from time to time. These are usually bricked basins with bottoms of loosely layered bricks, under which there is loose, easily water-permeable sand or slag. The sludge generally needs about 10 to 20 days to become compacted and it is then taken off. It provides a manure, rich in humus, which improves arable land, particularly of light soil. It is usually given free to local farmers.

In 1907, *Imhoff* built the first modern primary clarification plant in the form of the Emscher (Imhoff) tank. It consisted of two basins, one above the other. The intermediate partition was formed from sliding walls carrying a slit in the middle, through which the sediment entered the lower basin, the real digestion (septic) tank. The OMS- and Dywidag tanks are constructed similarly. The Kremer tank has a detached septic tank.

Water which has been only mechanically clarified is still putrescible and would unduly load the receiving water. It must thus be subjected afterwards to a biological process for breaking down the putrescible organic compounds.

Purification of wastewater has made some progress in the cities and communities in recent years. Nevertheless, no purification of wastewater is undertaken in about one-third of the communal sewage systems (W. Germany) and only every fifth purification plant is really adequate.

Complete biological purification is the aim in every case. This is accomplished by:

5. Biological Purification Procedures

1. *Percolation filters:* The clarified wastewater is distributed over slag material. It trickles through it, coming into contact with air and is thereby purified through numerous wastewater bacteria and microorganisms.
2. *Activated sludge process:* The clarified wastewater is agitated in cemented basins with the help of paddle wheels or blown-in air. This forms an active sludge, honeycombed with vast numbers of microorganisms which biologically clarify following wastewater. The settled sludge is drawn off in secondary settling tanks and is partly reused as "return sludge" in the aeration tanks.
Mineralisation occurs during the biological process. The permanganate consumption of the water thus sinks appreciably. A well-functioning biological purification removes almost entirely the plentiful ammonium ion amounts in the water clarified by settling. Nitrite and nitrate appear in their place.
3. *Sewage farms* (irrigation fields)
4. *Sludge-blanket filter* (Reaktivator process) or flocculation-vortex-layer (Accelator process)
With these, the sludge is propelled back through the wastewater, causing rapid precipitation of the organic matter in a reaction zone. A 90% reduction of the BOD of municipal waste water takes about 3 hours.
5. *Sewage lagoons* and *fishponds:* These can accomplish biological purification of mechanically clarified wastewater (pond manuring near Munich).
6. *Chemical purification of wastewater:* This has been practised little in Germany so far. The *Laughlin* process (colloidal precipitation with ferric chloride, lime and paper pulp) is used extensively in the USA.
7. *Suitable cultures of Scirpus lacustris:* Dr. *Käte Seidel* has succeeded in Krefeld in purifying receiving water and wastewater with the help of this plant; even phenol was removed.

Concerning 1: The water from sedimentation is brought on to the percolating filters (slag filters of 6 to 15 feet height) with the help of watering channels or rotating distributor arms (*Segner* slewing arms, also called sprinklers). The water is expelled from one side of the arms, imparting the rotating movement. The water forms thin jets so that considerable atmospheric oxygen comes into contact with it.

"Biological slime" forms on the percolating filters during the "running in" time (see also p. 61).

The biological percolating filters were formerly open but nowadays they are walled in, on account of the bother from flies. Adequate artificial aeration must then be ensured. The sewage organisms which effect breakdown of the dissolved organic matter require considerable oxygen. An average of 0.5 cubic metre (125 gallons) of wastewater can be treated in 24 hours by 1 cubic metre of percolating filter mass. This figure reaches 2 to 4 cubic metres of water with the modern high-rate sprinkling filters.

Concerning 2: The activated sludge process is being used nowadays instead of the trickling filter process. The water from the sedimentation clarification is led into long cement basins about 3 to 6 feet deep. Here it is vigorously agitated with paddle wheels or compressed air. Activated sludge is thereby formed. It is loose and packed with innumerable plant and

animal wastewater organisms. Part is taken away, part returned to the first tanks (sludge return). This sludge is especially active. It adsorbs colouring matters and other organic compounds and breaks them down (see *Höll*, Zbl. Bakteriol. II, Abt. *81*, 198 [1930]).
The wastewater spends about 6 hours in the sludge tanks. The activated sludge is still putrescible and contains much water. It cannot be dried directly on sludge drying beds and is usually returned to the digestion chambers of the sedimentation clarification plant for digestion. According to *W. Bucksteeg* and *H. Thiele*, sludge activity can be tested with TTC (triphenyltetrazolium) (GWF *100*, 916 [1959] and Korresp. Abwasser 1966, p. 9).
A special form of the actived sludge plant is the *"oxidation ditch"* which is particularly good for smaller communities. The clarification process with oxidation ditches needs a great deal of space, however, which is not the case in the further development to the " total clarification plant of Prof. *Kehr*" (see also *R. Nehrkorn* and *H. Reploh*, Ges.-Ing. *87*, 143–146 [1966] and *K. A. Möhle* and *K. Mudrack*, Die. Nieders. Gemeinde, 1968, p. 5).
D. Kehr has indicated new ways of "aerobic sludge stabilisation" (Korresp. Abwasser 1966, p. 10; ATV-Berichte *14* [1962]; and GWF *104* [1963]).
The *Mammut-Rotor* of the Passavant Werke, Michelbach, W. Germany, is suitable for larger activated sludge plants; the *Kavitator-process* of the Süddeutsch. Abwasserreinig. Ges. Ulm, W. Germany (SAG-activated sludge plant) for smaller plants.

Concerning 3: Sewage irrigation with the sedimentation-treated water is the oldest procedure. It is still used considerably in Germany but has now been economically superceded. In narrowly restricted regions it is replaced now by spray irrigation of sewage. Where technically feasible, unclarified wastewater can be sprayed. 1 hectare (2.5 acres) of sewage land suffices for pre-clarified sewage from 1000 people.

Concerning 4: The now *Accelator process* (firm of Lurgi, Frankfurt/Main) and *Reaktivator process* (firm of Bran & Lübbe, Hamburg) are flocculation procedures which accomplish in a very limited space extensive purification of all types of wastewater and yield a greatly thickened sludge. In the former process, powerful aeration can be achieved in addition through a Cyclator-system.
Little room is required by the *Accelator-* and *Cyclator-processes* and also by the *Pulsator process* (firm of Degremont) and *Flocculator process* (firm of Wabag, Kulmbach).
The *Aero-Accelator* process of the Lurgi firm of Frankfurt has given good results with wastewaters, such as those from dairies, which are difficult to treat (*A. Kannemeyer*, Molkerei-Z. *9*, No. 7 [1958]). Further developments are the *Simplex Highpower plant* with high-power centrifuges (firm of H. Koppers, Essen), the *Vortair-Aerator* of the firm of Lurgi, Frankfurt, and the *Koagulator Bauart Passavant* of the Passavant-Werke, Michelbach.
Complete clarification plants are built by the firms of Dorr-Oliver, Wiesbaden (fully biological small clarification plant "*Completreaktor*"); H. Koppers, Essen (Simplex); and A. Rieber GmbH, Reutlingen (Rieber), all of W. Germany.
The flocculation process has proved satisfactory abroad in hundreds of plants already, especially with chemical precipitating aids (lime, ferric chloride, aluminium sulphate), in drinking water conditioning and in treatment of wastewater.

6. The Third Sewage Purification Stage

Removal of phosphates from the purified wastewater has now become necessary as a third purification stage, since intense eutrophication of streams through phosphates could occur.

H. E. Klotter reported his experience in this connection in "Vom Wasser", 1964. He removed the bulk of the phosphate with aluminium sulphate. His investigations show that two-story clarification plants are especially unsuitable regarding phosphate release into effluents running into lakes, ponds and reservoirs.

H. Liebmann discusses extensively the problem of the third purification stage in his collective work "Münchener Beiträge zur Abwasser-, Fischerei- und Flussbiologie" (Munich, Vol. *12,* 56 [1965]).

In the Wahnbachtal reservoir, phosphate removal from the inflow has proved satisfactory. It was done by bringing the water to the best pH (between 6 and 7) with calcium hydroxide solution and using ferric salts and Wisprofloc as flocculation agents (*H. Bernhardt* and *K. Kremling,* Vom Wasser *35,* 160–226 [1968]).

High state subsidies are available in Sweden for clarification plants. These are still higher when a third purification stage is constructed; this acts as a stimulus to building. The simultaneous process of *E. A. Thomas* is that mainly used in Switzerland. It employs ferric chloride to precipitate phosphate as iron-return sludge process and is most successful in many communal clarification plants. More than 90% of the phosphate is eliminated.

7. Trade Wastewater

The compositions of trade wastewaters can vary greatly. This cannot be discussed in more detail here, where we are giving only some of the more frequently encountered types.

Acid-containing wastewater from factories for metal articles and steeping processes: These must at least be neutralised and freed from iron before being introduced into the receiving water. pH-recorders can be used for control.

See *R. Weiner* [59] concerning wastewaters of the metal industry; and *H. Liebmann* concerning liquors from galvanising works (Münchener Beiträge zur Abwasser-Biologie, Vol. 7, Munich, 1960). They are nowadays freed from poisonous metal salts with the help of ion exchange units also (*F. v. Ammon,* Lit. Ber. *17,* 248 [1969]).

The nickel dissolved in used nickel-plating baths can now be extracted economically by using cation exchangers. Both the bath and the nickel may then be reused (firm of Götzelmann, KG, Stuttgart, etc.)

Alkali-containing wastewater must likewise be neutralised.

Phenol-containing wastewater: Phenols and their derivatives occur in wastewaters from coke plants, gasworks, and plants for tar and wood distillation. They gain access to the receiving waters in these wastewaters.

Rough field determination of the phenols: Phenol-containing water yields a blue colour with a few drops of ferric chloride solution.

Phenol is usually determined according to *Hinden-Splittgerber* (p. 217).
The following ways of *purifying phenol-containing wastewaters* in sources such as those mentioned above, are listed here:

1. Expulsion by means of hot gases (hot air).
2. Extraction with solvents (benzyl acetate, Phenosolvan-process).
3. Adsorption with active charcoal (Lurgi process).
4. Ion exchangers (*H. Anders,* GWF *92,* 238 [1951]).
5. Bacterial degradation.
6. Biological degradation with fungi of the *Oospora* genus (*K. J. Palatý,* Prague).
7. Cultures of *Scirpus lacustris* in effluves and receiving waters (*K. Seidel,* Krefeld, W. Germany).

Cyanides in wastewaters from steel tempering and galvanisation plants must be taken out before the water is led into the receiving waters. This detoxification can be done with bleaching powder, sodium hypochlorite, chlorine gas, but only in strongly alkaline solution on account of formation of chlorine cyanide; or by blowing out from the acidified solution–but only under certain definite conditions.

Magdeburg P-procedure (Dr. *Nolte*): Since most industrial wastewaters do not contain enough phosphate and nitrogen, the activated sludge process can be carried out only after adding phosphate and ammonia. Under this condition, the Magdeburg activated sludge process worked out by Dr. *Nolte* has proved very good for purifying industrial wastewaters, especially those containing phenols. Ammonium phosphate is added, minimum amounts being 3 mg PO_4/l and 15 mg N/l in the wastewater. See p. 227 for supplementary phosphate removal.

Sulphite liquors: Sulphite liquors from factories for sulphite treatment of wood pulp contain calcium bisulphite and free sulphur dioxide together with large amounts of ligninsulphonic acids, sugar-like matter and fibrous materials. *These liquors are purified* in various ways. Best is evaporation of the liquor and ignition of the residue; or use of flotation- and precipitating agents, such as iron salts or aluminium sulphate (Deutsche Adka-Ges., Giessen/Lahn, W. Germany).

Coal washing waters: The Sedipur process of the BASF, Ludwigshafen, is satisfactory for their purification.

Brewery wastewaters: These are suitable for use in agriculture (irrigation or spraying). Even clarified brewery wastewater is bad for receiving waters; it contains large amounts of nutrient suited to wastewater fungi.

Wastewaters from starch works: These also may be considered for agricultural purposes. See in *Seyfried,* Veröff. Inst. für Siedl. Wasserwirtsch., TH. Hannover, No. 10 concerning purification procedures for wastewaters from starch production.

Wastewaters from laundries: These and other wastewaters containing large amounts of synthetic detergents should be treated with an anti-foam agent (e.g., Struktol ML 131, on a silicone base, furnished by the Chem. Fabrik Schill & Seilacker, Hamburg).

Wastewaters from the paper industry: These are difficult to purify but this can be carried out now by the Adka apparatus which functions according to the flotation principle (Deutsche Adka-Ges., Giessen/Lahn, W. Germany).

Field detection of sulphite: A strip of filter paper, moistened with a potassium iodide/starch reagent, is suspended above the acidified (with phosphoric acid) sample in a round-bottomed flask which is slightly warmed. Sulphur dioxide is shown by a blue coloration of the strip. If the sulphite content is high, the initial blue colour is rapidly bleached.

Iodometric determination of sulphite: The receiving water under investigation is filled into a 300 ml flask, as in the oxygen determination (p. 102). 10 ml of 0.1 N iodine solution and 3 ml of phosphoric acid are introduced on to the flask bottom. Regardless of running over, the flask is stoppered and the contents shaken. After 3 minutes, the contents are transferred to a beaker and titrated with 0.1N thiosulphate using starch indicator.

Interferences: Hydrogen sulphide and sulphides interfere and must be subtracted from the result. Organic matter also may interfere. In these cases, the acidified water can be gently warmed and the SO_2 or H_2S expelled by aeration. The consumption of 0.1N iodine solution by the water thus treated is evaluated and subtracted from the total value.

Calculation:

$$\text{mg } SO_3/l = \frac{\text{ml of consumed 0.1N iodine} \times 4.003 \times 1000}{\text{contents of the flask minus 13 ml}}$$

Arsenic, copper and zinc are determined as in drinking water.

Cyanide determination
According to *Stumm, Woker* and *Fischer* (Schweiz. Z. f. Hydrologie, 16 [1954] and *K. Knie* and *H. Gams* (Österr. Wass.-Wirtsch., Vol. 8, No. 10 [1956]), *poisonous cyanides* can be determined highly accurately in concentrations down to 10 µg/l through the reaction of bromine cyanide with benzidine. If complex iron cyanide compounds are present, the distillation with tartaric acid, described in the literature, yields incorrect results. Thiocyanates are determined as well by this method.

Procedure for the Austrian method: To 10 ml of water under investigation are given in succession, shaking after each addition: 1 ml of buffer solution (82 g of sodium acetate + 60 ml of distilled water, then 60 ml of glacial acetic acid added and the volume made up to 200 ml); 0.5 ml of saturated bromine water; 0.5 ml of 2% arsenite solution (8g of As_2O_3 + 4.4 g of anhydrous sodium carbonate, boiled with 48 ml of distilled water and solution made up to 242 ml); 6 ml of mixed reagent (1 g of benzidine hydrochloride dissolved in 15 ml of 0.2M HCl by warming, and 15 ml of pyridine added; prepared freshly each day); 4 ml of acetone. Colorimetry is then carried out.
A very convenient field method for cyanide determination is that with the Hellige comparator (*F. Hellige,* Freiburg im Breisgau, W. Germany), using chloramine solution and a barbituric acid/pyridine mixture.
The pyridine/pyrazolone method has been introduced as a standard procedure in the USA. *F. Dietz* (Vom Wasser *34,* 202–208 [1967]) reports on this.

In concentrations of even 0.1 mg CN/l, cyanide is fatal to fish (*R. Czensny,* Abh. a. d. Fisch., Lief. 2).
According to *K. Knie,* the limit of tolerance in river water is 10 μg CN/l.
In the International Standards, 0.2 mg CN/l has been fixed for drinking water.
H. Knöpp has worked out a new procedure, the "A-Z-Test" for testing the toxicity of waste waters (Deutsch. Gewässerkundl. Mitt. *5,* 66–73 [1961 [).
Serial investigations with ion-specific method: Using the ion-specific electrodes of the Orion Research Inc. (Cambridge, USA), serial investigations of cyanide ion can be carried out with considerable saving of time (see also p. 22).

E. Some Comments on Boiler Feed Water, Boiler Water and Cooling Water

1. Boiler Feed Water

a) General

Every cubic metre of boiler feed water introduces large amounts of salts into the boiler. These would constantly accumulate through evaporation if the water were not previously treated or removed from time to time.
In the simplest case of evaporators and flame-heated boilers with only a few atmospheres of pressure, a too extensive enrichment of salts can be prevented by taking out part of the boiler water and replacing it with boiler feed water.
Boiler water is checked with a hygrometer. The older hygrometers still use the Baumé scale. 1° Bé corresponds to a density of 1.007. Best is the "special hygrometer 123d" which gives intervals of 0.005. The boiler water of low pressure boilers should be diluted if the density exceeds 1.01. Any concentrating of salts is dangerous with high power and high pressure boilers; sludge depositions also cause danger and interference. The salt content of the boiler water of high pressure boilers should therefore not exceed 0.6° Bé at 20° C. A method of control is provided by the temperature-compensated density measuring instrument of the Chemische Werke, Hüls, W. Germany (*R. Neubauer,* Grosskesselbes. No. 2/3 [1948]). In recent times, the salt content of the condensate is checked continuously, e.g., through conductivity measurements.
Boiler feed water for modern high pressure boilers must always be pretreated. Chief sources of trouble are hardness-salts. There are two principal ways of elimination: Either softening with calcium hydroxide solution, followed by removal of the residual hardness with trisodium phosphate (p. 184); or continuous total demineralisation with Permutits or Lewatits (p. 185).
The *cation-anion exchangers* offer the only way to attain regular operation with very high pressure boilers. Permissible degrees of hardness of boiler feed water are 0.05° dH total

hardness for boiler pressures up to 40 atmospheres; and 0.02° dH for boiler pressures from 40 to 100 atmospheres. The total (residual) hardness for very high pressure boilers should be below 0.01° dH.

Boiler feed water must be free of iron (maximum value of 0.02 mg Fe/l and 0.00 mg Fe/l for very high pressure boilers) and free of mineral oil also.

Calcium sulphate (gypsum) is a particularly disadvantageous component of the water in boiler operation, since its inverted temperature/solubility relation means that it separates at the hottest points. The overheating due to gypsum boiler scale may lead to a boiler explosion or at least to rupture of the boiler wall.

Enough soda must thus be added to waters of high non-carbonate hardness to eliminate calcium sulphate as calcium carbonate which is removed from time to time with the boiler sludge (this is necessary even for low pressure boilers). Trisodium phosphate also has proved good in these cases without leading to such an increase of caustic soda concentration as occurs in the soda process. The *trisodium phosphate process* (Chem. Fabrik Budenheim near Mainz) can, however, be used also at higher pressures for successful removal of calcium sulphate from the feed water.

b) Silica in Boiler Feed Water

Silica in the water likewise leads to very resistant scale formation in the boiler. Silicate and gypsum scale are the most resistant of all and have the lowest heat conductivity. The boiler walls at these spots can thus suffer considerable overheating which may lead to damage or even explosion, as mentioned above.

The *silica* must therefore also be removed for higher boiler pressures. This is best done by heating the feed water with magnesium oxide, either in the boiler itself, or, better, in special processing plants beforehand. According to the *Hoechst procedure,* silica removal and decarbonisation can be performed at the same time using *dolomite hydrate.* The upper limits of silica concentrations (in mg SiO_2/l) are 5 for low pressure boilers, 0.1 for high pressure boilers and 0.01 for very high pressure boilers.

In high pressure boilers working at over 50 atmospheres, silica is deposited also on the turbine blades. This reduces the turbine performance. The silica layer must be laboriously removed with dilute sodium hydroxide solution. Silica is thus removed before feeding, as stated. This is best done with Permutit or Lewatit. Glass bottles should not be used for water samples.

Silica can be completely removed with Permutit ES, Lewatit ES and HL, Wofatit L 165 and Duolite A 101 D (firm of Benkiser, Ludwigshafen). Residual silica amounts to only 1–10 μg/l. The silikometer (firm of Bran & Lübbe, Hamburg) may be used for control.

Very pure water acts as an acid at the high temperatures of the steam boiler. As is known, the pH value falls with increase in temperature. The pH of the boiler feed water and in the boiler water should be in the alkaline region.

Since the iron boiler walls are attacked even by untreated boiler water at the high temperature prevailing in the boiler, a definite amount of sodium hydroxide must always be present, expecially when small quantities of oxygen are still present in the boiler water. At least

200 mg NaOH/l should be present for boiler pressures below 50 atmospheres; for high pressure boilers, the concentration of NaOH should not exceed 100 mg/l, however.
Too much NaOH leads to *caustic embrittlement* of the boiler and to tears in the rivets. This latter can be prevented by adding sodium sulphate to the boiler water. Formerly, a threefold excess of sodium sulphate over sodium hydroxide was customary; this is adhered to less rigidly nowadays since the addition of alkali has become replaced by phosphate addition.
The "caustic soda number" was a guide formerly used. Its value had to lie between 400 and 2000.
Caustic soda number =

$$\text{mg NaOH/l} + 0.222\,(\text{mg Na}_2\text{CO}_3\text{/l} + \text{mg Na}_2\text{SO}_3\text{/l}) + 0.036\,\text{mg P}_2\text{O}_5\text{/l}$$

The "alkalinity number" is used today. It is obtained simply by multiplying the p-value (p. 92) by 40.

c) Oxygen in Boiler Feed Water

The last *traces of oxygen* also must be removed from the feed water for very high pressure boilers. This is carried out by adding sodium sulphite, by boiling (thermal degassing)[1] or by vacuum degassing, attaining 0.03 mg O_2/l. In recent times, dithionite ($Na_2S_2O_4$) and hydrazine have been used. The latter undergoes thermal decomposition in the circulation water, rendering it alkaline.
4 mg of SO_2 are needed to combine with every 1 mg O_2 in the chemical procedure. Boiler feed water should not contain more than 1 mg SO_2/l in excess; and boiler water not more than 10 mg SO_2/l at pressures below 60 atmospheres.
Treatment with sulphite usually follows a thermal treatment. Thermal degassing removes other gases, such as CO_2, as well as oxygen. This is necessary for high pressure boilers.
A new procedure for taking out the last traces of oxygen is filtration through the redox exchange resin Duolite S 10. It can be regenerated (with sodium dithionite solutions) like other exchange resins (firm of G. Benkiser, Ludwigshafen/Rhein).
The oxygen content may be controlled electrochemically according to *Tödt,* or by the oxygen recording plant of the Chlorator GmbH, Grötzingen, W. Germany.

d) Hydrazine Determination

2 ml of sulphuric acid (1:1) and 20 ml of dimethylaminobenzaldehyde reagent are added to 100 ml of water sample, yielding a yellow or red colour. This is estimated after 10 minutes by measurement at 458 nm in a spectrophotometer, or in a filter photometer, against a blank solution.
The reagent is prepared by dissolving 20 g of p-dimethylaminobenzaldehyde in 1250 ml of isopropanol. It is commercially available as "Bayer reagent".

1 Suitable for this are the Doppel-Entgaser and Umlauf-Entgaser of the Firm of Philipp Müller, Stuttgart-Güglingen.

Maximum p-values for boiler water

Boiler pressure (atmospheres)	Maximum p-value
up to 20	15
up to 40	12
up to 64	3
up to 80	1
up to 125	0.5
up to 160	0.1

Maximum values for total salt content in boiler water

Boiler pressure (atm.)	Salt content (mg/l)
up to 20	4000
up to 40	3000
up to 64	2000
up to 80	1500
up to 125	750
up to 160	500

Maximum phosphate values in boiler water

Boiler pressure (atm.)	Phosphate (mg P_2O_5/l)
up to 40	30
up to 64	25
up to 80	15
over 80	10

These data claim in no way to be complete. Processing of boiler feed water has become a vast domain of its own. Reference is therefore made to the specialist literature (see the list on p. 264).

The technical progress–high pressure boilers up to 250 atm.–would be inconceivable without the great advances made in water chemistry.

Only soft water with a total hardness of 4.0° dH at the most may be used as *feed water for locomotives*. The salt content of the feed water should be below 400 mg/l. The degree of alkalinity (p. 91) must not exceed the residual hardness if the latter is less than 0.5° dH. Addition of polymeric phosphates to the locomotive feed water is advantageous, even when the water has been previously decarbonised.

2. Cooling Water

Cooling water must not bring about any deposits of scale or of looser nature on the cooling surfaces. It should therefore have a low hardness, especially a low carbonate hardness. The latter ought not to exceed 5° dH in recooling. Against this, the water should not be so soft as to attack the pipes, cooling surfaces and tank materials. Carbonate hardness must be distinguished from "apparent carbonate hardness" (p. 179).

In recooling, used chiefly in chimney coolers and finally in cooling ponds, the salt content should not increase through evaporation of large amounts of water to a value of over 3000 mg/l. Suitable amounts of fresh water must thus be added always and also part of the water containing the increased salt concentration must be removed.

Further, cooling water should contain neither iron nor manganese or, at least, not more than 0.3 mg Fe/l. Deposits occur otherwise in the pipes, clogging them. In time, the cooling surfaces become insulated too. Water which contains iron and manganese can also be made available as cooling water by treatment with suitable polyphosphates. This is quite economical with circulated cooling water. The cooling water must not come into contact with the air in this case. 4 mg polyphosphate must be added for every mg of Fe.
No more than small amounts of organic matter should be present in the cooling water or else sludge separates. The upper limit is a permanganate consumption of 25 mg $KMnO_4$/l.
Hard waters have to be softened. This is best performed with the *Bücher* process, using quick lime (p. 183). Polyphosphate addition suffices with waters of medium carbonate hardness (up to 6° dH); best is sodium tripolyphosphate. The water temperature must not then rise above 65° C or else orthophosphate is formed. This gives rise to deposits of scale and sludge. The polyphosphates prevent precipitation of the hardness-formers (formation of scale) and also dissolve up deposits of scale (see also p. 188).
Algae and plankton should be filtered from the cooling pond before the water is returned in the circulation. Their growth can be prevented in the cooling pond through copper salts. See p. 55 concerning microsieves.
K. Schilling has developed a portable apparatus with directions for study of boiler water and boiler feed water. It is supplied by the firm of Franz Bergmann KG, Berlin-Zehlendorf.

F. Water for Brewing Purposes

Water for brewing must be hygienically without shortcoming. Everything that has been said about drinking water applies thus to water for brewing. Water good for the one purpose is generally good for the other but this is not necessarily always the case. Various mineral components of the water which play no particular part in drinking water have a considerable influence on the quality of the beer. The salts in the water, hydrogen carbonates of calcium, magnesium and the alkali metals, raise the pH of the wort which in turn reduces the extract. Colour and taste of the beer are thus influenced unfavourably. Water containing little calcium is needed for light beers. There are three principal types of beer, according to the nature of the water used for brewing:

1. *Pilsen*	no carbonate hardness	no non-carbonate hardness
2. Munich	carbonate hardness of less importance	high non-carbonate hardness
3. *Dortmund*	high carbonate hardness	high non-carbonate hardness

Kolbach and *Schwabe* found that the calcium concentration of the water must be 3½ times a great as the magnesium concentration to counteract the acidity-reducing effect of the bicarbonates. In the science of brewing one calculates with the *"lime value"* which is given by adding the calcium hardness to half the magnesium hardness. 3.5° calcium hardness are

equal to 1° of total alkalinity. By dividing the lime value with 3.5, the "balanced alkalinity" is obtained. Subtraction of this from the total alkalinity yields the "residual alkalinity". This "residual alkalinity" is an important factor in brewing. If the calcium- and magnesium content is high compared to the total alkalinity, a low "residual alkalinity" is yielded nevertheless. If the residual alkalinity exceeds 5°, the water should be conditioned for preparation of light beers.
More than 50 mg MgO/l as magnesium sulphate in the water for brewing is bad for the beer quality and sometimes provokes diarrhoea of the consumer. A common salt content of over 500 mg/l is harmful to light beers; of over 800 mg/l, harmful to dark beers.
A high content of nitrate (over 25 mg/l) interferes with the fermentation since nitrites are formed. Nitrites must not be present in brewing water, apart from the fact that they are a sign of polluted water (p. 69). Nitrites are powerful yeast poisons.
During boiling of the wort, the bulk of the water for brewing becomes sterilised; yet the foodstuff regulations do not permit the use for beer manufacture of water which is not in perfect condition. Besides, part of the brewing water is often used, without boiling, for diluting the wort.
Water for brewing should be clear, colourless and odourless as well as naturally having a fully satisfactory taste.
The processing of water for brewing is subject to certain administrative regulations which must be squared with the responsible customs and excise authorities, on the basis of the brewing duties. A brewery can therefore not carry out softening, decarbonisation or chlorination of the water for brewing without consideration. On the other hand, simply boiling the brewing water is disadvantageous for various other reasons.
Carbonates and bicarbonates of potassium, sodium and calcium and also calcium sulphate can, however, be added to the water for brewing as desired, without hindrance.
Decarbonisation with calcium hydroxide solution is the best and cheapest method for normal waters. One must only be sure that no overdosage of calcium occurs. The firms of Hager & Elsässer of Stuttgart-Vaihingen, and Philipp Müller of Stuttgart-Güglingen have special experience in the field of decarbonisation of brewing water, in particular for obtaining pure water completely free of magnesium.
Waters containing much magnesium carbonate and alkali carbonates require special treatment: E.g., for magnesium hydrogen carbonate by strong overheating, followed by treatment of the clarified water with carbon dioxide and then blending with 2 to 3 parts of the untreated water.
Waters containing humic substances must always be processed for brewing purposes (flocculation with aluminium sulphate, p. 52).
Softening with Permutit or Lewatit is unsuitable for brewing water, since the sodium introduced by the exchange is just as unfavourable as calcium and magnesium. Only decarbonisation with calcium can yield a good brewing water, possibly followed by a complete demineralisation (p. 187). If municipal tap water is used in a brewery, it must be remembered that the customary decarbonisation fails where phosphates have been added.
Water containing iron ($>$ 0.3 mg Fe/l) interferes during brewing (damage to yeast) and yields cloudy beer. Removal of iron by closed aeration (p. 175) is thus advisable. Iron would be precipitated at the same time during any decarbonisation with calcium.

A high silica content of the brewing water acts unfavourably on the clarification of the beer.
The pH value of the brewing water is important for the action of enzymes. The pH value of processed water must be adjusted to ca. 7.0, and the p-value to 0 (see also *K. W. Naumann*, "Das Brauwasser").

VIII. Mineral Water and Medicinal Water

A. Mineral Water

The question whether water from a natural or artificially tapped source is to be termed mineral water or medicinal water is answered in W. Germany by the "Verordnung über Tafelwässer" (Table Water Order) of 11/2/38, or the "Begriffsbestimmungen für Kurorte, Erholungsorte und Heilbrunnen" (Definitions of Spas, Convalescence Resorts and Medical Springs) of 1/1/65.

The Table Water Order[1] contains the following definitions:

Table waters are:

1. Mineral waters
2. Waters, low in minerals
3. Artificial mineral waters.

Concerning 1: "*Mineral waters* are natural waters from natural of artificially tapped sources, which contain at least 1000 mg of dissolved salts or 250 mg of free carbon dioxide in one kg." The *limiting value* of 1000 mg/l is derived from the total of all anions and cations evaluated during the complete analysis, and not from the residue on evaporation. The former is usually higher than the latter, since the heating during evaporation causes partial decomposition of the bicarbonate. Half of the determined hydrogen carbonate is thus added to the residue on evaporation to obtain value limits. The term mineral water can thus be applied also to waters which contain rather less than 1000 mg of residue from evaporation of 1 kg (see also *K. E. Quentin,* Z. Anal. Chem. *146,* 18–26 [1955]). Occasional values below the limit of 1000 mg/kg should therefore not lead directly to objections (*W. Fresenius,* "Der Naturbrunnen", p. 153 [1963]).

Iron may be removed from mineral waters by aeration and this works also with those containing considerable amounts of CO_2 (performed by, e.g., the Berkefeld Filter-Ges., Celle, W. Germany). The removal of hydrogen sulphide by aeration is permitted too for mineral waters (done by, e.g., Neue Continental Bauges., Hanover-Linden).

The use of ozone in removal of iron and of sulphur would appear also to be permitted.

According to § 7 of the Table Water Order, the removal of iron and of sulphur must be declared on the label of the bottle in the same kind of writing as the name of the source. This applies equally to impregnation with CO_2, i.e., "freed from iron and carbon dioxide added" or "freed from sulphur and carbon dioxide added". These mineral waters can be termed "natural mineral water".

"Säuerlinge" (acidulous springs) are mineral waters with a natural content of at least 1000 mg dissolved free CO_2 per kg. They may be subjected to no alteration except addition of further CO_2 (impregnation).

1 Reichsges.Bl. (Reich Law Gazette) I, p. 199 [1938].

"Sprudel" (bubbling or effervescent waters) are acidulous springs which essentially bubble forth from natural or artificially tapped sources under natural carbon dioxide pressure. The term "Sprudel" is allowed also for a mineral water which has been bottled under addition of carbon dioxide, even when it has been freed of iron or sulphur by aeration.
Concerning 2: *Waters, low in mineral content* and which are obtained from natural or artificially tapped sources, are also termed table waters. Apart from impregnation with CO_2, these waters may not be altered in any way.
According to the new Fremdstoff-Verordnung (Decree covering foreign matter) of 19/12/59 in W. Germany, silver salts (chloride or sulphate or the sodium silver chloride complex) may be added in amounts up to 1 mg Ag/l, without declaration, to table waters (Bundesgesetzblatt [Federal Law Gazette] 1959 Part 1, No. 52). (this has been modified in the meantime).
"Sole" (Brine) is a natural water, rich in salt, with a minimum salt content of 14 g/kg, principally sodium chloride.
Mineral waters, the salt contents of which have been raised by abstraction of water, are also termed brines (discharge from graduation works or evaporating pans).
In the new definitions for *medicinal springs* from 1965 (see below), the value limit for brines is adapted to that of the Table Water Order (see below). The limit was about 10% higher in the old definitions. It is now thus 14 g/kg.
The Table Water Order does not apply to all mineral waters but only to those destined to be table waters.
The investigation of mineral water is largely considered in the section on drinking water. The next section contains some supplementary information on waters of very high mineral content.

B. Medicinal Water

Medicinal springs are "natural or artificially tapped water or gas sources which possess curative properties as a result of their chemical composition, their physical properties (e.g. thermal springs) or according to experience" (*M. Abt* in "Der Naturbrunnen", *11,* 242–246 [1961]). *K. E. Quentin* has expounded in detail the legal aspects of mineral and medicinal water in a commentary on the foodstuff law of *W. Zipfel* (Munich, 1967).
The new "Definitions of Spas, Convalescence Resorts and Medical Springs of the German Watering Place Society (Deutscher Bäderverband)", issued on 1/7/68, contain as definition of medicinal waters: Natural medicinal waters come from medicinal springs which emerge naturally or have been artificially tapped. They must have medicinally authenticated curative, alleviating or preventive (health-maintaining or health-promoting) properties. Their chemical and physical properties are to be checked continuously through medicinal water analyses and control analyses.
The categories of water listed below constitute most of those admitted as *medicinal waters* (as long as their properties or components, e.g. tar, bitumen, phenols etc., do not exclude their use for curative purposes):

Group A: Waters containing more than 1000 mg of dissolved solid mineral material in 1 kg. These medicinal waters are more precisely characterised through those ions which are present in concentrations of 20 mval % or more of the total cation or anion concentrations. The prevailing cations and then anions are enumerated in descending order in this characterisation.

The medicinal waters of Group A are divided into 4 chief groups, according to the prevailing anions, namely, into chloride waters, hydrogen carbonate waters, carbonate waters and sulphate waters. The old terms, such as muriatic, saline, earth salt sources etc., are eliminated by the definitions. One distinguishes now between:

1. Chloride-Waters
   a) Sodium-chloride-waters (frequent) (formerly "muriatic sources")
   b) Calcium-chloride-waters (formerly "earth-muriatic sources")
   c) Magnesium-chloride-waters (rare)
   Sodium chloride waters with more than 5.5 g sodium and 8.5 g chloride ion per kg can also be referred to as brines.
2. Hydrogen carbonate-waters (formerly "alkaline sources")
   a) Sodium-hydrogen carbonate-waters (rare)
   b) Calcium-hydrogen carbonate-waters (frequent)
   c) Magnesium-hydrogen carbonate-waters
3. Carbonate-Waters (rare)
4. Sulphate-Waters
   a) Sodium-sulphate-waters (formerly "saline sources")
   b) Magnesium-sulphate-waters (rare) (formerly "bitter sources")
   c) Calcium-sulphate-waters (frequent)
   d) Iron-sulphate-waters (rare)
   e) Aluminium-sulphate-waters (very rare)

Group B: Waters containing less than 1000 mg of dissolved solid mineral material in 1 kg, but possessing a minimum amount of particularly active constituents.

The medicinal waters of Group B are divided into:
1. Iron-containing waters with at least 10 mg Fe/kg (formerly, wrongly, "steel sources")
2. Arsenic-containing waters, with at least 0.7 mg As/kg (corresponding to 1.3 mg/kg of hydrogen arsenate, $HAsO_4$)
3. Iodine-containing waters, with at least 1 mg I/kg
4. Sulphur-containing waters, with at least 1 mg/kg of titratable sulphur
5. Carbon dioxide-containing waters with at least 1000 mg CO_2/kg
6. Radon-containing waters with at least 18nC/l = 50 Mache units (ME)
7. Radium-containing waters, with at least 10^{-7} mg Ra/kg (dropped from the definitions of 1965)

Group C: Waters, which, independently of their content of mineral material, possess a natural temperature exceeding 20° C (thermal waters or hot springs). Thermal waters of low mineral content are referred to also as *akratothermal waters.*

Group D: Waters of normal natural temperature and low in mineral content, fulfilling none of the above conditions but established as medicinal waters through special clinical expert opinion.

If a medicinal water fulfils several conditions, the characterisation according to Group B is prefixed, e.g., iodine-containing sodium-chloride-water.

The characterisation for waters containing carbon dioxide and for thermal waters is added as suffix, e.g. sodium-chloride-acidulous water or sodium-calcium-chloride-sulphate-thermal water.
With medicinal waters of multiple character, e.g., one which is a brine and acidulous water at the same time and also a thermal water, it is left to the chemist to decide, with the help of his specialist knowledge, which is the most important property to use for naming it: e.g., iron-containing thermal brine, or thermal brine.
Thermal waters in Germany always come from a greater depth. The temperature in the interior of the earth increases 1° C for every 30 m (100 feet) depth *("geothermal gradient").* The limits given for Group B apply, according to the new definitions, not only to the source (as customary formerly) but also to the place of consumption, i.e., for the drinking-room or bathing rooms. Valuable constituents (e.g. iron, CO_2, H_2S) may suffer partial loss of the storage, conveyance and introduction into the bath are unsuitable. This should be avoided with the help of detailed investigations from tapping to consumer site. At any rate, samples for comparison have to be taken nowadays at two places at least (see *K. E. Quentin,* Heilbad u. Kurort, *9,* 61 [1957], *W. Müller,* Z. f. angew. Bäder u. Klimaheilk. *1,* 1953, *E. Komma,* Heilbad u. Kurort *12,* 46–48 [1960], and *A. Kastel, M. Klusacek* and *W. Weis,* Heilbad u. Kurort *12,* 81–83 [1960]).
According to the new definitions, medicinal waters may be delivered for curative purposes only when no constituent of any sort is added or removed. Any processing or modification whatever is forbidden. The sole exception is warming for drinking or bathing purposes.
No material may be added to or removed from medicinal waters for bottling. The term "pure, natural bottling without treatment" has been fixed. The term "natural bottling" is applied to waters to which carbon dioxide is added up to 1 atmosphere at 20° C, including the CO_2 already present. This addition of carbon dioxide is, however, permitted only when iron or other materials precipitate during storage (this must be proved and confirmed with a specialist opinion).
Protection of medicinal sources is subject to special water legislation, since these sources are irreplaceable natural possessions. Our medicinal sources are being endangered more and more nowadays by wastewater, waste products and effluents from power stations. The new definitions aim at guaranteeing that our sources remain unaffected and pure, through continual hygienic-chemical and bacteriological examination. The bacterial number of medicinal waters should not exceed 20/ml.
In addition to this, yearly control analyses must establish whether the concentrations of the most important components of the medicinal water correspond to those of the major medicinal water analysis. Some spring waters, especially shallow springs, show fluctuating composition; this can often be eliminated by technical measures (*W. Müller,* Vom Wasser *18,* 27–42 [1950–51]).
Other sources, in particular those coming from a great depth, remain fairly constant (*W. Fresenius,* Schrift d. Deutsch. Bäderverband *8* [1952]). The author knows of medicinal springs which have stayed chemically constant to an extraordinary extent for decades and which also have a completely constant temperature.
Regular observation of the spring delivery is naturally important for the management of the resort, especially for the periodic and intermittent springs. Even a temporary overstraining

of the sources of medicinal springs may have an unfavourable influence on the chemical and physical composition (*W. Müller*, loc. cit.).
Most medicinal waters have strongly corrosive properties and this endangers the source tappings and pipes (*C. Genser*, Schrift. d. Deutsch. Bäderverband *8*, [1952]). Alterations to the tapping and estate of the medicinal source must thus be prevented by regular technical and chemical investigations. Plastic, "Hagusta" and "OBO"-pipes are being increasingly used in watering places (*W. Wevelmeyer*, Heilbad und Kurort 3129 [1957]).
Standard procedures for investigation of medicinal waters have been developed by the Analytical Commission of the Deutscher Bäderverband (German Union of Watering Places) and these have been incorporated in the definitions (see above) as an appendix. These standard analyses are reproduced below.

1 Standard Analyses

a) Major Medicinal Water Analysis

There must be a major medicinal water analysis of every balneologically used medicinal water. It must be renewed every 20 years in the large spas, where more than 10,000 guests take the water or more than 100,000 baths are prepared.
Likewise, the major analysis must be repeated every 20 years in bottling concerns which supply over 300,000 bottles of medicinal water yearly.
The major medicinal water analysis must comprise at least:

a) General Information
   Name and address of the institute conducting the investigation
   Name of the representative
   Date of sampling and of the investigation on the spot by the institute commissioned.
   Characterisation of the sampling spot through its general site and altitude.
   Description of the geological situation, based on existing documentation
   Information about the depth, cross-section and lining of the borehole or the tapping otherwise, other technical information connected with the tapping
   Delivery or capacity in litres/min
   Weather and preceding meteorological conditions
   Other data and observations.
b) Perception tests of odour, taste, coloration and clarity at the moment of sampling, 8 hours later (if possible) and after arrival in the laboratory, then quoting the time elapsed from sampling up to this moment.
c) Physical and physico-chemical investigation
   Temperature in °C at the sampling spot, quoting the air temperature and pressure (millibars)
   Density at 20 °C
   pH-value, determined electrometrically at the source
   Radioactivity: Content of radon (Rn) and radium (Ra)
   Spectroanalytical investigation
d) Quantitative chemical investigation, following qualitative tests
   Content of dissociated components (ions) in mg/kg, in mval/kg and in mval %
   Content of undissociated components in mg/kg and mmole/kg, and of dissolved gases in mg/kg, mmole/kg (0°, 760 mm Hg)
   Sum of solid dissolved and total dissolved matter in mg/kg, of cations and anions in mval/kg, and residue on evaporation after drying at 180° C. Analysis of freely escaping gases.

e) Content of medicinal value-determining constituents at the place of use
f) Bacteriological and, if necessary, microscopic examination
g) Characteristic

The major medicinal water analysis may be published only in one of the three following styles:
a) Complete style, intended for scientific purposes
b) Intermediate style (mainly intended for larger prospectuses), consisting of:
Name and address of the institute conducting the investigation
Date of sampling
Temperature of the water
Radioactivity
Quantitative chemical investigation as under d)
Characteristic
c) Miniature style (intended for labels and smaller prospectuses), consisting of:
Name and address of the institute conducting the investigation
Date of sampling
Columns of milligram contents
Columns of millival % contents
Content of undissociated constituents and dissolved gases in mg/kg
Characteristic

b) Minor Medicinal Water Analysis

A minor medicinal water analysis must be made every 20 years for all those medicinal waters with which drinking cures are conducted at the spa or which are sent away as medicinal or table water; and which are used for preparing baths or supplying inhalation rooms etc (provided of course that no major medicinal water analysis is required for these water, l.c.)

The minor medicinal water analysis must comprise at least:
a) General information:
Name and address of the institute conducting the investigation
Name of the representative
Date of sampling and of the investigation on the spot by the institute commissioned
Characterisation of the sampling spot through its general site and altitude of the source
Short description of the geological situation, based on existing documentation
Information about the depth, cross-section and lining of the bore hole or the tapping otherwise; other technical information connected with the tapping
Delivery or capacity in 1/min
Meteorological conditions
Other data and observations
b) Perception tests of odour, taste, coloration and clarity at the moment of sampling, 8 hours later (if possible), and after arrival in the laboratory, then quoting the time elapsed since sampling.
c) Physical and physico-chemical investigation
Temperature in °C at the sampling spot, quoting the air temperature and pressure (Millibars)
Density at 20°C
pH-value, determined electrometrically at the source
d) Quantitative chemical investigation, following qualitative tests
Content of K^+, NH^+, Ca^{2+}, Mg^{2+}, Mn^{2+}
NO_2^-,NO_3^-,Cl^-,$SO^2{}_4^-$,HCO_3^-
H_2SiO_3,CO_2

If necessary, the contents of other constituents, characteristic of the source, are given, e.g., As, I, Rn, titratable sulphur.
Quotation of the components and sum total

e) Content of medicinal value-determining constituents at the place of use
f) Bacteriological investigation
g) Characteristic

The minor medicinal water analysis must be published only in one of the three styles, as the major analysis (see above).

c) Control Analysis

A control analysis is necessary every year for each medicinal water for which a major water analysis is carried out. Control analyses every 3 years are performed on all other medicinal waters.
The control analysis should comprise the most important components of the particular medicinal water.

d) Hygienic Investigation

1. A hygienic investigation must be made every year on all medicinal waters which are drunk, inhaled or used in curative baths, and on all water used for making up and for rinsing in bottling of medicinal water.
2. The hygienic investigation must be begun at the sampling spot and consists of the hygiene chemical, microscopic and bacteriological examination as well as the indispensable inspection of the site. This comprises examination of the hygienic nature of the source, processing, conveyance and delivery of the medicinal water (including the rooms for bathing, inhalation and drinking) and of the dilution and rinsing water.
Iron-containing acidulous waters possess bactericidal activity as a result of their high iron content. Sometimes much too favourable results are thus found, e.g., as in Bad Elster (W. Germany) (*J. Lenk,* Dissertation, Dresden 1970).
3. A hygienic investigation should be undertaken as far as possible in collaboration with the medical officer. It should be carried out always by the same institute if that is possible, in order to guarantee constancy of the checking.

2. Explanatory Comments of the Medicinal Water Analysis

If one accepts that drinking water analysis requires some experience, then this applies with special force to medicinal water analysis. Here, it is often necessary to determine accurately fractions down to a thousandth of a milligram in waters of mineral content as high as 100–300 g/kg and more.
It would go beyond the scope of this book to attach an accurate guide to the minor and major medicinal water analysis. The water chemist has sometimes, however, to decide whether a water is to be termed a mineral water or medicinal water. Some tips about the special features

of mineral water analysis are consequently given here, the more necessary because there is a serious shortage of summarising presentations in this domain. All standard works (*R. Fresenius* [4]) have been out of print for a long time and are also difficultly obtainabe even in libraries. We possess only the valuable work of the Zürich professor *Dr. Gübeli-Litscher,* published in Innsbruck in 1948. Volume VIII of the Handbuch der Lebensmittelchemie, on mineral waters and medicinal waters *(K. E. Quentin* and *W. Fresenius),* is appeared in 1970[1].

It is customary in medicinal water analysis to carry out the determinations four, or at least three, times in order to obtain good average values.

Some of the special features of mineral waters (e.g., the density, content of carbon dioxide) have made it advisable to take weighed samples (not volumes as with drinking water analysis) and express the results as mg/kg.

It is often noticeable that some constituents are present in the highly concentrated brines in amounts that greatly exceed their normal solubilities. Alkali chlorides increase appreciably the solubilities of calcium sulphate and calcium carbonate, for instance (see *L. Rothmund,* Löslichkeit und Löslichkeitsbeeinflussung, Leipzig 1907).

In the *major medicinal water analysis,* a much larger *sample* is taken than in drinking water analysis. At least 50 kg are taken, plus a number of 1- to 2-litre bottles which are filled with the sampling tube (p. 11). Even for the minor medicinal water analysis, larger amounts of the spring water are needed, usually about 5 bottles of 1–2 l, besides samples for H_2S,O_2, etc. Samples of thermal water are best obtained free of air bubbles in plastic bottles since these expand on filling and contract on cooling.

Carbon dioxide must usually be expelled from *acidulous waters* before the titrations; this has been mentioned already under the individual determinations in the drinking water part of this book.

The *residue on evaporation* is now measured for mineral water through drying at 180° C.

1. Determinations of Cations

a) Potassium and Sodium

These alkali metals are nowadays best determined by flame photometry. The values for potassium are almost always very accurate. The test solutions must, however, contain exactly the same amounts of minerals as the water under investigation[2]. The determination of sodium is not always so accurate, especially with mineral waters of high concentration with predominating high Ca- and Mg-concentrations; these must be allowed for in any case. Usually, however, one can manage by precipitating the sodium with uranyl zinc acetate followed by determination of the precipitate gravimetrically or by dissolving it in hydro-

1 See also H. Vogt, Lehrbuch der Bäder- und Klimaheilkunde, Berlin (1940); and W. Amelung u. A. Evers, Handbuch der Bäder- und Klimaheilkunde, Stuttgart (1962) concerning the chemistry of medicinal waters.

2 see p. 35.

chloric acid and titrating with Titriplex (EDTA). Potassium can be determined gravimetrically with tetraphenylborate instead of the perchlorate method (except with brines, p. 135).

b) Lithium

At least 5 kg of the mineral water under investigation is acidified with hydrochloric acid and largely evaporated down. This is best done for waters of high mineral content in a porcelain dish on a water-bath; towards the end, especially with waters containing calcium sulphate, warming from above with infrared radiators helps prevent bumping.
The residue is transferred without filtering into a beaker and an amount of barium chloride corresponding to the sulphate present is added to the solution at the boiling point. The precipitate is filtered after it has settled out. The completeness of precipitation can be tested by cautiously adding barium chloride and sodium sulphate solutions alternately.
After most of the acid has been removed with ammonia or by evaporating in an open basin, alkaline earths and heavy metals are precipitated from the hot solution with milk of lime. When the precipitate has settled out, completeness of precipitation is checked with calcium hydroxide solution. If this is so, one filters again and precipitates excess calcium in the hot filtrate, using ammonium oxalate. The precipitate is usually voluminous. It is well washed with water containing ammonium oxalate. The filtrate is tested for completeness of precipitation. Hydrochloric acid is then added to it and the solution is evaporated to dryness. The last traces of moisture are removed in a drying oven at 130° C. After it has cooled, the dry residue is finely powdered and immediately warmed with absolute alcohol. The solution is filtered from the insoluble sodium and potassium chlorides and the filtrate evaporated in a platinum basin. Ammonium salts are expelled by heating the residue. With brines, the calcium oxalate precipitate sometimes contains much NaCl; in this case, the precipitate is dissolved in hydrochloric acid and re-precipitated with ammonium oxalate in ammoniacal solution; the NaCl would otherwise interfere later.
The residue consists essentially of lithium chloride. It is immediately powdered and treated with 25 ml acetone and 1 drop hydrochloric acid (d = 1.19). After mixing several times, it is filtered into a platinum dish and the filter + residue washed with three 5 ml portions of acetone.
The solution is evaporated on the water-bath and the residue cautiously ignited so as to destroy the last amounts of organic material. A few drops of sulphuric acid are added and the product is fumed off and weakly ignited. The final residue of lithium sulphate is weighed after it has cooled. 1 mg of it corresponds to 0.1263 mg Li.

c) Flame-photometric Determination of Lithium

After interfering elements have been eliminated, lithium can be determined very accurately by flame photometry according to the basic studies of *W. Schuhknecht;* it is done at 670.8 nm. The preparation of standard and buffer solutions is given on p. 136 under potassium (For further details, see *W. Schuhknecht* and *H. Schinkel,* Z. Anal. Chem. *194,* 161–183 [1963]).

d) *Calcium and Magnesium*

The instructions given in the section on drinking water (p. 116) suffice in most cases for determination of calcium and magnesium. The method with Complexone (EDTA) is generally inapplicable to mineral waters.

e) *Iron*

In waters rich in iron, the iron is best determined by titration with 0.01N $KMnO_4$, since it is always present in the divalent form in mineral waters containing hydrogen carbonate.
The water to be investigated is filled into a weighed standard flask (500 ml or 1 l capacity) with spherically widened neck, using the sampler tube (p. 11). After reweighing, 10 ml of 20% sulphuric acid and 15 ml manganese sulphate/phosphoric acid are immediately added.
The manganese/phosphoric acid reagent is prepared by adding a solution of 50 g crystalline manganese sulphate in 250 ml water to a mixture of 250 ml phosphoric acid (d = 1.30) and 150 ml water + 100 ml sulphuric acid (d = 1.84).
The carbon dioxide liberated from the hydrogen carbonate in the water generally suffices to prevent oxidation of the ferrous ions.
Titration is carried out, swirling the solution cautiously, with 0.01N potassium permanganate to a permanent pink colour. A blank is performed under identical conditions, with the same amount of distilled water.
Calculation: 1 ml 0.01N $KMnO_4$ is equivalent to 0.5585 mg Fe. The blank must be subtracted from the value obtained. If As (III) is present, a subtractive correction must be made of 4 mval Fe per mval $HAsO_4$.
The instructions under "drinking water" are suitable for waters containing little iron. For brines, reference solutions must be prepared with corresponding salt concentrations.

2. Determinations of Anions

The most important anions can generally be determined in the same way as described in the section on drinking water; mineral water was treated there as a special case where necessary. Further details are given below only for brines and mineral waters of high salt concentration.

a) *Chloride*

Volhard's method is more accurate than Mohr's method for waters containing over 1000 mg Cl/l. 25.0 g of sample are acidified with nitric acid and 10.0 ml of 0.1N silver nitrate (or correspondingly more) are added. The solution is heated until the silver chloride has completely flocculated. The silver chloride in this form behaves indifferently when thiocyanate is subsequently added (*V. Rothmund* and *A. Burgstaller,* Z. Anorg. Chem. *63,* 330 [1909]).

After cooling and without filtering, 2 ml of a saturated (in the cold) solution of ferric ammonium sulphate, acidified with nitric acid, are added and the excess silver is back titrated with 0.1N ammonium thiocyanate to a red coloration. The end-point can be determined more precisely in a potentiometric titration. 0.01N $AgNO_3$ can be used towards the end of the titration, yielding thereby very accurate results.
Brines must be suitably diluted or 0.5N solutions of silver nitrate and ammonium thiocyanate are used. In this latter case, however, it is best largely to separate the heavy precipitate from the solution by washing (without filtration).
Bromides and iodides are naturally co-titrated; this must be allowed for in the calculation.

$$1 \text{ ml } 0.1\text{N } AgNO_3 = 3.5457 \text{ mg Cl}$$
$$1 \text{ ml } 0.5\text{N } AgNO_3 = 17.7285 \text{ mg Cl}$$

b) Sulphate

In connection with sulphate determination on mineral waters of high salt content, it must be noted that the check on complete precipitation is especially important (p. 48); it is done on the filtrate after concentrating and washing the barium sulphate precipitate with hot distilled water.
Silica is removed by fuming off with HCl and filtering the solution of the residue in dilute HCl.

c) Boric Acid

H. Baron's test for *detecting boron* in plant material can be used as a *preliminary test* (Z. Anal. Chem. *143,* 339 [1954]). It is a colour (blue) reaction with 1,1'-dianthrimide (Merck) in sulphuric acid (2.5 ml of the concentrated water sample + 12.5 ml conc. H_2SO_4 + 5 ml dianthrimide-sulphuric acid[3]). A quantitative method is possible when the mixture is allowed to stand for 5 hours at 70° in a drying oven and parallel experiments are made with standard solutions; the filter S 61.5 with the Elko II colorimeter is used.
Procedure: 2 to 5 kg water are evaporated to a small volume after adding a little purest sodium carbonate. This small residue is filtered and the filtrate evaporated further. It is then acidified with hydrochloric acid and poured into absolute ethanol. This mixture is filtered, the filtrate rendered alkaline with sodium hydroxide and the alcohol distilled off. The residue is brought to dryness in a crucible and ignited. The ignition residue is taken up in hot water plus added ammonium carbonate and the solution filtered.
Silica is removed by adding an ammoniacal zinc solution to the filtrate, evaporating it to dryness and igniting. The residue is taken up with water and the solution freed from zinc silicate and zinc carbonate by passing in carbon dioxide.
0.1N HCl is added to the filtrate after concentration, until the methyl orange colour change occurs; CO_2 is expelled by boiling for 10 minutes under reflux.
Boric acid is determined in the cooled solution by adding 10 g of mannitol and titrating with

3 0.05 g 1,1'-dianthrimide in 100 ml sulphuric acid (d = 1.84).

0.1N NaOH to the red of phenolphthalein indicator (1 ml of 0.0375% solution).
1 ml 0.1N sodium hydroxide = 4.383 mg HBO_2.
Boric acid can be determined colorimetrically with quinalizarin in a micromethod; 10 ml of water sample is mixed with 20 ml conc. sulphuric acid and 1 ml quinalizarin solution (0.01% in 90% H_2SO_4) (*Gmelin,* Boron volume).

d) *Iodide and Bromide*

P. Höfer's instructions are followed for mineral waters of high salt concentration (Ges. Ing. *74,* 224–226 [1953]); evaporation is not carried out.
To 100 g water are added 6 ml of phosphate-common salt solution (100 g of each of monosodium phosphate, $NaH_2PO_4 \cdot 1H_2O$, disodium phosphate, $Na_2HPO_4 \cdot 12H_2O$ and sodium pyrophosphate, $Na_4P_2O_7 \cdot 10H_2O$, and 35 g sodium chloride, all in 1400 ml water). Regardless of any precipitation, 6 ml of a 0.5N hypochlorite solution (14.0 g Cl/l) are added and this solution is slowly warmed to ca. 90° C, taking 5 to 10 min. and shaking frequently, this temperature must not be exceeded. 6 ml of 2N formic acid are then added, causing the precipitated alkaline earth phosphates to dissolve. Chlorine in the atmosphere above the liquid is removed by suction with a rubber pressure bulb.
The solution is cooled under the tap, 1 ml 1N potassium iodide is added and ca. 0.5 ml 10% HCl to bring the pH to 3.0. After 10 minutes, the liberated iodine is titrated with 0.01N thiosulphate. 1 ml 0.01N thiosulphate is equivalent to 0.2116 mg iodine.
For the bromine determination, this solution is acidified with a further 10 ml of 10% hydrochloric acid and retitrated with the thiosulphate after 10 minutes. 1 ml 0.01N thiosulphate is equivalent to 0.1333 mg bromine. A blank is necessary since the added potassium iodide may contain free iodine. Four replicates are needed and 2 or 3 blanks run simultaneously. A solution of 0.1489 g potassium bromide in 1 l distilled water is used as standard; 1 ml of it contains 0.1 mg bromine.
Interferences: Nitrites interfere but they are hardly found in medicinal springs.
Some mineral springs contain a high amount of iodine (iodine springs), e.g. the Adelheid spring in Bad Heilbrunn with 23.21 mg iodine/kg; the Wilhelmina spring in Bad Wiessee with 34.78; and the iodine spring in Bad Tölz with 41.5 (all in Germany).

K. E. Quentin reports on the *determination of carbon dioxide* in waters rich in it (Heilbad und Kurort, 1967).
An approximate determination of the free carbon dioxide in acidulous waters can be made with the "Karat" apparatus of *P. Haertl* (Firm of Weidemeyer & Co., Kassel, W. Germany).

3. Determinations of Trace Elements

a) *Dithizone Method*

Trace elements are those found in very small amounts in living organisms and which possess a biological activity: some are important for life. They have thus been termed also "inorgan-

ic vitamins". Amongst them are iron, copper, zinc, cobalt, manganese and iodine. The so-called subsidiary trace elements, lead, mercury, silver, tin and arsenic have no detectable biological activity. Little is known of the part played by molybdenum in balneology, although it certainly has a powerful biological activity. Suitable methods for quantitative determination of molybdenum traces in water are lacking.
The author's experience shows that the dithizone method is the best for quantitatively determining some metal trace elements provided that it is correctly used and that all reagents and vessels are specially treated. Since trace elements play an important part today in mineral waters too, the special application of the dithizone method to mineral water analysis is described here[4].

Cleaning of Vessels
All vessels destined for use must be cleaned first with alcohol and then with warm nitric acid. They must then be shaken with dithizone solution (concentrated, in chloroform or carbon tetrachloride) in acid and alkaline medium. Complete purity is checked by shaking the vessels with some dithizone-purified water (see below), previously brought to pH 9 with purified ammonia[5], and then with 0.5 ml of a 25×10^{-6}M dithizone solution in CCl_4 (see below). The carbon tetrachloride phase must remain colourless or else the vessel has to be cleaned again. Later it is enough to shake several times only with dithizone-CCl_4 solution.

All reagents and the (doubly) distilled water also must be extracted with dithizone solution and with ammonia (1:200) until there is no blank.

Dithizone which has entered the aqueous solution, recognised through the yellow colour of the solution, is removed by extraction with chloroform.
Blanks and checks on inhibitors are absolutely essential at every step of the analysis. If a blank shows a positive result, reagents and vessels must again be extracted with dithizone solution and washed afterwards with chloroform and carbon tetrachloride. Brown glass and polyethylene vessels are unsuitable for samples and reagents. Only glass stoppered vessels of Jena glass are suitable; rubber stoppers should never be employed.

Trace elements, especially lead and zinc, may evade detection if the samples of mineral water stand in the sampling bottles; this is as a result of adsorption on the walls of the vessel or through separation of iron-containing precipitates. Freshly taken water samples must thus be used for detection of traces or the sample must be acidified when it is taken.

The dithizone solutions in carbon tetrachloride, mentioned below, are stable for only 24 hours; they must be shielded from light during this period.
Hydrogen sulphide delays or prevents reactions of the heavy metals (see also *E. B. Sandell*, Col. Determination of Traces of Metals). It must therefore be removed before analysis with

4 The standard work of *G. Iwantscheff* "Das Dithizon und seine Anwendung in der Mikro- und Spurenanalyse", Weinheim 1958 is referred to. The author is grateful to the writer of this book for the information in this domain. Determination of cation traces without error is possible only when the quoted analytical procedure is adhered to.
5 Pure dilute ammonia is prepared by isothermal diffusion into dithizone-purified water, e.g. placing dishes of ammonia and of distilled water in an empty desiccator. Enough ammonia has passed into the water in 24 hours.

dithizone. Phosphates, also have an unfavourable influence on the dithizone extraction; they seldom occur in interfering amounts in medicinal and mineral waters.

b) Dithizone-Analysis

Adherence to the following working procedure enables 1 to 3 μg amounts of the trace elements to be determined.

Determination of copper

200 to 500 g of mineral water are treated in a separating funnel with 3 ml of a 10% solution of hydroxylamine hydrochloride (Merck 4616) and 1N hydrochloric acid to bring the pH to 2. The mixture is shaken up after each addition and then shaken with 0.5 ml of dithizone solution for 2 min. For normal mineral waters, the dithizone solution contains 12.8 mg/l in carbon tetrachloride (Merck products 3092 and 2208, respectively), i.e., $5 \cdot 10^{-5}$M; a 10^{-4}M solution is used for brines.

If the colour changes from green to red-violet, more than traces of copper are present. Extraction with dithizone reagent is then repeated until the green colour persists for 2 min. The pH of 2 must be maintained since lead is co-extracted at a higher pH and interferes; it then evades determination later.

Copper can nevertheless be present in traces even if there is no visible colour change during the first extraction. In this case, the excess of dithizone is removed after washing with 0.5 ml 0.5N HCl[6]; this is done washing the CCl_4 layer with dithizone-purified distilled water and shaking with ammonia (1:200) for 10 sec. If copper is present, the red-violet colour then appears in the CCl_4-phase. It is extracted once more with 5 ml 0.5N HCl[6] and the coloration measured photometrically at 550 nm, comparing with standard solutions treated in the same way. Should the CCl_4-layer change colour during extraction, the combined layers are made up to 10 ml and treated with distilled water and ammonia as above.

With brines, extraction with the more concentrated dithizone solution is followed by treatment of the CCl_4-phase with 6N hydrochloric acid, to decompose the copper dithizonate; the latter goes into the acid layer. The aqueous layer is separated and evaporated and treated anew with hydroxylamine and HCl to bring to pH 2; it is then extracted with the $5 \cdot 10^{-5}$ M dithizone solution as given above. The CCl_4-layer is separated, excess dithizone washed out of it and the solution submitted to colorimetric determination (see above).

An orange or yellowish colour occurring in the first extraction can indicate the exceptional interference of mercury. It must be masked by addition of 1 ml 1% potassium iodide solution before the copper dithizonate is extracted.

Determination of lead and zinc

Iron, present in most waters, is retained in solution in the aqueous phase from acidic extraction of copper, by treating with 2N Rochelle salt solution and then with 1% ammonia to adjust to pH 7.0. The solution is extracted one or more times with $5 \cdot 10^{-5}$M dithizone/CCl_4 solution, as above. A violet-red or pink-red coloration in the dithizone/CCl_4 layer indicates

6 Extraction with 0.5N HCl has two purposes in the copper determination: 1. to remove interfering ions; 2. to prevent the copper dithizonate from change due to longer action of the alkaline solution (washing solution of 1:200 ammonia).

lead and zinc. Extraction with more dithizone reagent is then carried out until the green dithizone colour no longer changes or vanishes. The aqueous layer is separated and discarded.
The separated CCl_4-phase is made up to 10 ml, washed with 10 ml distilled water and shaken twice with 5 ml 0.5N HCl for 1 min; the colour of the organic phase changes back to green and the lead and zinc dithizonates pass into the aqueous layer. This is separated and washed twice with pure CCl_4.
The washed aqueous layer is made up to 20 ml and divided into two parts. Part I is tested for lead, Part II for zinc.

Part I (lead determination) is rendered alkaline (pH 8–9) with 1N dithizone-purified ammonia and 6 ml of a 10% KCN solution is added (KCN, Merck 4967 is lead-free). This solution is shaken for 1 min with 0.5 ml $5 \cdot 10^{-5}$M dithizone/CCl_4. A pink colour appears if lead is present. The solution is then shaken with further 0.5 ml amounts of the dithizone reagent until the green colour persists. Whether no colour change occurs or merely a dirty green coloration is observed, the separated CCl_4-phase is made up to 10 ml and shaken with two 5 ml portions of 0.5% KCN solution for 10 sec to mask zinc, mercury and any entrained copper. The pink solution is washed once with distilled water and evaluated photometrically at 520 nm, comparing with the colours from standard solutions.
Only bismuth yields a red colour with dithizone under the conditions given; it is scarcely to be expected in waters, however. In cases of doubt, the red extract can be shaken with a few ml of 0.1N Michaelis buffer solution (pH 4.6, see below); this decomposes the lead dithizonate (green colour of the dithizone solution) and only the red colour of bismuth dithizonate remains.
Preparation of the Michaelis 0.1N buffer solution: 13.6 g sodium acetate and 6.0 ml acetic acid are made up to 1 l with doubly distilled water.

Part II (zinc determination) is treated with 6 ml of a 10% aqueous sodium acetate solution and its pH adjusted to 4.5–4.8 with dilute (1N) ammonia. After adding 0.5 ml of a 50% sodium thiosulphate solution, the solution is shaken with 1.5 ml of $5 \cdot 10^{-5}$M dithizone as above. If zinc is present, the colour turns pink to blue-violet. Further shaking with dithizone solution is then carried out until the CCl_4-phase stays clear green. The united CCl_4-layers are made up to 10 ml and, provided that no change of colour is observed, washed with 1% sodium acetate, with ammonia (1:200) and finally with distilled water until the aqueous phase remains colourless. The CCl_4-layer stays pink to blue-violet during this and can be subsequently determined photometrically at ca. 540 nm.

Determination of silver
The pH of a further sample of the mineral water (200 to 500 g) is brought to 3–4 with N H_2SO_4 (without adding hydroxylamine). The solution is extracted one or more times for 1 min with 0.5 ml of $5 \cdot 10^{-5}$M dithizone/CCl_4 in a separating funnel until the last dithizone portion remains green. Silver is shown by a golden yellow colour.
The united organic phase is made up to 10 ml with CCl_4 and this is extracted two or three times with 5 ml amounts of 0.5N H_2SO_4 and finally with water. Excess dithizone is then removed by shaking with 0.01N ammonia. The golden yellow organic phase is separated,

washed with water and evaluated colorimetrically at ca. 460 nm, comparing with standard solutions, treated in the same way.
Even if the CCl_4-phase shows no visible colour change on extraction, the extraction with sulphuric acid and with ammonia, as described above, is carried out and the photometric measurement made.
Should the green dithizone phase undergo no colour change, the separated sample of water is tested immediately for mercury after acidifying anew with sulphuric acid (see below).
On account of their high NaCl-content, brines are brought to pH 5–6 in the determination of silver with dithizone. The end-point of the extraction would probably be a pink coloration in this case (presence of zinc and lead). These elements can be removed from the organic extract by washing twice with distilled water and 0.5N H_2SO_4. When this yields a purely green colour, no silver can be present or it would give a yellow-green to yellowish coloration. This can be pursued further with silver test solutions. Any mercury in the brines can be extracted at pH 1 before the silver. If copper is present also, it is separated beforehand at pH 1.5 through extraction with stronger dithizone solution; the silver is masked by the high chloride content of the brine at pH 1.5. The silver can then be extracted by adjusting the pH of the aqueous brine sample to 5–6; the end-point of this extraction would be shown by a green colour or, in presence of Zn or Pb, by pink. The usual procedure (with removal of excess dithizone) could then be continued by shaking twice with 0.5 N H_2SO_4.
A pure yellow colour may be derived from mercury too; it reacts with dithizone before silver does. Whether it is due to mercury or to silver can be ascertained by shaking the extract obtained for silver determination in the procedure with 0.5N HCl. This decomposes the silver dithizonate but leaves the mercury dithizonate unchanged. Silver in the HCl-solution can then be re-extracted with dithizone at pH 5–6.
A violet-tinted mixed colour may be observed at the end of extraction should the mineral water contain much copper. The combined extract is then shaken 1 min with 10 ml 0.5N hydrochloric acid. After reducing the pH of this acid solution to 5–6, silver can be extracted from it without interference.

Determination of mercury
The pH of another 300–500 g sample of mineral water is brought to 1.0 with 2N H_2SO_4. It is extracted for 1 min with 1 ml amounts of $5 \cdot 10^{-5}$M dithizone/CCl_4 until the organic phase remains green. When more than traces of mercury are present, the dithizone solution turns orange-yellow. If not, the excess of dithizone must be removed by washing with ammonia; this applies also to the former case, when the last dithizone solution which stays green is added to the previous yellow-coloured extracts.
If the mineral water contains much copper and silver, it is advisable, before addition of ammonia, to extract the united CCl_4-layer with 10 ml 0.5N sulphuric acid and to increase the acidity beforehand to 0.2N. Interference by copper can be avoided also by prior addition of formic acid (*Woebling* and *Steiger,* Z. Angew. Chem. *46,* 279 [1933]).
The combined extracts are made up to a suitable volume with CCl_4 and shaken once with 5 ml 0.5N HCl. Excess dithizone is removed by shaking twice with 10 ml amounts of ammonia (1:200). The remaining organic layer is washed by shaking with a few ml acetic acid, and is evaluated photometrically at 485 nm.

Mercury dithizonate is highly light-sensitive[7]. Work must thus be carried out in the dark, or else the mercury escapes quantitative determination. The dithizonate can be desensitised by adding 10 ml 2N acetic acid. Alternatively, the mercury must be extracted very rapidly. If mercury is masked in a parallel test by adding some potassium iodide (1 ml 1% solution), then a negative coloration test provides further identification evidence of mercury.
1 to 3 μg of the mentioned trace elements can be detected with the procedure given.

Detection of Beryllium: Quinalizarin (1,2,5,8-tetrahydroxyanthraquinone) has proved to be a most sensitive reagent for beryllium, in mineral water investigations also. N NaOH is added to 10 ml of a suitably concentrated sample of the mineral water until present in 0.3N strength, 0.4 ml of a 0.05% quinalizarin solution (0.3N in NaOH) is then added. A colour change from violet to cornflower blue occurs if beryllium is present. A roughly quantitative estimation is possible with the help of standard solutions. The limit of detection is about 0.15 μg Be.
Heavy metals, including iron, do not interfere. Interference of magnesium can be eliminated by adding ammonia. With mineral waters rich in iron, this metal is precipitated along with the calcium and adsorbs all the beryllium present in solution; the precipitate can be dissolved in HCl and the beryllium extracted from this solution with butyl acetate. It can then be colorimetrically determined as above or also with Chromazurol S (0.05% solution with 0.2% gum arabic).
These tests are quoted here since beryllium is important technically.
A further highly sensitive reagent for beryllium is Beryllon II; very small amounts of beryllium yield a red-violet to blue-violet, or light blue colour (higher concentrations) with a 0.01% solution of the reagent in HCl. The usual addition of Rochelle salt and 5% Trilon B (EDTA) solution prevents interference by iron and other heavy metals (*F. D. Krivoruchko*, Russ. Z. f. Hygiene [1965], Chem. Abs. *63*, 2302).
The scheme for analysis of drinking water is valueless for balneology; an example for a minor medicinal water analysis is thus given below.

c) Example of a Minor Medicinal Water Analysis (Temple Spring, Bad Salzschlirf)

1 kilogram of the water contains:

Cations	Milligrams	Millivals	Millival%
Sodium ion (Na^+)	2945.0	128.06	74.43
Potassium ion(K^+)	131.0	3.351	1.95
Ammonium ion(NH_4^+)	0.38	0.021	0.01
Calcium ion(Ca^{2+})	493.7	24.64	14.32
Magnesium ion(Mg^{2+})	191.6	15.76	9.16
Iron(II) ion(Fe^{2+})	6.0	0.214	0.12
Manganese ion(Mn^{2+})	0.28	0.01	0.01
		172.0	100.0

7 It turns pole grey-violet.

Anions			
Chloride ion(Cl^-)	4715.8	133.00	77.34
Sulphate ion(SO_4^{2-})	863.1	17.97	10.45
Hydrogen carbonate ion(HCO_3^-)	1281.0	21.00	12.21
Hydrogen phosphate ion(HPO_3^{2-})	0.03	0.00	0.00
		172.0	100.0
		Millimoles	
Undissociated components	15.34	0.19	
Silicic acid(H_2SiO_3) Total:	10643.23		
Gaseous components			
Free carbon dioxide(CO_2)	1700.0	38.63	
Free hydrogen sulphide(H_2S)	0.00	0.00	
Total:	12343.23		

Nitrite ion (0.000 mg/kg) and nitrate ion (below 0.1 mg/kg) were not detectable.
Characteristic: Sodium-chloride-acidulous spring.
In addition, general information is given about the spring, concerning site, depth, piping, capacity and discharge; and also the temperature of the spring, its reaction and pH-value, density and residue on evaporation.

Water as a Material

The investigation of water by the chemist, bacteriologist and biologist always concerns only that which is present in dissolved, colloidal or undissolved, living or non-living form. No book devoted to the study of water has hitherto dealt with the material, water, as such.
Water is, however, the most remarkable material on our earth. All its properties, even the fundamental properties, are peculiarities and exceptions. Everyone knows, for example, that water does not contract but expands on freezing (about 11%); this is because its maximum density is at + 4° C and not at 0° C. This has the important consequence for life in waters, that ice floats on the water and that our lakes do not freeze to or from their bed.
Further, water has the largest specific heat. The warmth taken up during the summer period by the large volumes of water is thus slowly released in winter; large amounts such as the seas consequently function as temperature compensators.
Water is the liquid with the highest dielectric constant. Its extreme insulating power, due to this, enables the ions of opposing charge, cations and anions, to co-exist in salt solutions.
It is, moreover, the best solvent for all types of substance. Water therefore never occurs in nature in really pure form, i.e., without any trace of dissolved minerals, gases or organic materials. Its good solvent power is due to the pronounced dipole character of the water molecule. The molecule has a positively and a negatively charged end and these poles are especially widely separated. When water molecules approach one another, the positive pole of one attracts the negative pole of another; molecule aggregates are thereby formed which,

depending on temperature, have up to 8 H_2O units. The higher the temperature, the fewer such associates are present in the water.

The bipolar properties of water play a prominent part also in corrosion in aqueous medium.

A striking peculiarity is the molecular structure of water in which the two hydrogen atoms form an angle of 105° at the oxygen atom. One had assumed formerly that the water molecule was linear, (H-O-H), but it is known with certainty today that the centre of gravity of the water molecule does not lie in the middle. This gives water its dipole moment which imparts to it many extraordinary properties. The dipole moment of water is, moreover, especially large. The aggregated state of water stands out amongst those of analogous hydrogen compounds, likewise its viscosity and surface tension. The viscosity can be altered in various ways through electrolytes, and the surface tension through surface-active materials such as soaps and tensides.

Another speciality of water is the formation of colloidal systems. The colloids, e.g. humus colloids, have definite electric charges. These colloids can be discharged through various added materials. Practical advantage is taken of this in, e.g. coagulation with aluminium sulphate (after having established the zeta-potential).

Some years ago it was discovered that naturally occurring water does not consist alone of normal "H_2O" but that a small proportion of heavy hydrogen and oxygen atoms are present. Ordinary hydrogen of atomic weight 1 (protium) has two isotopes: the heavy hydrogen atom, deuterium, of atomic weight 2; and the superheavy hydrogen atom, tritium, of atomic weight 3. Two heavy isotopes of oxygen of atomic weight 16 are also known; they have the atomic weights 17 and 18. All these hydrogen and oxygen isotopes are present in greater or lesser amounts in almost all naturally occurring waters (ground water, surface water and rainwater). There are 18 of them and 12 ions can be roughly determined in the water. *Deuterium* has gained enormous technical significance in recent years as heavy water, D_2O, for cooling reactors in atomic power stations. The D_2O concentration must be continuously more than 99%; this can be determined by infra-red spectroscopy (e.g., with the PMQ apparatus of the firm of Zeiss, Oberkochen, W. Germany). The deuterium content of naturally occurring waters varies as a result of the differing temperatures of evaporation and condensation of normal and heavy water; the content is highest in ocean water and near coasts.

K. Höll has reported on this in connection with medicinal springs (Heilbad und Kurort, *17*, 136–139 [1965]).

Water has a powerful internal tension. Its cohesive properties are demonstrated by the very varied drop formations which result when drops of water from a fine dropping jet land on an especially clean dish. The characteristic drop forms for many waters are compiled very lucidly in *T. Schwenk*'s illustrated book, "Bewegungsformen des Wassers", Stuttgart 1967 (the attempts to explain these in the book are, however, open to doubt).

Although theoretical discussions have deliberately not been included in this book, these brief comments on the material properties in general are intended to complete the overall picture.

IX. Reagents for Investigation of Drinking Water

In the chapter "Investigation of Water", reference is made to the following directions for preparation of many reagents. One must adhere rigidly to them. Other reagents are denoted with an asterisk (*). This means that they are based on the prescriptions given in the German Pharmacopoeia. Any desired concentration may be used if there is no asterisk.
The purity and effectiveness of all reagents must be checked continuously (p. 20).

Reagent No. 1 Active charcoal
100 g active charcoal (e.g. vegetable charcoal) are mixed with 2 l water and 40 ml sodium hydroxide and the whole boiled briefly. The product is filtered and freed from sodium hydroxide by washing on the filter with dilute hydrochloric acid and water. It is then dried in an oven at 100°.

Reagent No. 2 Aluminium hydroxide
125 g alum are dissolved in 1 l of distilled water and aluminium hydroxide is precipitated by adding ammonium hydroxide. It is washed by decantation with distilled water until the water is free from chloride, nitrate and ammonia.

Reagent No. 3 Ammonium chloride solution for lead determination
20 g purest ammonium chloride and 2 g dilute acetic acid are diluted to 100 ml with distilled water. Ammonium chloride often contains traces of lead which must be established in a blank. The solution may be freed from lead by shaking with talc.

Reagent No. 4 Standard solution of ammonium ion
2.966 g purest ammonium chloride are dissolved in water and the volume made up to 1 l. 10 ml of this "Solution I" are diluted to 1 l. 1 ml of this "Solution II" contains 0.01 g NH_4^+/l. 1 ml of Solution II, diluted to 100 ml with ammonium ion-free water (Reagent No. 10), yields a solution of 0.1 mg/l.

Reagent No. 5 Barium chromate solution
5 g of barium chromate are dissolved in 50 ml of 25% hydrochloric acid and the solution made up to 1 l. The final solution can be kept some time but may have to be filtered.

Reagent No. 6 Standard solution of lead ion
0.1830 g lead acetate (analytical grade) is dissolved in 3% acetic acid and the solution made up to 1 l with this; or 0.1599 g lead nitrate (dried at 100° C) + 1 ml nitric acid are diluted to 1 litre. 1 ml of each solution contains 0.1 mg lead ion.

Reagent No. 7 Chlorine reagent for determination of chlorine number
200 ml 0.1N sodium hypochlorite and 100 ml 1N sodium hydroxide are diluted to 1 l with distilled water (Reagent No. 10). The activity is determined by standardisation.

Reagent No. 8 Reagent for detecting chlorine
1 g of colourless dimethyl-p-phenylenediamine hydrochloride is dissolved in 50 g conc. hydrochloric acid and the solution diluted to 250 ml.

Reagent No. 9 Reference solution for chlorine
0.115 g methyl red is dissolved in 5 ml 1N sodium hydroxide plus 95 ml distilled water. This stock solution is diluted further by a factor of 100.

Reagent No. 10 Purest distilled water, free from ammonium and nitrite ions, carbon dioxide and organic materials
Barium hydroxide and potassium permanganate are added to ordinary distilled water and the solution is distilled through an especially clean condenser of Jena glass.

Reagent No. 10a DPD-reagent
0.11 g diethyl-p-phenylenediamine sulphate, 2 ml of 10% sulphuric acid and 2.5 ml of a 0.8% solution of the disodium salt of ethylenediaminetetraacetic acid are dissolved in a little distilled water and the solution made up to 100 ml. The reagent can be kept some time in the dark.

Reagent No. 11 Diphenylamine-sulphuric acid
Only completely white diphenylamine may be employed. 0.085 g of it is mixed with 190 ml chemically pure sulphuric acid (1:3 vol) in a 500 ml graduated flask. Pure concentrated sulphuric acid is then added with shaking, causing the mixture to warm up. The diphenylamine melts and goes into solution. The solution is filled up to the neck of the flask with conc. sulphuric acid. After cooling, it is made up to the mark with conc. sulphuric acid, bearing in mind the calibration temperature. If the sulphuric acid contains nitrogen compounds and no pure acid is available, the impurities can be removed by previously adding 0.5% ammonium persulphate and heating.
Should the prepared diphenylamine solution appear bluish from traces of nitric acid (e.g. from the laboratory atmosphere), it can be rendered usable by heating to 115° C on a sand-bath. If the sulphuric acid contains nitric acid, a little common salt should be added before heating. The reagent can be kept for a longer period.

Reagent No. 12 Reference solution of iron
1. 0.901 g ferric alum (purest potassium ferric sulphate) is dissolved in 10 ml hydrochloric acid and distilled water in a 1 l graduated flask. The solution, which has become warm, is subsequently made up to the mark, at the correct temperature. 1 ml = 0.1 mg Fe.
2. 0.7022 g ferrous ammonium sulphate, $FeSO_4 \cdot (NH_4)_2SO_4 \cdot 6H_2O$ (analytical grade) is dissolved in 50 ml 25% sulphuric acid and the solution diluted to 1 l with distilled water. It is then oxidised as required with 0.1N $KMnO_4$. 1 ml of original solution contains 0.1 mg Fe^{2+}. The solution can be diluted 1:10 as needed.

Reagent No. 12a Ferrous ammonium sulphate solution
1.106 g ferrous ammonium sulphate ($FeSO_4 \cdot (NH_4)_2SO_4 \cdot 6H_2O$) is dissolved in freshly boiled distilled water + 1 ml 25% sulphuric acid. After it has cooled, the solution is made up to 1 l. 1 ml solution contains 0.1 mg Fe^{2+}.

Reagent No. 13 Potassium hexachloroplatinate reagent
2.49 g potassium hexachloroplatinate (K_2PtCl_6) and 2.016 g crystallised cobaltous chloride, $CoCl_2 \cdot 6H_2O$, are dissolved in distilled water + 200 ml hydrochloric acid (d = 1.19) and the solution made up to 1 l. The solution contains 1 mg Pt/ml.

Reagent No. 14 Potassium palmitate solution according to *Blacher*
0.1 g phenolphthalein is dissolved in 500 ml 95% ethanol in a litre flask and 300 ml of distilled water are added. 25.6 g purest palmitic acid (Merck) are dissolved in this mixture by warming on the water-bath and adding 0.7% alcoholic potassium hydroxide until the solution is pale pink-red.
The potassium hydroxide solution is prepared by warming 7 g potassium hydroxide in a little alcohol and subsequently making the solution up to 1 l with alcohol at the calibration termperature. If the colour is too strongly red, it should be just removed by adding dilute hydrochloric acid and the just visible pink-red formed by adding the potassium hydroxide solution again. The solution is then ready for dilution to 1 l. Any palmitate separating out on cooling can be redissolved by warming the flask. *Blacher*'s solution can be prepared in propyl alcohol also; no separation then occurs in the cold, which is advantageous for field analyses in winter.
Blacher's solution may also be obtained commercially (Merck).
The potassium palmitate solution is standardised with 20 ml 0.1N barium chloride solution (12.215 g $BaCl_2 \cdot 2H_2O$/l) which is diluted to 100 ml with boiled, distilled water. Phenolphthalein indicator is added and the solution titrated to pink-red (pH ca 8.8).
The firm of Merck now provides standard calcium solutions which facilitate standardisation of the palmitate solution.

Reagent No. 15 Caramel solution for colour determination
1 g cane sugar is dissolved in 50 ml distilled water and 1 ml sulphuric acid (1 + 2) is added. This solution is boiled for exactly 10 minutes. 1 ml 33% sodium hydroxide is then added and a further 10 minutes boiling carried out. After cooling, the resulting solution is made up to 1 l. 1 ml = 1 mg caramel.

Reagent No. 16 Reference solution for copper
(for colorimetric determinations). 1.964 g purest crystallised cupric sulphate is dissolved in 500 ml distilled water. 1 ml = 1 mg Cu.

Reagent No. 17 Kieselguhr standard
Purest kieselguhr (Merck) is powdered in an agate mortar and sieved through fine silk ("Müllerseide") (No. 20). The sieved product is suspended in water and any deposit which forms is discarded. The suspension is evaporated to dryness and the residue powdered anew. 1 g of this final product is suspended in 1 l distilled water yielding a solution of turbidity degree 1000. 1 ml = 1 mg kieselguhr.
This standard turbidity must be renewed often, at least every one or two months. Other dilutions may be prepared as needed.

Reagent No. 18 Methyl orange indicator
1 g methyl orange is dissolved in 1 l hot water. Any product subsequently separating is filtered off.

Reagent No. 19 Reference methyl orange solution according to *Haase* and *Gad*
0.115 g methyl red is dissolved in 5 ml 1N NaOH; 5 ml 0.01N sodium hydroxide are added and the solution is diluted to 100 ml. This solution is diluted 1:100 with distilled water.

Reagent No. 20 Molybdate-sulphuric acid solution
A 10% solution of ammonium molybdate and a 50% sulphuric acid solution are mixed in the proportion 1:3. This mixture can be kept only 8 days under light exclusion; the two components are thus stored separately.

Reagent No. 21 Naphthol reagent according to *Zambelli*
3 g α-naphthol are dissolved in 100 ml 80% acetic acid.

Reagent No. 21a Naphthylamine reagent
0.2 g naphthylamine is dissolved in 10 ml glacial acetic acid plus 40 ml distilled water and the solution diluted with 250 ml distilled water.

Reagent No. 21b Nessler's reagent
100 g mercuric chloride and 70 g potassium iodide are dissolved in as little water as possible. A solution of 160 g sodium hydroxide in 500 ml water (Reagent No. 10) is added with stirring and the final solution made up to 1 l with water.
Excess of mercuric chloride increases the sensitivity and excess of potassium iodide decreases it. Precise work is thus necessary. The qualities of purchased reagents vary somewhat for this reason. In comparisons with standard solutions, these differences are unimportant; but new calibration curves are needed each time a new reagent is used. Nessler reagent, (Merck 9028) must be used with the Hellige comparator. Nessler's reagent cannot be kept for long.

Reagent No. 22 Reference solution of nitrate
0.1635 g of purest potassium nitrate or 0.137 g sodium nitrate is dissolved in distilled water, 1 ml chloroform is added and the solution made up to 1 l with distilled water. 1 ml contains 0.1 mg NO_3^-.

Reagent No. 23 Reference solution of nitrite
0.150 purest sodium nitrite or 0.185 g potassium nitrite is dissolved in water, 1 ml chloroform is added and the total volume made up to 1 l with distilled water.
1 ml = 0.1 mg NO_2^-.

Reagent No. 24 Phosphate buffer
46 g potassium dihydrogen phosphate (KH_2PO_4) and 24 g trisodium phosphate ($Na_3PO_4 \cdot 12H_2O$) are dissolved in distilled water and the volume made up to 1 l.

Reagent No. 25 Reference phosphate solution
0.377 g disodium hydrogen phosphate ($Na_2HPO_4 \cdot 12\ H_2O$, analytical grade) or 0.141 g potassium dihydrogen phosphate (KH_2PO_4) is dissolved in distilled water and the volume made up to 1 l with distilled water. 1 ml contains 0.1 mg PO_4. This solution is diluted 10 or 100-fold with 3% sodium chloride solution; it cannot be kept.

Reagent No. 26 *Riegler*'s reagent
2 g pure sodium 1,4-naphthionate and 1 g purest ß-naphthol are shaken vigorously with 200 g water and the solution is then filtered. The reagent solution is colourless and can be kept for longer periods in dark bottles without suffering change.

Reagent No. 27 Soap solution according to *Boudron* and *Boudet*
150 g lead soap are softened by warming on the water-bath and triturated with 40 g potassium carbonate to form a homogeneous mass. This is extracted with absolute alcohol and the solution filtered from undissolved residue. The alcohol is distilled off and the residual potassium soap is dried at 100° C on the water-bath. 20 g of this potassium soap are dissolved in 520 ml of 56% alcohol by warming, the warm solution is filtered if cloudy and allowed to cool.
This solution is filled into the special, tap-less burette ("hydrotimeter") to the mark above the zero. 40 ml of the barium nitrate solution described below are placed in the special glass-stoppered bottle and the soap solution is added until a lather is formed. If the burette reading is less than 22, the soap solution is too concentrated and must be diluted witz 56% alcohol until precisely 22 burette degrees are equivalent to the 40 ml of barium nitrate.
For this standardisation, 0.574 g pure barium nitrate, dried at 100°, is dissolved in distilled water and the volume made up to 1 l. 100 ml of this solution contains barium equivalent to 22 mg calcium carbonate and 40 ml of the solution contain barium equivalent to 8.8 mg calcium carbonate. The solution thus has a hardness of 22 French degrees = 12.3 German degrees.

Reagent No. 28 Rochelle salt solution
100 g of chemically pure Rochelle salt (Tartarus natronatus pro analysi) are dissolved in 200 g distilled water by warming, and 10 ml of Nessler's reagent are added. After standing for 1 to 2 days, a yellow precipitate of an ammonium salt separates. The clear solution is siphoned off and can be kept in the dark.

Reagent No. 29 Sodium hydroxide/carbonate
100 g crystallised sodium carbonate and 50 g sodium hydroxide are dissolved in 300 ml distilled water by warming. Any further necessary water is added and the solution then centrifuged. The carbonate must be free from ammonia or the reagent mixture must be boiled for a longer time.

Reagent No. 30 Sodium carbonate solution for determination of carbon dioxide
1.2045 g of purest, dried (at 160°–180° C) sodium carbonate is dissolved in boiled water in a litre flask. After cooling (the flask is closed with a soda-lime tube), the contents are made up to 1 l. According to *Czensny,* 100 g sodium sulphate are addes previously. The same solution can be obtained if 22.8 ml 1 N or 228 ml 0.1 N sodium carbonate solution are diluted to 1 l with boiled distilled water. 1 ml corresponds to 0.5 mg CO_2, i.e., the use of 100 ml water corresponds to 5 mg/l of free CO_2.

Reagent No. 31 Sulphanilic acid reagent
10 g sulphanilic acid are dissolved in 100 ml conc. sulphuric acid (d = 1.84).

Reagent No. 32 Tolidine reagent
90 ml distilled water are brought to the boil, 10 ml conc. hydrochloric acid (d = 1.19) are added and then 0.1 g o-tolidine (analytical grade). Boiling is continued until the tolidine

has completely dissolved, the solution is then cooled and made up to 100 ml. The reagent can be prepared also from 0.135 g o-tolidine hydrochloride which is more easily soluble. The reagent keeps for 6 months, provided it is protected from light. It must also be protected from frost, since o-tolidine separates irreversibly at O° C.
The solution must not be stored in bottles with rubber stoppers.

Reagent No. 33 Stannous chloride
1 g stannous chloride ($SnCl_2{\cdot}2H_2O$, analytical grade) is dissolved in 10 ml hydrochloric acid, if necessary with gentle warming. A granule of metallic tin is added and the solution is diluted with 40 ml distilled water. The reagent has to be freshly prepared each time since it keeps for only a few hours. Solid stannous chloride also cannot be kept long; only clear, dry crystals may be used to prepare the reagent.

Reagent No. 34 Zirconium-alizarin reagent
0.3 g zirconiumoxychloride ($ZrOCl_2 \cdot 8H_2O$) is dissolved in 50 ml distilled water in a one-litre flask.
A solution of 0.07 g sodium alizarinmonosulphonate in 50 g water is slowly added to the zirconium oxychloride solution, swirling the contents of the flask. The resulting solution becomes clear on standing some minutes.
112 ml fuming hydrochloric acid ($d = 1.19$) are diluted to 500 ml with distilled water; and 37 ml sulphuric acid are added to 400 ml distilled water and the volume made up to 500 ml with distilled water. These acid solutions are mixed after cooling and the mixture poured into the alizarin-zirconium solution in the one-litre flask. The final solution is mixed well and is ready for use an hour later after changing colour. It is stored in a refrigerator but must be newly prepared every 2 to 3 months.

Reagent No. 35 Cerium reagent
120 g cerium (III) chloride (analytical grade) are dissolved in 100 ml previously boiled distilled water, 3 ml 25% hydrochloric acid are added and the mixture is boiled for 2 hours. After cooling (the flask is closed with a tube containing sulphite), the volume is made up to 1 l with boiled distilled water.

X. Covering Document for Water Samples

Sender-in of the sample ....
Name of well or water works ....
Day and hour of sampling ....
Air temperature .... Water temperature ....
Purpose of the investigation ....
Is investigation regular? ....
Or is there a particular reason? ....
(suspicion of pollution or failure of purification plant)
Purpose for which the water is used? ....
As drinking water, service water or for definite technical aims? ....
definite technical aims? ....
Is the water supply individual or central? ....
Or has the water been taken directly from an open stretch of water (lake, pond, river or stream) or from a spring? ....

A. Investigation of Individual Water Supply

Individual water supply (with hand pump)? ....
Is there a drilled well? ....
How deep is the ground water level? ....
Does the ground water level fluctuate with the time of year? ....
And the water level in the well? ....
Or is there a dug well? ....
a) a closed dug well with wooden cover? ....
or iron cover? ....
Is the well footing satisfactorily covered? ....
b) an open dug well (cistern, draw well)? ....
Does the rim of the well project above ground level? ....
Is the dug well bricked or does it consist of concrete rings or have wooden sides? ....
What is the base of the well made of (stone, cement, gravel or soil)? ....
How deep is the well? ....
When was the well built, cleaned or repaired? ....
Does the water level of the well depend on the water level of neighbouring stretches of water? ....

Or on rainfall? ....
Are there polluted influents? ....
Or is the water collected on roofs? ....
Or is surface water present? ....

B. Further Details of the Well Vicinity

How far is the well from the farm? ....
Are there near the well: lavatories, manure heaps, liquid manure pits, seepage pits, stables, stable run-offs, sewers, foul water gutters, drainage ditches, contaminated brooks, village ponds, refuse tips, cemeteries, camping sites, gravel quarries, sewage plants, mineral oil stocks, works with radioactive wastes or waste waters (distance away)? ....
Are there moors or swamps near the well (distance away)? ....
Are salt deposits, brine springs or graduation works in the neighbourhood? ....
Is the well at a higher spot and can the water run away from the well on the surface? ....
Or does the ground slope down to the well? ....
Is the temperature of the water constant or does it vary with the air temperature at the time? ....
Does the appearance of the water change at times? ....
Does it become cloudy or coloured at times? ....
Is there a run-off gutter for well water which runs over, or does it seep through immediately near the well? ....
Is the well surrounded by earth (grown or raised) or has trodden loam been applied? ....
Has the well been in continuous use? ....
Is the well situated in the flood area of a water course and what is its nature? ....
What is the surrounding ground like? Are there ravines or cracks in it near the well? ....
Have there been local cases in the last few years of illnesses which can be carried by drinking water (e.g. typhoid, paratyphoid)? ....

C. Central Water Supply

What is the type of delivery of the water (shallow or deep well; vertical or horizontal well or reservoir water delivery with a strainer)? ....
What is the daily well delivery? ....
Further questions as under B.

Bibliography of the Chemical Part

The bibliography titles are reprocuded here from the German edition. They are followed by translated titles for the benefit of those whose knowledge of German is limited. It is hoped that, with their help, such readers will be able to decide easily if a given source is of value to them.

Literatur vom chemischen Teil

Trinkwasser-Untersuchung und Beurteilung (Mineralwasser)

1. Deutsche Einheitsverfahren zur Wasser-, Abwasser- und Schlammuntersuchung. Herausgegeben von der Fachgruppe Wasserchemie, Weinheim, 3. Aufl. im Erscheinen seit 1965.
2. Die amerikanischen Einheitsverfahren zur Untersuchung von Wasser und Abwasser. Deutsche Übersetzung von *F. Sierp,* München (1951).
3. Einheitliche Anforderungen an die Trinkwasserbeschaffenheit und Untersuchungsverfahren in Europa. (Schriftenreihe d. Ver. Wasser-, Boden- u. Lufthygiene [1960], *H. Kruse.*)

3a. European standards for drinking water. WHO, Genf 1961.

3b. Internat. Standards for Drinking Water, WHO, Genf, 2. Ed., Genf 1963.

4. Handbuch der Lebensmittelchemie. J. Schormüller, Bd. VIII. S. W. Souci u. K. E. Quentin. (Wasser u. Luft), Berlin 1969/70.
5. *H. Beger,* Leitfaden der Trink, und Brauchwasserbiologie. Stuttgart (1951). 2. Aufl. von *D. Lüdemann* (1966).
6. *O. Gübeli-Litscher,* Chemische Untersuchung von Mineralwässern. Innsbruck (1948).
7. *H. Klur* und *W. Olszewski,* Untersuchung des Wassers an Ort und Stelle. 8. Aufl., Berlin (1943).
8. *Lunge-Berl,* Chemisch-technische Untersuchungsmethoden. Bd. II, Teil 1, Kesselspeisewasserpflege von *A. Splittgerber* (1923).
9. Die Untersuchung von Wasser. 4. Aufl., E. Merck AG, Darmstadt (1966).
10. *Ohlmüller-Spitta,* Untersuchung und Beurteilung des Wassers und Abwassers. 5. Aufl. von *W. Olszewski* und *O. Spitta,* Berlin (1931).
11. *J. Tillmanns,* Untersuchung von Trink- und Brauchwasser. Halle 1915.
12. *Zimmermann,* Photometrische Metall- und Wasseranalysen. 2. Aufl., Stuttgart (1961).
13. Vattenkemiska Analysmetoder. *L. Karlgren,* Uppsala 1962.

Wasserhygiene

14. Hygienische Leitsätze für die Trinkwasserversorgung. Veröffentl. a. d. Geb. d. Med. Verw., 38 Berlin (1932).
15. Leitsätze für die Trinkwasserversorgung DVGW-Schrift (1949).
16. Trikwassergüte – DVGW-Aussprache – Tagung (1954).
17. Leitsätze für die zentrale Trinkwasserversorgung, Berlin (1959), DIN 2000.

18. Leitsätze für die Einzel-Trinkwasserversorgung, Berlin (1959), DIN 2001.
19. Das Lebensmittelgesetz, Textausgabe (1959). B. Behrs Verlag, Wiesbaden.
20. Verordnung über den Zusatz fremder Stoffe bei der Aufbereitung von Trinkwasser (Trinkwasseraufbereitungsverordnung). Bund.Gesetz-Bl. I, S. 762–763 (1959), S. 479 (1966).
21. Einheitliche Anforderungen an die Trinkwasserbeschaffenheit und Untersuchungsverfahren in Europa. *H. Kruse* – Schrift.-Reihe Ver. Wa-, Bo-, Lufthygiene Nr. 1 a (1960).
22. *H. Kruse,* Wasser. Hannover (1949).
23. *W. Kumpf, B. Dietrich* u. *J. Kumpf,* Radioaktive Substanzen und Wasser. 2. Aufl, München (1969).

Wasserversorgung

24. DIN-Taschenbuch B 10, Bd. 12, Wasserversorungsnormen.
25. Die Wasseraufbereitung als wissenschaftliche Aufgabe und ihre praktische Ausführung. DVGW-Sonderheft.
26. Taschenbuch für das Gas- und Wasserfach. 3 Teile, Teil III: Zentrale Wasserversorgung, herausgegeben vom DVGW (*E. Naumann*), München (1963).
27. Rohrnetzschäden – ihre Ursache, Feststellung und Verhütung. DVGW-Aussprachetagung (1961).
27a. Wasseraufbereitung in Hausanlagen. DVGH-Broschüre. Von *K. E. Oehler* 1965.
28. *G. Giebler,* Chemische Wasserstatistik der Wasserwerke in der Bundesrepublik und West-Berlin. München (1959).
29. *H. Bahlsen,* Das Wasser. 2. Aufl., München (1954).
30. *E. Bieske,* Handbuch des Brunnenbaues. 3 Bde. Berlin 1965.
31. *E. Bieske,* Bohrbrunnen. München (1953).
32. *J. Brix, H. Heyd, E. Gerlach* u. *K. Hünerberg,* Die Wasserversorgung. 6. Aufl., München (1963).
33. *C. Dahlhaus* u. *H. Damrath,* Wasserversorgung. 5. Auflage, Stuttgart 1967.
33a. *Fr. Dosch,* Untersuchungen zur hygienischen Prüfung eines gedeckten Trinkwasserbehälters. Wien 1964.
34. *G. M. Fair* u. *J. C. Geyer,* Wasserversorgung und Abwasserbeseitigung. Deutsche Bearbeitung *B. Dietrich* u. *J. Morris,* München (1962).
35. *J. Gandenberger,* Fernwasserleitungen. Weinheim (1951).
35a. *E. Gerlach* u. *K. Hühnerberg,* Die Wasserversorgung, München 1963.
36. *G. Giebler,* Chemische Wasserstatistik der Wasserwerke, 3. Aufl., München (1959).
37. *L. W. Haase,* Werkstoffzerstörung und Schutzschichtbildung im Wasserfach. Weinheim (1951).
38. *J. Hentze,* Wasserbau. 2 Teile (1958 u. 1960).
38a. *E. Hömig,* Metall + Wasser. 3. Aufl., Essen 1970.
39. *O. Jetsch,* Aufbereitung des Wassers. Berlin-Ost (1956).
41. *H. Liebmann,* Trinkwassergewinnung aus Oberflächenwasser. München (1958).
42. *J. Mutschmann* u. *F. Stimmelmayr,* Taschenbuch der Wasserversorgung. 4. Aufl., Stuttgart (1965).
43. *H. Schneider* u. *Chr. Truelsen,* Die Wassererschließung. Essen (1952).
44. *H. Schneider,* Vertikalbrunnen – Horizontalbrunnen.
45. *H. Stooff,* Chemische und physikalische Fragen der Wasserversorgung. Schriftenreihe Wa. Bo. Lu. (1949).

Oberflächenwässer, Abwasser

46. *H. Brix, K. Imhoff u. R. Weldert,* Die Stadtentwässerung in Deutschland. Jena (1934).
46a. *K. R. Dietrich,* Die Abwassertechnik, Heidelberg 1939.
47. *W. P. Dunbar,* Leitfaden für die Abwasserfrage. 3. Aufl., München (1954).
48. *W. Hosang,* Stadtentwässerung. 5. Aufl. (1969).
49. *W. Husmann,* Praxis der Abwasserreinigung. 3. Aufl., Berlin (1969).

50. *K. Imhoff,* Taschenbuch der Stadtentwässerung. 22. Aufl., München (1969).
51. *G. Jordan, M. Manthey-Horn, F. Meineck, P. Saner* u. *R. Schmidt,* Die chemische Untersuchung von Abwasser unter Berücksichtigung der Schlammanalyse und der Vorflutuntersuchung. Berlin (1941). Kl. Mitt. Ver. f. Wasser-, Boden- u. Lufthygiene, 17. Jahrg.
52. *D. Kehr,* Wasser- und Abwasserprobleme bei der Planung von Industriebetrieben. Zbl. f. Industriebau, Bd. I (1955/56).
52a. *O. Klee,* Reinigung industrieller Abwässer B 6 Chemie Monogr. 1970.
53. *E. Koschare,* Stadtentwässerung und Abwasserbehandlung. Köln (1961).
54. *R. Lautrich,* Der Abwasserkanal. 3. Aufl., Hamburg (1970).
54a. *H. W. Lieber,* Grundlagen und Praxis der Behandlung von Abwässern der Metallindustrie Saulgau 1968.
55. *H. Liebmann,* Handbuch der Frischwasser- und Abwasserbiologie. München (1951) (2 Bde.).
56. *H. Liebmann,* Münchener Beiträge zur Abwasser-, Fischerei- und Flußbiologie. München (1953 u. 1965) (14 Bde.).
57. *R. Maucha,* Hydrochemische Methoden in der Limnologie. Stuttgart (1932).
58. *F. Meinck, H. Stooff,* u. *H. Kohlschütter,* Industrie-Abwasser. 4. Aufl., Stuttgart (1968). 1969.
58b. *W. Pürschel,* Abwasser-Behandlung, Wasserwirtschaft. Westermanns Fachbücher f. Ing. Braunschweig.
58c. *W. Schäperklaus,* Lehrbuch der Teichwirtschaft. 3. Aufl. Berlin 1967.
59. *F. Sierp,* Die gewerblichen und industriellen Abwässer. 3. Aufl., Stuttgart (1967).
59a. F. Strather, Berbereichemie, Berlin 1967 (enth. Gerberei-Abwasser).
59b. *W. Triebel,* Leitfaden für den Betrieb von Kläranlagen, Frankfurt/M. 1965.
60. *R. Weiner,* Die Abwässer in der Metallindustrie. 2. Aufl., Saulgau (1961).
61. *G. Witzel,* Das Wasserhaushaltsgesetz. 5. Aufl., Berlin (1964).
62. Veröffentlichungen des Instituts für Siedlungswasserwirtschaft der Technischen Hochschule Hannover (Prof. Dr. *D. Kehr*) (20 Bde).
63. Stuttgarter Berichte der Siedlungswasserwirtschaft von Prof. Dr. *Pöpel* (14 Bde).
63a. *K. R. Dietrich,* Die Abwassertechnik, Hamburg (1970).

Kesselspeisewasser

64. Analysenverfahren für den Kraftverkehrsbetrieb. Ver. d. Großkesselbesitzer, Essen. 2. Aufl.
65. *K. Dorfner,* Ionenaustauscher. 2. Aufl., Berlin (1964).
66. *R. Freier,* Kesselspeisewasser, Kühlwasser. 2. Aufl., Berlin (1963).
66a. *H.-D. Held,* "Kühlwasser". Vulkan-Verlag, Essen (1970).
66b. *H. E. Hömig,* Physikalische Grundlagen der Speisewasserchemie. 2. Aufl., Essen 1966.
67. *S. Leik,* Das Wasser in der Industrie und Haushalt. Dresden (1949).
68. *H. Reichelt,* Vom Rohwasser zm Hochdruckdampf. Halle (1949).
69. Richtlinien für die Aufbereitung von Kesselspeisewasser und Kühlwasser. Ver. d. Großkesselbesitzer, Essen, 4. Aufl.
70. *A. Splittgerber,* Kesselspeisewasserpflege. Leipzig (1936). 2. Aufl., H. Ullrich, Berlin 1963.
71. *K. Stoll,* Das Wasser im Kesselhaus. München 1969.

Weitere Literatur in den einzelnen Abschnitten

Periodische Werke über das Gesamtgebiet

72. *Literaturberichte* über Wasser, Abwasser, Luft und Boden. Im Auftrage des Vereins für Wasser-, Boden- und Lufthygiene. Berlin-Dahlem, von Prof. Dr. *F. Meinck,* Verlag Gustav Fischer, Stuttgart. 18 Bände, von 1950–1970.
73. *Vom Wasser,* Ein Jahrbuch für Wasserchemie und Wasserreinigungstechnik, herausgegeben im Auftrage der Fachgruppe Wasserchemie von Prof. Dr. *W. Husmann.*
73a. Fortschritte der Wasserchemie, Akademie-Verlag, Berlin-Ost.
74. Mitteilungen d. Ver. d. Großkesselbesitzer, Essen.
75. Schriftenreihe GWF: Wasser u. Abwasser (13 Bde.).

75a. DVGW-Arbeitsblätter.
75b. DIN-Normen, herausgegeben vom Deutschen Normenausschuß, Beuth-Vertrieb GmbH, Berlin 15.

Zeitschriften

76. Das Gas- und Wasserfach – GWF, Verlag R. Oldenbourg, München.
77. Gesundheitsingenieur, Fachzeitschrift für Städtehygiene. Verlag R. Oldenbourg, München.
78. Wasser und Boden, Verl. Lindow, Hamburg 55.
79. Wasser und Abwasser, Verl. Winkler u. Co., Wien.
80. Österreichische Wasserwirtschaft, Springer-Verl., Wien.
81. Gas, Wasser, Wärme Fachzeitschrift Wien.
82. Wasser, Luft und Betrieb, Krauskopf-Verlag, Mainz.
82. Wasser + Abwasser, Verlag Nord-Süd-Werbung, München 15.
84. Water and Water Engineering, Verlag Loxley Brd., London.
85. Water Research, Pergamon-Verlag, Oxford u. London.

Translated Titles

Investigation and Assessment of Drinking Water (Mineral water)

1. German Standard Methods for Investigation of Water, Wastewater and Sludge (Mud); published by the specialist group, Water chemistry, Weinheim, W. Germany (3rd Edition since 1965)
2. American Standard Methods for the Examination of Water and Wastewater (the reference is, in fact, to a German translation)
3. Standard requirements of drinking water quality and procedures of investigation in Europe
4. Handbook of Food Chemistry, Volume VIII
5. Guide to the biology of drinking and service water
6. Chemical investigation of mineral waters
7. Investigation of water on the spot
8. Chemico-technical methods of investigation, Vol. II, Part 1, care of boiler feed water
9. The investigation of water
10. Examination and assessment of water and wastewater
11. Examination of drinking and service water
12. Photometric analysis of metals and water
13. Chemical analysis methods for water

Water Hygiene

14. Hygienic guiding principles for drinking water supply
15. Guiding principles for drinking water supply
16. Drinking water quality – DVGW discussion at a conference
17. Guiding principles for central drinking water supply
18. Guiding principles for individual drinking water supply
19. The Foodstuff statute, published text
20. Decree on the addition of extraneous materials in conditioning of drinking water (Drinking water processing law)
21. (same as 3.)
22. Water
23. Radioactive substances and water

Water Supply

24. DIN pocket book B 10, Vol 12, standards for water supply
25. Water conditioning as a scientific task and its execution in practice. DVGW special issue

26. Pocket book for Gas and Water Department. 3 parts, part III: central water supply, issued by the DVGW
27. Damage to pipeline networks, its cause, establishment and prevention. DVGW discussion conference
27a. Water treatment in domestic plants. DVGW prospectus
28. Chemical water statistics of water works in the Federal German Republic and in West Berlin
29. Water
30. Handbook of well construction, 3 volumes
31. Drilled wells
32. Water supply
33. Water supply
33a. Investigations for testing hygienically a covered drinking water tank
34. Water supply and removal of wastewater (original in English)
35. Long distance water pipe lines
36. Chemical water statistics of water works
37. Material destruction and formation of protective layers in water departments
38. Water engineering, 2 parts
39. Conditioning of water
40. Water in industry and the household
41. Obtaining of drinking water from surface water
42. Pocket book of water supply
43. Opening-up of water supplies
44. Vertical wells – horizontal wells
45. Chemical and physical questions in water supply

Surface water, Wastewater

46. Town drainage in Germany
46a. Sewage technique
47. Manual of wastewater questions
48. Town drainage
49. Sewage purification in practice
50. Handbook of town drainage
51. The chemical analysis of wastewater, including sludge analysis and examination of receiving waters
52. Water and wastewater problems in planning industrial establishments
53. Town drainage and treatment of sewage
54. The sewer
54a. Fundamentals and practice of dealing with wastewaters of the metal industry
55. Handbook of the biology of fresh water and wastewater
56. Münich contributions to the biology of wastewater, fisheries and rivers
57. Hydrochemical methods in limnology
58. Industrial wastewaters
58a. Textbook and handbook of wastewater technique
58b. Treatment of sewage, water economy
58c. Textbook of fish culture
59. Industrial and trade wastes
59a. Chemistry of tanning (including tannery wastewater)
59b. Manual for clarification plant establishments
60. Wastewaters in the metal industry
61. The law on water resources policy
62. Publications of the institute for water economy of settlements (Technical College of Hannover) (20 volumes)

63. Stuttgart reports on the water economy of settlements (Prof. Pöpel) (14 volumes)
63a. Wastewater Technique

Boiler feed water

64. Analytical procedures for power stations
65. Ion exchangers
66. Boiler feed water, cooling water
66a. Physical bases of feed water chemistry
67. (same as 40.)
68. From raw water to high pressure steam
69. Directives for processing of boiler feed water and cooling water
70. Care of boiler feed water
70a. Cooling Water

Further literature is mentioned in the individual sections

Periodicals covering the whole domain

71. Literature reports on water, wastewater, air and soil. Commissioned by the society for water-, soil- and atmospheric hygiene
72. Vom Wasser, a yearbook for water chemistry and technique of water purification, issued as an assignment of the specialist group for water chemistry
72a. Progress in water chemistry
73. Communications of the Union of large boiler owners
74. Written issues of the GWF: Wasser u. Abwasser (Water and wastewater) (13 volumes)
74a. Working sheets
74b. DIN-standards, issued by the German Standards Group

Journals (translations of the names)

75. Gas and Water Department-GWF
76. Sanitary Engineer, specialist journal for municipal hygiene
77. Water and Soil
78. Water and Wastewater
79. Austrian Water Economy
80. Gas, Water, Heat specialist journal, Vienna
81. Water, Air and Factory
82. Water and Wastewater

Part 2

The Bacteriology of Drinking Water

by

Helmut *Peter*

I. General Remarks about Bacteriological Investigation Procedures

Bacteria are unicellular forms of life. Since 1857 *(Nägeli)* they have been classified in the plant kingdom as Schizomycetaceae. They are distinguished from unicellular forms of animal life by having no cell nucleus; according to *Piekarski, Robinow* and others, the functions of a cell nucleus are taken over in bacteria by nucleoids, which, as nuclear equivalents, are non-uniformly distributed in the plasma of the bacterial cell. *R. Müller* [*90*] accordingly suggested that a special bacterial kingdom should be delimited alongside the plant and animal kingdoms. This proposal is supported by the extraordinary high number of bacteria–a handful of soil contains, according to *Thimann* [130], as many bacteria as there are human beings on our earth–and also by the multiplicity of types corresponding to this high number. There has been no lack of attempts to classify systematically this diversity, as in botany since Linnaeus' time. Classification according to morphology has now only historical significance (This is into cocci [spherical], bacteria [rod-shaped] and spirillae [screw-formed]). The grouping of bacteria into order, sub-order, family, tribe, genus or species is dictated not only by the form, variable within limits, but also by biochemical performance in anabolic and catabolic metabolism, and by serological features. To *Bergey* [14] and his co-workers is due the credit of having unified the work of *Lehmann* and *Neumann* [75] with new microbiological knowledge and having created a new systematic classification of bacteria which is employed internationally. Since *Bergey*'s death, it has been kept up with the present state of knowledge through a large commission of co-workers. In accordance with *Bergey*'s "Manual of Determinative Bacteriology", the protophyta are divided into 3 classes:

1. Schizophyceae – blue-green algae
2. Schizomycetaceae – bacteria
3. Microtatobiotes – rickettsia and viruses

The Schizomycetaceae are subdivided into 9 orders; the most important for water bacteriology, in so far as they are medically interesting, is that of the Eubacteriales. This order consists of 13 families which, in turn, are subdivided into tribes, genera and species.

Creatures from the class of Microtatobiotes (rickettsia and viruses) also play a part in medicinal water bacteriology. They constitute microbial life which is too small to be visible in the light microscope and which possesses the common characteristic of being able to multiply only in living cells. Particularly interesting are the viruses of bacteria, the bacteriophages. The detection of viruses is technically especially demanding and exceeds the domain of normal bacteriology. In contrast, bacteria are detected through classical

procedures of microscopic investigation, cultivation on artifical nutrient media, and animal experiments.
Direct microscopy of unstained bacteria is predominantly used in the hanging drop to establish the motility of the bacteria. As a result of the low light refraction of bacteria, it is inferior to the light microscopy of stained specimens. Staining is sometimes performed with basic aniline dyes, such as fuchsin, gentian violet or methylene blue in order to visualise the external form of the microorganisms. Staining is carried out also in combined working operations, aimed at delineating the chemical substrates of the bacteria. Gram staining may be mentioned amongst these combined staining procedures; it utilises a reagent for a magnesium ribonucleate present in "gram-positive" bacteria. Another is the Ziehl-Neelsen staining, which enables the wax-like substances in mycobacteria, so-called "acid-resistant" bacteria, to be stained.
Gram staining is most important in the systematic classification of bacteria. It depends on retention of basic aniline dyes in the interior of the bacterial cell. These are then converted into dye-iodine complexes by steeping in Lugol's solution. The subsequent treatment with alcohol removes the complexes from gram-negative bacteria which are thereby decolourised. The dye-iodine complexes in gram-positive bacteria cannot pass through the cell wall and so the bacteria retain their dark blue colour. The gram-negative bacteria are visualised by contrast staining. Examples of both types of bacteria are tabulated below:

Gram-positive	Gram-negative
Staphylococci	*E. coli*
Streptococci	Coliform bacteria
Actinomycetes	*Salmonella*
Lactic acid bacteria	*Shigella*
Diphtheria (Loeffler's) bacteria	*Pseudomonas* and *proteus* species
Anthrax bacilli	Vibriones
Yeasts	Spirillae and spirochaetes
Clostridia in young cultures	

Details for carrying out gram-staining are on p. 307.
Protein degradation products are used as basic substrate for cultivating bacteria on artificial media. They are added to the cultures as meat extracts, peptones or amino acids. Depending on the purpose of the investigation, protein materials, carbohydrates, indicators or special substances are added to aid growth of the bacteria to be cultivated and to inhibit as far as possible the growth of others.
A distinction is made between solid nutrient media and liquid nutrient solutions, which are mostly based on meat broth. Nutrient media are rendered solid by adding gelatinising substances to the broth. Agar-agar and gelatine are those most frequently used but other materials, such as silica gel or serum, solidified by heating, find routine application. It was only after *Robert Koch* had introduced solid culture media that it became possible to obtain individual colonies, as formed from the single bacterium on the jelly, and hence pure bacterial cultures. Almost every bacterial species requires special nutrient media in order to attain its best growth. These are given in the discussion of those bacterial species which are important in water bacteriology.

Experiments with animals serve for diagnosing certain pathogenic microorganisms. In water bacteriology it can play a part in the detection of tubercular bacteria in wastewater.
It is already evident from this brief account that there are numerous requirements, including comprehensive equipment and apparatus, for work with bacteria. If accurate results are sought, this work cannot be pursued as a side-issue. Further, bacteriology of water has been extended in recent years to include the detection of bacteriophages and viruses. Special knowledge is absolutely essential in this domain.
In accordance with the law for prevention and combating of infectious diseases (Bundesseuchengesetz–Federal Epidemic Decree–of 18/7/61), permission of the controlling authorities is necessary to be able to cultivate pathogenic microorganisms which, through the use of modern selective nutrient media, is playing an incomparably greater role than in earlier decades; this permission may be given only when there are definite qualifications. Experiments with animals also may be carried out only with special permission of the highest authority for protection of animals, usually the Ministry for Agriculture and Forestry (of the Land in Germany). These public injunctions make the demands clear which are asked of the investigator.

II. The Bacteriological Working Place

All apparatus, glass articles and chemicals needed for the bacteriological investigation of water must be sterilised, so that no secondarily introduced bacteria interfere with or ruin the work. Glass articles, such as sample bottles, test tubes, pipettes, flasks, petri dishes, funnels and burettes, are sterilised in the hot air steriliser. Sterilisation means destruction of all microorganisms, including the resistant spores. To accomplish this, the application of dry heat, hot air, at 170° C for 30 minutes is necessary, reckoning from the moment at which the temperature is attained.
Hot air sterilisers are available commercially in all usual sizes from many firms. In apparatus with static hot air, regions of temperature deviating by up to 20% from the desired value may occur. Apparatus with circulating hot air is thus preferable.
Test tubes, open flasks, burettes etc. are plugged with pulp before sterilising. The stoppers can be home-made or purchased in customary sizes. Test tubes are stored in rectangular baskets of galvanised iron wire and sterilised. Pipettes are placed in cans of sheet copper. All articles to be sterilised must be packed so that they cannot become infected again by atmospheric microorganisms after sterilisation. Liquid media are sterilised in autoclaves. These pressure apparatus work with saturated water vapour. An excess pressure of 1 atmosphere for 30 minutes at 120° C is necessary to destroy heat-resistant spores.
Various chemicals can be broken down, especially at these temperatures, and therefore the temperature of boiling water is used when such culture additives are present; this water is prepared in the Koch digester. Fractional sterilisation, tyndallisation, is used for liquids for which even this temperature is too high; this is rarely employed in water bacteriology, however. Liquid cultures are stored in test tubes or bottles, and solid culture media in petri dishes.
Plastic vessels are used mostly nowadays for this, and are discarded after being used once. They are sterilised already by the manufacturing firm and are delivered in a sterilised packing so as to be usable immediately without further preparation. Sterilisation of the glassware needed and preparation of the culture media are best performed in a special room. This should have as minimum equipment, a digester, an autoclave, a hot-air steriliser, a rough balance and a semi-analytical balance, shelves for chemicals and glassware, benches for work and bunsen burners. An electric refrigerator is indispensable for storing ready prepared culture media and solutions.
The real bacteriological working laboratory should have room for three incubators, adjusted to temperatures of 22° C, 37° C and 44° C. Since the room temperature often exceeds 22° C, the first-named incubator must be equipped with automatic water cooling. As a makeshift solution, an incubator without water cooling can be placed in a cellar room, if it has been ascertained with certainty that, even on the hottest days, the temperature in it never

rises above this 22° C value. The microscope is the most important instrument on the bacteriological place of work. Modern microscopes are equipped with condensers which enable the investigation to be carried out in bright field and dark field illumination and with the phase contrast procedure. This last-named method in particular, introduced into microbiological practice by *Köhler* and *Loos*, inaugurated by *Zernicke* [*66*], permits observation of unstained structures of microorganisms, important for water bacteriology. For details, reference must be made to the comprehensive specialist literature. Another indispensable item on the place of bacteriological work is the magnifying glass, used for counting bacterial colonies. The binocular magnifier of the firm of Zeiss has rendered us good service. Counters, e.g., that of *Wolffhügel*, are other aids for this purpose. Modern apparatus, such as the bacteria counting device of *Dannhofer*, can appreciably ease the weary work of counting colonies. The incubated petri dish is placed on a desk-shaped case, in which a counter and means of illumination are built and which is equipped with a lens of adjustable position.

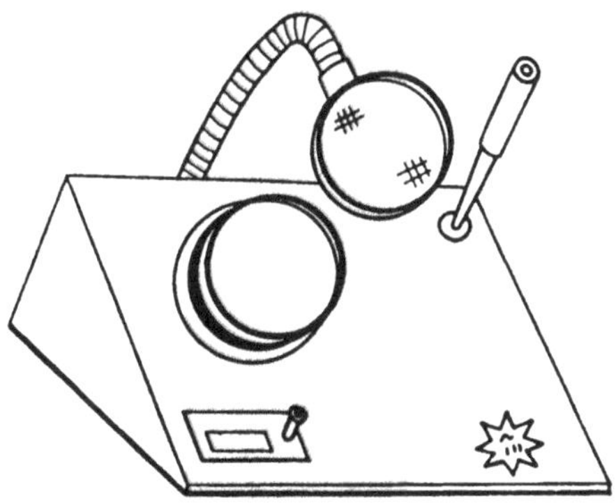

Fig. 1 Bacterial counting apparatus

Devices for producing a vacuum in suitable containers (*Zeissler's* anaerobic jar, *Knorr's* anaerostat, etc.) are necessary for cultivating anaerobic bacteria which is possible only in a milieu containing little or no oxygen. Suitable vacuum pumps are furnished by various firms.

Special apparatus have been developed for using membrane filters; these are discussed in more detail in the section on the membrane filter procedures (p. 317).

The equipment needed for bacteriological investigations is therefore quite extensive. It will be augmented with further apparatus in accordance with the demands of the individual investigator. But all apparatus is merely his aid and tool. He himself determines when they are to be used profitably and his knowledge and his experience are decisive.

III. Water and Disease

Drinking water should be free of pathogenic organisms and such constituents which are liable to be detrimental to human health. This prime hygienic requirement is, unfortunately, not always fulfilled. Consequently, the history of epidemics is largely identical with the history of drinking water epidemics. There is hardly a pathogenic germ which is unable to attack humans also via infected drinking water. It was, above all, the great cholera plague in Hamburg at the end of the 19th century that made the public aware of the connection between drinking water and epidemic. *Binding* depicted this plague, which caused immeasurable human suffering and misery, in his famous story "Der Opfergang". It was later possible to calculate the material damage brought about by the epidemic wave. The erection of a new drinking water plant, which the city of Hamburg had not carried out through false economy, cost 22.6 million marks in 1893, after the epidemic had subsided; *Gottstein* [*42*] estimated the material damage of the plague to be 430 million marks! The relation between the cost of a water works and the harm of an epidemic, from whatever organisms it may spring, would be even more unfavourable in our modern, highly industrialised society. Perhaps administrative officials would find it illuminating and useful to read *Binding's* story today and calculate corresponding possible costs for their town.

It is principally typhoid epidemics through infected drinking water which have attracted attention in our time, after the second world war. Only those epidemics are mentioned here however which are conspicuous through the larger number of affected cases; there were many others. Thus 400 people caught typhoid in Westerode in 1946, 16 of which died (*Harmsen* [*49*]). In the same year, 425 cases of typhoid were recorded in Neu-Ötting, due to infected drinking water (*Rimpau* [*104*]); in the same place, in 1948, 750 victims were recorded (*Seeberger* [*118*]). The typhoid epidemic in Hagen, in 1955, was caused by infected dairy water; there were 600 cases, of which 237 patients had typhoid and 205 were shown to have paratyphoid (*Kruse* [*71*]). Fresh in the memory is the typhoid epidemic in Zermatt in the winter of 1963, which was also traced to infection of drinking water. In many epidemics, e.g., in Hagen, not just a single *Salmonella* species was found to be the pathogenic organism, but several *Salmonella* species which had contaminated the water with a mixed infection.

Even *Shigella,* provokers of bacterial dysentery, which were formerly thought to be too sensitive to be able to endanger people via water, have been established as the cause of waterborne epidemics; this is a result of the refinement of bacteriological diagnosis. *Diosi* and coworkers [*31*] described a water epidemic in Romania which was caused by *Shigella boydii. Nikodémusz* [*91*] and *Saad* and co-workers [*107*] found *Shigella flexneri* in Hungary as pathogen of a dysentery epidemic which had been transmitted by drinking water.

Publications have multiplied in the last few years of cases in which pathogenic strains of *Escherichia coli* in drinking water caused intestinal complaints of babies and infants. Thus *Monnet* [*84*] found the *E. coli* types 0111 and 055 to be the cause of drinking water epidemics in various parts of France. The serotype 0124 was found several times in Hungary as a pathogenic polluter of drinking water. This work of *Lanyi* [*73*] and *Borian* [*15*] completes this selection of a few examples only from recent times. *Ulbricht* [*134*] detected *Dyspepsia coli* bacteria in about 0.1% of the water samples studied; most frequent occurrence was the especially virulent species 0127, and less often, 055.

Gertrud Müller [*87*] investigated serologically for 2 years the occurrence of *Dyspepsia coli* among all coli bacteria isolated from drinking and surface water. She found these organisms to comprise 0.4% of the *E. coli* isolated from wells, 0.5% of those from ship tanks and 0.1% of those from surface water.

Probably pathogenic coli species bring about infection more often via water than hitherto assumed. This likelihood ought thus to be considered in all unexplained intestinal affections, especially in childrens' homes, babies' cots, etc.

This list of infectious diseases which can be transmitted by water–salmonella infections, bacterial dysentery and illnesses due to coli bacteria – is by no means exhausted. Mention may be made, too, of tuberculosis (its pathogenic agent is often found in large amounts in the waste water from sanatoria, which has not been disinfected properly), plague, anthrax and tularaemia. The agents of the last-named disease gain access to the water through infected rodents, especially rats and mice, and this passes the infection on to humans. We ourselves experienced a tularaemia epidemic in the East, where we were able to identify *Bact. tularense* in the well water or sludge. Along with *Brucella, Leptospira* can also be transmitted through water. According to *Kathe* [*58*], not only rodents of the Muridae family but also large animals, such as horses, cows or pigs, can be sources of infection by *Leptospira* which, depending on the provoker type, manifest themselves clinically as Weil's disease, swamp fever, canine distemper, etc. *Mochmann* [*83*] described infection of drinking water by *Leptospira,* leading to 126 cases of illness in one case (Portugal) and 390 in the other (Yugoslavia). According to *Ramkow* [*101*], the Yugoslav epidemic in 1938 had 8 fatal cases.

Not only bacteria but their toxins also are able to exist in water and be detrimental to man and animal. *Schoop* [*114*] referred to a larger calamity of loss of chickens where botulism toxins could be detected in the water. Degradation products of bacteria, which, as pyrogens, can cause fever when introduced parenterally into humans, have often been found in water. The institute for research on pharmaceuticals, in Munich, detected pyrogens in the drinking water of the town N.; we had previously found large numbers of bacteria in the water as a result of wastewater entering after a flood catastrophe; the water was immediately adequately chlorinated (*Schlemmer* [*112*]). The bacteria were destroyed by the action of the chlorine but their degradation products could still be detected days later in experiments on animals. Normal humans certainly suffer no damage from these pyrogens in the drinking water but an uneasy feeling remains when one thinks of babies whose intestinal wall is more permeable than that of adults.

The aetiological agents of maladies due to viruses are frequently eliminated by man and animal and can gain access to drinking water via wastewater if conditioning is inadequate. The poliomyelitis virus is still found in receiving water months after epidemics have subsided and, according to *Reploh* [*103*], can even be regularly detected in the wastewater of large towns in which a larger number of humans live who function as virus carriers. Hepatitis epidemics of considerable magnitude have been initiated by polluted water. In New Delhi, in 1955–56, 30.000 people became ill through infected drinking water (*Ruys* [*106*]). *MacLean* [*80*] reported the detection of adenoviruses (APC-viruses) in baths water which caused sore throat and inflammation of the eyes among patrons of swimming pools; and of Coxsackie viruses which led to eczema with blisters. ECHO-viruses also were isolated from wastewater by *Pittler* and *Höpken* [*94*], *Primavesi* [*98*] etc. *Knocke* [*61*] and others succeeded in detecting reoviruses also. Apart from adeno- and reoviruses, all viruses belong to the group of enteroviruses which play the largest part in the hygiene of water and wastewater. It seems as if the detection in water of viruses pathogenic to man belongs today already to the routine work of the virologists active in this field.

Humans do not appear as a rule to become infected and then ill by directly taking up the aetiological agent present in the water; rather, at least with typhoid and paratyphoid, a decisive part is played by enrichment of the bacteria on crockery, food storage articles or the foodstuffs themselves, which act as culture media for the bacteria. *Steiniger* [*127*] pointed out that the water of the Flensburg harbour contains about 50–65 *Salmonella* per ml. Bathers who swallowed about 50 ml of water during bathing, did not become infected.

A fact of interest in connection with epidemics is that, immediately after infection of the drinking water, people are taken ill with a so-called "water illness"; this is non-specific and is evidently due to the great general increase in bacterial numbers in the polluted water. The real epidemic breaks out only after an incubation period of 14 to 21 days. This water illness as a forerunner of the epidemic could be observed particularly well in the great typhoid epidemic in Hanover in 1926, which led to 2423 cases of illness, of which 282 were fatal (*R. Müller* [*89*]). It is discussed today whether the so-called water illness is not an infection with Coxsackie- or ECHO-viruses, the incubation times of which are shorter than those of *Typhus abdominalis* (*Primavesi* [*98*]).

Not only water contaminated with bacteria or viruses can harm human beings; blue algae, protozoa and worms are also able to infect humans via water. Among protozoa may be mentioned especially the genus *Entamoeba*. In warmer climates, *Entamoeba histolytica* and *Entamoeba tetragena* can lead to severe human amoebic dysentery via drinking water (*Beger* [*12*]). Fully grown intestinal worms have not so far been detected in drinking water but the contamination of drinking water with eggs of worms has often been reported. Thus the eggs, capable of existing for a long time, of the roundworm, *Ascaris lumbricoides,* can be washed from manure heaps into leaky wells; or the eggs of *Ankylostama duodenale* be transmitted to humans through infected drinking water at higher temperatures.

One must utter a very vigorous warning against regarding epidemics borne by drinking water as historical events and believing that they are characteristic only of times of distress, such as after the two world wars; and that they can no longer occur nowadays, with the highly perfected notification duty of infectious diseases and a public health service which functions excellently. Maintained caution and permanent checking of all water-obtaining plants are needed as a result of the ever increasing contamination of surface water which, in many places where there is no suitable ground water, is utilised in conditioned form for drinking water purposes. Any slackness in this sphere can and will exact its retribution.

Among the chemical noxae which can endanger human health, those substances which can be present in the water as decomposition products of organic compounds are distinguished from those which have gained access to the water as soluble poisons through special circumstances, chance or inadvertency. Nitrates are worthy of special mention in the first group. If present in too high a concentration, they lead to *methaemoglobinaemia* of babies and infants (*Sattelmacher* [*108*]). Only children which have been fed with dried milk preparations dissolved in water can have this malady. Its symptoms are diarrhoea, blue appearance of the skin and increased difficulty of breathing which can lead to cessation of respiration and death (*Delfmann* [*29*]). It is assumed that the babies can become ill only when the upper sections of the alimentary canal are inhabited by bacteria which can reduce nitrate to-nitrite. Coliform bacteria appear to have a special rôle here. It is anyhow noticeable that methaemoglobinaemia occurs in particular when a well water is used which is intensely contaminated by bacteria. Occurrence of the illness is favoured by the acidity of the infant gastric juice (lower than that of adults) and by the weaknesses in the ferment system of the red blood corpuscles, which is still present in the first months of life (*Teerhaag* and *Eyer* [*132*]).

Simon [*123*] reports that an inquiry at all German childrens' hospitals revealed at least 695 cases of methaemoglobinaemia of babies between 1956 and 1967, traceable to an increased nitrate level of the water. *Werner, Thal* and *Wuttke* [*138*] state that especially babies of up to three months of age are endangered because they still possess predominantly blood corpuscles with foetal haemoglobin (HbF) which is easily oxidisable. In cases of infant methaemoglobinaemia one should, however, devote attention not only to the water but also to the nourishment as a possible cause; for example, spinach in particular contains appreciable amounts of nitrate which can be converted into nitrite by bacterial action if the air is unfortunately infected (*Borneff* [*17*], *Lodenkämper* [*78*]). *Sattelmacher* [*108*] refers in his monograph to work of *Popp* who detected considerable amounts of nitrate-reducing bacteria in dried milk preparations used for preparing food for babies.

Different authors quote very variable values for the permissible limits of nitrate concentration in drinking water. *Gärtner* and *Staak* [*39*] suggest that the possibility of methaemoglobinaemia should be pointed out when nitrate values are 40–100 mg/l and the danger of it when nitrate values exceed 100 mg/l. *Zimmermann* [*142*] considers that the limit should be at mg NO_3/l, a value required in the draft for a drinking water statute, from the Federal German ministry of health. *Büsing* [*17*] believes a limit of 20 mg/l to be necessary. Even though *Wedemeyer* [*136*] has described cases of infant methaemoglobinaemia in which the well water had nitrate contents well below the official value of 50 mg/l, this 20 mg/l limit will scarcely be realisable in practice. These cases show that occurrence of nitrate is not the only condition for the appearance of this illness. As a precautionary measure, *Sattelmacher* [*108*] demands that hospitals where the nitrate content of the water exceeds 50 mg/l should have sterile methylene blue solutions in readiness for intravenous injection; and that if the water has more than 100 mg NO_3/l, water of low nitrate content should be available in bottles or tins.

Among inorganic elements which can be transmitted by water to lead to human illness, lead with its poisonous action has been known since ancient times. Lead can be taken up from water which has attacked lead water pipes or tanks soldered with lead, or run into rainwater cisterns from lead roofs. Apart from poisoning of wells, arsenic in water which can lead to human melanosis or hyperkeratosis is usually determined by the geology.

Fluorides, which can be present in excessive amounts in the drinking water especially from volcanic regions, interfere in the enamel formation of the growing teeth of young people. On the other hand, fluoride is added to drinking water which contains none, as a protection against tooth decay. Over 40 million inhabitants of the USA drink fluoridated water. The fluoridation of drinking water in Kassel is a known case in Germany; since 1952,0.7 mg F/l has been added to the drinking water in the suburb of Wahlershausen (*Hugelmann* [*55*]). The effects of fluoridation have generally found a positive echo in the USA (*Kruse* [*70*]) but opinions differ in Germany. In opposition is the feeling that it is not the task of the public water supply to provide the consumer with materials which are said to prevent illness [*5*]. And the establishment of a precedent for adding other substances, e.g. medicaments, is feared. This latter objection is not wholly without reason when one reads that, at a recent symposium in London, it was discussed whether the over-population problem could not be solved by adding to drinking water the hormones contained in the anti-baby pill; or the population calmed down during high political tension by adding sedatives (*Regau* [*102*]).

Hettche [*54*] gives the *urochromes* as further chemical compounds which can impair human health; he regards them as the cause of endemic goitre. After uptake by humans, urochromes in the drinking water can combine with the copper present in the serum, which effects transport of iodine to the thyroid gland. Urochromes occur only in wells which have been polluted by human and animal excrement and the water of which shows other pollution indicators also. *Hettche's* ideas have not not remained without being contested (p. 47).

Borneff and *Knerr* [*16*] detected carcinogenic substances in water, coming from tar, soot and asphalt (p. 78). Detergents also may be among carriers responsible for transporting these substances from the surface of the earth to the ground water. However, only waters which are in any case hygienically objectionable can be impaired by carcinogenic substances.

IV. Inspection of the Site

The hygienist is a doctor and, like every other doctor who begins the examination of his patients with their previous case histories, he also begins with anamnestic enquiries for judging occurrences of water. Thus the age of the well system is important, and significant also is the question of the ground water level. Of decisive importance in the assessment can be the fact, whether the wells dried up during years of considerable drought and whether the water level falls sharply when larger amounts of water are taken, or rises after periods of rain. If rivers, streams or ponds are near, it is important to know whether the water level in the well corresponds with that of the surface waters. Inspection of the site is of prime importance for the hygienist, just as inspection of the sick person is for the examination of a patient. The classical ideas of the hygienists at the turn of the century are not out-dated as far as inspection of the site is concerned, but are as valid as on the day they were put into print. Thus *Gärtner* [*37*], the patriarch of water hygiene, wrote that the decision about the sanitary state of a water cannot be made in most cases in the laboratory; very important data are acquired there but the deciding factor is knowledge of the circumstances; one must take a look to be able to assess a water and subsequent investigation provides auxiliary help. In all cases, therefore, the local survey has precedence over other procedures of investigation. Every examination should be preceded by a test of the geological, hydrological and management factors, establishment of the origin and nature of the water and ascertaining data on its conditioning, piping and storage. Just as the laboratory results give the practising doctor valuable hints but cannot replace the diagnosis, in the same way the laboratory findings must support and supplement the impression from inspection of the site in deciding whether a water source is hygienically in order. It is expedient to carry out the local survey in accordance with a definite scheme. The questionnaires given by *Höll* on p. 6 and p. 262 of this book have been very useful to us.

Whether, after an official inspection of the site, the chemical or the bacteriological water investigation should take precedence seems an idle and pointless dispute. It is better to institute correctly both these procedures of investigation, using critical judgment; their results are often not parallel. Thus, in theory, a chemically flawless water is fully capable of sheltering epidemic-provoking germs. And a water which comes from the ground beneath old settlements can contain nitrite or ammonia and yet be bacteriologically in order. The bacteriological investigation of newly bored wells often promises little success as long as the well is subjected to external contamination; in this case, the chemical findings offer quicker hints about the water quality to be expected. On the other hand, the bacteriologist is responsible in all questions of hygiene which concern a potential danger to humans; it would be absurd to establish chemical indicators for microorganisms leading to epidemics. But to require of him data about the dangers of corrosion to pipes or something similar is to ask too much. Bacteriologist and chemist have the same aim, namely, to supply the population with drinking water in perfect order; both discover repeatedly that their knowledge and perception are not without gaps. It is also not possible to draw conclusions about the hygienic quality of the water from a single bacteriological finding. This solitary result is like a momentary picture and circumstances may change rapidly again. *Büsing's* suggestion [*21*] of differentiating between the assessment of hygienically flawless water and water to which there is no bacteriological objection is sensible and complies with practical needs.

Soil which has developed naturally has several million bacteria per mg in its upper layers. Thousands are still found at a depth of 1 metre whereas there are scarcely any at 3 to 4 metres (10 to 14 feet) depth. When surface water percolates into soil, the loss of bacteria is greater, the deeper the water gets into the soil. This depends on the adsorption of bacteria on soil particles, on killing of bacteria through change of milieu and chemical matter, on the activity of protozoa which are able to destroy the vegetative bacterial forms, on the action of antibiotically active substances from other microorganisms, and on numerous other factors (*Knorr* [*64*]). In the surface soil layers, the water is purified in an aerobic environment but the purification processes in deeper layers are anaerobic. The "struggle for existence" between the bacteria begins in this heterotrophic zone. First victims are the microorganisms alien to the surroundings, but above all those coming from the intestines of man and animals [*4*]. The nature of the soil is just as important for the natural purification of surface water as are physical conditions or electrostatic forces. Ground water is more endangered in fissured soil of large pore diameter than in soil of small pore size. Good protection for the ground water is provided by top layers of water-impermeable material, mostly loam or clay strata. If these are broken through, however – and this occurs more frequently than formerly as a result of the building and constructional works in our industry-dominated times – all the bacteria in the soil or surface water can get into the ground water and then, via wells, reach human consumption. The vast majority of bacteria is harmless to human beings and thus unimportant from a hygienic standpoint. A large bacterial number indicates, however, that a water is not pure ground water; the possibility must always be reckoned with, that pathogenic organisms can gain access to the ground water along with the non-pathogenic species. The duration of sojourn in the soil, i.e., the time during which a single bacteriological or chemical contamination is broken down, is quoted as 60 days (*Knorr* [*63*]). Such a singly polluted water is no longer dangerous for humans after this time.

However, even when ground water in perfect order is pumped, the water can become infected in the water works or pipe system, leading to outbreaks of an epidemic. It is not correct to assume that a pressure of 6 to 7 atmospheres prevails in the pipe system and that through this alone, no surface water can enter in the event of a broken pipe. Every water vacuum pump demonstrates the contrary. Suction can be created through transverse narrowing in the mains, eddy formation in the dead water spaces, formation of cavities in the mains as a result of considerable consumption of water (fire engines); infected surface water can then penetrate into the pipe system (*Kirchbach* [*59*]).

If the conditions are unfavourable, all the microorganisms in the soil or surface water can therefore get into the ground water. It is thus doubtful whether one can speak of a specific bacterial flora of the water. Just in the last few years, great efforts have been made to single out bacteriological indicator forms and bacterial associations for particular ground- and surface water (*Oberzill* [*92*]) and to differentiate between the ubiquitous microorganisms and those bacteria and bacterial associations which are characteristic of particular places; yet it appears that the goal being able to establish unambiguously specific native bacterial flora of ground water has not been attained.

This microbiology of the water and soil which does not follow medicinal aspects offers points of view which can be important for the knowledge of past and future.

Concerning the past, *Dombrowski* [*32*] succeeded in isolating living bacteria from salt samples of primary rocks which must be as old as the salt in which they were enclosed, i.e., ca. 180 million years. *Dombrowski* assumed that the water contained in their protein compounds had been withdrawn and that the bacteria would keep indefinitely in this non-metabolising state. Their living functions awaken when water is added. Living bacteria could also be obtained from rock salt from Canadian Middle Devonian, about 320 million years old. The oldest living creatures which we know could be isolated as bacteria, capable of life, from Lower Cambrian salt, probably the oldest salt deposit on earth with its 500 to 600 million years age. These researches have opened a new biological chapter in which the question "what is life?" can be introduced.

Immense tasks await this non-medical branch of water- and soil microbiology in the future because certain indicator bacteria may eventually be able to furnish clues to deposits of materials useful in the economy of energy (*Schwarz* [*117*]).

The activity of bacteria, saprophytically present in water, can have desirable and undesirable consequences in the economical and technical sense.

Amongst the positive achievements of bacteria may be mentioned briefly the degradation of protein to amino acids, the breakdown of amino acids to ammonia through the action in unison of various groups of microorganisms, the conversion of ammonia to nitrate (nitrification), the fission of carbohydrates to alcohols and organic acids (lactic, propionic, formic acids etc.) and the oxidative performances of microorganisms. Without them, there would be no harvest and no production of bread, cheese and other foodstuffs. Those interested in this branch of microbiology may refer to *Thimann's* work "The life of bacteria" [*130*].

Less appreciated are the capsules and sheaths of many water bacteria which consist of a mucous substance. Slime formation can often be traced to iron- and manganese bacteria, sulphur bacteria and aerobically growing bacilli of the *Mesentericus* group (*Beger* [*11*]). These are not only unaesthetic when they are hanging down from taps, but can also block pipes and thus interfere appreciably with the water works function.

V. Sampling

The purpose of the bacteriological examination of water is, from the hygienic aspect, to establish if the water is pure. Purity in the bacteriological sense means that the water is suffering no influx of surface water, which is always more or less contaminated, and which is shown by an increase in the total bacterial number; that the water is free from bacteria of the Coli-Aerogenes group and other bacteria which are to be regarded as obligate or facultative in testinal inhabitants of man or animal; and that it is free of pathogenic organisms of all types.

How often bacteriological water examinations should be carried out depends on the local situation. It is determined by the local medical officer who is also responsible for providing the population with drinking water in good order. The WHO [72] has suggested for European countries the investigation frequencies in the table below:

Number of persons served by the water supply	Longest interval between single investigations	Minimum number of samples taken per month from the distributing system
Below 500	1 year	
500–1.000	6 months	
1.000–2.500	3 months	
2.500–5.000	2 months	
5.000–10.000	1 month	1 sample
10.000–15.000	1 month	2 samples
15.000–20.000	1 month	4 samples
20.000–50.000	2 weeks	1 sample per 5.000 inhabitants
50.000–100.000	4 days	1 sample per 5.000 inhabitants
over 100.000	1 day	1 sample per 10.000 inhabitants

These guide values must naturally be regarded as minimum requirements which should be supplemented by further examinations in accordance with local conditions. The general directions of the Federal German Ministry of Health, not yet published, provide for analogous frequencies of investigation.

Correct sampling is a condition for regular execution of the bacteriological investigation. Very small errors when taking the water sample can influence the whole investigation unfavourably and result in incorrect data. The investigator himself ought therefore to take the sample if at all possible; only the bacteriologist himself is fully acqainted with the necessary conditions and the required activity and knack have become so much a part of him that errors can be avoided. Examination of samples which one has not taken oneself is always risky and it should not be undertaken. Delivery or sending in a false water sample to mislead the investigator deliberately is, alas, also not rare. Usually samples which one has not taken oneself are not accompanied by the indispensable parallel inspection of the site. Such an inspection performed a month previously may yield information which has appreciably changed in the meantime.

Even the time at which the sample is taken can be decisive for the investigation. In periods of frost, the surface soil layers are frozen and therefore impermeable to contamination; bacteriological examination can then yield a too favourable result. It is similar during a long spell of drought, during which the transport agent, water, is lacking which can carry the contamination from the surface in broken ground into the ground water. These ways of influencing the bacteriological results vanish after rainfall or melting of snow. On the other hand, just this examination during prolonged drought can give important information, since the ground water level falls and water can be drawn from levels which normally are not used for water catchment.

The amount of water which has to be taken during sampling is also important. It is dictated by the nature of the water and the aim of the investigation. When few or no microorganisms of the coliform group are expected in a water, a volume of 100 ml is enough for the so-called coli test. However, this can be said only with reservation; statistical inquiries have shown that there is a possibility of only 63% that a coli bacterium, present in the water, will really be detected in one 100 ml sample. This probability increases to 86% when two 100 ml samples are investigated, and to 95% with three [72]. The real number of coli bacteria present can therefore be given only when several samples are investigated, although the determination of the coli titre in 100 ml of water satisfies the practical demands if samples are taken frequently from known waters. Since this test has to be carried out at various temperatures and with different, decreasing amounts, and since the total bacterial number in the water is a routine determination also, 250 ml of water are the minimum for a sample. The bottle itself must be of low-alkali glass and have a ground glass stopper which should be further protected by not too thin aluminium foil folded over it like a cap. The bottle must not be filled right up to the top so that it can be well shaken before the investigation; the bacteria are thereby distributed as uniformly as possible in the water.

The bottle must naturally be sterilised and also on the outside, if it is to be dipped into the water under investigation. In order to prevent re-infection with atmospheric microorganisms or bacteria from the investigator's hand etc., the surface of the bottle is protected by wrapping in paper before the sterilisation procedure.

No generally valid rules can be laid down for the site of sampling. Here, too, it is a matter of the local circumstances, which dictate whether one samples from the well itself, from the water conditioning plant or from the taps. The last named is the simplest when the water is flowing continuously from the tap. The opening of the bottle is then carefully heated round ("flamed") with an alcohol burner, the bottle held beneath the running stream of water and, after filling, the mouth of the bottle is again heated with the burner flame.

If water is not running all the time from the tap (which is nearly always the case), the tap is first cleaned with a cottonwool pad soaked in alcohol and then heated with the flame of a small soldering lamp. This heating must be continued until all the water in the tap has been driven out by the water vapour pressure and the tap is dry. Leaking water taps cannot be sterilised. After the outlet of the tap has been sterilised, the tap is opened so that a uniform

but not too vigorous stream emerges. This should rinse out all microorganisms which may be situated on the inner wall of the water pipe or on washers in the tap. Running the water, which should continue for at least 15 minutes, better 20, prevents sampling of stagnant water which was standing in the pipe. One proceeds similarly with wells which are equipped with a pump. The pump outlet is sterilised here, too, and water is pumped off for 20 minutes. One must watch that the pumped off water cannot flow back into the well and also that, if the water level of the well is low, pumping is not continued so far that the water remainder, often mixed with churned up mud, is sampled for the investigation.
When the well possesses no pump, water must be drawn directly but under no circumstances with any drawing-buckets to hand or similar non-sterile device. Various appliances have been proposed for taking water samples from wells or surface water in a bacteriologically unobjectionable way. Those due to *Hettche* [*53*], *Kraus* [*67*] and *Laun* [*74*] may be quoted here. Each investigator will choose from the water sampling devices developed by specialists one which appears especially suited to his need. *Beger's* sampler [*10*] has proved most suitable for us. It consists of a cylindrical container of galvanised metal, the upper part of which can be removed and carries a reel with metal cable. The real sampling apparatus is attached to this cable. A plate holding the glass stopper of the sampling bottle can be raised or lowered with a screw. The device has the advantages of being easily manipulated and that it can be sterilised as a whole. A claimed disadvantage is that particles of dirt, which are often present on the surface of the water, are sampled at the same time when the apparatus breaks through the surface; this can be countered by making the space between the glass stopper and bottle neck as small as possible and by thrusting it through the water surface quickly and smartly.

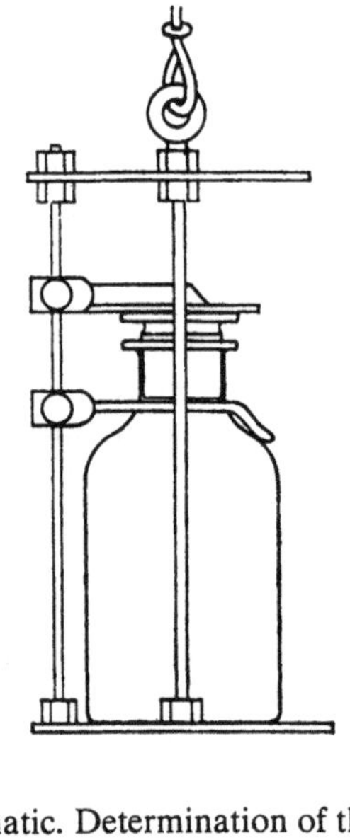

Fig. 2 Beger's water sampling device

Sampling of untapped sources is problematic. Determination of the total bacterial number with them seldom yields a true picture; evaluation of the coli test may be useful.

The outlet of tapped sources cannot usually be sterilised and, when they are running continuously, this is unnecessary. Only the neck of the bottle is then heated with the flame and the bottle is held in the stream of water.

In larger towns, the drinking water is usually provided from several water works with various wells. The consumer thus gets a mixed water, made up of individual waters of differing qualities. The water must then be sampled separately from the individual wells and the mixed water, used by the consumer, taken from taps in the mains; sampling from the extremities of pipes should be avoided.

According to *Spitta* [*126*], it is advisable in water treating plants to investigate both the raw water and the clean water as it leaves the plant. If possible, these samples should correspond, i.e., the sample of pure water should be taken after a time which corresponds to the passage through the plant of the raw water from which the first sample was taken.

If a water has been chlorinated, any microorganisms which may still be present in the water must be protected from subsequent action of the chlorine during transport of the sample. This can be done by adding sodium thiosulphate to the water bottle before sterilisation. 0.25 ml of 0.01N sodium thiosulphate solution is added to bottles of 250 ml content, 0.5 ml to those of 500 ml content. The same amounts of 1% sodium sulphide solution ($Na_2S.9H_2O$) should be added before sterilisation to water which has been treated with silver.

The samples of water must be investigated as soon as possible after having been taken. The maximum interval between sampling and investigation is given as 6 hours. This is not chosen arbitrarily but is derived from the rate of multiplication of bacteria in liquid culture media. When nutrient solutions are inoculated with bacteria and the bacterial count is determined at various time intervals, the usual curve found is that in Fig. 3. The horizontal part, A, is the latent or lag phase; B is the acceleration phase; C is the logarithmic phase. It is evident that the water samples must be examined before the acceleration phase is reached. If samples cannot be investigated very soon, it is advisable to cool the bottles in fragments of ice during transport.

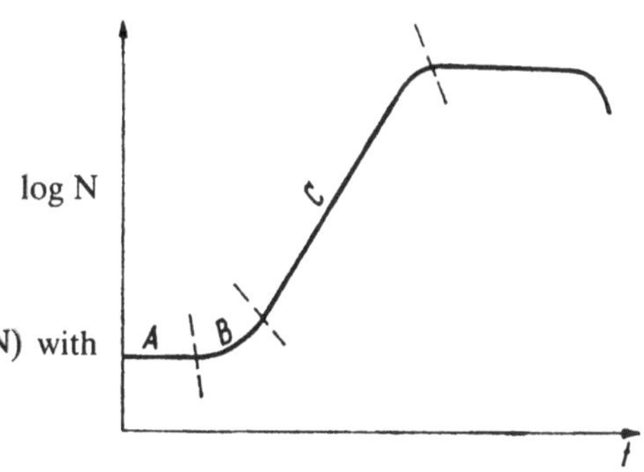

Fig. 3 Relation of log (bacterial number, N) with time, t

Reduction of the water temperature prevents too fast a multiplication of the bacteria. Sensitive microorganisms can, however, suffer a "cold shock" during the cooling, which may unfavourably affect their subsequent cultivation (*Knorr* [*62*]). The soonest possible investigation of the water samples is therefore advisable in every case. Examination in the laboratory is nevertheless always preferable to that outside or under makeshift conditions, since it is better suited to work under sterile conditions.

VI. Determination of the Bacterial (Plate) Count (Number)

The determination of the bacterial number aims at taking in all bacteria living in the water, those native to the ground water and those which reach the water via an alimentary route. The bacterial count is always quoted as the number in 1 ml of water.

The determination of the bacterial count is problematic because water is inhabited by bacteria of the most widely varying types. These flourish at differing optimum temperatures. The bacteria alien to the water are cultivated best at 37° C. They are termed thermophilic and come from the upper soil layers or from surface water. Some non-thermophilic bacteria grow best at 22° C and others at still lower temperatures. A compromise is thus necessary for the incubation temperature in routine investigations; temperatures must be chosen at which those bacteria for which they are not the optimum values, at least do not suffer harm. It is similar in the choice of the degree of acidity of the culture medium. Bacteria which grow anaerobically cannot flourish in culture media or nutrient solutions which contain oxygen or to which atmospheric oxygen has access. The selection of the culture medium itself is the most important thing in determining the bacterial number. There is no doubt that media based on meat broth, customary in medical microbiology, offer fewer microorganisms the possibility of development than do culture media based on further degraded breakdown products of proteins. The bacterial numbers are significantly higher on these oligotrophic culture media than on the nutrient substrates made from the usual meat broth (*Schwarz* [*117*]).

Wolters and *Schwarz* [*139*] quote the so-called "P-agar" for measurement of the bacterial count. Using an incubation temperature between 15 and 20° C, for 8 to 14 days, it yields a plate count 5 to 20 times larger than that obtained with the standard methods (gelatine or poured agar plates). P-agar consists of 1 g of peptone, 0.1g K_2HPO_4,1 l of sterilised local or tap water and 13 g agar-agar. The salts which are added are also important. Bacteria which live in sea water require salt concentrations differing from those needed by freshwater organisms. For each water a special culture medium really needs to be developed. This can be attempted by preparing the culture medium, not with distilled water but with the water under investigation, the individual chemical composition of which ensures the best conditions for development of its native bacterial flora. Deliberations of this sort are, however, not practicable for routine investigations.

The nature of the gelatinising substance is also important for the bacterial growth. More bacteria develop on gelatine culture media than on those prepared from agar. The bacterial counts on culture media based on silica gel lie between those on gelatine and on agar media, according to literature data.

The essential factors playing a part in determination of the bacterial count are thus milieu, temperature and culture medium. Should these questions not be solved in the most satisfactory way, then doubt is cast likewise on a further condition for evaluating the bacterial number with cultures. One assumes that each individual bacterium is suspended in the water. These single bacterial cells then multiply on the solid culture medium by division. The second generation divides on the same spot and this continues until, from the single bacterium, a colony is formed which can be seen with the naked eye or with a magnifying glass. The total number of these colonies is counted and termed the bacterial (or plate) count (or number). But bacteria are not always isolated in the water; many cluster together and then form only a single colony through growth on a solid medium. One tries to separate these bacterial conglomerates beforehand by energetically shaking the water sample. Other bacteria adhere, partly as bacterial associations, to very small particles of dirt in the water; these associations also germinate later to form only one colony.

The times needed by bacteria to form a colony vary widely. Appreciably more colonies are found after incubation for 7 days than for 1 day.

Schäfer [*111*] demands that the incubation should be basically extended to 7 days and that directives laying down a shorter time should be revised. The appraisal of some water sources would then change. Here, too, the bacteriologists who have to answer the questions of hygiene must settle for a compromise. Too long an incubation time delays publication of the results of the investigation, which may be too late to enable the necessary hygienic measures to be adopted. The routine determination of the bacterial count is thus a quite arbitrary method, which by no means includes all the bacteria living in the water. This is, however, not vital from the medical point of view because the doctor has only little interest in the native bacterial flora which is not pathogenic for humans. He is more interested in a method which permits detection of penetration into the ground water of surface water, which can carry pathogenic organisms. For this purpose, the determination of the bacterial count on poured gelatine plates, one of the oldest bacteriological methods of investigation, has proved useful for decades. Decisive alone is that the method is uniform in all the laboratories which work with it and that it is carried out with culture media prepared by a uniform procedure. Comparable results can be obtained only by standardisation. The samllest deviations, also in the technique of carrying out these methods of investigation, can frustrate the attainment of this goal.

According to the official directives of the Reichsgesundheitsamt (German Health Office) at the time (5/8/1940), nutrient gelatine had to be prepared by a uniform procedure (*Hallmann* [*46*]). Glassware for this includes flasks, sterilised test tubes, funnels and measuring vessels; and chemicals include Liebig's meat extract (Probacid I of the firm of Maggi or Standard I Bouillon of the firm of Merck may be used as substitutes), peptone of the firm of Witte, gelatine for bacteriological purposes (e.g. from Merck), analytical grade sodium chloride and N NaOH and Na_2CO_3 solutions.

10 g of meat extract, 10 g of peptone and 5 g of sodium chloride are dissolved in 1 l of water in a glass flask and heated in a digester for 30 minutes. 100 g of gelatine are added to 900 ml of this liquid after cooling. The gelatine requires about 30 minutes at ambient temperature to swell. The gelatine is then dissolved by heating in the digester for at most 30 minutes, stirring continuously. The pH of the hot liquid is adjusted to 7.0 with sodium hydroxide solution and then brought to 7.4 with sodium carbonate solution. The pH-values are measured with a potentiometer. After cooling to 40° C, the gelatine product is clarified by adding egg albumin; this may be the clear white from a hen's egg,

or crystalline material (Albumin ovi siccum) which is triturated to a paste with water (1.5 g/20 ml of water) and then homogeneously mixed into the gelatine by tilting the flask.
The gelatine is reheated to 100° in the digester and filtered, either directly on the steam bath or with the help of a hot water jacketed funnel. Any necessary pH adjustment is performed and 10 ml portions of the gelatine are then filled into sterile test tubes with a burette. These are then re-sterilised in the digester by heating once to 100° C for 10–15 minutes. The tubes are cooled as fast as possible in cold water or by placing in a refrigerator. They are stored in a cool room, protected from light, best standing in a wooden box which can be firmly closed. If the Standard I Boillonpulver of the firm of Merck is used instead of the meat extract, peptone and common salt do not have to be added.
The ready culture gelatine should be slightly yellowish and transparent. It must not be soft at temperatures below 26° C nor become fluid when heated for 3 minutes in a water bath at 30° C.

Various firms have introduced ready-prepared gelatine on to the market in recent years. The relevant amount of the powder is dissolved in water, sterilised in autoclaves (30 minutes at 120° C) and filled into test tubes. This eases considerably the preparation of bacteriological nutrient solutions and culture media. Their quality satisfies the directives of the country in which they are manufactured. Especially good for preparation of culture gelatine have proved the products of the firm of Oxoid (Nährboden und Chemie GmbH, Wesel) and the firm of Merck.
The culture gelatine can be advantageously obtained ready filled if modest amounts are to be used. Amongst German firms which sell this ready-usable culture gelatine may be mentioned the Biologische Arbeitsgemeinschaft in Lich/Oberhessen and the Membranfiltergesellschaft in Göttingen.
Application to poured gelatine plates should take place as soon as possible after the water samples have been taken, as has been said several times. The work should be performed in a place protected from the influence of dust and draughts. Small amounts of water are taken from the water sample under investigation in the bottle which is previously well shaken. Sterilised pipettes are used, as far as possible without sucking. Their sizes should be such that 1.0,0.5 and 0.1 ml samples can be accurately measured. The amount of water taken for investigation is governed by the nature of the sample. 1.0 and 0.5 ml amounts are needed from pure well or ground water. Smaller amounts, e.g. 0.5 and 0.1 ml, suffice with water where the local conditions lead to expectation of contamination. It may be necessary to dilute the water under examination with sterilised, physiological sodium chloride solution, in order to obtain accurately still smaller amounts. For this, 1 ml of the water under investigation is pipetted into 9 ml of the sterile salt solution in a sterile test-tube. This is well mixed, yielding a dilution factor of 10 (1 + 9). Dilution of 1 ml of this new solution with 9 further ml of sterilised physiological sodium chloride solution provides a dilution factor of 100, i.e., 0.1 ml of itcontains only 0.001 ml of the original water. The mouth of each test tube is heated round with the Bunsen burner ("flamed") after taking out the pulp stopper, and each pipette is used only once. The water sampled with the pipette is emptied, without blowing out, into a sterilised petri dish, during which the cover of the dish is briefly lifted on one side. 10 ml of gelatine are then added, after liquefying at 35° C and cooling to 30° C. Water and gelatine are then mixed to homogeneity by carefully tilting and rotating the petri dish. The gelatine plate is allowed to set as quickly as possible, best in a refrigerator.
At least two gelatine plates must be prepared from every water sample. After the gelatine solidifies, the plates are placed in the incubator where they are left for 48 hours at 22° C.

The time is extended to 72 hours for chlorinated water, so as to include also the microorganisms damaged by the chlorination, which grow more slowly.
After incubation, the number of colonies which developed form the bacteria is ascertained. Should there be only a few, this is done by direct counting; the counted colony is marked on the rear side of the dish with a wax pencil. Larger numbers of colonies are counted with the help of an apparatus: *Wolffhügel's* colony counter is probably the most used in Germany. Fundamentally it consists of a glass plate in which a grid is etched, of 1 cm^2 sized square areas. The four middle squares and those on the four diagonals (Fig. 4) are subdivided into nine smaller squares (3 × 3). The petri dish, which has an inner diameter of 9 cm, is placed on the counting apparatus so that the four central fields are exactly in the middle of the plate. On each of the four sides there are then two more of the subdivided squares within the dish. If the water sample has a high number of bacteria, it is then enough to count the colonies which have grown above the twelve fields arranged as a cross. A petri dish of 9 cm diameter has a usable surface area of 63.6 cm^2. The number of colonies counted in the twelve 1 cm^2 squares must then be multiplied by 63.5/12 = 5.3, to yield the total number of colonies present on the whole plate.
There are numerous other counting apparatus and devices but space forbids describing them in this treatment. Each investigator will familiarise himself with a procedure which especially appeals to him.
Very small colonies can often not be taken in with the naked eye. Against this, fine air bubbles may be considered to be colonies. A magnifying glass is thus often advantageous. A counting lens on a tripod can be placed directly on the glass plate in *Wolffhügel's* counting

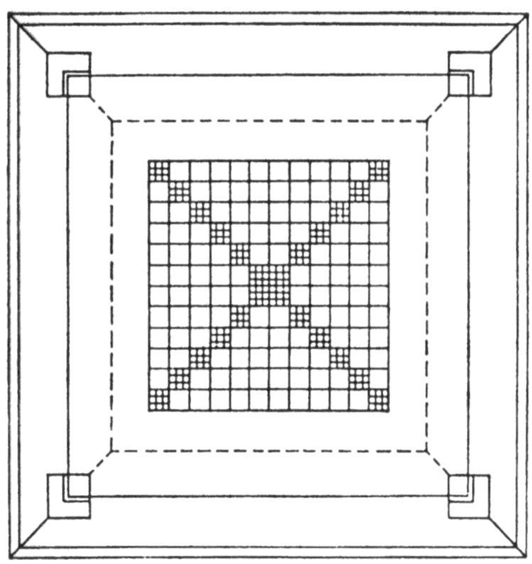

Fig. 4 Wolffhügel's counting device

apparatus. The binocular magnifier of the firm of Zeiss has also proved useful in counting bacterial colonies. The lenses should have a magnification power of about 6, to achieve uniform conditions in this feature too (*Spitta* [97]). Nowadays, counting with microscopes, formerly much used and discussed, is carried out only in special cases.

Conversion to the 1 ml base by multiplication is especially fraught with error with gelatine plates prepared from water dilutions. Consequently, counting is carried out on that gelatine plate prepared from the largest amount of water sample. Fictitious accuracy is avoided by expressing the bacterial count per ml with only two significant figures. 162 colonies is thus recorded as 160, but 48 as 48 (American Standard Methods [*2*]).

Water plates with culture gelatine are used as the standard procedure for determining the bacterial number in the water. Gelatine culture media have the disadvantages, however, that incubation is not possible at temperatures higher than 25° C, since the gelatine liquefies; and that many microorganisms can liquefy the gelatine by certain secreted ferments. These organisms capable of liquefying gelatine come mostly from the upper soil layers and their detection is therefore often a welcome clue to the origin of the water. If they occur in excessive amounts, however, the whole surface of the gelatine liquefies and colony counting is no longer possible.

Various procedures have been quoted to prevent complete liquefaction of the gelatine culture media, after the first appearance of gelatine-liquefying bacteria (which can mostly be recognised in due time by inspecting the plates after 24 hours of incubation). *Kruse* [*68*] proposes that a small crystal of $KMnO_4$ be laid in the gelatine-liquefying colony. *Spitta* [*126*] inhibits its further, too rapid growth by spotting with a lunar caustic pencil (silver nitrate). The dilution of water samples, already described, can also suppress the influence of gelatine-liquefying bacteria.

Nevertheless one often wishes to use a gelatinising medium other than gelatine, if only to be able to incubate at 37° C and thus determine the thermophilic bacteria which come from surface soil layers or directly from humans or animals. Agar-agar has proved the best of these materials since the early days of bacteriology, even though it shows inhibiting influences on the development of the pure water bacteria.

The *nutrient agar* is also filled in 10 ml amounts into test tubes and is prepared according to the instructions below:

3 g of meat extract, 5 g of peptone and 15 g of acid-free agar-agar are left standing with 1 l of distilled water for several hours in a flask. The mixture is then boiled in a digester until the ingredients have gone into solution (ca 1 hour). Distilled water is added to make up to 1 litre and the pH value adjusted to 7.2 with N sodium hydroxide. The nutrient agar can be clarified by filtration or sedimentation if it is allowed to cool slowly in tall cylinders after the boiling. The sedimented material in the bottom part of the cylinders can be cut off after solidification. The agar is then liquefied again and filled into test tubes which have been sterilised by 30 minutes at 120° C in autoclaves. This nutrient agar for bacteriological water investigations is lower in nutrient (more oligotrophic) than other media commonly used in bacteriology.

Nutrient agar differs from nutrient gelatine in that it liquefies and sets at higher temperatures. It becomes liquid only after being maintained a longer time at ca 100° and sets at about 40° C. If plates are prepared from nutrient agar and water, care must be taken that the microorganisms are not damaged by too hot agar. The agar in the tubes is left on the water bath at 100° C until it is liquid and is then cooled to 45° on a water bath. Pouring and evaluating the agar plates is performed as described for gelatine plates.

Agar-agar is imported from the Far East. During the world wars and in the likewise difficult post-war periods, it was hard to obtain, if not utterly impossible. There was no lack of attempts to replace it by other gelatinising substances.

Nutrient media prepared from *silica gel* proved so good that they have retained an established place in water bacteriology after the periods of emergency just mentioned. They have been incorporated into the German Standard Procedures [*30*]. Nutrient media prepared from silica gel have the advantage that they are prepared at room temperature and no damage to heat-sensitive bacteria thus occurs. They can be incubated at any desired temperature and are generally not liquefied by gelatine-liquefying bacteria. Further advantages of silica gel plates are their complete transparency and good buffering. Certain experience is needed for their preparation, however. According to the German Standard Procedures [*30*], this is done as follows:

First the water glass solution is prepared by adding 120–150 ml of water glass (Liquor natrii silicici DAB [= German Pharmacopoeia 6] 12–15%) to 880–850 ml of distilled water, free from copper, previously boiled and cooled. The solution is transferred to a sterilised burette, the tip of which dips into a water glass solution to prevent its becoming choked, and the opening of which is fitted with a soda-lime tube. The acidic nutrient broth is prepared by boiling for 30 minutes in the digester a mixture of 10 g of meat extract and 10 g of peptone in 1 l of distilled water, and then filtering. 50 ml of the water glass solution, 50 ml of the nutrient broth and 100 ml of distilled water, plus 1 ml of bromothymol blue indicator solution (40 mg in 100 ml of distilled water) are titrated with phosphoric acid (d = 1.013) to the colour change to blue-green. A volume equal to twice this titration amount is then removed from the nutrient solution and the same volume of conc. phosphoric acid (85%, d = 1.153) is added to it. This acidic nutrient solution can be kept as a stock for a longer period under sterile conditions. The poured plates are prepared by mixing rapidly equal volumes of the water glass solution and of the acidic nutrient solution in a sterilised flask and pouring into petri dishes already containing the amounts of water to be investigated (pipetted amounts of 1.0, 0.5 and 0.1 ml). Water and silica gel nutrient medium are mixed by careful tilting of the dish. Gelatinisation occurs in a few minutes and incubation is then begun.

Limiting Value ("Grenzzahl")

The maximum total number of bacteria which is permitted in a drinking water is termed the "limiting value" ("Grenzzahl").

The value of this drinking water standard has been fixed at 100 in Germany for many decades; this means that a maximum of 100 bacteria may be present in 1 ml of a drinking water. The conception of limiting value is disputed just as hotly as the figure of 100.

It is not without interest to establish how this number was arrived at. According to *Spitta,* the Imperial Health Office (Kaiserliches Gesundheitsamt) demanded in 1899 that a surface water after purifying filtration through sand, should yield a filtrate containing not more than 100 bacteria per cm^3. No data were given about the degree of contamination of the surface water. When one remembers that surface water can contain millions of bacteria per ml, this reduction to only 100 detectable microorganisms in the filtrate is naturally a satisfactory result. The concept of a limiting value of 100 bacteria thus had quite defined assumptions but since that time it has been applied to every drinking water and presumably can no longer be banished. A part was probably played also by the desire to assess water with a number which saves all subsequent reflection and by the fact that the number 100 is easily remembered. It is regrettable that this wrongly used numerical value, which is surely too high for a pure ground water, should still be valid today and, as far as I know, is included in the draft of the drinking water regulations of the Federal German Ministry of Health.

Spitta [*126*] writes rightly that the "limiting value" is really only a means of rescuing those who have entered but little into the problems of water hygiene, and that its application by such investigators is doubly dubious; that whoever has occupied himself more with the subject, will–apart from officially established cases–tie himself down as little as possible to limiting values and will try to fashion an opinion by weighing up and assessing mutually the individual data.

It is perhaps suitable to approach this problem from another side, namely, from the standards which have been proposed for mineral water. *Seeliger* [*121*] reports on an international symposium, held at Evian in 1959, where a maximum of 10 bacteria per ml after 3 days incubation was suggested, although it was admittedly an anti-bacterially treated mineral water. *Zimmermann* [*143*] demanded a maximum of 20 bacteria per ml of mineral water but *Gehrig* [*41*] considered 30 to be tolerable.

It is not quite understandable why other numerical values are applied to drinking water than to mineral water, since drinking water is more often drunk and used than mineral water; if drinking water represents pure ground water, it should not really harbour appreciably more microorganisms than mineral water which usually comes, however, from greater depths

It is objected that the limiting value of 100 is suitable because, as a result of increasing water consumption, considerable amounts of processed surface water must be used as drinking water. This argument is certainly unsound, for it is exactly in a treated and then chlorinated water that the bacterial number should be much lower than in a naturally occurring water.

VII. Determination of *E. Coli* and of Coliform Bacteria

In the determination of the total bacterial count of the water, the nature of the bacteria detected, i.e., their classification according to family, sub-family, genus and species, is fairly immaterial. Information is sought in other procedures of investigation of particular bacterial genera and species, about whether the water contains bacteria from the human or animal intestine which represent danger of infection for human beings. *August Gärtner* [*38*] formulated this in a classical way "where human faecal bacteria are able to get, there typhoid, dysentery and cholera bacteria will be found".

Mochmann [*82*] states that a human eliminates 50 x 10^{12} bacteria daily. Only some of these can be cultivated and, according to *Haenel* [*48*], the majority consists of anerobically growing bacteria. Amongst these predominate bacteria of the type of *Lactobacillus bifidus,* which, however, can be cultivated only on special nutrient media and under strictly anaerobic conditions. They play no part in water bacteriology on account of the difficulties of breeding.

Only about 10% of aerobic bacteria are present in human excrement, including a bare 2% of coli bacteria and less than 0.5% of enterococci. These bacteria are obligate residents of the human intestine. The enterococci settle more in the upper intestinal sections, the coli bacteria in the lower. If the figures can be accepted as correct, a human eliminates about $5 - 100 \times 10^{10}$ coli bacteria per day. Since neither man nor animal is often particularly careful with waste products, the coli bacteria are widely distributed. They are always taken as proof that the soil or objects in or on which they are detected have been in contact with faeces. This conclusion is valid only when these bacteria are specific to faeces. Since the beginning of the bacteriological era, an immense literature has developed about this question. *Spitta* [*126*] wrote as long ago as 1931 that the coli question of water bacteriology had been studied nearly 40 years already and had still found no conclusion and peace. It would appear, however, that, as a result of a stricter classification and extension of knowledge about the biochemical performances of the coli bacteria and related species, there are enough features distinguishing for practical purposes faecal-specific coli bacteria and other bacteria. The Enterobactericeae Subcommittee worked out a system for enterobacteria for Bergey's manual of 1963. These are bacteria which occur in the intestine and we have modified the system according to a table in the work of *Weber-Schutt* [*135*] (Table 1).

The bacteria of this family are short, gram-negative rods which under unfavourable culture conditions can form chains and filaments and occasionally also polar bodies. Capsules are seldom formed from *E. coli,* often from *Aerobacter* and regularly from *Klebsiella.* Most coli strains are motile, most *Aerobacter* forms are non-motile and *Klebsiella* is always non-motile.

The enterobacteria demand little of the nutrient substrate. They grow on nutrient gelatine and agar in aerobic and anaerobic milieu and form convex, circular colonies, often shaped like vine leaves, on gelatine; *Aerobacter* colonies are usually slimy and those from *Klebsiella* are always so, tending to flow together during longer incubation.

Table 1

Bacterial Species	General Distribution	Significance of detection in water concerning faecal contamination
Escherichia coli	Faeces of man and warm-blooded animals	proof
Aerobacter aerogenes *A. cloacae*	Ubiquitous, soil, water, plants, etc.	suspicion
Klebsiella	Faeces, respiratory tract, otherwise as *Aerobacter*	suspicion
Citrobacter intermedium	Ubiquitous, intestinal saprophytes of cold-blooded animals	suspicion
Citrobacter freundii	Predominantly intestinal saprophytes of cold-blooded animals	suspicion
Proteus vulgaris	Ubiquitous putrefaction bacterium, occurring also in the intestine	suspicion
Salmonella typhi and other *Salmonella*	In the intestine, faeces and urine of man and animals; cause of typhus and paratyphus	proof
Shigella dysenteriae and other *Shigella*	Intestine and faeces of man; cause of dysentery	proof

B. coli and *Aerobacter* species can ferment various types of sugar, yielding gas and acid. This property enables them to be distinguished from other bacteria on differentiating nutrient media. Lactose in particular is rapidly fermented by *E. coli, Aerobacter* and *Klebsiella,* in contrast to the *Proteus, Salmonella* and *Shigella* species.

Most coli strains form colonies with a typical metallic lustre on the fuchsin-containing Endo medium; this is stable even when the plates are kept for a longer time. A likewise metallic, brassy shimmer is characteristic of *E. coli* colonies on other differentiating nutrient media, such as MacConkey's agar (much used in England) and the eosin methylene blue agar.

Few families of bacteria show so many variants and mutations as the enterobacteria. This is shown in the form of the colonies, the biochemical properties (ferment weaknesses, ferment losses) and antigenic efficacy. Serological procedures for separating into various species have only limited application. As a result, one must fall back on their biochemical performances to distinguish the individual species; fermentation of lactose is of major importance here. All those species which attack lactose only slowly and protractedly have been termed paracolon bacteria.

All gram-negative, nonspore-forming, rod-shaped bacteria which ferment lactose into gas and acid in less than 48 hours at 37° C are classified as "coliform bacteria"; for classification as real *E. coli*, degradation of lactose at 44° C within 48 hours is a requirement *(Kruse* [72]). In addition, genuine *E. coli* should be able to form indole from peptone water, to yield a positive methyl red test and negative Voges-Proskauer reaction, and not to grow when citrate is the sole source of carbon. These four tests are often referred to collectively as the "IMViC" tests, after the initial letters of the individual tests. This gives 16 combinations which display relatively many intermediate stages in the scheme given by *Paw. Wilson* and co-workers undertook additional classification, so that the reaction picture is practically complete. We have taken *Wilson*'s scheme from *Schubert*'s work [*116*] and modified it somewhat (Table 2).

Table 2

	Indole formation	Methyl red test	Voges-Proskauer test	Citrate growth
Coli type I	+	+	–	–
Coli type II	–	+	–	–
Intermediate type I	–	+	–	+
Intermediate type II	+	+	–	+
Irregularis type I	+	+	–	–
Irregularis type II	–	+	–	–
Irregularis type III	–	+	–	–
Irregularis type IV	–	+	–	+
Irregularis type V	–	–	+	–
Irregularis type VI	–	–	+	+
Irregularis type VII	+	–	–	+
Irregularis type VIII	–	–	–	–
Bact. cloacae	–	–	+	+
A. aerogenes type I	–	–	+	+
A. aerogenes type II	+	–	+	+

Wilson extended the IMViC reaction by including testing for gas formation in the MacConkey broth at an incubation temperature of 44° C, and including gelatine liquefaction. *Habs* and *Müller* [45] propose that hydrogen sulphide formation should be tested also in a routine procedure. *E. coli* variety I and the irregularis variety I have the same IMViC result but can be distinguished through gas formation at 44° C, shown only by the former. *Citrobacter freundii* has the same IMViC picture as the intermediate variety II according to *Jawetz, Melnick* and *Adelberg* [57] but as the intermediate variety I according to *Hallmann* [47]. It can be distinguished from these intermediates through detection of hydrogen sulphide formation. In the above scheme, *Bact. cloacae* shows the same reaction picture as variety

I of *Aerobacter aerogenes*. It differs from the latter, however, in that it liquefies gelatine within a week at an incubation temperature of 22° C (*Habs* and *Müller* [*45*]). Bacteria with the IMViC formula of + − − −, absent from the above table, are counted among genuine coliform bacteria by *Hans Schmidt* [*115*]. That different authors have different opinions is hardly surprising since the individual bacteria can often display variable biochemical behaviour.

Fermentative abilities can be weakened or lost during prolonged sojourn in soil or water; the bacteria can occasionally regain these abilities by means of longer cultivation in a medium which is the most suitable for them. A scheme of this sort must not be taken too rigidly. Important is that the genuine faecal coli bacteria, corresponding to the *E. coli,* variety I and the irregularis variety I (*Schubert* [*116*]), can be distinguished through the IMViC tests from forms which, although present in the intestine, can be found also elsewhere. Only the real coli varieties permit the conclusion that the water in which they are detected has entered into contact with faecal matter. However, not only *E. coli* but also typhoid and paratyphoid bacteria are eliminated with faeces from healthy bacillus carriers, germ carriers and sick persons; only the faecal-specific bacteria thus permit the conclusion that there is a danger of infection through the water in which they are detected. Determination of the age of a water pollution by *E. coli* can be hygienically significant. Should it be established in cases of low amounts of *E. coli* that bacteria have been carried a long time in the ground water, it can be concluded that aetiological agents have probably died off during this period. A low coli value could thus then be tolerated. If, on the other hand, the pollution of the water with *E. coli* has recently occurred, all measures must be taken immediately to locate the origin of the pollution and to eliminate it. *Selenka* [*122*] has developed a method of determining the age of *E. coli,* which allows those coli bacteria older and those younger than 5 days to be distinguished. Details of this interesting but rather laborious procedure must be consulted in the original work.

The other bacteria of the enterobacteriaceae family come from the intestine but also from surface soil layers. They naturally should not occur in a pure ground water but they do not prove directly that faecal contamination has occurred. To regard them as faecal indicators would mean an excessive broadening of the coli concept. According to *Habs* and *Langeloh* [*44*], detection of coliform bacteria in water should be a first hint about endangerment of the water; this must then be confirmed or contradicted through aimed further investigations. Their experience showed that when a water in which coliform bacteria have been detected, is repeatedly investigated, sooner or later genuine coli bacteria can often be detected.

E. coli occurs only in the intestine of man and warm-blooded animals

Naturally it can now and then be transferred to other living creatures and be a temporary inhabitant of their intestine; but it is primarily an obligate intestinal resident of humans and warm-blooded animals.

Cold-blooded creatures also eliminate bacteria with their faeces which, through the results of biochemical reactions, are often indistinguishable from the coli bacteria of warm-blooded animals. It is, however, hygienically important to differentiate these "cold-blooded coli" from those bacteria which are eliminated only from warm-blooded creatures; only in the excrement of warm-blooded animals are to be found bacteria which are pathogenic to humans, whereas the faecal eliminations of cold-blooded creatures contain no aetiological agents which are transferable to humans (as far as we know). As a means of distinction, the fact can be utilised that the warm-blooded coli bacteria can grow at temperatures above 37°, up to 46° C, and can generally break down dextrose and lactose at these temperatures with formation of gas and acid. The cold-blooded coli bacteria do not normally possess this ability but can multiply and ferment only at temperatures up to 37°. In practice, one therefore tests the water samples at 37° for *E. coli* and, in a parallel investigation, incubation is performed at 45° C. This is the *Eijkman* procedure. According to the original instructions of *Eijkman,* the incubation temperature should be accurately 46°, and fluctuations of temperature of ± 0.5° not possible. Such changes of temperature cannot be fully avoided even with well regulated thermostats, however. A temperature of 45° C has thus been suggested for the *Eijkman* procedure *(Spitta [126]).* Other authors recommend 44° C for incubation and consider it important to incubate in a water bath and not in an incubator; the former has a more constant temperature and ensures that the material under investigation warms up more quickly *(Schubert [116]).*

The composition of the *Eijkman* nutrient solution has also been criticised; it is only slightly buffered and any streptococci which are incubated at the same time might bring about a not negligible lowering of pH; this could harm the coli bacteria (*Burkhardt* and *Legler* [*23*], *Heinrich* [*50*]).

One cannot conceal the fact that the question of the cold-blooded coli has not yet been settled finally. Thus *A. Müller* [*85*] considers that these organisms come from warm-blooded animals and exist in the cold-blooded animals as transition forms which can be termed "passage coli".

However this may be, *the Eijkman procedure has shown itself in practice to be an indispensable criterion for distinguishing those coli bacteria which originate from warm-blooded animals (and are thus genuine faecal indicators) from other bacteria; routine investigation without it is now unthinkable.* In its basic conception it is therefore justifiably included in the proposals of the European Bureau of the WHO for standard procedures of investigation, and in the German standard procedures.

Quantitative determination is the aim of the investigation of *E. coli* and coliform bacteria. This can be performed by measurement of the coli titre or coli test, i.e., the smallest amount of water in which *E. coli* can be detected, or, how many coli bacteria are present in 100 ml of the water under investigation.

The example below illustrates how the quotation of the coli titre, i.e., the smallest volume of water in which *E. coli* can be detected, can often be confusing to those who are not permanently occupied with these things:

If *E. coli* is detected in 100 ml of water (i.e., the coli titre is 100), this implies that in smaller amounts of water, e.g. 10 ml or 1 ml, no *E. coli* can be detected. The number of coli bacteria in the 100 ml of water is therefore between 1 and 9 in this example–for, if more than 10 were present in the 100 ml of water, one could be detected in the tenth part of this volume, i.e., in 10 ml. If the coli titre is given as 1 ml, then of course *E. coli* is present also in 10 ml and 100 ml of water, but not in the tenth part, 0.1 ml. This yields a coli content of at least 1 but less than 10 per ml.

Various volumes of water are treated in order to establish the coli titre. The membrane filter procedure is used nowadays for quantitatively measuring *E. coli*; it has displaced the older methods such as evaporation procedures or plate pouring procedures. More details are given later about the membrane filter method (p. 317) and the method for determining the coli titre is discussed below. Several nutrient solutions and media are necessary for this and should be prepared for routine investigations according to the directives of the German standard procedures [*30*].

A 10% lactose-peptone solution is used for preliminary culture of *E. coli* and the coliform bacteria. 100 g of Witte peptone and 50 g analytical grade sodium chloride are boiled in 1 litre of water in a digester for 1 hour. 100 g of lactose (purest, for bacteriological purposes) are then added. After boiling for a further 20 minute period, the hot solution is filtered and filled into test tubes or smaller glass flasks which are again sterilised by 20 minutes in the digester. This 10% lactose-peptone solution is used for the investigation of 100–5 ml amounts of water. The water under examination is treated with $\frac{1}{8}$ of its volume of the 10% nutrient solution, so that the end product is about 1% in lactose-peptone.

A 1% lactose-peptone solution is used in investigating amounts of water of 1 ml and less. 100 ml of the 10% lactose-peptone solution is mixed with 900 ml of sterilised distilled water. The resulting solution is filled into Durham fermentation tubes; these are closed with pulp plugs and sterilised again for 20 minutes in the digester.
Solid differentiating nutrient media are needed to identify *E. coli* which has become enriched in the preliminary culture. Eosin–methylene blue is most commonly used in the USA but in Germany, the *Endo medium,* introduced by the Japanese bacteriologist in 1904, has not been displaced by the many other nutrient media which have been developed in the intervening years. It is prepared as given below:

1 litre of 3% nutrient agar with pH of 7.4–7.5 in liquefied and still warm state is treated with 10 g of lactose and 5 ml of a concentrated alcoholic fuchsin solution, prepared from 10 g of basic fuchsin and 90 ml of ethanol. Freshly prepared sodium sulphite solution (10 g of $Na_2SO_3 \cdot 7H_2O$ in 90 ml of distilled water, dissolved by heating but without boiling) is then added to this mixture until the liquid nutrient medium still just appears faintly pink but becomes colourless on cooling after having been poured into petri dishes. About 20 ml of the sulphite solution are needed normally.
The 3% nutrient agar serves also as basic substrate for most other solid agar nutrient media. It is prepared by dissolving 10 g of meat extract, 10 g of peptone and 5 g of sodium chloride in 1 litre of distilled water and adding 30 g of powdered agar-agar. Filamentous agar strips are best cut into smaller pieces the day before and left overnight in water to swell. The mixture is then boiled until the agar dissolves and the pH-value adjusted to 7.3–7.5 with N sodium hydroxide solution. After clarification, most simply by sedimentation, as described in the preparation of nutrient agar for water plates, the agar product is reliquefied and filled in usable amounts into tubes which are sterilised in an autoclave.

The nutrient solutions and/or indicators below are required to carry out the IMViC tests:
1. The formation of indole depends on fission of the amino acid tryptophan, which is then degraded to indole by "indole-positive" bacteria. Tryptophan must therefore be present in the nutrient solution.

It is prepared by boiling 10 g of peptone, 10 g of analytical grade sodium chloride and 0.1 g of tryptophan in 1 litre of water for 1 hour in a digester. The pH is brought to 7.1–7.3 with N NaOH, the solution filtered and filled in 5 ml amounts into test tubes which are sterilised on the following day by heating a further hour in the digester. The Standard II Bouillon of the firm of Merck can also be used as substrate for the test of indole-forming capacity of the bacteria.

The *indole detection* itself is performed by adding 10 drops of a p-dimethylaminobenzaldehyde solution to the incubated culture. The reagent is prepared according to Kovács' modification:

5 g of dimethylaminobenzaldehyde are dissolved in 75 ml of amyl alcohol on the water bath at 60° C and then 25 ml of hydrochloric acid (chemically pure, d = 1.19) are added. The reagent is allowed to stand 6–7 hours until its colour changes to yellow. It is stored in an amber glass bottle.

2. The *methyl red test* shows the powerful acid formation by *E. coli*, which yields a pH value of below 5.2 after 4–5 days incubation; this contrasts with *A. aerogenes*. The nutrient solution needed for this is prepared from:

5 g of peptone, 5 g of dipotassium phosphate (K_2HPO_4) and 5 g of glucose; these are dissolved in 900 ml of distilled water and boiled for 20 minutes on the water bath. The mixture is made up to 1 litre, the pH-value is adjusted to 7.0–7.2 and the product is filled in 5 ml amounts into sterilised test tubes. These are sterilised on the three following days by heating for 20 minutes in a digester. The methyl red indicator is made by dissolving 0.04 g of methyl red in 60 ml of ethanol and making the volume up to 100 ml with distilled water.

3. The *Voges-Proskauer test* is carried out with the same glucose-peptone solution used for the methyl red test. The test is based on detection of acetoin (acetylmethylcarbinol) which is formed as an intermediate in the breakdown of carbohydrates by *A. aerogenes* and related strains. The *Barrit* reagent is required to detect acetoin; it contains 6 g of α-naphthol in 100 ml of ethanol and 40% potassium hydroxide *(Schubert [116])*.

4. Only *A. aerogenes* and related strains and not *E. coli* can multiply in a nutrient solution which contains citric acid as the sole source of carbon. *Koser's citrate broth* has the following composition:

1.5 g of sodium ammonium phosphate ($NaNH_4HPO_4 \cdot 4H_2O$)
1.0 g of potassium phosphate (KH_2PO_4)
0.2 g of magnesium sulphate ($MgSO_4 \cdot 7H_2O$)
3.0 g of sodium citrate ($Na_3C_6H_5O_7 \cdot 5.5H_2O$)
are dissolved in 1 litre of distilled water by warming. The pH of the solution is adjusted to 6.8 if necessary with 0.1N NaOH or 0.1N HCl. 5 ml amounts are filled into sterilised test tubes which are then heated in an autoclave for 20 minutes.

Determination of the Coli Titre (Test)

Before determining the coli titre, one must decide what amounts of water are to be investigated. Since drinking water must contain no coliform bacteria–and hence naturally no *E. coli*,– in 100 ml, this amount has to be investigated in any case. Examination of 50, 10 and 1 ml amounts of normal drinking water is also recommended. If pollution is expected, volumes smaller than 1 ml, i.e., 0.5, 0.1 and 0.01 ml should be studied.

Each water sample must be investigated in duplicate since incubation is performed at both 37° and 44° C. Water amounts of 10 ml and more are investigated in *Eijkman* fermentation tubes or erlenmeyer flasks with Durham tube fitment. Water volumes of less than 10 ml are studied in ordinary test tubes with gas collecting tubes. The 10% lactose-peptone solution is first put into the vessels for investigation, using a volume equal to one-eighth of the water volume to be taken. Thus 12 ml of 10% lactose-peptone solution are used with 100 ml water sample, 6 ml with 50 ml, 1.2 ml with 10 ml etc.

10 ml of the 1% lactose-peptone solution are used in the investigation of 1 ml and less of wa-

ter. The tubes containing the required amount of lactose-peptone solution can be resterilised in the digester but this is usually unnecessary if work has been carried out under sterile conditions. The water amounts to be studied are then added to the tubes, taking precautions to maintain sterility. One sample is incubated at 37°, the other at 44° C. After 24 and 48 hours it is tested whether there is turbidity or gas formation in the tubes. When the contents of the fermentation tube remain clear without change, the investigation is concluded: *E. coli* and coliform bacteria were not detectable in the amount of water taken for investigation.

Using a platinum loop, sterilised by heating to red heat, an Endo plate is inoculated from those tubes with turbid contents, regardless whether there has been evolution of gas (since this can be prevented by accompanying bacteria even when genuine coli bacteria are present); the inoculation is such that the development of single colonies can be reckoned with. The nutrient media are incubated for 24 hours at 37° C and then inspected for colonies which are suspected as coming from coli bacteria.

E. coli grows in dark red, flat colonies, showing the typical fuchsin lustre which is also stable when the nutrient medium stands for a longer time and which colours the medium red around the colony. The other bacteria form colourless to pink colonies which generally display no metallic sheen. *A. aerogenes* develops principally in hemispherical colonies, usually without the fuchsin lustre, but it can also develop like the coli bacteria. After some time, the colonies can become lighter coloured and show formation of slime. The coliform bacteria tend to behave like *E. coli* but they do not redden the nutrient medium so strongly as a rule.

Before questionable colonies are further differentiated in the IMViC test, a gram-stained preparation ought to be inspected microscopically to make sure that it is really a question of gram-negative, nonspore-forming rods.

The *indole detection* part of the IMViC test is then carried out by inoculating the tube containing the tryptophan nutrient solution with a small portion of the colony, using the platinum loop; this is incubated for 24 hours at 37° C. 1 ml of indole reagent is then added and the mixture shaken gently. The indole reagent, which, as a result of its low specific gravity, covers the nutrient solution, turns red if indole is present. The reaction is accelerated by gently warming.

A tube containing glucose-peptone-phosphate solution is inoculated in the same way for the *methyl red test* and the *Voges-Proskauer test,* and incubated for 4 to 5 days at 37° C. 2 ml of the nutrient solution are then removed with a sterile pipette to a test tube and treated with 1 ml of the Barrit α-naphthol reagent and 0.4 ml of 40% KOH solution. A positive test is shown by the appearance of a wine-red colour after some minutes; no change, or a faint greenish colour is negative.

The methyl red test is performed on the remainder of the glucose-peptone-phosphate culture. 5 drops of methyl red solution are added to the incubated nutrient solution. A positive result, i.e., acidification of the substrate to a pH-value below 4.4, is shown by a red colour.

The *citrate test* is carried out by very lightly inoculating the tube with *Koser*'s citrate broth. Great care must be taken that no other material (particles of nutrient medium) get into the citrate medium. After 48 hours incubation at 37° C, the liquid is inspected for turbidity through growth of bacteria. If so, this is a positive result; a non-turbid nutrient solution, i.e., no growth of bacteria, is negative.

The bacteria in question can then be classified as in the scheme on p. 299, according to the results of the different parts of the IMViC quartet.

Many modifications of these tests have been proposed. Thus *Schubert* [*116*] performs the methyl red and Voges-Proskauer reactions after only 48 hours incubation; this saves considerable time in making the diagnosis. He and other authors consider the incubation temperature to be too high, claiming that more unambiguous results are obtained at 30° C. For years there has been discussion–sometimes vehement–about the composition of the *Eijkman* nutrient solution and the use of Endo plates. The culture medium of *Murae* for differentiating *Aerobacter* from *Escherichia* appears to represent a step forward according to *Braune* [*19*] whose work includes the prescription for preparing this medium. But we believe that, in this domain adapted to practical needs, we should restrict ourselves to the methods of investigation quoted here; through the use of procedures as closely similar as possible, results are obtained which can be compared most satisfactorily.

The use of tablets, provided by the Biologische Arbeitsgemeinschaft, Lich/Oberhessen, W. Germany, enables the Voges-Proskauer and indole tests to be performed rapidly.

A Voges-Proskauer test tablet is dissolved in 1 ml of distilled water. The tube is then inoculated either with a thick culture suspension or with a culture mass, using a platinum loop. After incubating at 37° C for 4 hours, 0.6 ml of a solution of an α-naphthol tablet (1 tablet dissolved in 5 ml of 95% ethanol) and 0.2 ml of a 40% aqueous KOH solution are added. A cherry red colour after shaking vigorously is a positive test.

For the indole test, 1 indole tablet is dissolved in 1 ml of distilled water. The solution in the tube is inoculated with a thick culture suspension or culture mass, then incubating for 4 hours at 37° C. A solution prepared from 1 tablet of p-dimethylaminobenzaldehyde dissolved in 5 ml of amyl alcohol-conc. HCl (3 + 1) is poured over the incubated solution. A red ring of colour is a positive test for indole.

It is therefore possible to carry out two of the IMViC tests within 4 hours and, if the Voges-Proskauer test is positive, to eliminate from consideration coliform bacteria which do not belong to the genuine faecal coli bacteria.

The test of breakdown of urea can also be performed quickly with test tablets.

According to the German standard procedures, the results of the investigation are quoted as follows:

Example: Using lactose-peptone solution, a coliform bacterial titre of 1.0 ml was found by incubation at 37° C and a coli titre of 10.0 ml by incubation at 44° C. Distinction of *E. coli* was done according to the IMViC test. In this example, objection must be made to the water, since neither *E. coli* nor coliform bacteria should be present in 100 ml of drinking water.

Another way of expressing the coli number or the number of coliform bacteria must be mentioned. This is the method in the English-speaking countries, using the MPN ("*M*ost *p*robable *n*umber").

This procedure has been proposed by the WHO in the "International Standards for Drinking Water" for routine investigation of coliform bacteria.

For the "Presumptive Test", tubes containing lactose broth and fitted with a gas collecting attachment, are inoculated with various amounts of water (depending on the expected degree of contamination of the water) and incubated for 24 hours at 37° C.

If the test leads to turbidity and gas formation, the "Confirmed Test" is carried out. Two tubes containing brilliant green bile broth or MacConkey broth are inoculated from the tube showing the positive Presumptive Test. One tube is incubated for 48 hours at 37° C, the other for 24 hours at 44° C.

If this confirmatory test is also positive (formation of gas and acid), the "Completed Test" is performed. In this, Endo or eosin methylene blue agar plates are streaked with material from the tubes of the Confirmed Test. The colonies yielded are cultivated further to accomplish the final diagnosis of *E. coli* or coliform bacteria. It must be established with certainty through gram staining, which should be carried out on a sample from an inoculated and incubated nutrient agar slant, that it is really a gram-negative, nonspore forming rod.

The most probable coli number in 100 ml of water can be calculated, or read from tables when the number of positive tubes in the Presumptive Test is known. The tables for various

Table 3

Number of tubes giving a positive reaction			MPN index	95% Confidence limits	
3 of 10 ml each	3 of 1 ml each	3 of 0.1 ml each		Lower limit	Upper limit
0	0	1	3	< 0,5	9
0	1	0	3	< 0,5	13
1	0	0	4	< 0,5	20
1	0	1	7	1	21
1	1	0	7	1	23
1	1	1	11	3	36
1	2	0	11	3	36
2	0	0	9	1	36
2	0	1	14	3	37
2	1	0	15	3	44
2	1	1	20	7	89
2	2	0	21	4	47
2	2	1	28	10	149
3	0	0	23	4	120
3	0	1	39	7	130
3	0	2	64	15	379
3	1	0	43	7	210
3	1	1	75	14	230
3	1	2	120	30	380
3	2	0	93	15	380
3	2	1	150	30	440
3	2	2	210	35	470
3	3	0	240	36	1300
3	3	1	460	71	2400
3	3	2	1100	150	4800

amounts of water which are taken, and the various numbers of tubes are given in the work cited above. Table 3 is printed by kind permission of the WHO. In our opinion, it is especially well adapted to practical requirements, since it is based on 3 tubes with 10 ml, 3 with 1 ml and 3 with 0.1 ml of water and thereby is fairly certain of covering an average consumption of material.

The lactose broth has the eomposition:

meat extract	3g	up to 1 litre of distilled water
lactose	5g	
peptone	5g	

The ingredients, without the lactose, are dissolved on the water bath. The lactose is then added and the volume made up to 1 litre with the distilled water. After sterilisation, the pH-value should be between 6.8 and 7.

The *brilliant green bile broth* is prepared as follows: 10 g of peptone and 10 g of lactose are dissolved in 500 ml of distilled water. 20 g of dehydrated oxgall (Merck Nr. 3756), dissolved in 200 ml of distilled water and with pH adjusted to 7–7.5, are added and the volume is made up to 975 ml with distilled water. The pH of the resulting solution should be 7.4. After adding 13.3 ml of a 0.1% aqueous solution of brilliant green in distilled water, the final volume is brought to 1 litre with distilled water. This solution is filtered through cottonwool into tubes which are sterilised.

MacConkey's broth has the following composition:

commercial sodium taurocholate	5g
lactose	10g
peptone	20g
sodium chloride	5g
distilled water	up to 1000 ml

The lactose is added after the other ingredients have been dissolved and the solution is filtered through filter paper. The pH of the filtrate is adjusted to 7.4. 1 ml of a 1% alcoholic solution of bromocresol purple or 5 ml of a 1% aqueous solution of neutral red is (are) added and the mixture is filled in 5 ml amounts into Durham fermentation tubes. These are sterilised by heating at 115° C for 15 minutes in an autoclave or in a digester for 30 minutes on three successive days.

According to the original directions, the solutions below are required for gram staining:

a) Carbol gentian violet solution

saturated alcoholic gentian violet solution	10.0 ml
acid. carbol. lique fact.	1.0 ml
aqua dest.	100.0 ml

b) Dilute Lugol's solution

iodine	1 g
potassium iodide	2 g
aqua dest.	5 ml

The mixture is triturated in a mortar until homogeneous and then diluted with distilled water to 300 ml

c) Acetone spirit

acetone	3 ml
96% ethanol	97 ml

d) Dilute carbol fuchsin solution

saturated alcoholic fuchsin solution	10.0 ml
acid carbol. lique fact.	5.0 ml
aqua dest.	100.0 ml

This solution is diluted 1:10 with distilled water. It does not keep and cannot be used when coloured precipitates form.

In gram staining, the air-dried preparation is first fixed by heat. The carbol gentian violet solution is then dropped on to it through a small filter funnel, left 3 minutes and then poured off. Lugol solution is added and immediately poured off to remove coloured residues. The Lugol solution is again dropped on and left to stain for 2 minutes. It is then poured off and washing with the acetone-alcohol performed until no more clouds of dye wash off. After washing with water, contrast staining is carried out (30 sec. with dilute carbol fuchsin) and then rinsing with water.

VIII. Pathogenic Coli Strains

In 1928, the childrens' doctor *Adam* isolated coli bacteria from babies' stool. He considered them to be the cause of infant enteritis. But the exact differentiation of pathogenic coli strains became possible only after serological research on the various antigenes of *E. coli;* these strains play an increasingly large part in water bacteriology. Most coli varieties can be subdivided in over 100 groups through their thermostable O-antigene which is built up of polysaccharides and is linked to the bacterial body. Only about 6 of these enter into consideration as the cause of toxic infant enteritis. They belong to the O-groups 128, 127, 111, 86, 55 and 26. The coli bacteria possess K-antigenes as well as these different O-antigenes. These are thermolabile and their substrate is the capsules of the bacteria. The K-antigenes are divided into three groups, A, B and L, depending on their resistance to heat.

All K-antigenes mask the O-antigenes; the former must be destroyed by heating the coli bacteria for 2½ hours before the O-antigenes can be treated with an anti-O-serum.

Some time ago the Behringwerke brought polyvalent anti-O-K-sera on to the market. These react also with living coli bacteria and permit a rapid preliminary test for pathogenic coli strains.

Some of the suspected colony is taken off with the platinum loop and mixed to homogeneity on a glass slide with 1 drop each of the polyvalent anti-O-K-sera I and II. The slide is tilted carefully for two minutes to ensure good mixing. If there is no visible agglutination after this time, the strain concerned contains none of the corresponding K-(B)-antigenes. If agglutination occurs, testing with monovalent anti-O-K-sera is carried out to elicit to which serological group the coli strain belongs. A not too thick suspension in physiological sodium chloride solution of a pure culture of the strain in question is prepared and heated 30 minutes on the water bath at 100° C to destroy the B-antigenes. The agglutination reaction is repeated with this suspension, using specific saturated O-sera.

IX. Detection of Enterococci and *Clostridium Welchii*

Although it is generally possible to distinguish faecal-specific coli bacteria from coliform bacteria and bacteria of the *Aerobacter* group, bacteria are found over and over again which deviate from the characteristic type. It is then difficult or even impossible to decide whether the water has been faecally contaminated or has been affected by inadequately filtered surface water. These difficulties inspired the quest for further control bacteria in the water which are characteristic for faecal contamination. The search for enterococci in the water as a test of faecal pollution has yielded good results especially by Anglo-American authors but also increasingly by German investigators.

The enterococci belong to the sub-family of the streptococci which, according to Bergey's manual, are classified under the lactobacillaceae. Streptococci are spherical- or oval-shaped bacteria linked together in chains of a greater or lesser number of units and which are gram-positive. They grow on solid culture media as small discus-shaped colonies and can display various haemolysis forms on agar plates containing added blood. Their classification according to the nature of the haemolysis and their biochemical behaviour represents a large advance in clinical bacteriology. A final definitive classification was first accomplished by *Lancefield* in 1928, however; with the help of serological methods, he detected in extracts of streptococci substances which were specific for groups and types. Most streptococci possess a group-specific carbohydrate antigene which permits their classification. The enterococci belong to the serological group D and, according to *Seelemann* [*93*], can be divided into six different types according to their biochemical behaviour:

Streptococcus faecium; Str. durans; Str. glycerinaceus; Str. liquefaciens; Str. zymogenes; and Str. bovis.

These individual types could be of interest in water bacteriology since various authors have recently been professing the opinion that the type *Streptococcus faecium* is of animal origin whereas *Streptococcus glycerinaecus (faecelis)* comes from humans (*Slanetz* and *Bartley* [*125*], *Balteanu* and co-workers [7]). The two types are distinguished through their biochemical effects. The latter (*S. glycerinaecus)* breaks down glycerol but strains of the former (*S. faecium*) do not attack this alcohol. It does not yet appear certain if the distinction between these two types can be utilised to yield an unambiguous assertion about origin, the more so because *Knothe* [*65*] has been able to detect *Streptococcus faecium* also, in no small amounts, in human urine samples. It is important for water bacteriology that the enterococci form a serologically homogeneous group of obligate intestinal inhabitants which possess certain common cultural properties. Thus they can multiply in nutrient media of augmented sodium chloride content (up to 6.5 %), withstand pH changes of the medium up to 9.6 and temperatures up to 63° C and coagulate and reduce 0.1 % methylene blue solution. This capacity for resistance to external influences makes their detection in water particularly suitable. Whether the detection of enterococci should be included in the routine investigation of a water may be undecided; but in cases of doubt, investigation of these faecal-specific microorganisms can effect a significant enlargement of our diagnostic scope. Enterococci can be technically most easily detected through the membrane filter procedure which is discussed later in this book. Nevertheless, the classical methods of investigation with preliminary culture of definite amounts of water in liquid nutrient media followed by transfer to indicator nutrient media hold an established position.

Guthof [43] carries out the preliminary culture in a special broth containing sodium azide and sodium chloride. *Lorenz* [79] on the other hand, regards as adequate the enrichment of enterococci in the usual *Eijkman* lactose-peptone solution. The preparation of this solution is given above under the heading of determination of coli titre (p. 302) and its use here in routine investigations has the advantage that the same medium can be taken for preliminary culture in detection of both coli bacteria and enterococci. From the tubes which have been incubated at 44° C and show cloudiness, material is transferred to Endo plates or other indicator nutrient media containing lactose in order to detect. *E. coli* and coliform bacteria; to isolate enterococci, material is transferred to nutrient media which contain inhibitors for other bacteria that do not, however, affect enterococci in the given concentration.

Sodium azide, potassium tellurite and common salt are the principally used inhibitors of this sort. We should like to forego discussion of the individual chosen nutrient media quoted for breeding enterococci and to restrict ourselves to giving two special media which have proved valuable in practice.

Schäfer [110] has used a nutrient medium in which the additives to 100 ml of 2% liquefied agar medium of pH value between 8 and 9 are:

1.0 g sodium taurocholate
0.5 ml ferric chloride, $FeCl_3$, solution (10%)
2.5 g NaCl
3.5 ml potassium tellurite solution (1%)
0.1 g aesculin

The nutrient medium is sterilised for 10 minutes in the digester.

G. Müller [86] modified *Slanetz's* nutrient medium for detection of enterococci on membrane filters; it is prepared from:

4.0 g glucose
8.0 g potassium dihydrogen phosphate
0.8 g sodium azide
0.2 g triphenyltetrazolium chloride

These ingredients are dissolved in 1 litre of water without heating, then mixed with the same amount of liquefied (but not too hot) 2% nutrient agar and poured into petri dishes.

This nutrient medium, given for membrane filter procedures, is too soft for inoculation and spreading with the platinum loop; it can, however, be made harder without difficulty and loss of value by increasing the agar concentration.

The enterococci grow on these nutrient media as colonies of 1–2 mm size, characterised through blackening on *Schäfer's* medium. This blackening is due to decomposition of the aesculin into glucose and aesculetin through ferments of the enterococci. The aesculetin then reduces the ferric chloride present. Through reduction of the triphenyltetrazolium chloride in the modified *Slanetz* medium, the enterococci yield light pink to dark brown colonies. The growth of most other bacteria is so inhibited that they cannot multiply on these nutrient media. Suspect colonies must nevertheless be further differentiated in order to be regarded with certainty as enterococci. The biochemical properties of the enterococci can be utilised here for identification, namely: growth on blood agar containing 6.5% of added sodium chloride; heat resistance up to 60° for 30 minutes; fission of sorbitol and raffinose. One should convince oneself through a gram preparation that streptococci are really present before embarking on these investigations, however. Detection of enterococci in routine laboratory work is facilitated by an acid-glucose broth in powder form for preliminary culture, provided by the firm of Merck.

There are variable data about the relation between the number of enterococci and the number of coli bacteria found in the water. Most authors who have carried out comparative investigations agree, however, that the number of enterococci in drinking water and wastewater does not exceed the number of coli bacteria, but is less, according to the water. This seems quite natural since the enterococci are eliminated in small amounts from humans and warm-blooded animals and their survival ability in the soil and water appears to lie between that of the typhoid and coli bacteria (*Guthof* [*43*]).

All these facts underline the primacy of investigating *E. coli;* detection of enterococci may assist this valuably in certain cases.

Detection of *Clostridium Welchii.*

The detection of *Clostridium welchii (Clostridium perfringens, Welch-Fränkel* gas gangrene bacillus) in water has been repeatedly given in the literature of recent years likewise as a possible sign of faecal contamination *(Seeliger* [*121*]). *Cl. welchii* is certainly present in large amounts in the eliminations of man and warm-blooded animals but it is also detectable in considerable amounts in the soil, both manured and unmanured. *Zeissler* [*141*], probably the most recognised experit on anaerobic organisms, was thus able to write that Fränkel's gas gangrene bacillus is ubiquitous, probably the most widely distributed microorganism which exists at all. In view of this widespread distribution, we cannot regard *Cl. welchii* as faecal-specific. *Clostridium perfringens* which has been detected in the water can of course come from the intestine of man or animal but it can also come from the soil- and enter the ground water with surface water where there are not enough protective strata. Its appearance in the water is thus an indication of inadequate filtration of surface water and demonstrates in any case that the water is polluted. Detection of *Cl. welchii* is, however, no proof of a faecal infection of the water. *Perelli* [*93*] considers the detection of *Cl. welchii* as an indicator of an older contamination and the detection of *E. coli* as a sign of more recent contamination.

Cl. welchii can thus function only as a general indicator of pollution; investigation of it in water can be quite useful and complement the results of other procedures of investigation.

The *Clostridium* bacteria belong to the Bacillaceae family and differ from other spore-forming rods in that they can grow only under anaerobic conditions or when the oxygen pressure is greatly reduced. *Cl. welchii* is not so sensitive to oxygen as are other bacteria of this genus, however. *Cl. welchii* organisms are large, unwieldy rods with rounded corners, gram-positive in fresh cultures but more or less gram-negative after some days. They have no flagella and are therefore non-motile. They form spores only on nutrient media free of carbohydrates. Capsules can be formed in the animal body or on nutrient substrates, rich in protein. On glucose-blood-agar plates under exclusion of oxygen, *Cl. welchii* forms raised, round, brilliant colonies of about 2 mm size; they appear strawberry red at first, changing via grey and brown to olive green when oxygen gains access. Further cultural characteristics of *Cl. welchii* are formation of gas and acid in nutrient solutions which contain carbohydrate, and decomposition of sterile milk with formation of gas and a cheesy product. *Cl. welchii* is most simply detected in water with the help of the membrane filter. Accordingly, the method of detection is quoted when the membrane filter procedure is described below (p. 322).

X. Detection of Viruses and Bacteriophages

Viruses and bacteriophages are characterised through being ultramicroscopic, i.e., invisible with the light microscope; and through being able to multiply only in living cells. Human or animal cells serve as hosts to viruses; bacteria are attacked by the phages, which multiply in them.

The classification of viruses according to clinical considerations i.e., which diseases they cause, has proved to be unsatisfactory since the same symptoms of illness can be brought about by various viruses, and the symptoms can manifest themselves differently, depending on the host organism and its reaction. Moreover, most virus infections are without clinical symptoms.

Accordingly, classification is preferred today based on chemical criteria and morphological distinctions.

Viruses consist of a nucleic acid, surrounded by a protein mantle. The nucleic acid is either deoxyribonucleic acid (DNA) or ribonucleic acid (RNA).

The DNA-viruses comprise 4 classes:

1. Papova-viruses; 2. Adeno-viruses; 3. Herpes-viruses; 4. Smallpox viruses.

The RNA-viruses are likewise divided into 4 classes:

1. Picorna-viruses with the enteroviruses (polio-, Coxsackie- and Echo-viruses and probably hepatitis viruses); 2. Reo-viruses; 3. Arbo-viruses; 4. Myxo-viruses.

In water and wastewater microbiology those viruses play the principal part which are most resistant to chemical and physical influences; these are adeno-viruses, entero-viruses and reo-viruses.

Probably the infectious hepatitis virus is the most important in water hygiene; but it cannot yet be detected with the methods which are available today. Entero-viruses and vaccination viruses are found so often in wastewater that virologists attach the same importance to detection of viruses in drinking and wastewater as to detection of bacteria in the bacteriological domain. The question arises whether this is justified if one accepts the accuracy of data from the American Public Health Association [*1*], according to which the ratio of relative frequencies of occurrence of viruses to coliform bacteria in wastewater is 1:65,000. An additional factor is that detection of a virus infection of the water can be accomplished only with a significantly larger outlay of technical aids and time expenditure than the bacteriological investigations require.

Viruses have not so far been detected in treated municipal water in the USA. This may be because they were not present, or because the technique of virus detection is not yet sufficiently developed.

One is attempting the world over to perfect the methodological conditions. *Borneff* reported on a symposium in Cincinnati in 1965 devoted to these problems [*17*]. The report states that considerable progress has been made in concentrating the viruses in the water and also in transferring the concentrate to cell cultures. Concentration is effected by dialysis of the water against polyethylene glycol; by ultracentrifugation and polyphosphoric acid precipitation (*Pittler* and *Höpken* [94]); and by filtration. *Gärtner* [40] developed alginate filters for retaining viruses, through which was filtered the water which had been rendered free of bacteria by membrane filtration; the filters were then dissolved in citrate solution. These alginate filters are now manufactured industrially by the Sartorius-Membranfilter GmbH. The solution of the filter was then inoculated (in 0,5 ml amounts) into cultures of monkey kidney and hela cells (*Primavesi* [98] and *Wüstenberg* [140]).

It cannot be our aim here to give further details about detection of viruses in water. As things are at present, this remains the concern of specialist institutes; there are so far no routine determinations. This does nothing to alter the fact that the microbiology of water is developing in a virological direction and that the virological investigation of the water will become part of the routine investigation when there is further progress in acquiring scientific information.

Bacteria function as hosts to bacteriophages. According to our present knowledge, phages multiply in the living bacterial cell by injecting nucleic acid into it when they encounter bacteria which possess receptors specific for the particular phage. The phage DNA brings about the formation of enzymes in the bacterium which modify the bacterial metabolism in such a way that phage DNA is formed instead of the bacterial DNA; with its protein coating, this provides new phages. This process of formation of new phages is repeated until the bacterial cell bursts and the phages are liberated. Most bacterial species can function as hosts for several types of bacteriophages; the latter are, however, specific for particular bacteria. Investigations of phages serve generally in microbiology as a model for genetic problems in the molecular domain and have led here to substantial new information [6].

The detection of phages also has practical value in the microbiology of water. Phages are often more resistant to environmental influences than are their host bacteria. If phages are detected in water, this shows that the host bacteria must have been present in the water, even when they are no longer alive.

Carstens [25] quotes the example of a river in which typhoid bacteria and typhoid phages were found just below the point of entry of wastewater. Two and a half miles further downstream, only typhoid phages were detected and further away still, neither the bacteria nor the phages. *Rische, Keiger* and *Stempel* [105] infected a receiving water with adapted phages which do not occur in nature. Over a stretch of about 20 miles they were able to establish only a dilution of the phage amount but no influence on them of the self-purification capacity of the river. In certain cases, therefore, the search for bacteriophages can supplement that for pathogenic germs and coli bacteria, and be important in enquiries about epidemics.

Lenk's [77] work on the Berlin lakes showed that the investigation of *Salmonella* phages yielded more specific results than the search for the *Salmonella* themselves. The ratio between *Salmonella*-positive tests and phage-positive tests was 2:14.

The detection of phages has been introduced also into routine diagnostics. The demand was made at the Evian symposium, already mentioned, and reported by *Seeliger* [121], that mineral water, treated anti-bacterially, should be free from coli and *Shigella* phages as well as fulfilling bacteriological requirements of perfection.

Sturdza and *Russu-Pandelescu* [129] *detect phages in water* by first mixing 30 ml of the water under investigation with 30 ml of nutrient broth in an erlenmeyer flask. To this is then added 0.2 ml of a broth culture (3 hour) of the bacteria, the phages of which are being looked for; this is incubated for 16–18 hours at 37° C. The flasks are then heated for an hour on the water bath at 58–60° C. A drop of the mixture is then transferred with the loop to fresh growths (ca 17 mm = ¾" diameter) of the test organisms on agar plates. The result of 6–8 hours incubation (if necessary with a further 16–18 hours storage in a refrigerator) is noted. Phages, if present, show themselves as sterile spots or circular spaces (taches vierges) in the bacterial mats. The test organisms must be chosen correctly because there are bacteria which are resistant to phages or which are reduced to prophages by infecting phages and are thus immune to the action of phages.

The number of phages in a sample of water can be determined by adding increasing amounts of the water to suitable amounts of nutrient broth and then proceeding as described above. *Sturdza* and *Russu-Pandelescu* suggest mixing 30, 10, 5, 1.0, 0.1, 0.01, and 0.001 ml of water with the same volumes of broth; a parallel control series is made up but not subsequently inoculated with the test bacteria.

A shortcoming of this procedure for detecting phages in water is, according to *Lenk* and *Ackermann* [*76*], their observation in testing for typhoid vi-phages or *Salmonella* phages, that other *Salmonella* species suffered attack; this makes it technically difficult to isolate specific typhoid-vi phages and leads to considerable time demands. According to the same authors, the temperature of 58–60° is too high, since some types of phage are inactivated by heating at 60° for an hour. They therefore warmed the water-broth-test strain mixture for an hour at 56°. Thermal damage to the phages is avoided by passing the mixture through membrane filters; these retain the bacteria but absorb the phages only to an insignificant extent (*Uecker* [*133*]).

It has moreover proved advantageous to use the so-called blue plate instead of an ordinary agar plate for testing the action of phages (*Rättig* and *Hofmann* [*110*]). This blue plate contains dextrose and marine blue.

The colour indicator plate is prepared as follows:

To 500 ml of liquefied meat broth-peptone agar, containing 6 g of dextrose, are added 4 ml of a 2% aqueous marine blue solution. After mixing, the product is poured into petri dishes. The marine blue solution should be prepared a few days in advance and filtered before use. The blue dishes are dried for 2–3 hours at 37° C before using and inoculated as fresh as possible.

We ourselves have generally had good experience of *Sturdza* and *Russu-Pandelescu's* technique in searching for coli phages in waters.

XI. The Membrane Filter Procedure

Most of the methods for bacteriological water investigation so far described date basically from the beginning of this century. They have indeed been improved and refined in the course of the years but essentially our ideas and methods do not deviate from those which our fathers and grandfathers developed. This is remarkable when one regards the rapid progress made in all domains of medicine; but it is certainly no disadvantage since it simply demonstrates the correctness of the knowledge gained at the beginning of the bacteriological era. The sole exception is the membrane filter procedure because it has been brought to its probably final state of perfection only in recent years. A special treatment of this method of investigation and its possibilities of application to bacteriological investigation of water thus appears justified. This treatment is based in many essential features on the work of *Daubner* [*28*], compiled in a monograph. It is regrettable that this unique and valuable monograph has not yet been translated into world languages and has not been published outside the home country of the author.

The use of semipermeable membranes of animal tissue was described as long ago as the 18th century. *Fick* was the first to prepare artifical filters [*35*] (1855), from collodion; these were studied later above all by *Bechold* [*19*] (1906). After further work by various authors, *Zsigmondy* and his co-worker *Beckmann* engaged in the preparation of membrane filters during the war years 1914–1918 and patented their methods in 1918. This established the foundations of the manufacture of such filters, taken over by the Sartorius A.G., now their subsidiary company the Sartorius-Membranfilter GmbH of Göttingen. Membrane filters were prepared also abroad for widely varying purposes, e.g. in the USA, England, USSR and Czechoslovakia. In 1932, *Barsow* [*8*] published his studies on detection of coli bacteria in water using membrane filters; these were then further used and developed in all countries after the second world war. Membrane filter procedures have been standard methods in the USSR since 1950 and in the USA since 1960.

Nitro- or acetylcellulose are the basic materials for preparation of membrane filters. These are dissolved in organic solvents, in particular acetone; with the help of a pouring device the solutions are brought on to exactly horizontal glass plates, or on to an endless polished metal belt. The solvent evaporates gradually, leaving a paper-thin skin with many tiny pores. The rate of evaporation can be varied by change of air humidity and temperature. This rate, and the concentration of the cellulose product and the solvent, govern the ultimate pore diameter which can thus be varied between 10μ m and 5 nm.

A uniform pore diameter is one of the requirements of membrane filters for use in microbiology. The mean pore diameter can be calculated from the rate of flow of distilled water through them under a definite pressure; for this, the Hagen-Poiseuille formula is used, which expresses the relation concerning passage of liquids through capillaries. *Henneber-*

ger and *Crodel* [52] describe a special apparatus for this purpose, with which very accurate results can be obtained.

The maximum pore diameter can be determined by the bubble pressure method, in which the membrane filter is stretched in a special apparatus and submitted to air pressure from underneath. The pressure needed to force the first air bubbles through the filter into the water above it is a measure of the maximum pore size. This last-named can be calculated from formulae which take account of the surface tension of the water, atmospheric pressure, etc. The bubble pressure method can be used for determination of maximum pore diameters down to 0.3 μm only, because the necessary pressures become otherwise too high. Smaller pore sizes are determined with standard sols.

The imperviousness of the filters to bacteria is decisive in microbiological work. It can be investigated by filtering a suspension of a group of bacteria and then testing the filtrate, collected under sterile conditions, for the presence of these microorganisms using a culture. Filters supplied by the German industry are, however, tested by the works so as to fulfil all requirements concerning porosity and calibration; only in special cases, therefore, must one carry out one's own calibration.

Filters of average permeability come into consideration for water bacteriology. For detecting *E. coli* and coliform bacteria, the filter type SM 11406, of mean pore diameter of 0.6 μm is used; for determining total bacterial number, the filter type SM 11407, of mean pore diameter 0.3 μm.

The filters can be sterilised by boiling in distilled water. They are placed in a beaker of diameter about 1 cm larger than that of the filters and sterilised water is poured over them. A few boiling stones are added to prevent bumping. The water is then boiled for about 20 minutes, avoiding too hectic boiling which causes the filters to roll about. Membrane filters can also be sterilised in digesters or autoclaves. For this purpose, the autoclave packings of type SM 13706 or SM 13707 have proved suitable; they have a colour indicator which shows when sterilisation is complete.

Chemical sterilisation of membrane filters can be performed by placing in 1% nipagin (methyl-p-hydroxybenzoate) solution for 15–20 minutes or exposing to formaldehyde vapour in a desiccator for 24 hours. The disinfecting solution must be removed from the filters by passing sterilised distilled water through several times. Membrane filters from the Göttingen suppliers can be obtained also which have been sterilised with gaseous ethylene oxide; they can be used directly.

Coarser impurities can be removed from water containing suspended matter by preliminary filtration through a filter which is permeable to bacteria. A simple assembly permits this pre-filtration to be carried out with the bacteriological filtration itself in a single working operation.

The most varied apparatus have been developed for work with membrane filters; which is used depends on the purpose of the investigation. The stainless steel filtration apparatus, SM 16201, for reduced pressure of 50 mm, has proved satisfactory for bacteriological investigations of water. It consists of a filter plate with sintered metal filter and stopcock, and a graduated funnel-shaped attachment which can be joined to the filter plate with a lever mechanism giving watertight closure of the filter resting on the sintered metal. The apparatus can be combined with an auxiliary unit for pre-filtration. It is made of stainless steel which facilitates sterilisation. The funnel-shaped attachment has a capacity of 500 ml. A cylindrical upper part of 40 ml capacity can be attached when smaller volumes of water are to be investigated. The filtration apparatus is mounted on a Witt jar or filtering flask by means of a pierced rubber stopper. The flask is joined with a rubber tube to a water pump or, if several apparatus are used at the same time, to an electric laboratory pump.

The apparatus can be sterilised in an autoclave or hot-air steriliser. This method is preferred when work is to be carried out under really sterile conditions. It is sufficient for routine investigations to saturate a wad of cottonwool with methylated spirit, moisten the inside of the funnel and the filter plate plus sintered filter and then to ignite this with a burner. The burning alcohol is drawn through the fritted filter. This method does not respect the strict requirements for sure disinfection but, according to *Kruse* [*69*], has been shown to be adequate for daily practice.

After the filter and filtration apparatus have been sterilised, the filter is laid on the fritted plate using sterilised forceps, the upper part is joined to the filter plate and the stopcock is closed. The apparatus is mounted on the filtering flask, which is joined with vacuum rubber tubing to a water pump. Larger amounts of water (e.g. for detection of coli in water) are measured into the funnel, smaller amounts (less than 20 ml) are added with a pipette to 20 ml of sterile water previously placed in the funnel, and then mixed. This procedure is necessary with smaller amounts of water to attain as uniform a distribution as possible of the bacteria on the filter surface. The vacuum is then laid on, the stopcock beneath the filter plate is opened and the water drawn through the membrane filter. The bacteria remain on the surface of the membrane filter. By spirally swirling sterile distilled water from above downwards, bacteria which may still be adhering to the inner wall of the funnel attachment are washed on to the filter. The stopcock is closed after filtering, the funnel attachment is taken off, the filter held at its outside edge with sterilised forceps, removed from the sintered metal base and rolled out on to the culture medium. Air bubbles must not be entrapped between filter and culture medium since the nutrient substrate cannot diffuse through them.

To determine the bacterial count, 10 and 1 ml of water, or perhaps smaller amounts, are filtered, according to the nature of the water. The filtration apparatus can be dispensed with (*Daubner*'s procedure) with water volumes of 1 ml or less. The membrane filter is then placed on a pad of sterilised, ordinary filter paper in a petri dish and the water is added to the filter with the help of a pipette; the membrane filter must not be flooded over. 1 ml of water is sucked through the membrane filter by the filter paper pad in about 1 minute. Nutrient gelatine used for determining the bacterial number in the membrane filter procedure should contain not more than 10% of gelatine. If nutrient agar is used, it should have only 1% agar addition so that the nutrient media remain as soft as possible and the nutrients can diffuse more easily through the membrane filter. Silica gel nutrient media and liquid nutrient solutions, with which sterilised, ordinary filter paper is saturated, can be used also. The colonies which develop on the surface of the membrane filter are counted after incubation for 24 and 48 hours at 22° C and at 37° C. The amount of water to be filtered should be chosen so that as far as possible, not more than 150 colonies can form on the filter. Bacterial beds are easily formed with a larger number of colonies and these merge and render counting impossible. The bacterial colonies which develop on the surface of the filter mostly appear colourless and translucent and thus easily escape notice. To eliminate the error which would ensue, the incubated membrane can be stained, or dyes or indicators can be added to the nutrient media. The first possibility is most simply realised by laying the membrane filter for a minute on to normal filter paper which has been soaked in a 0.01% solution of methylene blue. When the membrane filter is then transferred to a filter paper moistened with distilled water, the filter itself is decolorised whereas the colonies retain their blue colour. As an additive in the second possible method, *Bucksteeg* and *Thiele* [*20*] recommend 1% triphenyltetrazolium chloride solution. *Daubner* uses one tenth of this concentration. This substance is reduced to a red formazan, yielding reddish to dark red stained colonies on the membrane filter; these are easily determined.

Colonies on membrane filters are appreciably smaller than those habitually encountered on the solid nutrient media of medical bacteriology. This shows that the bacteria on membrane filters do not enjoy the best conditions for multiplication which one otherwise endeavours to give them.

Attempts are often made to count bacteria directly in the water, using a microscope. Investigations have been carried out since 1929 (*Cholodny* [*26*]) with the membrane filter procedure also, aiming at direct microscopy; this has not been adopted as a routine procedure. This is probably because pure water has only few bacteria, which then lie too far from one another on the filter to enable counting to be exact; and because polluted water carries with it particles of dirt and detritus which can be confused with microorganisms.

The distinction between living and dead bacteria also is impossible with routine procedures. The direct method has found much successful application in hydrobiology (*Beling* and *Maier* [*13*] and *Jannasch* [*56*]) and can be advantageous in special cases in medical microbiology too in order to obtain a rapid idea of the microbial flora of a water sample; the method is accordingly briefly described here:

The sterilised filter is fitted into the sterilised filtration apparatus and the water under investigation is filtered. The amount taken is again dictated by its degree of contamination and should be such, that, under optimum conditions, not more than 30 microorganisms are to be seen in the field of view of the microscope. After the water has been filtered, the bacteria are fixed on the membrane filter by slowly filtering 5 ml of 5% formalin solution. A final washing is carried out with sterile distilled water. The membrane filter can then be stained in the filtration apparatus. This is done by adding dropwise 5 ml of dye solution to the membrane filter and leaving it there for the desired time. The staining time is 30 minutes with erythrosin (3% solution in 5% aqueous phenol) or 5 minutes with methylene blue (10 ml of a saturated aqueous solution plus 90 ml of 5% aqueous phenol solution). The staining dye solution is then sucked off and rinsing is carried out with distilled water until the membrane filter shows only a weak colour shade. Alternatively the filter paper can be removed from the apparatus and laid on sheets of cardboard or ordinary filter paper in a petri dish, which have been soaked in the solution of the dye. The time of action is the same as that used within the apparatus. The membrane filter is then placed on filter paper, saturated with distilled water, and this is repeated until the membrane filter appears only feebly stained. The membrane filter is dried and cut into suitably sized pieces which are put on to glass slides on which a drop of cedar wood oil has been placed. The oil expels all the air from the filter pores within a few seconds, making the filter transparent. A further drop of the cedar wood

oil is added to the piece of the filter, a very thin cover glass laid on it and the slide examined microscopically.
The American Standard Methods recommend filtering 500 and 100 ml of water, when it is obviously not contaminated, for detection of *E. coli* by means of the membrane filter; for well water, which may be polluted, they recommend 10 ml, 1 ml or 0.1 ml of the water. Investigation should be carried out on diluted samples of water which is more strongly polluted.
For detection of coli, *Beling* and *Maier* [*13*] propose the water amounts:

tap water	250–500 ml
drilled well water, best twice filtered separately	10–100 ml
dug well water, best twice filtered separately	5– 50 ml
surface water, best twice filtered separately	1– 10 ml
wastewater must be diluted considerably and filtered in various dilutions	

First the smaller water sample and then the larger are filtered with strongly polluted water samples; the filtration apparatus does not then have to be resterilised.
After filtering, the filters are placed on Endo nutrient media which, however, have a somewhat different composition from those used in classical procedures of investigation.

1 g of lactose (dissolved in 7 ml of distilled water by warming), 0.13 g of sodium sulphite (Na_2SO_3, likewise dissolved in a few ml of distilled water) and 2.5 ml of filtered fuchsin solution (10 g of basic fuchsin in 90 ml of ethanol) are added to 100 ml of liquid nutrient agar which contains 1–1.5% agar instead of the usual 3%. After mixing well, it is poured into petri dishes.

After incubating for 18–20 hours at 37° C, the dark red colonies with fuchsin lustre are counted on the membrane filter. These colonies form red points on the underside of the membrane filter. Doubtful colonies can be further differentiated by transfer inoculation on to normal Endo nutrient media. The occasional light red colonies also should be further investigated by transfer of pure cultures to Endo agar; they may develop to *E. coli* or colonies of coliform bacteria. The membrane filter can also be cut in half with scissors after filtration. The one half is placed on Endo agar for membrane filters, and the other transferred to a tube containing 1% lactose-peptone solution. This latter is then incubated at 44° C and further treated as is described under the *Eijkman* procedure.
A liquid Endo nutrient broth is proposed in the American Standard Methods. 3 ml of it is dropped on to a sterilised absorption pad of filter paper in a petri dish; after filtration, the membrane filter is laid on this and incubated. The nutrient broth has the composition:

tryptone or polypeptone	10.000 g
thiopeptone or thiotone	5.000 g
casitone or trypticase	5.000 g
yeast extract	1.500 g
lactose	12.500 g
sodium chloride	5.000 g
dipotassium hydrogen phosphate	4.375 g
potassium dihydrogen phosphate	1.375 g
sodium lauryl sulphate	0.050 g
sodium deoxycholate	0.100 g
sodium sulphite	2.100 g
basic fuchsin	1.050 g

These substances are dissolved by warming in 1 litre of distilled water containing 20 ml of ethanol.

The nutrient solution must not be boiled nor submitted to steam under pressure. It is stored in a refrigerator and must not be more than 4 days old when used.
This and other nutrient solutions have the advantage that the nutrient substrates are better able to penetrate the membrane filter than are firmer nutrient media, and also that the membrane filter is incubated with a small amount of nutrient solution. It is thus more economical than other methods.
The membrane filter procedure can be applied with advantage also for *detection of enterococci* in water. There are fewer enterococci than coli bacteria in water and, according to a rule of thumb of *Daubner,* ten times as much water must be taken for detection of the former than of *E. coli. Slanetz, Bent* and *Bartley* [*124*] give a special medium for cultivation of enterococci on membrane filters. It has been included in the American Standard Methods since 1960 and has the composition:

tryptose	20 g
yeast extract	5 g
glucose	2 g
dipotassium hydrogen phosphate	4 g
sodium azide	0.4 g

These substances are dissolved in 1 litre of water and the pH is adjusted to 7.2. 10 g of agar are added and the mixture is heated until the agar has dissolved. After cooling, 0.1 g of 2, 3, 5-triphenyltetrazolium chloride is added, which has been dissolved previously under sterile conditions in a few ml of distilled water. The nutrient medium is poured into petri dishes.

Neither *E. coli* nor staphylococci nor *Proteus, Sarcinae* nor *Pseudomonas* organisms can grow on this nutrient medium; but enterococci yield red to purple tinged colonies of 0.5–1 mm diameter. Other bacteria can, however, show these colours through reductive ferments. Consequently, the test for enterococci must be carried out by transferring to classical culture media and employing usual microbiological methods. Membrane filters are incubated with the *Slanetz* culture medium for 48 hours at 37° C.
Clostridia are detected on membrane filters by heating the water sample for 30 minutes at 80° C on the water bath so as to kill off the vegetative bacterial forms. From each water sample, 100, 50 and 10 ml amounts are then filtered through membrane filters and the filters placed in a sterile petri dish. They are held on the bottom of the dish with flamed forceps and covered with 15 ml of liquefied *Wilson-Blair* agar for anaerobic cultivation; the medium is not poured on to the membrane filter but on to the bottom of the dish so that the filter slowly becomes inundated. After the culture medium has solidified, the petri dish is sealed with plasticine to render it air-tight, and incubated for 24 hours at 37° C. An anaerobic jar is usually not absolutely necessary. If one is aiming at really anaerobic conditions, however, cultivation should be performed in an evacuated vessel (*Zessler*'s anaerobic jar; *Knorr*'s anaerostat); sealing of the petri dishes is then superfluous.
The *Wilson-Blair* culture medium for anaerobic cultivation is prepared as follows, according to *Daubner* (care not to confuse with the *Wilson-Blair* nutrient medium for cultivation of *Salmonella*!):

1. 20 g of sodium sulphite in 80 ml distilled water.
2. 20 g of glucose in 80 ml of distilled water.
3. 0.8 g of ferrous sulphate in 9.2 ml of distilled water.

Each of these solutions is boiled and filtered. A mixture is made of 20 ml of 1., 10 ml of 2. and 2 ml of 3. This is added to 200 ml of nutrient agar (2.5%) with a pH-value of 7.5. 15 ml amounts of the liquid nutrient medium are filled into test tubes and heated for 15 minutes at 120° C in an autoclave.

Clostridium welchii grows in this medium, reducing the sulphite to sulphide which then reacts with the ferrous sulphate to yield black ferrous sulphide. Typical black coloured colonies are thus formed, which are fished out and identified with the customary cultivation procedures, if necessary after microscopic examination.

XII. Bacteriological Examination of Baths Water

Baths water is a priori not drinking water. However, the recommendations of the Federal German Ministry of Health, passed to the German Land Governments on 27/1/64, include the requirement that the water used for filling open-air and closed baths should have the same bacteriological and chemical-physical quality as drinking water. It thus appears desirable to discuss here the bacteriological investigation of baths water; sampling and examination show some special features compared to those of drinking water. The benefits to health of bathing and swimming, in that they further circulation, respiration and metabolism, are generally well known. Swimming is the best form of deportment gymnastics. It is therefore necessary, especially in the present time of sedentary occupations and increasing leisure periods of all sections of the population, to provide adequate bathing and swimming facilities. Equally important, nevertheless, are the right layouts for baths and measures to ensure good water quality; this averts certain dangers and prevents the good intention from having the opposite outcome.

The Medical Officer of Health has the responsibility for hygienic flawlessness of closed and open-air baths, which he superintends as part of his duty inspections. The control of water quality is usually left to a bacteriologist, however, engaged for the water investigation. A few significant facts are quoted here in this connection:

Every bather introduces bacteria and organic substrates into the baths water. According to *Klosterkötter* [*60*], between 3 million and 3 milliard (3×10^9) bacteria are found on the skin surface, in the nose and throat region and in the region of the anus and vulva, depending on the personal hygiene of the individual. The organic contaminants consist of rubbed-off skin cells, mucous secretions, perspiration, skin fat, particles of excrement, urine, cosmetics and ointments. About 0.5 g has been estimated per bathing guest. The baths water is thus permanently exposed to re-infection, which cannot be removed merely by change of water; continuous filtration (circulation) and disinfection (chlorination) of the baths water are necessary.

Gertrud Müller [*88*] points out that whereas the danger to health from drinking water is characterised by the chain of infection intestines–water–intestines, that from baths water is characterised by this and also by the sequences nose and throat region–water–nose and throat region, and skin–water–skin. With baths water, the search for indicator bacteria (as for *E. coli* and enterococci in drinking water) is thus not sufficient. Bacteria, pathogenic to humans, coming from the nose and throat region or the skin, must be included in the investigation. Especially useful here are the coagulase-positive staphylococci. According to studies of *Gertrud Müller,* quoted above, they occur independently of *E. coli* and enterococci, are more resistant to chlorination than coli bacteria and come without doubt from bathing humans. According to *Pulverer* [*99*], the test of the ability of pathogenic (to humans) staphylococci to coagulate plasma (plasma coagulase reaction) is replacing the procedures formerly used for distinguishing staphylococci, such as haemolysis (solution of the blood colouring matter, haemoglobin), formation of colouring matter and other biochemical feats.

Bacteria are by no means as uniformly distributed in the baths water as we assume they are in drinking water. A surface film forms on chlorinated baths water, consisting of secretions from the skin and mucous membranes of the bathers. The bacterial count of this surface film can be a thousand times greater than that in the deep water. *Heinrich* and *Gröschel* [*51*] explain this through the surface layer conditions which lead to lowered surface chlorine concentration as a result of gas exchange, and through increased chlorine consumption by organic substances. These differences do not exist in natural baths. This situation has consequences for the sampling and investigation of the baths water.

Samples should be taken both at the entry point of fresh water and at the exit point of the used water. Samples must be taken, too, from the surface, which has particularly extensive contact with the bathers, and from the depth, so as to be able to control the function of the circulation. According to the directives of the Federal German Health Office, the water in artificial baths must be continuously chlorinated–an excess of 0.3 mg free Cl/l must be present at the water outflow. The water sample bottles must therefore be treated with sodium thiosulphate immediately after sampling. As said already on p. 289, 0.25 ml of a sterile 0.01N sodium thiosulphate solution is added to 250 ml bottles, and 0.5 ml to those of 500 ml content.

The bacteriological investigation of the baths water comprises the determination of the bacterial count and the coli titre, and the detection of coagulase-positive staphylococci.

The bacterial count is determined with poured agar plates, as described in Chapter VI of this contribution. The agar plates with baths water samples are incubated at 37° C, in contrast to gelatine plates with drinking water samples, which are incubated at 22° C; this allows the bacteria of human origin to be determined better. The bacterial count has only a limited information value with baths water because sampling is especially susceptible to chance. It cannot be omitted, however, since the directives of the Federal German Ministry of Health demand drinking water quality.

The bacterial count is interpreted differently in the individual Countries.

According to *Coetzee* [*27*], the total bacterial number in swimming baths in the USA must not exceed 200 per ml; in England, the bacterial count should not exceed 10 per ml in 75% of the samples and 100 per ml in 25% of the samples.

The Federal German Health Office gives no figure for the bacterial number in natural swimming waters but it must be less than 100 in swimming pools *(Carlson* [*24*]*)*.

The coli titre is determined by the ***Eijkman*** procedure, as in Chapter VII of the bacteriological section of this book; or, as in the English-speaking countries, according to the MPN (most probable number) method (p. 305). The investigation of *E. coli* in baths water thus shows no special features in comparison with that in drinking water.

The Federal German Health Office recommends for natural swimming waters that the *E. coli* titre should not exceed 0.1 ml, i.e. a maximum of 10 *E. coli* bacteria should be detectable in 1 ml of water. A titre of 0.01 ml can be tolerated in 20% of the samples if, after careful checking of the local conditions, no hygienic dangers through wastewater discharge are present in the immediate vicinity (*Carlson* [24]). In constructed swimming pools on the other hand, *E. coli* must not be detectable in 100 ml of the water.

This distinction between natural swimming waters and constructed baths corresponds to reality when one thinks of the many summer baths of smaller districts, which are highly useful to the country population, especially the youthful part.

The inclusion of staphylococci in the investigation of baths water is of high value at present. *Pulverer* [99] detects staphylococci by filtering 200 ml of baths water through a membrane filter and incubating the filter for 24 hours at 37° C on the *Chapman* selective culture medium for staphylococci. Suspect single colonies are directly transferred to a blood agar plate; the total amount of culture material is scraped from thicker colonies on the filter surface, extracted for 6 to 9 minutes with the same amount of toluene and then spread on to blood agar plates. The plasma coagulase activity is determined for colonies where doubt reigns.

This is done by rubbing culture material into a drop of rabbit plasma on a glass slide. Virtually instantaneous agglutination is a positive result and formation of a uniform turbidity is negative. If the result is unclear, the real plasma coagulase reaction is carried out. A loop of culture material is inoculated into 0.5 ml plasma in a tube, which is inspected after 2 or 24 hours in an incubator at 37° C. Positive and negative controls must be conducted at the same time. Staphylococci are termed coagulase-positive which caused total coagulation of the plasma.

Chapman agar can be obtained from the firm of Merck and from the Biologische Arbeitsgemeinschaft (biological working association) as a dehydrated culture medium.

Gertrud Müller [88] has not had this good experience with Chapman agar. She filters 100 ml of water through a membrane filter, incubates the filter in peptone broth and transfers from this to blood agar plates.

We ourselves have been carrying out detection of staphylococci for years, especially in the investigation of kineto-therapeutic baths, and, after enrichment, spread on to blood agar plates; these are available anyhow in the laboratory and we fight shy of burdening the routine operations with further selective culture media.

Favero and co-workers [34], in connection with assessment of staphylococci occurrence in baths water, state that there should be less than 100 staphylococci in 100 ml of water. This number seems to be too high, according to *Gertrud Müller*'s and our own findings, when one relates it to the strict requirements of the bacterial number and coli titre in chlorinated baths water.

Since no staphylococci are found in baths water which has been adequately disinfected, the detection of staphylococci ought to be regarded as a criterion for insufficient treatment of the baths water.

XIII. Detection of Pathogenic Bacteria in the Water

Before the membrane filter procedure was introduced, detection of pathogenic bacteria in water was difficult and often failed. In membrane filters is a film material which permits bacteria to be concentrated from larger amounts of water. Its use with special culture media opens up new prospects in the search for pathogenic microorganisms in water. Even here, though, it is not always possible with certainty to detect pathogenic bacteria in a water which has caused human illness. For one thing, the number of pathogenic bacteria in a contaminated water is usually very much smaller than the number of coli bacteria and other microorganisms in this water. The duration of life of pathogenic bacteria in water is restricted and dependent on many factors which cannot be estimated. Even *Salmonella* can multiply in water under definite conditions, but this depends on the nutrient substances, the accompanying flora and fauna and the temperature. *Steiniger* [*128*] established the presence of especially many *Salmonella* in contaminated sewer water and when the water was polluted with blood-containing waste from slaughter houses. On the other hand, many bacteria are destroyed in surface water by protozoa, predominantly ciliates, rotifers and flagellates; these require temperatures of 10 to 25° C and, according to *Beger* [*12*], appear to prefer alimentary bacteria to the real water bacteria. Another reason why it is difficult to detect the aetiological agents of illnesses, e.g., typhoid, in an infected water, is the time of incubation of the infection maladies. In the case of typhoid, this interval between taking up the germ and the outbreak of the illness is about 14 days. And at first, after that, uncharacteristic symptoms are encountered. Further time elapses thus before the diagnosis is certain, the illness is reported to the competent health office and suspicion is created that perhaps a polluted well was the origin of the illness. From the time of infection of the sick person, at least 4 weeks will then have passed before the water is tested for *Salmonella.* In this time, the germs can long since have disappeared, by dying off or being washed away. This is frequently the case when the water is polluted only once.

Since these natural circumstances cannot be altered, it will not be possible to perfect the methods of investigation still further so as always to be able, with certainty, to detect in water germs causing illness. A negative result of detection of an aetiological agent therefore brings no proof. One would ask too much of bacteriological methods to demand of them information about bacteria in a water which infected a human weeks earlier.

If, despite these difficulties, it has been possible during the last two decades to carry out detection of pathogenic bacteria in water ever more frequently, this is due to improvement of the methods of investigation through the introduction of membrane filters, and to further perfection of the nutrient media and solutions needed. In 1957, *Seeliger* [*120*] made a compilation of the *Salmonella* strains found in German water and wastewater in the previous year. 758 strains with 38 different serotypes were isolated. Paratyphoid B predominated, isolated 289 times. It appears to be especially resistant to outside influences. Then followed *S. montevideo* (63 times) *S. münchen* (48), *S. typhi murium* (44), *S. thompson* (31) and *S. bareylly* (30).

This compilation shows that, whereas formerly the cultivation of *Salmonella* from water was an almost unique event, isolation of these organisms belongs nowadays practically to routine diagnostics in water bacteriology.

In the *investigation of Salmonella in water,* as large as possible a volume of water is passed through a membrane filter. The larger the chosen volume, the greater is the chance of detecting *Salmonella.* Since the samples are often highly polluted, several filters are connected in series and pre-filtration is made use of. For this latter purpose, the auxiliary filtration unit is mounted on the filter plate carrying the bacteria-retaining membrane filter and screwed on tightly. The filter base plate of the auxiliary unit is sterilised with the flame and introduced. Membrane filter V is laid on to it and the funnel screwed on. The water to be examined is filled into the funnel and sucked through in the usual way. Before all the liquid has passed through the pre-filter, it is rinsed with a few ml of sterile distilled water; this is repeated several times. After having filtered, the membrane filter V is discarded. The bacteria-retaining membrane filter is then placed directly on a suitable selective culture medium, or the bacteria on this filter are enriched in a suitable pre-culture.

The culture medium quoted by *Wilson-Blair* has been proving satisfactory for years for the direct culture procedure of membrane filters on solid nutrient media in the investigation of *Salmonella.* As with most culture media, this nutrient substrate has undergone many modifications. Below is given *Schlirf* and *Seidenstücker*'s [*113*] method of preparation, based on the procedure of *Lovrekowitsch* and *Messerschmidt* [*81*]:

4 mixtures are first made up:

A. 50 g of glucose, dissolved in 250 ml of hot distilled water in a litre flask
B. 50 g of disodium hydrogen phosphate ($Na_2HPO_4 \cdot 2H_2O$), dissolved in 250 ml of hot distilled water in a litre flask.
C. 50 g of bismuth ammonium citrate, dissolved in 250 ml of hot distilled water in a litre flask.
D. 25 g of sodium sulphite, dissolved in 250 ml of hot distilled water in a 2-litre flask.

A is added to B yielding mixture I. C is added to D, yielding II. I is then added to II in the 2-litre flask and boiled. This part of the procedure is important and should be arranged so that the mixture begins to boil after 5 minutes. Boiling begins as gentle bubbling at a temperature of 101 to 101.5° C, and should be maintained using a medium flame for 7.5 minutes at the same temperature (controlled with a thermometer). To the still hot solution are added:

50 ml of a freshly prepared 13% solution of ferrous ammonium sulphate
25 ml of a 1% aqueous solution of brilliant green.

This liquid, termed the basic mixture, is filled hot in 50 and 100 ml amounts into flasks which can be tightly closed. As far as possible, air bubbles should be avoided. The flasks are stored in the dark at 17–18° C. The basic mixture should ripen for at least 8 weeks before use. It can be kept for 3 to 4 months.

A 100 g amount of the basic mixture is shaken round in the flask to mix in any deposit and is added to 500 ml of liquefied nutrient agar which is prepared with casein-peptone instead of peptone. The product is poured into petri dishes, avoiding air bubbles. The plates are left open for 30 minutes to dry. The bismuth sulphite nutrient medium should be freshly prepared and used after keeping for 3 days at the most (its efficiency deteriorates with age).

Since the preparation of this culture medium is somewhat tedious, other special media have been given for routine diagnosis of *Salmonella*. Two are quoted here: *Leifson*'s deoxycholate-citrate-lactose-sucrose-agar; and the *Salmonella-Shigella* (S–S) agar. These are efficient culture medium combinations. Both have been subjected to many variations.

The *Leifson* agar is prepared according to the prescription:

Nutrient agar with 1% Witte peptone	1000.00 ml
2% solution of neutral red	1.25 ml
lactose	10.00 g
solution A	50.00 ml
solution B	50.00 ml

Solution A consists of:

sodium citrate ($Na_3C_6H_5O_7 \cdot 2H_2O$)	17 g
sodium thiosulphate ($Na_2S_2O_3 \cdot 5H_2O$)	17 g
ferric citrate in flakes	2 g
distilled water	100 ml

(the solution must not be sterilised)

Solution B consists of:

sodium deoxycholate	10 g
distilled water	100 ml

(the solution should be sterilised for 1 hour at 60° C)

The *Salmonella-Shigella* nutrient agar (S–S agar) has the composition below, according to the American prescription as quoted by *Hallmann* [46]:

5.0 g meat extract	20.0	g	agar-agar
5.0 g peptone	0.33	g	brilliant green
10.0 g lactose	0.025	g	neutral red
8.5 g sodium citrate	100.00	ml	ox gall
1.0 g ferric citrate	1000.00	ml	distilled water

After the ingredients have been dissolved, the medium is boiled for 2–3 minutes and poured into petri dishes.

After the water under investigation has been filtered, the membrane filters can be directly laid on these culture media and incubated.

It is, however, usually better to enrich the *Salmonella*, often present in only meagre amounts in the water, on a *liquid preliminary culture* which inhibits the growth of bacteria other than *Salmonella*. Amongst other nutrient solutions, *Preuss*'s tetrathionate-crystal violet-digestion broth [97] can be used here with success. It is an extension of *Kaufmann*'s sodium thiosulphate-bile broth and is prepared as follows:

A mixture of 100 ml of a nutrient broth of pH 6.5 and 0.5 ml of a 0.1% aqueous solution of crystal violet is sterilised in the water bath. After it has cooled, 2 g of potassium tetrathionate are added and the mixture, without being sterilised, is run in 5 ml amounts from a sterile burette into sterile test tubes.

The membrane filter through which the water under study was passed is rolled up with the help of sterile forceps and introduced into a test tube. This is incubated for 18–20 hours and then applied to one of the mentioned *Salmonella*-differentiating media. Seeding alone on to *Wilson-Blair* culture medium is not to be recommended since, according to *Hallmann* [47], some paratyphoid strains grow poorly on it after enrichment on the tetrathionate medium and any sample agglutination needed later is thereby often impeded.

In recent years, various authors have reported good results in the detection of *Salmonella* in water which they obtained through direct enrichment in the potassium tetrathionate broth mentioned above. *Popp* [*95, 96*] used a subtle combination of direct seeding on *Wilson-Blair* culture medium, membrane filtration, and enrichment in a ten times more concentrated potassium tetrathionate broth; the incubated enrichment medium also was subjected to membrane filtration and this again incubated in simple potassium tetrathionate broth.

Eikmanns [*33*] detects *Salmonella* in water by adding to 100 ml of the water 100 ml of a *Heinrich* and *Pulverer* [*144*] use a broth with the ten-fold concentration of potassium tetrathionate which contained 0.8% of a 1% solution of brilliant green to suppress the accompanying flora. The ten-fold concentrated tetrathionate broth was then diluted ten times with the water under investigation and the product of 18–24 hour incubation was applied to *Wilson-Blair* culture medium. The enrichment can be repeated by transferring a loop of the sediment from the first incubated enrichment to a further tube containing 10 ml of fresh enrichment solution, diluted 1:10 with tap water. *Gaase* and *Temme* [*36*] confirm *Heinrich* and *Pulverer*'s results and consider this technique for detection of *Salmonella* to be clearly superior to the membrane filter method.

4% potassium tetrathionate broth containing Metachrome Yellow (Alizarin Yellow GGW) (1:5000) to inhibit the growth of *E. coli, Aerobacter, Proteus* species etc. After incubation for 48 hours, it is applied to *Wilson-Blair* culture medium, as in *Schlirf* and *Seidenstücker's* modification given above.

Smaller amounts are best prepared from ready-made materials in order to circumvent preparation of the often complex culture media and the time-consuming subsidiary work of sterilisation. These ready-made preparations contain powdered agar with the relevant chemicals. The culture medium is then made by dissolving in a definite amount of distilled water (quoted in the instructions), filling into vessels and sterilising.

These ready-made nutrient media are commercially available in Germany from the firm of Merck amongst others, as well as from those mentioned above in the discussion of nutrient gelatine.

In addition to these dehydrated (powdered) nutrient substrates, "expendible culture media" have been gaining popularity in recent times for smaller scale investigations. These nutrient media are furnished ready for use in expendible petri dishes or tubes. The prospectuses of the firms concerned include all the usual and liquid media, ranging from simple nutrient agar or nutrient gelatine to S–S agar, at prices below the cost of self-preparation on a smaller scale. German suppliers include the Biologische Arbeitsgemeinschaft in Lich/Oberhessen and the Sartorius Membranfilter GmbH in Göttingen with whose products we have had good results.

The "nutrient cardboard discs" of the latter firm must be mentioned in this connection. They were developed by *Thon* and *Beling* [*131*] and possess quite decisive value in bacteriological water investigation. The discs contain the nutrients for bacteria, in dehydrated form. The polyethylene plastic bags which contain the discs with membrane filters in sterile packings, are opened with sterilised (flamed) scissors. The discs are laid in sterile petri dishes already containing 3.3 ml of pipetted sterilised distilled water. The water under investigation is then passed through the membrane filters in the way described and the filters are placed on the discs and incubated for 16–20 hours. A lattice is imprinted on each membrane filter provided with the nutrient cardboard discs, each square of which has a side of 3.1 mm. This network can significantly ease the counting of bacterial colonies. For calculating the bacterial number, it is important to know that the surface area of a square is 1/130 of the effective filtration area of the filtration apparatus. The Endo type discs are used for detecting *E. coli*, incubating 20 hours at 37° and 44 °; the azide type for detecting enterococci, incubating 24–30 hours at 37°. The total bacterial count is made on nutrient cardboard discs containing added triphenyltetrazolium chloride (standard type). Most bacteria break down this redox indicator through ferments to yield yellow to red colonies on the membrane filters after incubation. Coliform bacteria can be cultivated on a nutrient cardboard disc containing tergitol. These bacteria ferment lactose, and incubation, performed ad 37° and 44°, gives light or yellow, or red colonies with a yellow halo, respectively. Dark red colonies without yellow halo cannot be ascribed to coliform bacteria.

Salmonella can be isolated by placing the membrane filters on a nutrient cardboard disc which contains bismuth sulphite, according to the *Wilson-Blair* principle. After incubation for 24–48 hours at 37°, colonies are yielded with a black or brown halo of metallic lustre.

Culture media containing lactose are used in principle for separating bacteria which break down the lactose (predominantly bacteria of the coli-aerogenes group) from those which do not ferment it (*Salmonella* and *Shigella*). These last-named therefore grow on Endo-Leifson or S–S agar as colourless, transparent, somewhat glassy colonies, often surrounded by a transparent zone of nutrient medium. *Salmonella* reduce sodium sulphite contained in *Wilson-Blair* culture media, yielding hydrogen sulphide as end-product of the reduction process. This gives bismuth sulphide which has a black-brown to greenish sheen. All colonies which do not ferment lactose or which reduce sulphite must be taken from the medium for further inoculation and differentiation.

This is done using methods with which medical microbiologists are conversant and which can be further prosecuted only with very difficult serological working procedures. The non-medical man is not so familiar with these methods if only because he rarely obtains the necessary permission according to the federal epidemic decree (Bundesseuchengesetz) to carry out investigations of this sort. There thus seems little point in treating in this contribution specifically medical procedures of detection for differentiating aetiological agents.

In the attempt to detect promoters of disease in water, one always proceeds basically by filtering as large an amount of water as possible through membrane filters and then transferring the filters to an enrichment broth which is more or less specific for the species of bacteria which one is searching for. For example, if there is a suspicion that cholera vibriones in water could have started a cholera epidemic, the membrane filters (in this case with a smaller pore diameter than needed for detection of coli bacteria) are introduced into the officially prescribed enrichment broth of peptone water plus excess alkali, incubated for a suitable period and then transferred to inoculate the special culture media of *Dieudonné* or *Pilon.*

Suspicion that there are cholera vibriones in water is naturally a highly dramatic matter. The doctor who believes himself confronted with this will take his method of handling it from the relevant specialist literature rather than from the author of this section (who can describe it only from the same sources!).

Similar considerations apply to detection of other disease promoters in water, for which there is and can be no generally valid prescription. The knowledge and experience of the doctor must be supplemented here by the art of improvisation.

Much work and trouble may be involved in the individual case of establishing the bacterial count and the coli titre, differentiating the bacteria from the group of enterobactericeae and detecting enterococci and *Clostridia,* regardless of the method to be used. The doctor who carries out the bacteriological investigation and judges the water from the hygienic point of view is spared one worry, namely that someone else is going to take from him the responsibility associated with performing these investigations.

XIV. Bibliography of the Bacteriological Part

As with the chemical bibliography (p. 264), a version with translated titles is given here after the list taken from the German edition.

XIV. Literatur zum bakteriologischen Teil

[1] American Public Health Association: Standard Methods for the examination of water and wastewater. 12. Edition. New York (1966).

[2] Amerikanische Einheitsverfahren zur Untersuchung von Wasser und Abwasser (R. Oldenbourg, München (1951).

[3] *Anonymus:* Zur Frage der Trinkwasserfluoridierung. Neue Deliwa Z. *8,* 248 (1955).

[4] *Anonymus:* Neuere Auffassungen über Filtrationsvorgänge. Gas- und Wasserfach *97,* 503 (1956).

[5] *Anonymus:* Zur Frage der Trinkwasserfluoridierung. Städtehyg. *8/9,* 158 (1958).

[6] *Anonymus:* Dem Lebensrätsel auf der Spur. Dtsch. Ärzteblatt *16,* 1952 (1966).

[7] *Baltaneu, E.,* und Mitarbeiter: Beiträge zum Enterokokkenstudium als Hygieneanzeiger in verschiedenen Wasserquellen. Igiena (Buc.) *14,* 679 (1965), zitiert nach Zbl. Bakt. I. Ref. *205,* 341 (1967).

[8] *Barsow, K. H.:* Zur Methodik der Bestimmung coliformer Bakterien an Membranfiltern (russ.). Z. Mikrobiol. *4,* 42 (1932), zitiert nach *Daubner* [28].

[9] *Bechold, H.:* Kolloidstudien mit den Filtrationsmethoden. Z. phys. Chem. *60,* 257 (1907).

[10] *Beger, H.:* Leitfaden der bakteriologischen Trinkwasseruntersuchungen (Urban u. Schwarzenberg, München und Wien 1946).

[11] *Beger, H.:* Bakterielle Schleim- und Gallertbildung in Trinkwasseranlagen. Gesundh. Ing. *70,* 373 (1949).

[12] *Beger, H.:* Leitfaden der Trink- und Brauchwasserbiologie (Piscator, Stuttgart 1952).

[13] *Beling, A.,* und *Maier, K. H.:* Die Wasserversorgung und ihre Überwachung mit Hilfe der Membranfilter. „Vom Wasser" *XXI,* 118 (1958), Verlag Chemie GmbH Weinheim.

[14] *Bergey, D. H.:* Manual of determinative Bacteriology. 7. Aufl. Balliere, Tindall and Cox. London (1957).

[15] *Borian, A.,* und Mitarb.: Enterocolitisepidemie infolge Verunreinigung des Trinkwassers mit dem Colistamm 0124. Orv. Hetil (Budapest) *100,* 1075 (1959), zitiert nach Zbl. Bakt. I. Ref. *178,* 160 (1961).

[16] *Borneff, J.,* und *Knerr, R.:* Cancerogene Substanzen im Wasser und Boden. Arch. Hyg. *143,* 390 (1959).

[17] *Borneff, J.:* Tagungsbericht. Zbl. Bakt. I. Ref. *201,* 346 (1966).

[18] *Borneff, J.:* Virus im Wasser. Arch. Hyg. 150, *207* (1967).

[19] *Braune, J. F.:* Zur Technik der Differenzierung zwischen Escherichia coli und Aerobakter aerogenes. Zbl. Bakt. I. Orig. *164,* 338 (1955).

[20] *Bucksteeg, W.,* und *Thiele, H.:* Verbesserung der Keimzählungstechnik in der Wasserbakteriologie durch Anwendung von 2,3,5-Triphenyltetrazoliumchlorid. Zbl. Bakt. II. *111,* 6 (1958).

[21] *Büsing, K. H.:* Die Trinkwasseruntersuchung und Begutachtung nach den Grundsätzen der Hygiene. Neue Deliwa Z. *11,* 379 (1959).

[22] Bundesgesundheitsamt: Viruserkrankungen als Gruppenerkrankungen nach dem Baden in Hallenbädern. Arch. d. Badewesens *17,* 92 (1964).

[23] *Burkhard, F.,* und *Legler, F.:* Streptokokken im Wasser als Versagerursache beim Colinachweis mittels des Titrationsverfahrens. Arch. Hyg. *142,* 389 (1958).

[24] *Carlson, S.:* Zur Hygiene der Freibadegewässer und öffentlichen Schwimmbäder. Bundesgesundheitsblatt *9,* 169 (1966).

[25] *Carstens, E.:* Bakteriophages and their possible use in sewage purification. Inst. Sew. Purf. *5,* 467 (1963), zitiert nach Zbl. Bakt. I. Ref. *194,* 346 (1964).

[26] *Cholodny, N.:* Zur Methodik der quantitativen Erforschung des Planktons. Zbl. Bakt. II. *77,* 179 (1929).

[27] *Coetzee, O. J.:* Bakteriologische Kriterien bei Badewässern. Publ. Hlth. in South Africa, Juniheft 1963, zitiert nach Zbl. Bakt. I. Ref. *197,* 552 (1965).

[28] *Daubner, I.:* Membránovè Filtre a ich Pouzitie y Hydrobakteriologii (Vydavateslstvo Slovenskej Akadèmie, Bratislava 1960) u. Manuskript für eine Neuauflage.

[29] *Delfmann, L.:* Tödliche Methaemoglobinämie durch nitrathaltiges Brunnenwasser. Zbl. allg. Path. path. Anat. *100,* 268 (1959).

[30] Deutsche Einheitsverfahren zur Wasser-, Abwasser- und Schlammuntersuchung (Chemie GmbH, Weinheim (1960).

[31] *Diosi, P.,* und Mitarb.: eine durch Shigella bodyii verursachte Wasserepidemie in der Gegend von Tara Birsci. Microbiologica (Bukarest) *7,* 55 (1960), zitiert nach Zbl. Bakt. I. Ref. *179,* 137 (1961).

[32] *Dombrowski, H.:* Das Alter des Lebens. Bild der Wissenschaft *2,* 654 (1965).

[33] *Eikmanns, K. H.:* Salmonellennachweis in Oberflächengewässern. Arch. Hyg. *152,* 350 (1968).

[34] *Favero, M. S., und Mitarb.:* Die Verwendung von Staphylokokken als Indikator für Schwimmbadverunreinigungen. *Pub. Hlth. Rep. (Wash) 79,* 61 (1964), zitiert nach Zbl. Bakt. I. Ref. *197,* 555 (1965).

[35] *Fick, R.:* Über Diffusion: Pogg. Anal. *94,* 51 (1855), zitiert nach *Daubner* [28].

[36] *Gaase, A.,* und *Temme, N.:* Verbesserter Salmonellen-Nachweis im Oberflächenwasser. Z. Hyg. *150,* 286 (1965).

[37] *Gärtner, A.:* zitiert nach *Nietsch, B.:* Ziele und Methoden der bakteriologischen Wasseruntersuchung. Gas, Wasser, Wärme *14,* 76 (1960).

[38] *Gärtner, A.:* zitiert nach *Wüstenberg, J.:* Boden und Wasser in H. Gärtner u. H. Reploh: Lehrbuch der Hygiene (Gustav Fischer, Stuttgart 1964).

[39] *Gärtner, A.,* und *Staag, H. H.:* Über die Bedeutung und Beurteilung des Nitratgehaltes im Trinkwasser. Gesundh. Ing. *81,* 117 (1960).

[40] *Gärtner, H.,* und *Schnurbein, R. Frhr.:* Untersuchungsergebnisse zur Poliomyelitis-Virus-Anreicherung aus Wasser und Abwasser, Arch. Hyg. *148,* 183 (1964).

[41] *Gehrig, A.:* Über die Ergebnisse fünfjähriger bakteriologischer Kontrollen in Mineralwasserbetrieben. Arch. Hyg. *143,* 456 (1959).

[42] *Gottstein, A.:* Die Lehre von den Epidemien (Springer, Berlin 1929).

[43] *Guthof, O.:* Enterokokken als Indikator faecaler Verunreinigungen von Wasser, Kongreßbericht. Zbl. Bakt. I. Ref. *169,* 329 (1958).

[44] *Habs, H.,* und *Langeloh, H.:* Die Bedeutung der sogenannten atypischen Colibakterien für die Trinkwasserbeurteilung. Arch. Hyg. *142,* 476 (1958).

[45] *Habs, H.,* und *Müller, H.:* Zur Differenzierung der coliformen Bakterien bei der Trinkwasseruntersuchung, Arch. Hyg. *144,* 1 (1960).

[46] *Hallmann, L.:* Bakteriologische Nährböden. (Georg Thieme, Stuttgart 1953).

[47] *Hallmann, L.:* Bakteriologie und Serologie (Georg Thieme, Stuttgart 1955).

[48] *Haenel, H.:* Aspekte mikroökologischer Beziehungen des Makroorganismus. Zbl. Bakt. I. Ref. *176,* 1 (1960).

[49] *Harmsen, H.:* Probleme der Westeroder Wasserversorgung, Städtehyg. *2,* 261 (1955).

[50] *Heinrich, S.:* Versager des Colinachweises nach Eijkman. Kongreßbericht. Zbl. Bakt. I. Ref. *169,* 330 (1958).

[51] *Heinrich, S.*, und *Gröschel, D.*: Zur Frage der Bildung eines Oberflächenfilms auf dem Wasser von Schwimmbädern. Zbl. Bakt. I. Orig. *191*, 312 (1963).
[52] *Heneberger, G.*, und *Crodel, B.*: Untersuchungen an Membranfiltern. Zbl. Bakt. I. Orig. *160*, 7 (1953).
[53] *Hettche, O.*: Oberflächenwasser als Trinkwasser. Zbl. Bakt. I. Orig. *160*, 15 (1954).
[54] *Hettche, O.*: Der Kreislauf des Wassers und der Kropf. Städtehyg. *6*, 1 (1955).
[55] *Hugelmann, H.*: Die Fluoridierung des Trinkwassers in Kassel. Gas- u. Wasserfach *94*, 511 (1953).
[56] *Jannasch, H. W.*: Beitrag zur Methode der direkten mikroskopischen Untersuchung von Mikroorganismen auf Membranfiltern: Über die Aufhellung der Filter. Zbl. Bakt. I. Orig. *161*, 225 (1954).
[57] *Jawetz, E.*, *Melnick, J. L.*, und *Adelberg, E. A.*: Medizinische Mikrobiologie (Springer, Berlin u. Göttingen 1963).
[58] *Kathe, J.*: Borellien und Leptospiren in H. Reploh und J. J. Otto: Lehrbuch der medizinischen Mikrobiologie (G. Fischer, Stuttgart 1965).
[59] *Kirchbach, H.*: Unter welchen Umständen kann in Trinkwasserleitungen Schmutzwasser eindringen? Arch. Hyg. *140*, 130 (1956).
[60] *Klosterkötter, W.*: Hygienische Probleme bei der Umwälzung des Badewassers im Schwimmbecken. Arch. d. Badewesens *17*, 108, (1964).
[61] *Knocke, K. W.*: Nachweis von Reo-Viren in Abwässern. Kongreßbericht. Zbl. Bakt. I. Ref. *206*, 543 (1967).
[62] *Knorr, M.*: Zur hygienischen Beurteilung der Ergänzung und des Schutzes großer Grundwasservorkommen. Gas- u. Wasserfach *92*, 104 u. 151 (1951).
[63] *Knorr, M.*: Die Praxis der hygienischen Überwachung und Begutachtung der Trinkwasserversorgung. Zbl. Bakt. I. Orig. *160*, 1 (1954).
[64] *Knorr, M.*, *Haas, F.*, und *Knorr, B.*: Beobachtungen über das Versagen der Biologischen Sperre bei der Entkeimung des versinkenden Wassers. Arch. Hyg. *140*, 130 (1956).
[65] *Knothe, H.*: Über das Vorkommen verschiedener Streptokokkenarten der Gruppe D bei urologischen Infektionen. Arch. Hyg. *150*, 180 (1967).
[66] *Köhler, A.*, und *Loos, W.*: Das Phasenkonstrastverfahren und seine Anwendung in der Medizin. Natuwiss. *29*, 49 (1941).
[67] *Kraus, P.*: Ein handliches, hydromechanisches Probeentnahmegerät zur bakteriologischen Wasseruntersuchung. Zbl. Bakt. I. Orig. *172*, 458 (1958).
[68] *Kruse, H.*: Wasser, Darstellung seiner chemischen, hygienischen, medizinischen und technischen Probleme (Schmorl u. Seefeld, Hannover 1949).
[69] *Kruse, H.*: Das Membranfilter. Gas- u. Wasserfach *93*, 712 (1952).
[70] *Kruse, H.*: Zur Frage der Trinkwasserfluoridierung. Gas- und Wasserfach *94*, 511 (1953).
[71] *Kruse, H.*: Die Typhusepidemie in Hagen/Westf. im Jahre 1955. Gas- u. Wasserfach *99*, 736 (1958).
[72] *Kruse, H.*: Einheitliche Anforderungen an die Trinkwasserbeschaffenheit und Untersuchungsverfahren in Europa. Schriftenreihe des Vereins für Wasser, Boden u. Lufthygiene (Gustav Fischer, Stuttgart 1959).
[73] *Lanyi, B.*, und Mitarb.: Enterocolitisepidemie infolge Verunreinigung des Trinkwassers durch Bact. coli, Serotype 124: 72: 42. Orv. Hetil (Budapest) *100*, 1075 (1959), zitiert nach Zbl. Bakt. I. Ref. *178*, 160 (1961).
[74] *Laun, R. H.*: Das pneumatische Wasserentnahmegerät. Gas- und Wasserfach *92*, 200 (1951).
[75] *Lehmann, K. B.*, und *Neumann, R. O.*: Atlas und Grundriß der Bakteriologie. 4. Aufl. (G. F. Lehmann, München 1907).
[76] *Lenk, V.*, und *Ackermann, H. W.*: Typhus Vi- und Salmonellenaphagen aus Fluß- und Abwasser. Zbl. Bakt. I. Orig. *181*, 224 (1961).
[77] *Lenk, V.*: Über die Bedeutung des Nachweises von Salmonellaphagen und E. coli als Indikatoren für das Vorkommen von Salmonellen. Zbl. Bakt. I. Orig. *200*, 353 (1966).
[78] *Lodenkämper, H.*: Tagungsbericht. Zbl. Bakt. I. Ref. *201*, 346 (1966).

[79] *Lorenz, W.:* Ein einfacher Nachweis von Faecalstreptokokken bei der Trinkwasseruntersuchung. Kongreßbericht. Zbl. Bakt. I. Ref. *169,* 329 (1958).
[80] *Mac Lean, D. M.:* Verunreinigungen von Wasser mit Viren. J. Am. Water Works Ass. *56,* 585 (1964), zitiert nach Zbl. Bakt. I. Ref. *200,* 413 (1965).
[81] *Messerschmidt, Th.,* und *Adler, E.:* Zum Nachweis von Typhus- und Paratyphusbazillen ist der Wismutsulfitagar sehr viel leistungsfähiger als andere Nährböden. Med. Klin. *44,* 834 (1949).
[82] *Mochmann, H. P.:* Das Abwasser als Infektionsquelle. Wasserwirtschaft, Wassertechnik *5,* 187 (1955).
[83] *Mochmann, H. P.:* Leptospiren als Wasser- und Abwasserinfektion. Desinfektion u. Gesundh.wesen *49,* 30 (1957).
[84] *Monnet, P.* und Mitarb.: E. coli Typ 0 111 und 0 55 im Trinkwasser. Ann. Inst. Pasteur *87,* 347 (1954), zitiert nach Zbl. Bakt. I. Ref. *158,* 178 (1955).
[85] *Müller, A.:* Escherichia coli im Kaltblüterdarm. Zbl. Bakt. I. Orig. *162,* 32 (1955).
[86] *Müller, Gertrud:* Der Enterokokkennachweis bei bakteriologischen Wasseruntersuchungen. Zbl. Bakt. I. Orig. *182,* 318 (1961).
[87] *Müller, Gertrud:* Das Vorkommen von agglutinierenden E. coli-Typen in Trink- und Oberflächenwässern. Zbl. Bakt. I. Orig. *203,* 465 (1967).
[88] *Müller, Gertrud:* Der Wert des Kokkennachweises bei gechlorten Wässern. Arch. Hyg. *151,* 402 (1967).
[89] *Müller, R.:* Lehrbuch der Hygiene 2. Auflage (G. F. Lehmann, München–Berlin 1942).
[90] *Müller, R.:* Medizinische Mikrobiologie (Urban und Schwarzenberg, München–Berlin 1952).
[91] *Nikodesmusz, E.,* und *Ormay, L.:* Ergebnisse der bakteriologischen Untersuchungen im Laufe einer durch Trinkwasser verursachten Dysenterie- Epidemie. Arch. Inst. Pasteur Tunis *36,* 43 (1959), zitiert nach Zbl. Bakt. I. Ref. *174,* 287 (1960).
[92] *Oberzill, W.:* Neuere Erkenntnisse auf dem Gebiet der Chemie und Mikrobiologie des Wassers. Scientia pharmazeutica *25,* 89 (1957).
[93] *Perelli, O.:* Vergleichende Untersuchungen über 2 Verschmutzungsindikatoren von Wasser, das zu Trinkzwecken bestimmt ist. Igiene mod. *57,* 89 (1964), zitiert nach Zbl. Bakt. I. Ref. *197,* 540 (1965).
[94] *Pittler, H.,* und *Höpken, W.:* Isolierung von Enteroviren aus Abwasser. Kongreßbericht. Zbl. Bakt. I. Ref. *201,* 336 (1966).
[95] *Popp, L.:* Vorfluter, ein reichhaltiges Salmonellen-Reservoir. Zbl. Bakt. I. Orig. *166,* 90 (1956).
[96] *Popp, L.:* Der Nachweis von Salmonellen im Fluß- und Abwasser. Zbl. Bakt. I. Orig. *189,* 366 (1963).
[97] *Preuß, H.:* Über eine neue Tetrathionatanreicherung. Z. Hyg. *129,* 187 (1949).
[98] *Primavesi, C. A.:* Virologische Untersuchungen von Oberflächenwässern und ihre Ergebnisse. Arch. Hyg. *150,* 196 (1967).
[99] *Pulverer, G.:* Staphylokokken im Freiland. Zbl. Bakt. I. Orig. *199,* 469 (1966).
[100] *Raettig, J.-J.,* und *Hofmann, S.:* Die Vorteile der Farbindikatorplatte in der experimentellen Arbeit mit Bakteriophagen. Zbl. Bakt. I. Orig. *161,* 121 (1954).
[101] *Ramkov, M.:* Über eine Trinkwasserepidemie von Weilscher Krankheit. Z. Hyg. *140,* 556 (1955).
[102] *Regau, Th.:* Menschen nach Maß (Bechtle, München 1965).
[103] *Reploh, H.:* Zur Frage der Verbreitung der spinalen Kinderlähmung durch Abwasser. Wasser u. Boden *5,* 280 (1953).
[104] *Rimpau, W.:* Typhus-Wasserepidemie in Neuötting. Med. Mschr. I. *452* (1947).
[105] *Rische, H., Keiger, G.,* und *Stempel, J.:* Der Einfluß der Selbstreinigung eines Vorfluters auf Bakteriophagen. Gesundh. Ing. *82,* 210 (1961).
[106] *Ruys, Ch.:* Einige hygienische Gesichtspunkte hinsichtlich der Wasserverunreinigung. Water-Boden-Luft *47,* 82 (1957), zitiert nach Zbl. Bakt. I. Ref. *171,* 263 (1959).
[107] *Saad, F.,* und Mitarb.: Eine durch Trinkwasser verursachte Epidemie in der Stadt Satoraljauj-

hely, Volksgesundheitswesen. (ungar.) *43/5,* 145 (1962), zitiert nach Zbl. Bakt. I. Ref. *187,* 238 (1963).
[108] *Sattelmacher, P G.:* Methaemoglobinämie durch Nitrate im Trinkwasser. Schriftenreihe des Vereins für Wasser-, Boden- u. Lufthygiene (Gustav Fischer, Stuttgart 1962).
[109] *Sattelmacher, P. G.:* Gefahren durch Nitrate im Trinkwasser. Gas u. Wasserfach *104,* 1321 (1962).
[110] *Schäfer, W.: Über die Rolle des Colinachweises im Wasser.* Zbl. Bakt. I. Orig. *160,* 1 (1954).
[111] *Schäfer, W.:* Zur Bewertung des Keimgehaltes bei Trinkwasseruntersuchungen. Städtehyg. *16,* 14 (1965).
[112] *Schlemmer, F.:* Befundbericht des Arzneimittelforschungsinstitutes in München.
[113] *Schlierf, K.,* und *Seidenstücker, H.:* Der Wismutsulfitagar zur Züchtung der Typhus-Paratyphusbakterien. Zbl. Bakt. I. Orig. *157,* 356 (1952).
[114] *Schoop, G.:* Teichwasser als Seuchenverbreiter. Dtsch. tierärztl. Wschr. *57* (1950).
[115] *Schmidt, H.:* Grundlagen der spezifischen Therapie (Bruno Schulz, Berlin–Grunewald 1940).
[116] *Schubert, R.:* Die coliformen Bakterien in der Wasserdiagnostik und ihre Differenzierung. Z. Hyg. *142,* 476 (1955).
[117] *Schwarz, W.:* Wasserbakteriologie – Probleme und Möglichkeiten. Gas- u. Wasserfach *95,* 443 (1954).
[118] *Seeberger, H.:* Der Wasserversorgungsingenieur und die wiederholten Typhusepidemien in Neuötting. Gas- u. Wasserfach *90,* 587 (1949).
[119] *Seelemann, M.:* Biologie der Streptokokken (Hans Carl, Nürnberg 1954).
[120] *Seeliger, H. P. R.:* Salmonellosen in der Deutschen Bundesrepublik und Westberlin 1956 Übersichtsreferat der deutschen Salmonellazentrale. Zbl. Bakt. I. Ref. *169,* 193 (1958).
[121] *Seeliger, H. P. R.,* und *Werner, H.:* Hygiene der Heilwässer, Mineralwässer und Limonaden, Z. Hyg. *147,* 470 (1961).
[122] *Selenka, F.:* Weitere Untersuchungen zur Bestimmung des Alters von faecalen Verunreinigungen im Wasser. Arch. Hyg. *150,* 660 (1967).
[123] *Simon, C.:* Nitratgehalt des Trinkwassers und Möglichkeiten zur Vorbeugung der alimentären Methaemoglobinämie. Kongreßbericht. Zbl. Bakt. I. Ref. *201,* 346 (1966).
[124] *Slanetz, L. W., Bent, O. F.* and *Bartley, C. H.:* Use of membrane filter technique to numerate enterococci in water. Public Health Reports *70,* 1 (1955), zitiert nach *Daubner* [21].
[125] *Slanetz, L. W.,* und *Bartley, C. H.:* Ermittlung und seuchenhygienische Bewertung der faecalen Streptokokken im Wasser. Am J. Publ. Hlth. *54,* 609 (1964) zitiert nach Zbl. Bakt. I. Ref. *200,* 412 (1965).
[126] *Spitta, D., Ohlmüller-Spitta:* Untersuchung und Beurteilung des Wassers und Abwassers (Springer, Berlin 1931).
[127] *Steiniger, F.:* Aus der Biologie der T.P.E.-Keime im Abwasser und abwasserverunreinigtem Vorfluter. Der prakt. Desinfektor *45,* 287 (1953).
[128] *Steiniger, F.:* Praktische Gesichtspunkte zur Infektionsgefahr mit T.P.E.-Keimen aus dem Wasser. Desinfektion u. Gesundh.wesen *47,* 116 (1956).
[129] *Sturdza, S. A.,* und *Russu-Pandelescu, M.:* Über die Bedeutung des Bakteriophagennachweises bei Gewässeruntersuchungen. Städtehyg. *8,* 81 (1958).
[130] *Thimann, K. V.:* Das Leben der Bakterien (Gustav Fischer, Jena 1964, Deutsche Übersetzung).
[131] *Thon, D.,* und *Behling, A.:* Beitrag zur Methodik bakteriologischer Flußwasseruntersuchungen. Zbl. Bakt. I. Orig. *176,* 542 (1959).
[132] *Teerhaag* und *Eyer, H.:* Nitrate im Trinkwasser. Der öffentl. Gesundheitsdienst *20,* 1 (1958).
[133] *Uecker, W.:* Membranfilter in der Bakteriophagenfiltration. Zbl. Bakt. I. Orig. *168,* 194 (1957).
[134] *Ulbricht, A.:* Dyspepsiecolivorkommen in der Steiermark. Kongreßbericht. Zbl. Bakt. I. Ref. *201,* 346 (1966).
[135] *Weber-Schutt, G.:* Der routinemäßige Nachweis von E. coli bzw. coliformen Bakterien im Wasser. Zbl. Bakt. I. Orig. *194,* 245 (1964).

[136] *Wedemeier, F. W.:* Methaemoglobinämie des jungen Säuglings durch nitrathaltiges Brunnenwasser. Arch. Kinderheilkunde *152,* 267 (1956).
[137] Weltgesundheitsorganisation: International Standards For Drinking Water. 2. edition W.H.O. Genf, 1963.
[138] *Werner, U., Thal, W.,* und *Wuttke, W. D.:* Schwerste und letal verlaufende Methaemoglobinämien durch nitrathaltiges Brunnenwasser bei jungen Säuglingen. Deutsch. med. Wschr. *90,* 124 (1965).
[139] *Wolters, N.,* und *Schwarz, W.:* Untersuchungen über Vorkommen und Verhalten von Mikroorganismen in reinen Grundwässern. Arch. Hydrobiol. *51,* 500 (1956).
[140] *Wüstenberg, J.:* Entwicklung von Verfahren zum Nachweis pathogener Viren im Abwasser. Forschung im Geschäftsbereich des Bundesministers für Gesundheitswesen, Bonn 1969.
[141] *Zeißler, J.:* Die Gasödeminfektion des Menschen in Kolle, Kraus, Uhlenhut: Handbuch der pathogenen Mikroorganismen. Bd. IV, 2 (G. Fischer, Jena 1931).
[142] *Zimmermann, W.:* Zur Methaemoglobinämie. Kongreßbericht. Zbl. Bakt. I. Ref. *169,* 339 (1958).
[143] *Zimmermann, W.:* Zur Hygiene der Mineralwasserfabrikation. Arch. Hyg. *142,* 609 (1958).
[144] *Heinrich, S.,* und *Pulverer, G.:* Ein Beitrag zur Methodik des Salmonella-Nachweises im Abwasser, Flußwasser und Schlamm. Z. Hyg. *145,* 529 (1959).

Translated Titles

1. American Public Health Association: Standard Methods for the examination of water and wastewater. 12. Edition. New York (1966).
2. American Standard Methods for Investigation of Water and Wastewater (evidently German translation, published in 1951 by R. Oldenbourg of Munich)
3. Anonymous: Concerning the question of fluoridation of drinking water
4. Anonymous: Newer ideas about filtration processes
5. Anonymous: Concerning the question of fluoridations of drinking water
6. Anonymous: On the track of the mystery of life
7. Contributions to the study of enterococci as indicators of sanitary conditions in various sources of water
8. On methods of determining coliform bacteria on membrane filters (quoted from *Daubner* [*28*])
9. Colloid studies with filtration methods
10. Guide to bacteriological investigations of drinking water
11. Bacterial formation of slime and gel in drinking water plants
12. Guide to the biology of drinking and service water
13. Water supply and its control with the help of membrane filters
15. Enterocolitis epidemic as a result of contamination of drinking water with the coli strain 0124 (abstract from Zbl. Bakt. I., Ref. *178,* 160 1961)
16. Carcinogenic substances in water and soil
17. Conference reports
18. Virus in water
19. Concerning the technique of distinguishing *Escherichia coli* and *Aerobacter aerogenes*
20. Use of 2, 3, 5 -triphenyltetrazolium chloride for improving bacterial counting in water bacteriology
21. Investigation and expert judgment of drinking water on a hygienic basis
22. Federal German Health Office: Collective infection through viruses after bathing in closed baths
23. Streptococci in water as the reason for failure of coli detection through the titration procedure
24. Concerning the hygiene of water in open-air and public swimming baths
25. *Carstens, E.:* Bakteriophages and their possible use in sewage purification. Inst. Sew. Purf. *5,* 467 (1963), zitiert nach Zbl. Bakt. I. Ref. *194,* 346 (1964).

26. Concerning the methods of quantitative research on plankton
27. Bacteriological criteria for baths water
28. Membrane filters and their application in water bacteriology (original in Slovak)
29. Fatal methaemoglobinaemia through nitrate-containing well water
30. German Standard Procedures for investigation of water, wastewater and sludge
31. An epidemic in the region of Tara Birsci (Rumania), caused by *Shigella bodyii* in water: abstract from Zbl. Bakt. I. *179,* 137 (1961)
32. The length of existence of life
33. Detection of *Salmonella* in surface waters
34. The use of staphylococci as indicators of contamination of swimming baths
35. Concerning diffusion; quoted from *Daubner* [*28*]
36. Improved detection of *Salmonella* in surface water
37. Aims and methods of bacteriological investigation of water; quoted from Nietsch, B.
38. Soil and water, in "Lehrbuch der Hygiene", from *H. Gärtner* and *H. Reploh*; quoted from *Wüstenberg, J.*
39. Concerning the significance and assessment of the nitrate content of drinking water
40. Results of investigation of enrichment of poliomyelitis viruses from water and wastewater
41. On the results of five years of bacteriological control in mineral water organisations
42. Experience gained from epidemics
43. Enterococci as an indicator of faecal pollution of water (conference report)
44. The significance of the so-called atypical coli bacteria for the assessment of drinking water
45. Concerning differentiation of coliform bacteria in the investigation of drinking water
46. Bacteriological culture media
47. Bacteriology and serology
48. Aspects of microecological relations of the macroorganism
49. Problems of the Westeroder water supply
50. Failures in Eijkman's coli detection (conference report)
51. On the question of formation of a surface film on the water in swimming baths
52. Investigations on membrane filters
53. Surface water as drinking water
54. Water circulation and goitre
55. Fluoridation of drinking water in Kassel
56. Contribution to the method of direct microscopic investigation of microorganisms on membrane filters: on the transparence of filters
57. Medicinal microbiology
58. *Borellia* and *Leptospirae,* in: "Lehrbuch der medizinischen Mikrobiologie" of *H. Reploh* and *J. J. Otto*
59. Under what conditions can dirty water enter drinking water conduits?
60. Sanitary problems in the circulation of baths water in swimming pools
61. Detection of reo-viruses in wastewaters (conference report)
62. On the hygienic assessment of the supplementation and protection of large ground water occurrences
63. The practical side of the hygienic supervision and expert assessment of drinking water supply
64. Oberservations on the failure of the biological barrier during sterilisation of percolating water
65. Concerning the occurrence of various streptococcal species of Group D in urological infections
66. The phase contrast procedure and its application in medicine
67. A handy hydromechanical sampling apparatus for bacteriological investigation of water
68. Water, presentation of its chemical, hygienic, medicinal and technical problems
69. The membrane filter
70. Concerning the question of fluoridation of drinking water
71. The typhoid epidemic in Hagen/Westphalia in 1955
72. Standard requirements of drinking water nature and procedures of investigation in Europe

73. Enterocolitis epidemic as a result of contamination of drinking water through *B. coli,* serotypes 124:72:42; abstract from Zbl. Bakt. I. *178*, 160 (1961)
74. The pneumatic water sampling apparatus
75. Atlas and compendium of bacteriology
76. Typhoid Vi- and *Salmonella* phages from river and wastewater
77. Concerning the importance of the detection of *Salmonella* phages and *E. coli*, as indicators for the occurrence of *Salmonella*
78. Conference reports
79. A simple detection of faecal streptococci in investigation of drinking water (conference report)
80. Contamination of water with viruses
81. On the detection of typhoid and paratyphoid bacilli in the bismuth sulphite agar, much more efficient than other culture media
82. Wastewater as a source of infection
83. *Leptospirae* as water and wastewater infection
84. *E. coli,* type 0111 and 055 in drinking water (original in French), quoted from Zbl. Bakt. I, Ref. *158,* 178 (1955)
85. *Escherichia coli* in the intestine of cold-blooded creatures
86. Detection of enterococci in bacteriological investigation of water
87. The occurrence of agglutinating *E. coli* types in drinking and surface waters
88. The value of detection of cocci in chlorinated waters
89. Textbook of hygiene, 2nd edition
90. Medicinal microbiology
91. Results of bacteriological investigations carried out during a dysentery epidemic caused by drinking water, quoted from Zbl. Bakt. I., Ref. *174,* 287 (1960)
92. Newer knowledge in the domain of chemistry and microbiology of water
93. Comparative investigations of 2 pollution indicators for water destined for drinking purposes (original in Italian), quoted from Zbl. Bakt. I., Ref. *197,* 540 (1965)
94. Isolation of enteroviruses from wastewater (conference report)
95. Wastewater, an abundant Salmonella reservoir
96. The detection of *Salmonella* in riverwater and wastewater
97. Concerning a new tetrathionate enrichment
98. Virological investigations of surface waters and their results
99. Staphylococci out of doors
100. The advantages of colour indicator plates in experimental work with bacteriophages
101. Concerning an epidemic of Weil's disease through drinking water
102. Humans made to measure
103. On the question of propagation of poliomyelitis through wastewater
104. Typhoid water-borne epidemic in Neuötting, Bavaria
105. The influence of self-purification of a receiving water on bacteriophages
106. Some hygienic points of view regarding water purification, quoted from Zbl. Bakt. I., Ref. *171,* 263 (1959)
107. An epidemic in the town of Satoraljaujhely, Hungary, due to drinking water, (original in Hungarian), quoted in Zbl. Bakt. I., Ref. *187,* 238 (1963)
108. Methaemoglobinaemia through nitrates in drinking water
109. Dangers due to nitrates in drinking water
110. Concerning the part played by detection of coli in water
111. On evaluation of the bacterial count in investigations of drinking water
112. Report on the findings of the Institute for Research on Pharmaceuticals in Munich
113. Bismuth sulphite agar for cultivation of typhoid and paratyphoid bacteria
114. Pond water as a spreader of epidemics
115. Fundamentals of specific therapy
116. Coliform bacteria in water diagnostics and their differentiation
117. Water bacteriology – problems and possibilities

118. The water supply engineer and the repeated typhoid epidemics in Neuötting, Bavaria
119. Biology of the streptococci
120. *Salmonella* in the Federal German Republic and West Berlin in 1956. Reviewing report of the German Salmonella Centre
121. Hygiene of medicinal waters, mineral waters and lemonades
122. Further investigations of the determination of the age of faecal contaminants of water
123. The nitrate content of drinking water and ways of preventing alimentary methaemoglobinaemia (conference report)
124. *Slanetz, L. W., Bent, O. F.* and *Bartley, C. H.:* Use of membrane filter technique to numerate enterococci in water. Public Health Reports *70*, 1 (1955), zitiert nach *Daubner* [21].
125. Detection and sanitary significance of faecal streptococci in water
126. Investigation and assessment of water and wastewater
127. From the biology of *Salmonella* bacteria in wastewater and receiving waters contaminated with wastewater
128. Practical aspects of the danger of infection by *Salmonella* bacteria from water
129. On the significance of detection of bacteriophages in investigations of waters
130. The life of bacteria
131. Contribution to the methods of bacteriological investigation of river water
132. Nitrates in drinking water
133. Membrane filters in the filtration of bacteriophages
134. Occurrence of *Dyspepsia coli* in Steiermark, Austria (conference report)
135. The routine detection of *E. coli* or coliform bacteria in water
136. Methaemoglobinaemia of young babies through nitrate-containing well water
137. Worldheathyorganisation: International Standards For Drinking Water. 2. edition W.H.O. Genf, 1963.
138. Very serious and fatal methaemoglobinaemia of young babies through nitrate-containing well water
139. Investigations of the occurrence and behaviour of microorganisms in pure ground waters
140. The development of procedures for detecting pathogenic viruses in wastewater. Research in the province of the Federal Minister of Health, Bonn, 1969
141. Gas gangrene infection of humans in Kolle, Kraus, Uhlenhut, Handbook of pathogenic micro-organisms, Vol. IV, 2
142. Concerning methaemoglobinaemia (congress report)
143. On the hygiene of manufacture of mineral water
144. A contribution to methods of *Salmonella* detection in wastewater, river water and mud.

Part 3

The Biology of Water
by
Dietrich *Lüdemann*

I. Biological Water Investigations

In accordance with the layout of the book, the following pages will deal principally with the biological water analysis of receiving, drinking and service water; with internal interferences with water supply plants which are of biological origin; and with methods for their recognition and presently available ways of eliminating them.
Biological examinations are just as indispensable as chemical and bacteriological analyses for the complete assessment of a water, whether it is a receiving, drinking or service water. The opinion has often been advanced that biological investigations in water works, especially of pure water tanks etc., are wholly superfluous, since a regularly conditioned drinking water should contain no plant or animal organisms. It has been repeatedly confirmed in practice, however, that even in the most technically modern water works, microorganisms occasionally survive all the newest processing procedures and gain access to the "pure water side" of a water works.
It has been known for a long time that numerous living creatures encounter adequate conditions for life in the water mains system. This applies to certain bacterial zoogloea, iron- and manganese bacteria and fungi; and also to resistant algae, small aquatic crustaceans, worms (nematodes), rotifers, water-lice, etc.
Thienemann [1926] and *Liebmann* [1951] have summarised the possibilities and limitations of biological water analysis. According to this, biological water analysis is always applicable, whether harmful wastewaters are just being released or not. The living community of plants and animals yields an average picture of the nature of the water which has been flowing through for a longer period. It may sound paradoxical, but the quality of a water can be judged even when it is no longer encountered in the investigation. In contrast, the chemist, for example, must sample the wastewater surge directly.
This is also the reason why the biologist obtains an average value from one investigation in most cases. He can then profess an opinion, whereas the chemist and bacteriologist are dependent on sampling several times. A water can therefore be assessed biologically more quickly than chemically or bacteriologically – and, incidentally, more cheaply!
The biological water analysis permits demonstration of the presence of matter which is harmful to organisms, but is scarcely able to provide information about the nature and composition of this matter. So far, biological reference organisms or indicators are known for only a few materials, e.g. hydrogen sulphide, iron, common salt. Chemical analysis must therefore be employed here to identify the particular harmful material.
Further, biological analysis can never furnish quantitative values for the amounts of materials in question; the chemist is needed here, too.
May these few points suffice for our purpose.

II. Methods

At the moment, efforts are being made (in W. Germany) to choose the best and most useful of the many procedures, as "biological standard methods". This work is being done by a specialist commission, to which representatives from applied research belong.
The methods chosen will then be declared as binding and for general use, just as the chemical and bacteriological standard procedures have been for years.
The work of the commission has not yet been concluded, unfortunately. Merely a few of the physiological procedures in water investigation have appeared up to now. For this reason, only the most necessary apparatus and the absolutely indispensable examinations and ecological procedures for sampling and evaluating the results are mentioned below. The "biological standard procedures" will thus have to wait until the next edition.

A. Sampling

A biological water investigation is not committed to a fixed working plan, as is chemical analysis. Sampling, etc is thus governed largely by the problem posed. The investigation of biologically caused interferences in a water works naturally proceeds differently from, say, a wastewater control.
To the biologist undertaking the investigation should be left the decision where to take the samples and which biological material is most suitable for the examination; appropriate sampling is vital for the value of the biological analysis.

1. Biological Investigation of Waters

A thorough study of the documents and other papers before sampling usually provides the biologist with considerable data about the local circumstances at the place of investigation. Geological and topographical maps, for example, can help a great deal also in setting up a plan for investigation. Nevertheless, even the experienced biologist will always tour the stretch of water to be investigated, taking his time, so as to become acquainted with all local factors which can influence sampling and which appear neither on maps nor in the documents (influxes of wastewater and rain water, disposal of rubbish in the vicinity, deposits on the banks, to mention only a few examples).
The influence of weather has also to be considered: The samples should be taken at the normal water level, and not at high or low water. The season plays a significant part too. The

amount and the composition ot the plankton in summer differ from those in winter, to take the simplest example.

In order to judge the self-purifying capacity of receiving waters it is often extremely important to investigate the biological community at the most unfavourable time of the year, i.e., in late summer to autumn at low water, or in winter.

Depending on local circumstances, the first samples in a *study of receiving waters* are taken about 50–100 yards above the inlet point of the relevant wastewaters, so as to gain a picture of the uninfluenced biological community for comparison purposes.

Special attention is to be devoted to the influx spot itself and its immediate environs. Here are often found, for example, considerable fungus, augmented algae growth, putrefying banks of sludge, coloration and clouding of the receiving water, etc. The second sample is taken here so that any suspended living or non-living matter carried by the wastewaters may be sieved off with a plankton net and then examined.

Subsequent samples are taken at about 100 yards below the influx point, in order to determine the damage to organisms of the receiving water, by the wastewaters or any of their poisonous components. Depending on the size of the receiving water, further samples may be needed further downstream. This enables determination of the stretch of water up to the point where the contamination is no longer noticeable and the water has regained its normal composition.

The biologist will generally find the following samples to suffice in the case just described:

1. *Sievable suspended matter,* best taken from the middle of the stream. These samples contain plankton (plant and animal organisms freely suspended in the water) and non-living matter (e.g. detritus, residual plant tissue, remains of animals, turbidity-provoking matter of mineral origin, etc.) (Fig. 1)
2. *Growths* from the riparian zone, taken from higher water plants of the riparian vegetation, or from stones, posts, landing places, quay walls etc. At the same time te bank flora and the macroscopically detectable animals are to be noted so as include this later in the worked-out report on the results.
3. *Bottom samples* from the bed of the stream, for investigation of the sediment (sand, gravel, mud, etc.) and the organisms living in it.

B. Apparatus

In recent times, the methods of investigation have been refined and perfected through the development of new instruments. Accordingly there are available to the hydrobiologist numerous technically flawless devices for obtaining these samples. The practical worker will generally be content with more simple apparatus, however.

The methods given below are recommended for sampling the suspended matter (plankton and non-living matter) in water:

The simplest procedure for obtaining suspended matter is sampling with water drawing bottles, which are provided in various constructional forms by the relevant firms, or with a water pump.

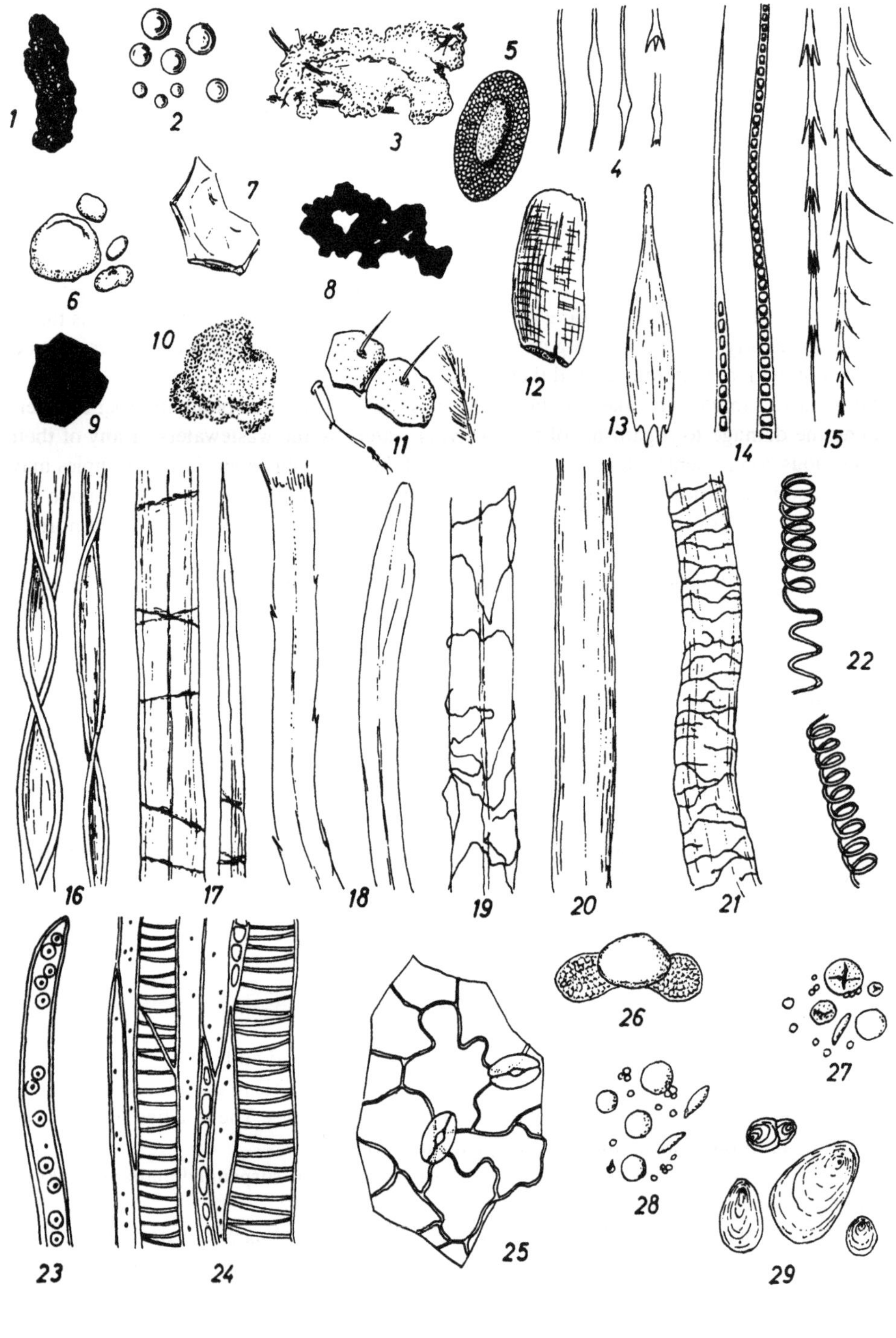

Among the advantages of both these methods is the fact that there is a direct relation between the volume of water drawn and the amount of plankton contained in it.
Only a few of the large number of water drawing vessels can be given here; they suffice completely for our purpose.

1. *Gad's* water drawing bottle

This simple apparatus is designed for water samples from modest depths. It has only a small diameter and so can be used easily for taking samples from water works wells, pipe lines, etc. (Fig. 2, 3)

2. *Ruttner's* water drawing vessel

This is probably the best known water sampler and enables samples to be taken also from greater depths of lakes. A thermometer inside the glass or plastic container indicates the water temperature at the same time. This apparatus is also used considerably by chemists, e.g. for samples from various depths for oxygen determination (Fig. 2, 4)

3. Horizontal water sampling bottle of *Jaag, Ambühl* and *Zimmermann*

This was developed especially for sampling flowing water. It can be used also for obtaining plankton samples from shallow brooks (Fig. 2, 6)
After sampling, the water is filled into suitable glass vessels (wide-necked bottles etc.). They are allowed to stand so that solid matter (including motile organisms which are killed off with a fixation agent such as 3% formalin solution) can settle. They are then transferred to glass slides by means of a pipette and examined microscopically.
A *disadvantage of the water drawing method of sampling* is stated to be the too small amounts of water; waters containing little suspended matter or a low plankton density thus yield a false "total biological picture".
These shortcomings can be eliminated for example by using a water pump which can withdraw as much water as desired from any depth. Electrically driven vane-type pumps are recommended, which exhibit uniform suction action and on which flow gauges show the amount of pumped water.
The water is then filtered with the help of a plankton net or suitably fine gauze. The plankton obtained in this way is fixed, as already mentioned, and can then be examined microscopically or kept for treatment later.
Plankton nets, principally of simple nature, are in general use in Germany for obtaining plankton. So-called "quantitative plankton nets" and "traps" are much used in special investigations in lakes and reservoirs (Fig. 2,1).
The simple net consists of a cone-shaped cut bag, ca. 35 cm (14") long, made of silk gauze, Perlon or Nylon. They are commercially available in various mesh sizes and grades. Thus

←

Fig. 1 Non-living consituents in water. 1. Fragment of manganese. 2. Oil drops. 3. Organic detritus. 4. Hairs from worms inhabiting the water. 5. Statoblasta of moor animals (*Plumatella*). 6. Grains of sand. 7. Chip of stone. 8. Soot. 9. Fragment of coal. 10. Flocculated iron oxide hydrate. 11. Remains of insects. 12. Muscle fibre of meat. 13. Butterfly scale. 14. Rat hairs. 15. Bird feather fragments. 16. Cotton fibres. 17. Linen threads. 18. Hemp fibres. 19. Crude silk fibres. 20. Artifical silk threads. 21. Sheeps' wool. 22. Spiral vessels. 23. Wood fibres of coniferous trees. 24. Wood fibres of decidous trees. 25. Upper skin tissue of a leaf. 26. Pine pollen. 27. Rye starch. 28. Wheat starch. 29. Potato starch.

Drawings: *Kowalsky*, after *Eichler*,
Gassner, *Kolkwitz*, partly originals.

the formerly almost exclusively employed Gauzes No. 25 and 27 have mesh diameters of 75 or ca. 40 µm, respectively.
Present-day nets of Nylon or Perlon gauze have an average mesh diameter of 50 m. The nannoplankton net, developed by *Höll*, is very suitable for investigating waters containing

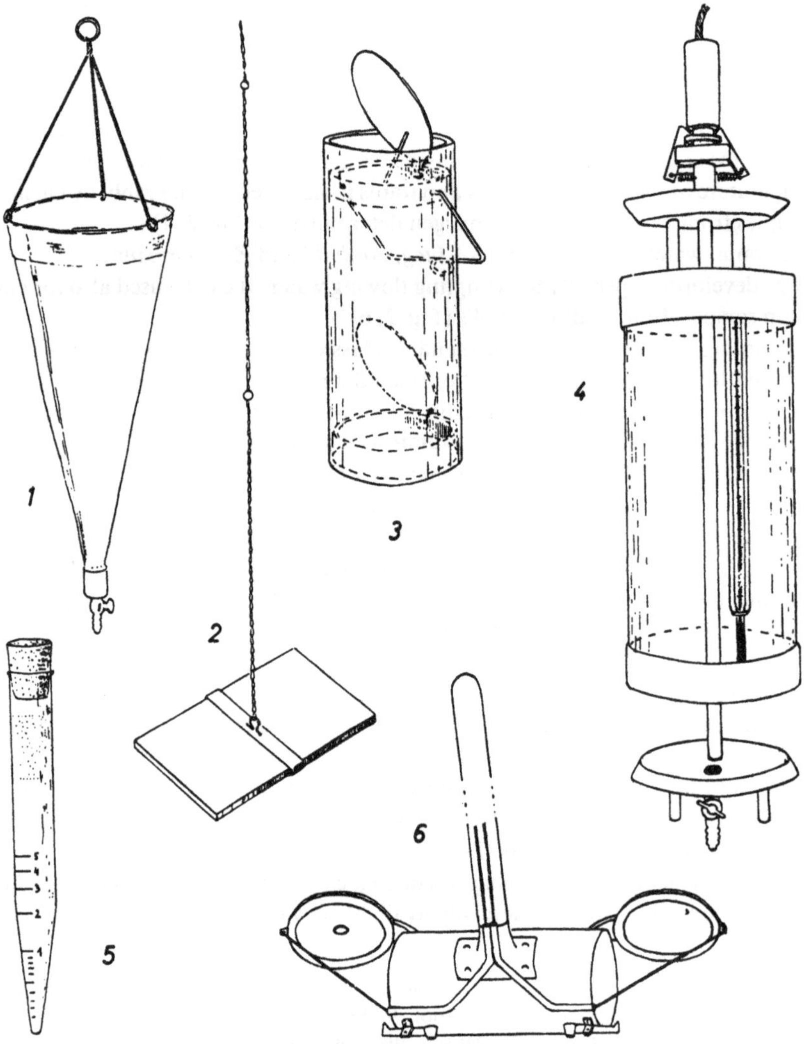

Fig. 2 Devices for sampling plankton. 1. Plankton net. 2. Visibility or sighting disc. 3. Gad's water-drawing bottle. The lid opens when water flows in. The handle is drawn folded down to illustrate the procedure. 4. *Ruttner*'s water drawing vessel. 5. Graduated plankton tube. 6. Horizontal water sampling bottle of *Jaag, Ambühl* and *Zimmermann.*

Drawings: *Kowalsky*, taken from *Liebmann*
partly originals.

little suspended matter and especially for biological control of drinking water. It is made of a new synthetic fibre and has mesh diameters of 28 and 10µm.
The net is drawn through the water for sampling. For more accurate investigations, however, definite amounts of water must be introduced with a vessel of known content. Standardised sampling beakers of 0.5 or, better, 1 litre volume, or 5- or 10-litre buckets are used for this.
25 or 50 litre amounts of raw waters are usually enough but 1 cubic metre (1000 litres, ca 250 gallons) is better for pure water.
After the net is removed from the water, the sieved suspended matter collects in the deposition beaker and is transferred to a graduated plankton glass by opening the tap (pinchcock, metal tap). The amount which has settled out after a definite interval can be read off. If the plankton is not to be examined immediately in its living condition, fixation is carried out as already described (Fig. 2,5).
It would go beyond the scope of this book to treat in more detail the possibilities and limitations of the plankton net method, often discussed in the literature. From reference experiments we know that only a small percentage of the phytoplankton present is retained by the net and that the net method gives no absolute values (*Bursche* [1953] and *Schwoerbel* [1966]). Despite this, experience shows that the method is usually adequate in practice (ascertaining the water quality, control of drinking water etc.).
Auxiliary equipment facilitates handling of plankton nets, sampling vessels and also "palisade scrapers", reed knives and similar apparatus. These can easily be attached with an adapter to an extensible pole, made of several metal parts which can be drawn out like a telescope to attain a length of up to 10 feet.

1. Membrane Filter Procedure

The membrane filter apparatus, present in every bacteriological laboratory today and in continuous use for routine investigations, were originally employed for sterilising liquids. The effective component of these filters is a membrane made from cellulose nitrate. Hydrobiologists have used the apparatus for obtaining sievable suspended matter in water (*Kolkwitz* [1924] and *Beger* [1966], for example).
A simple apparatus of the Membranfilterges. mbH, of Göttingen, made of stainless steel, is particularly good for control of drinking water and investigations of plankton. It comprises a filter plate, tap, stainless steel filter and a funnel-shaped top of 500 ml capacity. About 0.5 litre filters in 3 minutes when a water vacuum punp is fitted to the apparatus.
The Membranfiltergesellschaft mbH of Göttingen makes filters of various types and grades. There are available today for special investigations ultrafine filters with pore diameters of 35 down to 5µm. Coarser membrane filters with pore diameters of 500 to 1000µm suffice for plankton investigations.
Kolkwitz recommends transferring the residue on the membrane filter to a beaker by means of a jet of water or a rubber sheet. A quantitative microscopic investigation is then possible if the amounts of water have been measured. The moist membrane filter can also be inspected in incident or transmitted light with a microscope after placing a cover glass on it.

Investigations comparing the filtration powers of membrane-, sand- and plankton net filtration induced *Heynig* and *Ockert* [1962] to suggest the membrane filter method specially for routine biological control of drinking water–the more so because the fine pores retained very small organisms. The filtrate from one litre of water usually suffices for investigations of this sort.

Accurate values are obtained with the help of the membrane filter method even in quantitative investigations of phytoplankton (*Marvan* 1960). A flawless mode of work is a condition, of course; incorrectly chosen membrane filters with too large a pore size, or too powerful suction during filtration, etc., naturally lead to wrong results.

Sampling of plant growths and their investigation scarcely throws up any problems since quantitative evaluations are probably hardly required in practice.

Knives, scrapers and ladles can be used if necessary to dislodge the animal and plant growth colonies from stones, water frontage buildings, reed shafts and the walls of wells, filter plants and irrigators or percolators. The so-called "*palisade scraper*" finds general use. It consists of a metal frame with protruding metal edge and attached bag of coarse linen. Several models are available commercially (Fig. 3, 1, 2, 3).

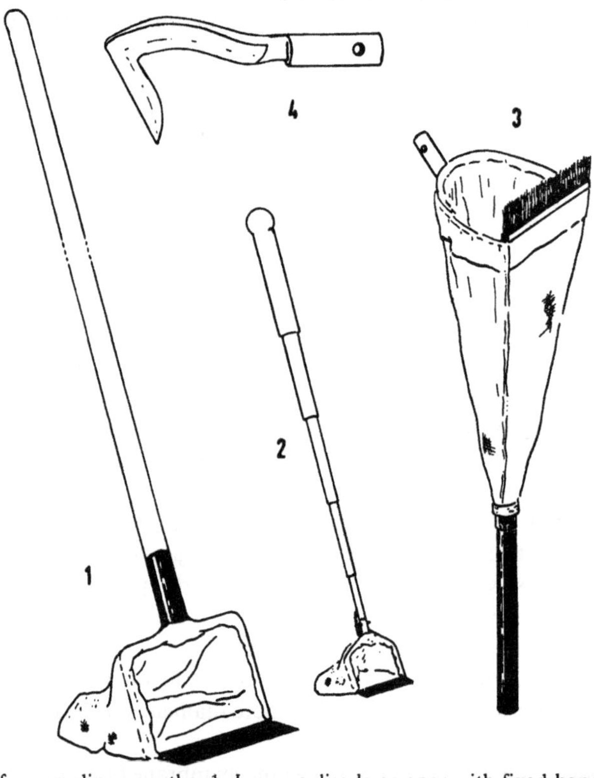

Fig. 3 Devices for sampling growths. 1. Large palisade scraper with fixed handle. 2. Small scraper with iron metal edge and fitment for an extensible handle. 3. Palisade scraper with brush and collecting bag. 4. Reed knife (3 and 4 also with an attachment for telescopic handle).

Drawings: *Kowalsky*, originals.

Aquatic plants can be cut off easily with the "reed knife". It has a sharp, metal blade in the approximate form of a hook and is fixed to a telescopic stick through an adapter (Fig. 3,4).
The collected material is inspected in a shallow dish, on the spot as far as possible, so as to be able to identify delicate organisms which might shrink during fixation. The material is transferred to specimen flasks and is preserved with a few drops of 3% formalin solution for more detailed preparation later.
If there is interest in examination of the living material, samples of growths may be preserved for some time at 4–6° C in a refrigerator without significant alterations occurring in the proportions of the species.

2. Investigations of Bottom Materials and Sludges

For the interests of practice, a number of simple and proved equipments are available for investigation of bottom materials and for their sampling from waters. Methodological difficulties are hardly to be expected, the more so because qualitative investigations predominate in drinking water biology.
Probably the simplest of these devices is the *sludge cup* (*Kolkwitz* [1907]). It can be used almost everywhere. It consists of a metal cup of about 250 cm³ capacity with a weighted bottom which contains a small hole to allow air to escape, and an iron ball which is joined to the side of the cup with two short chains. A ring attached to the iron ball is used for carrying a line. Fig. 4,6 shows a model modified for using a telescopic stick.
Purdy's *mud sampler* is the most suitable apparatus for taking larger samples of bottom materials. It has a cup of about 3 litres capacity. It is the device most used in N. America. The Ekman-Birge *dredge* is used extensively for hydrobiological investigation. It weighs about 5 pounds and has a cross-sectional area of 225–250 cm² (35 to 40 sq. inches). It enables quantitative evaluations of sediment samples to be carried out. The *Ekman* dredge is lowered open to the bottom; its jaws close with the help of a weight device (Fig. 4,2).
This type of dredge works well only in soft bottom deposits, however. If the deposits are hard or stony, the so-called *"mud shears"* of *Liebmann* or the *dredge* of *Van Veen* are better; these heavy devices with levering shear arms on the dredge jaws can attain considerable power (Fig. 4,1).
"Sludge core samplers" or *"sounding tubes"*, available commercially in various forms, are used to obtain samples of undisturbed soil profiles. Examples are the *tube plumps* of *Naumann* and of *Liebmann* and the *core samplers* of *Züllig* and *Ohle* (Fig. 4,3).
The tubes, made of metal, hard glass or plexiglas (closely resembling perspex) in the modern apparatus have holes bored at regular intervals. These are closed with Tesafilm (similar to Scotch tape) before use. After sampling, small samples can then be taken through the holes, corresponding to desired points in the sediment profile. The sludge core sampler of *Ohle,* provided by the firm of Hydrobios, Kiel, even has a plexiglas tube, divided along its length; the two parts are held together by clips with wing nuts. These are loosened after the sample is taken and the one half of the tube is removed; the sample lies in the other half, ready for further investigation.

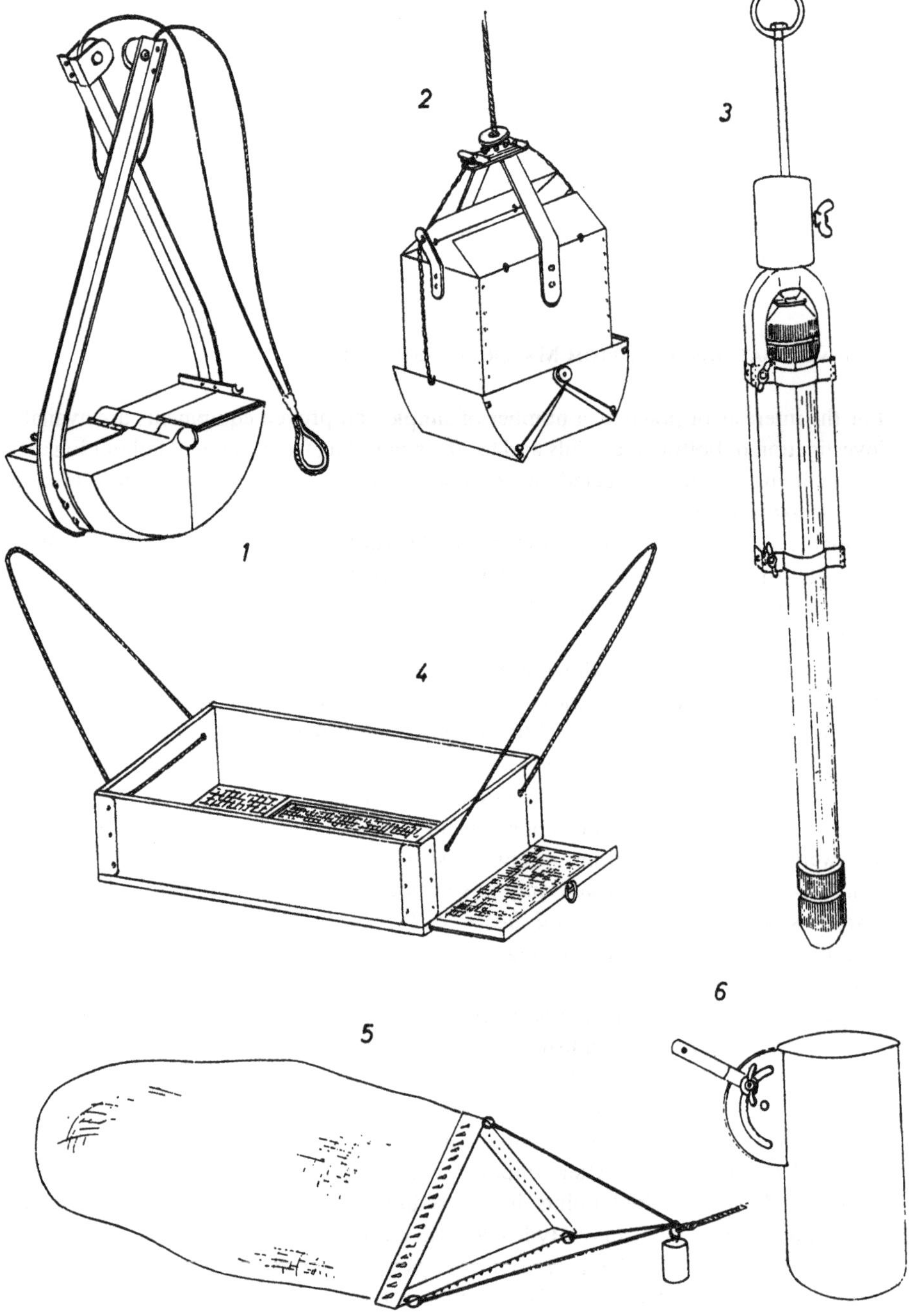
1
2
3
4
5
6

In use for a long time for qualitative purposes have been dredges, consisting of a heavy triangular iron frame on which a coarse gauze bag with a covering protective linen bag are fixed; an example is *Steinmann's dredge net.* Holes in the metal frame or rings attached to it carry a tow line. The foldable constructions of *Kolkwitz, Lundbeck, Fittkau,* etc., with rectangular framework function according to the same principle, like all dredges. They are drawn across the bottom on a long line, thereby sinking through their own weight into the sludge. The upper layer of sediment is then sampled in the net bag (Fig. 4,5).

After the dredge has been taken out, its content can be examined immediately or it can be fixed like all other biological samples, and stored in wide-necked flasks for subsequent study.

The *sludge sieve* of *Kolkwitz* with double sieve bottom (Fig. 4,4) finds considerable use for thorough washing out of the sample of sediment (mud, sand, etc.). It enables all larger organisms inhabiting the bottom to be picked out (Tubificidae, Chironomidae larvae, molluscs, etc.). According to *Schwoerbel* [1966], metal sieves of this sort are unsuitable for soft-bodied species (Turbellariae and certain other species of worm) since they are easily damaged during cleansing from mud. He thus recommends freeing the bottom sample from detritus by a cautious shaking movement of small portions in a bag made of gauze held in water. The organisms retained are either observed on the spot or they can be transported alive in wide collecting tanks for later examination or can be fixed in the usual way.

The organisms ascertained in bottom investigations are quoted either as numbers of each species or group; or as fresh weight of all the individuals of a species or group, per unit surface. Such evaluations seldom enter into consideration, however, for the usual practical investigations of water quality.

It may be stressed once more that the information value of a biological water analysis in an investigation of samples is much greater immediately after they are taken. Unfortunately the microscopic examination on the spot (a good travelling microscope is essential for this) cannot always be carried out as a result of the pressure of time needed for larger investigations of waters. On must remember, however, that many characteristic microorganisms, which are important as indicators for establishing categories of water quality, can scarcely be determined in the fixed state.

C. Ecological Procedures of Biological Water Analysis

Comprehensive presentations are to be found in *Helfer* [1916/1931] and *Liebmann* [1951] by those interested in the historical development of biological water analysis.

The early work of the Breslau botanist *F. Cohn* [1853] who was the first to draw up

←

Fig. 4 Devices for investigating bottom materials and sludges. 1. van *Veen* dredge, closed. 2. *Ekman-Birge* dredge, open. 3. Ohle mud proof stick. 4. *Kolkwitz* sludge sieve. 5. *Steinmann* dredge. 6. *Kolkwitz* sludge cup.

Drawings: *Kowalsky*, originals.

a system of biological water analysis, and that of his pupil *C. Mez* [1898] who drew special attention to the significance of protozoa in assessing a water, have been of particular importance for the development of the saprobic system described below.

1. Saprobic System

When water is contaminated with organic, putrescible matter, characteristic biological communities with typical indicator organisms are formed. In 1902, 1908 and 1909, *Kolkwitz* and *Marsson* published lists of plant and animal species which appeared suitable as "indicator organisms" for assessing the quality of a water. The conditions essential to life of these organisms may be modified in many ways through the pollution load; yet the following factors are decisive for the occurrence and zonal distribution of certain species:

1. Change of the oxygen situation, due to degradation of organic compounds.
2. Change of the chemical composition of the water due to formation of products of putrefaction.
3. Changes in the composition of the species, due to the modified living conditions.

The publications of *Kolkwitz* and *Marsson,* mentioned above, form the basis of the so-called *"saprobic system"* (sapros = decomposition, bios = life). In every assessment, it is not the single indicator form which is decisive, but always the whole biological community.

The principal degradation stages in the self-purification process of the water determine the division of the organisms into pure water and foul water species.

In accordance with the chemical classification of reduction, incipient oxidation and completed oxidation (mineralisation), the biological poly-, meso- and oligosaprobic zones can be distinguished. The mesosaprobic stage is subdivided into an α- and a β-section.

The indicator organisms can, following this classification, be ascribed to one or to two adjacent stages of the four:

1. oligosaprobic stage (scarcely polluted)
2. β-mesosaprobic stage (slightly polluted)
3. α-mesosaprobic stage (strongly polluted)
4. polysaprobic stage (very intensely polluted)

The fact that still today, in improved form, this saprobic system is being used successfully, despite all criticism, is proof that it is based on correct assumptions.

Thus at present the saprobic system drawn up by *Kolkwitz* and *Marrson* and revised by *Liebmann* [1947, 1951], is still in use. The revision was urgently necessary, for in the course of time (notably in recent years) new knowledge about the life requirements of plant and

⟶

Fig. 5 Typical organisms of the oligosaprobic stage. 1. *Cyclotella bodanica* (500). 2. *Tabellaria flocculosa*, a. girdle view. b. view of shell (500). 3. *Micrasterias truncata* (300). 4. *Euastrum oblongum* (250). 5. *Batrachospermum vagum var. ceratophyllum* (8). 6. *Ulothrix zonate* (1000). 7. *Fontinalis antipyretica* (0.5). 8. *Mallomonas caudate* (500). 9. *Strombidinopsis gyrans* (500). 10. *Notholca longispina* (100). 11. *Planaria alpina* (5). 12. *Holopedium gibberum* (12). 13. *Oligoneura rhemana*, larva (3). 14. *Perla bipunctata*, larva (2.5).
(Magnifications given in brackets)

Drawings: *Kowalsky*, after *Hustedt, Lindau-Melchior, Liebmann*, partly originals

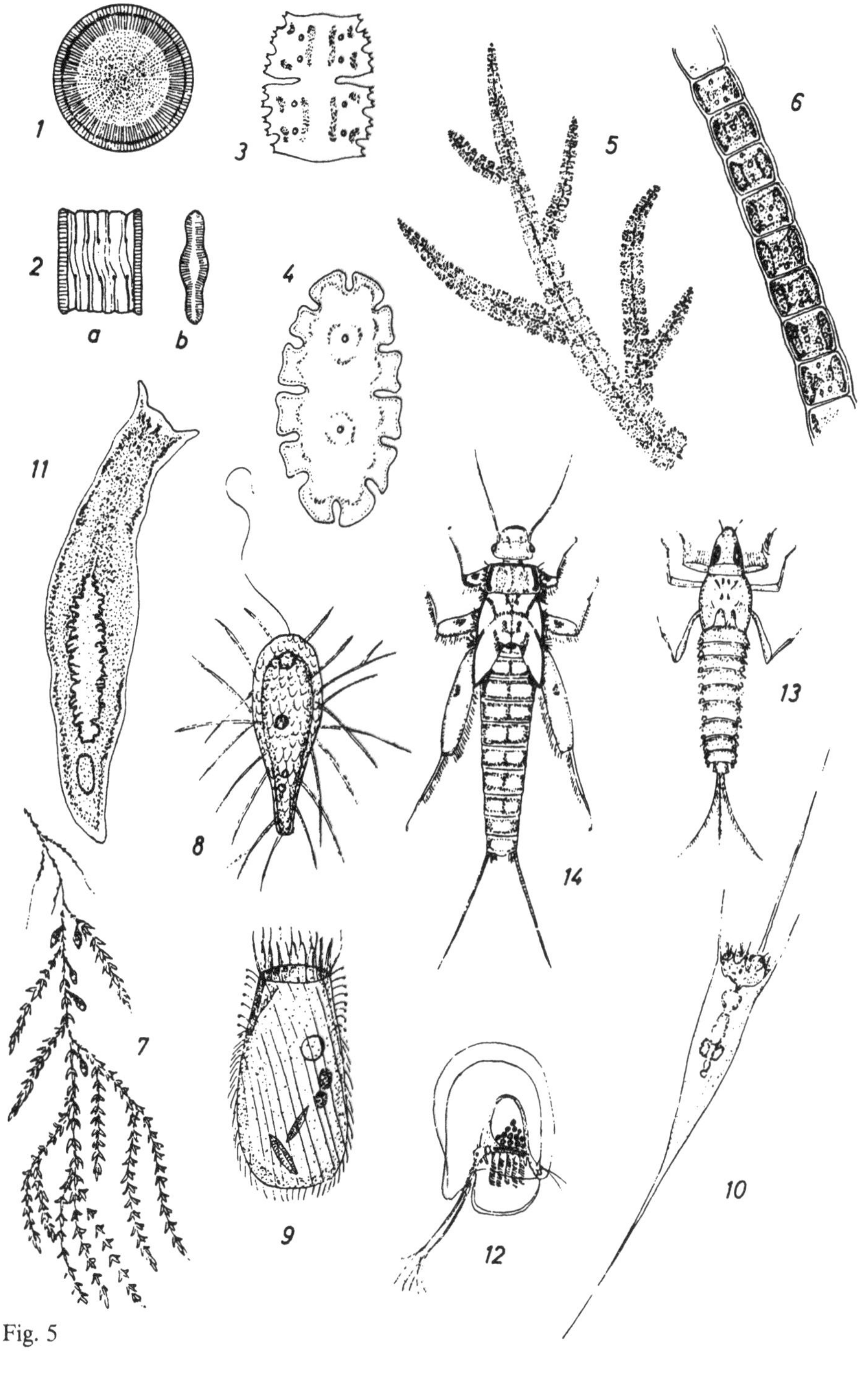

Fig. 5

animal organisms has appeared in numerous publications. *Liebmann* has largely considered all these new data in his *revised saprobic system.* The indicator organisms were chosen with a view to creating a system that could be used in all places and in all waters; *Liebmann* took particular notice of the protozoa, since as cosmopolitans they are suitable for judging waters. The four "classical" stages (oligosaprobic–polysabrobic) are retained in the revised system, although *Liebmann* refers to them as water quality stages I–IV.

2. Indicator Organisms

Since the work of *Kolkwitz* [1950] and of *Liebmann* [1951] contains detailed lists of the plant and animal indicator organisms, the individual stages of the saprobic system are briefly described here in respect of their biological circumstances.

1. Oligosaprobic stage (Water quality class I)

Decrease of the bacterial number to below 100 per ml is one characteristic of the pure water zone. For this reason, organisms which live on bacteria have largely disappeared. Typical, however, is the appearance of certain flat worms (Turbellaria), rotifers, snails, small crustaceans and a multiplicity of insect larvae (Epherimida, Perlidae, Trichoptera, etc.). Principal representatives of the plant organisms are diatoms (*Synedra, Surirella, Tabellaria, Cyclotella,* etc.), desmids (*Euastrum, Staurastrum, Micrasterias,* etc.) and various green and red algae (Fig. 5).

2. β-Mesosaprobic stage (Water quality class II)

The bacterial numbers are less than 10,000/ml of water. Many species of diatoms, Konjugatae and green algae are present and show their widest distribution in this stage. Higher aquatic plants also occur in numerous species. The dinoflagellates predominate among the protozoa. Snails, mussels, rotifers, small crustaceans and insect larvae are present too (Fig. 6).

3. α-Mesosaprobic stage (Water quality class III)

This stage has bacterial numbers of up to 100,000/ml. Among plant organisms predominate blue algae, diatomaceae and green algae and also some higher aquatic plants. A characteristic feature is the so-called "water (or lake) bloom", caused by immense growth of certain microorganisms. Many species of flagellates, ciliates and rotifers are present, but only few species of molluscs, crustaceans and insect larvae (Fig. 7).

⟶

Fig. 6 Typical organisms of the -mesosaprobic stage. 1. *Oscillatoria rubescens* (800). 2. *Melosira varians* (250). 3. *Tabellaria fenestrata* (250). 4. *Asterionella formosa*, a. typical colony (250), b. individual cell (800). 5. *Rhoicosphenia curvata*, a. girdle view, b. cell view (1000). 6. *Pediastrum boryanum* (250). 7. *Scenedesmus quadricauda* (750). 8. *Synura uvella* (300). 9. *Uroglena volvox* (330). 10. *Amoeba proteus* (60). 11. *Brachionus urceus* (30). 12. *Stylaria lacustris* (60). 13. *Daphnia pulex* (12). 14. *Chydorus sphaericus* (25). 15. *Ancylus fluviatilis* (6). 16. *Corethra plumicornis* larva (10).
(Magnifications given in brackets)

Drawings: *Kowalsky*, after *Hustedt, Huber-Pestalozzi, Engelhardt*, partly originals

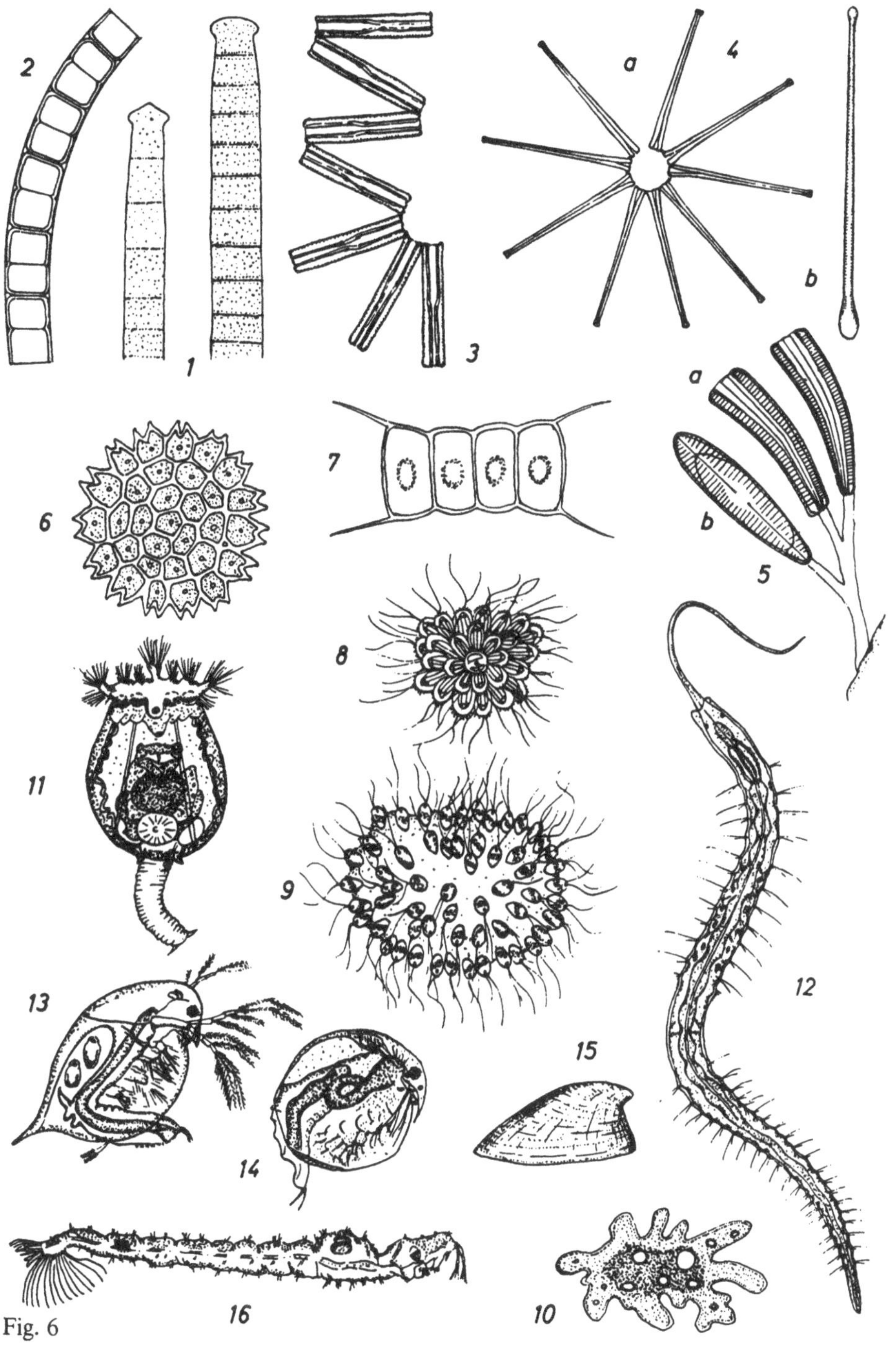

Fig. 6

4. Polysaprobic stage (Water quality class IV)
Typical of this stage are the extensive developments of bacteria. The bacterial numbers exceed 1 million per ml water. Only few species are encountered in this milieu of extreme conditions for life, then usually in very large numbers (flagellates, ciliates, Tubificidae etc.). Absent are diatomaceae, green algae, Konjugatae and higher aquatic plant forms. Likewise, no molluscs and crustaceans are to be found among the animal organisms (Fig. 8).

3. Assessment of the Results

Hitherto, the particular saprobic stage or class of water quality has been established on the basis of the number of indicator organisms, determined by biological-microscopic analysis, and the frequency of their occurrence. This type of assessment always requires an expert biologist with years of practical experience, however.
The lists of organisms, often covering many pages, are completely incomprehensible to the non-biologist. Attempts have thus been made recently by means of mathematical methods and graphs to acquaint those interested in water problems better with the results of the biological water analysis. Diagrammatic presentation as a coloured "map" of water quality has found extensive use also.
Brief reference will now be made to some of the best known procedures of this sort.
Based on his experience over the years, *Beger* [1952] created a type of indicator system which was specially adapted to the needs of water works. He, too, proceeded from *Kolkwitz*'s saprobic system, as is evident from the division into groups given below. He distinguished ten groups of indicator organisms, namely:

1. Pathogenic organisms and their carriers
2. Pure- and foul water indicators
3. Hydrogen sulphide indicators
4. Iron- and manganese indicators
5. Lime indicators
6. Salt indicators
7. Ground water and surface water indicators
8. Cold- and warm water indicators
9. Colour indicators
10. Odour and taste indicators

⟶

Fig. 7 Typical organisms of the -mesosaprobic stage. 1. *Oscillatoria formosa* (1000). 2. *Oscillatoria tenuis* (600). 3. *Cyclotella meneghiniana*, a. view from above, b. side view (1000). 4. *Stephanodiscus hantzschii, a. view from above, b. side view* (1000). 5. *Nitzschia palea* (1000). 6. *Navicula cryptocephala* (1000). 7. *Cosmarium botrytis* (500). 8. *Closterium leibleinii* (500). 9. *Anthophysa vegetans*, part of a colony (500). 10. *Colpoda cucullus* (300). 11. *Stentor coerulus* (100). 12. *Lionotus fasciola* (300). 13. *Charchesium polypinum*, a. structure of a colony (20), b. single individuals (200). 14. *Sphaerium corneum*, a. side view (1), view of a cell half (4). 15. *Stratiomys* species, larva (15). 16. *Asellus aquaticus* (5).
(Magnifications given in brackets)

Drawings: *Kowalsky*, after *Hustedt, Liebmann, Pringsheim*, partly originals

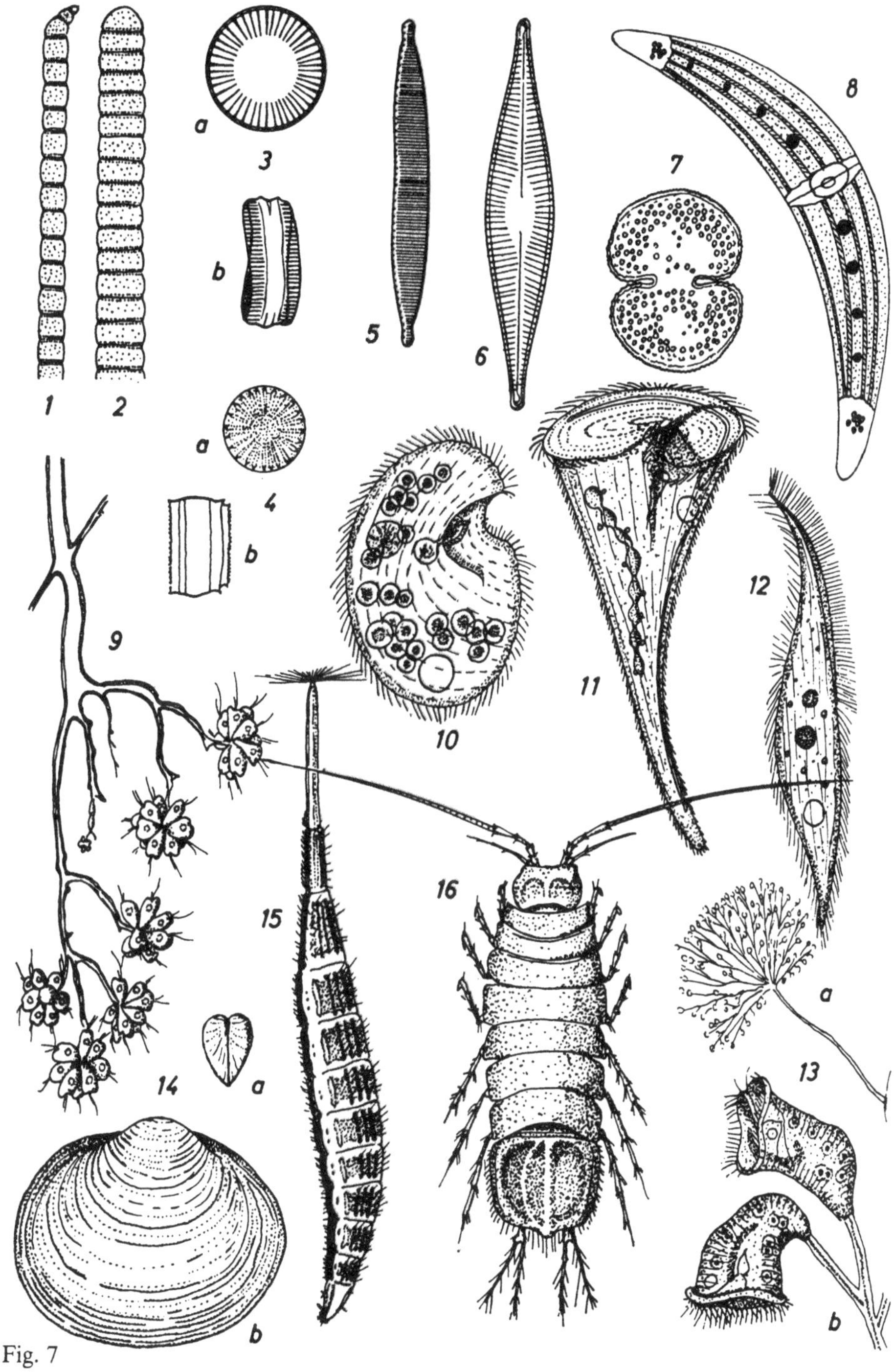

Fig. 7

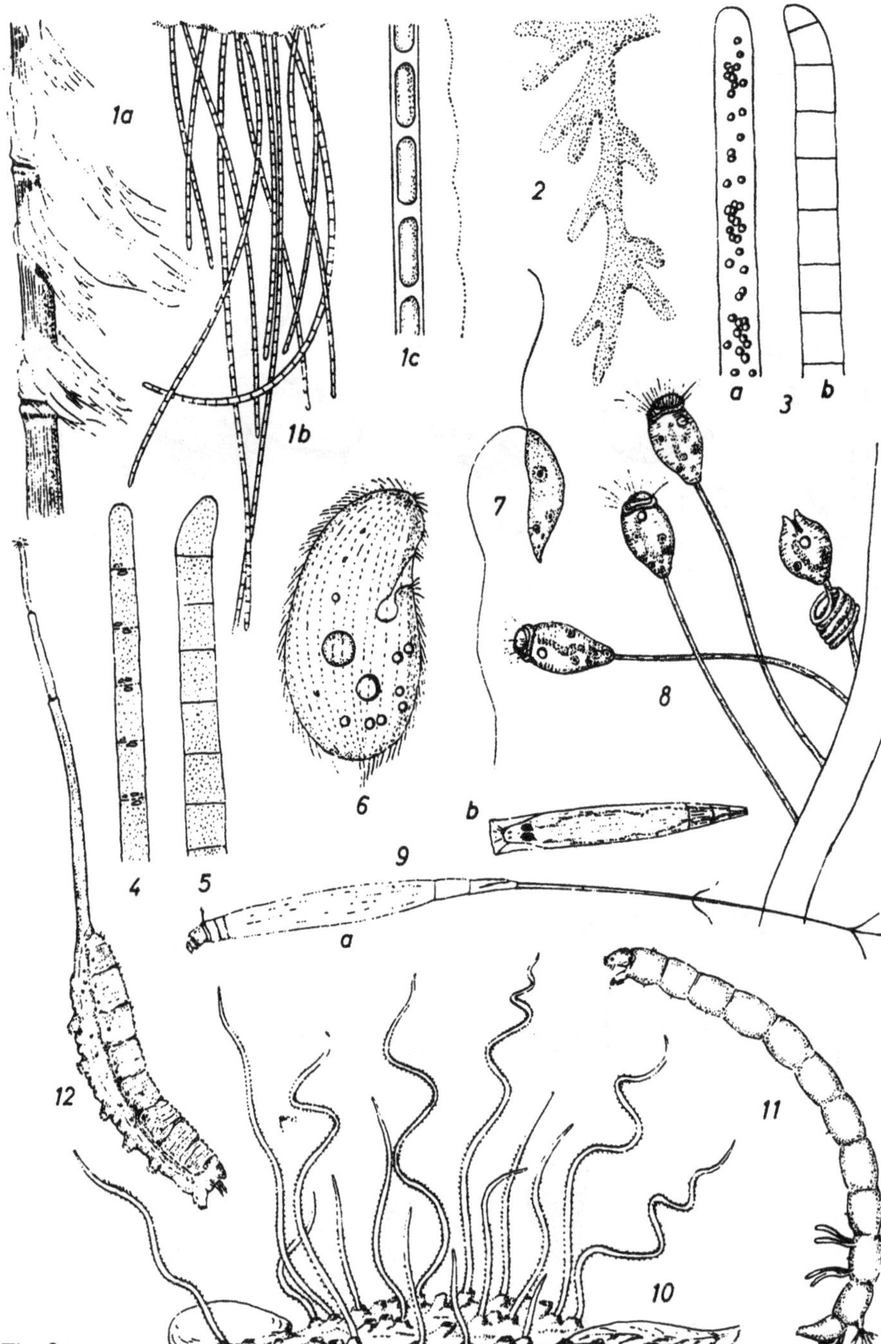

Fig. 8

The procedure of *Knöpp* [1955] is based also on the saprobic system. However, only macroorganisms are used for evaluating the results; their occurrence is calculated with the help of seven frequency degrees. The graph yielded with the help of a coordinate system yields a "longitudinal section of biological quality" of the stretch of water investigated. In this graph, the oligosaprobic and ß-mesosaprobic organisms are plotted above the abscissa and the α-mesosaprobic and polysaprobic indicator organisms are below it.

Kothé [1962] developed a simple method which takes special account of the so-called "species deficit" as an important criterion for investigation of waters. The saprobic system serves here, too, as starting basis, although only macroorganisms are used. The biological condition of the unpolluted water serves as reference standard, the "species standard". The "species deficit" is quantitatively expressed with the formula:

$$x\,\% = \frac{(A_1\text{-}A)\,.\,100}{A_1}$$

where A_1 = number of species at a reference sampling spot ("species standard") and A = number of species contained in the water in sampling spots below this. The values range from 0 to 100%, the highest value corresponding to the state of total destruction. The calculated values can also be plotted as curves in a simple coordination system. The abscissae express the relevant distance along the stream.

The method developed by *Pantle* and *Buck* [1955, 1956, 1959] is also widely employed. In this, the species found from evaluation of the samples are tabulated from a qualitative point of view (*Kolkwitz* saprobic system, also partly according to a personal classification) and a quantitative point of view (1 = occasional, 3 = frequent, 5 = numerous). The frequencies of occurrence of the relevant species (micro- and macroorganisms are calculated separately) are then added together for each saprobic stage. The saprobic index corresponding to every water section investigated can then be calculated according to a simple formula. The values can be plotted to give a graph showing the quality of the water along the stretch of water investigated; the graph can be in colour (e.g., a particular colour for each saprobic stage) to make it clearer. Striking findings, such as stretches of water without life, or zones of wastewater, may be emphasized by special marks.

Liebmann's suggestion for evaluating the results [1959]. has become known and is widely used in practice as the "Münchner Methode" (Munich method). The four classes of water quality are represented on a diagram with the colours:

←

Fig. 8 Typical organisms of the polysaprobic stage. 1. *Sphaerotilus natans*, a. growth on a reed stalk (0.5), b. bundle of bacterial threads (300), c. single thread surrounded with gel (2000). 2. *Zoogloea ramigera* (500). 3. *Beggiatoa alba*, a. in liquid of high H_2S-content, b. in liquid of low H_2O-content with transverse walls here visible (1000). 4. *Oscillatoria putrida* (1000). 5. *Oscillatoria chlorina* (1000). 6. *Colpidium colpoda* (250). 7. *Bodo putrinus* (2500). 8. *Vorticella microstoma* (190). 9. *Rotaria neptunia*, a. full length, b. drawn in (25). 10. *Tubifex tubifex*, colony in sludge (2). 11. *Chironomus thummi*, larva (7). 12. *Eristalis tenax*, larva (2).
(Magnifications given in brackets)

Drawings: *Kowalsky*, after *Kolkwitz*, *Wesenberg-Lund*, *Liebmann*, some originals

Quality class	I (= oligosaprobic stage):	blue
Quality class	II (= ß-mesosaprobic stage):	green
Quality class	III (= α-mesosaprobic stage):	yellow
Quality class	IV (= polysaprobic stage):	red

Shaded areas depict the intermediate stages which often occur. The colours used are:

Quality class	I/II:	blue-green
Quality class	II/III:	green-yellow
Quality class	III/IV:	yellow-red

As in the procedure of *Pantle* and *Buck,* just described, special features can be recorded on the coloured map by means of definite signs. For example, black points indicate places of especially intense "water bloom". Zones devoid of life are given as wavy lines and zones of destruction by colouring black the stretch of water concerned on the diagram (according to a proposal of *Pantle* and *Buck* [1955]). Black, slanting shading shows formation of sewage fungi. Regions where dead fish regularly appear are to be denoted by small crosses.

These few procedures will have to suffice. The aim was simply to indicate some ways of evaluating biological findings. Many methods have had to remain unconsidered in this book, and for the reasons that, despite some advantages, they have not become accepted, and, in other cases, that they require too much laboratory working time (e.g. almost all physiological procedures of water investigation) and are unsuitable for a practice which prefers simple methods.

Detailed compilations of procedures for investigation are to be found, for example, in *Beger* [1966], *Liebmann* [1959], *Liebmann* and *Reimann* [1964] and *Schwoerbel* [1966].

III. Biological Investigation of Drinking Water

Drinking water biology has been gaining importance steadily in recent times, due to the increasing tendency to condition surface water for drinking purposes. But in ground water works also there are many problems, not yet solved completely, which are associated with internal interferences (e.g. yellow deposits in wells through iron and manganese bacteria).

Practical experience shows, however, that the biologist is consulted only when pipes are clogged or filters blocked, or when the quality of the treated water is impaired by matter affecting colour, odour and taste, or when the consumer complains to the water works about small crustaceans, water-lice, (*Asellus aquaticus*) or worms emerging from the water taps. *Beger* [1954], *thus distinguishes the most important organisms for drinking water biology.*

1. in connection with hygiene
2. in connection with working techniques

The following conclusions are drawn:

Concerning 1: The investigations cover, for example, the detection of pathogenic organisms and of non-living and living pollution indicators, including organisms regarded as unaesthetic and animals which either can transmit pathogenic microorganisms or contribute to an increase in the bacterial number through their secretions.

Concerning 2: It is important to know the significance of certain organisms for the water works technique or for the individual processing stages. Are they necessary or are they pests which can cause internal upsets and must therefore be combated? If the latter is true, it must be ascertained which way these organisms have entered the water works and which measures are to be adopted in order to prevent this entry or reduce it to a minimum.

In investigations of this sort, the biologist can never limit himself to mere detection of the particular species. He has to consider the many mutual interactions between microbes, plant and animal organisms and the water supply plant, which represents an extreme region of life. *Rössner* [1966] points out in this connection that the appearance of troublesome organisms in drinking water hardly has anything to do with a real "contamination" of the water.

Procedures of investigation which originate from hydrography or wastewater biology are given in many specialist handbooks. They are scarcely suitable for biological drinking water control. The known *Sedgwick-Rafter* method (Standard Methods 1955), the power filter of *Beger* [1952] and the centrifugal filter of *Whipple* [1927], to mention only a few, have also not made a breakthrough in Germany. Even the frequently employed sediment investigation of samples from the water system can hardly be used. According to *Rössner* [1966], tap water which yields sediments already after standing for several hours, is to be regarded qualitatively as only poorly suited.

Occurrence, propagation and extensive development of definite species of organisms are

largely dependent on the ecological conditions of the various water resources (ground water, bank filtrate, surface water, etc.).
The biologist will thus take samples from the raw water and from the individual conditioning stages in order to obtain information about the origin and propagation of the species in the water works itself. The sievable suspended matter is obtained from the pipe lines concerned with the help of a fine mesh plankton net (e.g. the nannoplankton net of *Höll*); a palisade scraper suffices for growths in filter walls, etc.
As already stated, a mere few litres of water are not enough for a drinking water control; an average of 1 cubic metre (ca 250 gallons) is allowed to pass through the plankton net in the normal case of only low numbers of organisms.
Microscope slides can be suspended in the well to concentrate the iron and manganese bacteria which often occur in wells and pipe systems. The bacteria form colonies in adequate amounts on the slides during a definite exposure time and can then be examined microscopically. A profile of the qualitative and quantitative distribution of iron bacteria in a well may be obtained by fixing several slides to a thin line at definite intervals. According to our own studies, the zones of yellow deposition of a well can be located to within 10 cm with the help of this simple method.

A. Organisms in Plants for Obtaining and Conditioning Water

The "minor water supply" will only be touched on in this book. Water holes and "wall-less" moor wells are the simplest sources of water supply. These are pits of 3 to 5 feet depth, extending to the ground water. They are still to be found in this form in remote moorland regions (e.g., Emsland). The living plant and animal organisms found in them mostly have their habitat in surface waters. Peat mosses (*Sphagnum* species) and certain rotifers, water bears (*Macrobiotus* species), small crustaceans, etc. occur in moorland areas.
Dug wells are superior to the above water holes for obtaining water. A large number of species from both the open and closed dug wells with firmly fitting covers have been described. Some are genuine inhabitants of ground water, and some come from surface water. *Beger* [1966] and *Liebmann* [1951] etc. include detailed compilations.
Enough examples have become known of, e.g., certain protozoa species which have kept for years in closed dug wells without any loss of numbers. These include principally flagellates, ciliates, bottom fauna of the *Difflugia* and *Euglypha* genera. Occasionally fungus threads and spores of the genus *Fusarium* are encountered. Diatoms of the genera *Navicula, Microneis, Hantzschia*, etc. are regularly found in more shallow wells.
If the well yields iron-containing water, precipitated iron and iron bacteria of very many types are commonly observed. Genuine ground water is to be expected when the well is situated in meadows of valleys or in broken-up ground.
Drilled wells, known as abyssinian driven wells in their simplest form, represent a hygienically more satisfactory way of obtaining ground water than do dug wells.
The advantages of such wells are that their walls are impermeable and that they can attain depths which are inaccessible to the dug well. According to experience gathered up to now,

organisms cannot gain access from surface waters to this type of well. Only real ground water living creatures are encountered during development, e.g. Cyclopidae, the well shrimp *Niphargus* and the small snail *Lartetia.* This applies also in general to drilled wells with all types of woven filters. An exception constitute certain hygienically harmless water bacteria, including the iron and manganese bacteria. In microscopic analysis they are counted just as much among the usual content of the pumped water as are grains of sand, fragments of stone, hydrated iron oxide, particles of manganese, etc.

Spring water is widely used in mountain areas and also in the plains. The degree of purity of the water depends on its origin (genuine ground water, surface water etc.) and, above all, on the natural cropping out and tapping of the source. Cataract sources are the least contaminated; they descend more or less steeply into the valley and thus form no spring pools. All floating material is dragged with the water and no clogging occurs. The situation is much less favourable with the pool sources, where the water comes from below into a smaller or larger pit. This can sometimes become boggy and a smaller stream then runs from it.

A distinction must be made in the life of spring water between

1. the organisms which come from the source itself, swept out from underground, and
2. those plant and animal species which colonise the spring water after it has emerged from the ground.

Real ground water inhabitants belong to the first group. Species living above ground can also appear, however, when they enter through surface water connection with the source. To the second group belong the biologically peculiar communities of thermal springs, brines, sulphur- and iron springs and so-called acidulous waters. Their study has already become a biological specialist domain but it is unfortunately not possible to consider this within the scope of this book.

Undisturbed pool and swamp sources are characterised externally through luxuriantly growing beds of moss, mats of algae and higher aquatic plants (water cress, speedwell, *Stellaria uliginosa, Callitriche verna, Sium erectum,* etc.). Amongst animal organisms are found not infrequently crayfish (Gammaridae), small crustaceans (Cyclopidae, Ostracoda etc.), water mites (Hydracarina), flat worms (such as *Planaria alpina* which likes cold water). Sometimes even earth-worms forsake the moist earth for the region of the source, likewise *Haplotaxis gordiodes,* the real living domain of which is the moist layers of ground just above the ground water level.

Even though by no means all natural pool springs display rich vegetation and colonisation (sources in forest shade have usually few species), they nevertheless do not in the least fulfil the requirements of drinking water hygiene. It is known that earth worms, for example, can enormously augment the bacterial number of a water to which they have gained access. In one instance, up to 64 million microorganisms were found in 1 cm^3 of earthworm excrement.

B. Shortcomings in the Tappings of Springs

Living creatures appear not infrequently in the tappings and chambers of springs which have no connection with the essential animal world of ground and surface water. These include gnats, snails, spiders, frogs, toads, salamanders and mice, and even small species of birds have been observed. This is always a sign of constructional shortcomings (e.g. insufficiently secured ventilation shafts, leaks in the access to well chambers). Fine sieves are often missing in the spring tappings; these are intended to prevent ground water creatures from entering the pipe system on pumping.

C. Water Works Plants

The practical examples mentioned so far have concerned smaller plants for obtaining water. Large water supply units must count with the same troubles to some extent; but the situation is rather more favourable for them, since living organisms are generally retained by modern water processing, making use of accelators, plants for removing iron and manganese and the slow and rapid filters. Nevertheless, water works operation is often sensitively disturbed through considerable growth of particular organisms (iron- and manganese bacteria, unicellular algae, etc.).
Shallow and deep wells are used for obtaining water for water supply on a large scale. The ground water taken from these contains little nutrient so that only a few, highly specialised types of bacteria can live in it. The yellow deposits in the well are therefore largely caused by iron bacteria, for instance.
Some of the ca. 60 species of iron bacteria known at present are autotrophic, i.e., they require no organic carbon compounds. Others make use of organic carbon or inorganic carbon dioxide as source of carbon (mixotrophic species). The remainder are dependent on the supply of organic carbon compounds (heterotrophic species). Autotrophic species obtain their necessary energy from the oxidation of Fe^{2+} to Fe^{3+}. Some oxygen, even if very little, must be present in the water for the precipitation process to be possible.
Some classical disasters with iron bacteria are known from the literature (e.g. *Beger* [1966]). These were mainly due to the genus *Crenothrix* and led to extensive clogging of pipe systems and tanks. We know now from the diagrams in the publications at that time that the species *Leptothrix ochracea* and *Gallionella ferruginea* and the flagellate *Anthrophysa vegetans* took part as well as *Crenothrix* (Fig. 9).
Many observations and investigations in practice, and work on pure cultures have recently furnished new information about the conditions for life of iron bacteria. The following factors are decisive for their development:

1. the oxygen content of the water
2. the carbon dioxide content of the water

3. the content of dissolved Fe^{2+}
4. the pH-value

Iron bacteria flourish at a pH-value of about 5.4–7.2. A Fe^{2+} amount of 1.6 mg/l is wholly adequate for their prosperity. Laboratory experience shows that the upper limit for their growth is at 10–12 mg Fe^{2+}/l and that at 14 mg/l, growth in cultures ceases.

It has been established in addition that a higher redox potential prevails in the vicinity of *Leptothrix* coagulation. The increased rate of flow of the water in the neighbourhood of the well also has a favourable influence on the development of the iron bacteria.

All this knowledge is of great importance for practice. It may be mentioned in this connection that the collective term "iron bacteria" comprises all iron and manganese bacteria. Almost all bacteria which can precipitate iron are capable also of precipitating manganese.

In 1952, *Beger* ventured the opinion that there are many rod and coccal forms which escape microscopic determination but which play a significant part in the removal of iron and manganese from a water. Our own studies, carried out over many years on deep wells of W. German water works confirm this. It applies with particular force to the processes leading to yellow deposits in the wells.

Using his nannoplankton net, *Höll* has been able several times to isolate in water works 10 μm sized, colourless iron bacteria; these could not be determined more precisely but they mineralised when introduced into a suitable iron-containing nutrient solution (written communication, 1966).

Recently, specially constructed underwater cameras or apparatus for underwater television have been successfully used for checking developing or existing yellow deposits and corrosion in wells and pipe systems (*Schneider* [1962], *Schultz* [1963] etc.).

Hydrogen sulphide, often present in ground water, either disappears already during iron or manganese removal, or it must be taken out by a special degassing process. This can be by aeration of the water in open plants or by atomising, percolation or spraying.

Such plants are characterised by massive development of sulphur- and of iron bacteria. The sulphur bacteria, e.g. the well-known *Beggiatoa alba,* form chalky coloured deposits on the walls. Below them are layers of black, greasy putrefied sludge which is rich in iron sulphide.

There is naturally much plant and animal life in open water works plants, the permanent forms of which have gained access to the water through the aeration or atomising processes. Detected have been diatoms, green algae, mosses, ferns, flagellates, ciliates, rotifers, various species of worm and insect larvae, especially Chironomidae.

1. Slow Filters

The processing of larger amounts of water by filtration of surface waters through sand is still irreplaceable today.

Several groups of factors are involved in slow sand filtration which should provide a filtrate free of suspended matter and low in bacterial content:

1. the mechanical sieving action of the sand

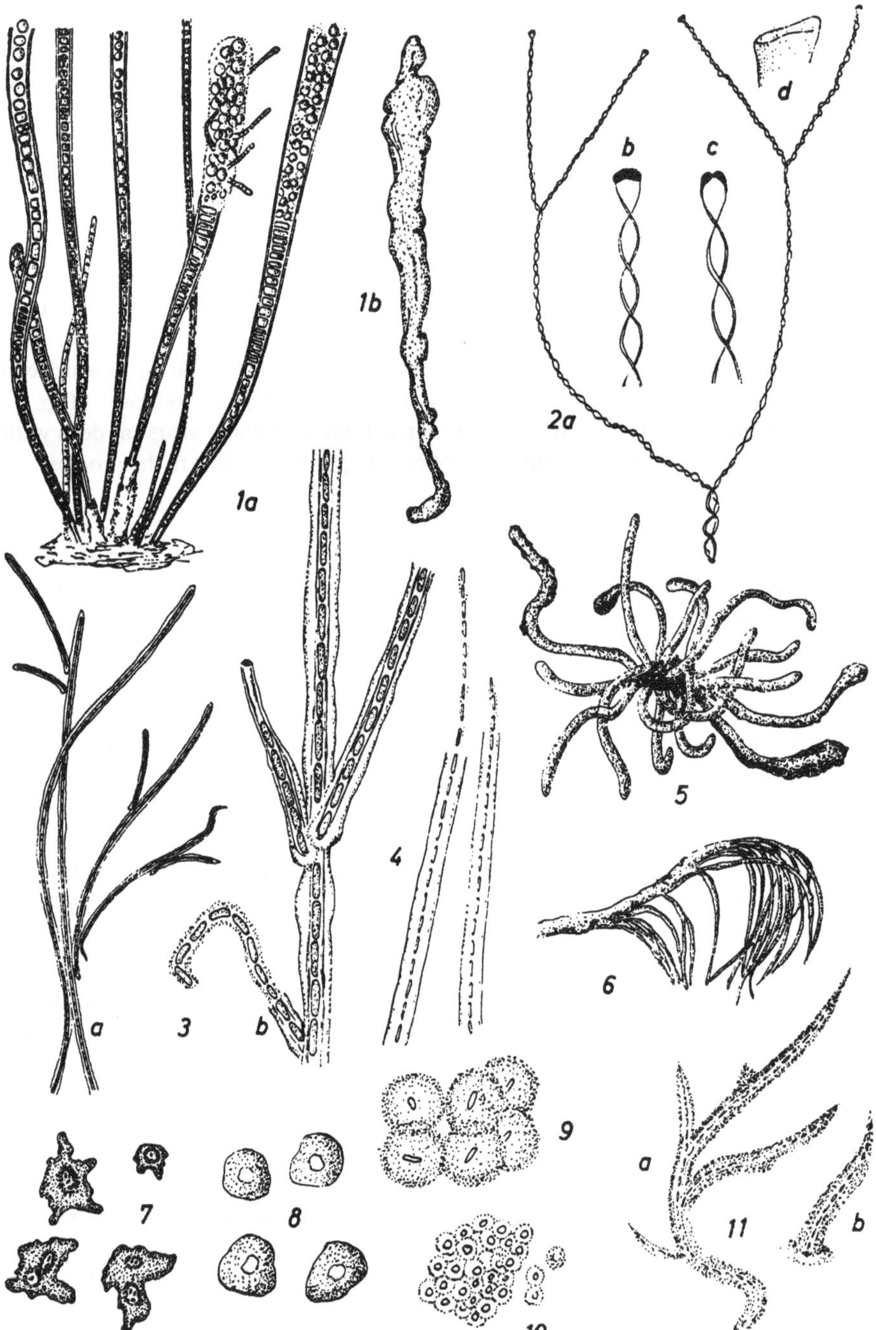

Fig. 9

2. chemical changes
3. adsorptive processes in the boundary layers
4. biological processes in which bacteria, algae, protozoa and also higher forms of life take part.

In our climate, certain diatoms dominate in winter and mostly green algae and some blue algae also, in summer. The animal species are principally representatives of the so-called mesopsammon fauna i.e., animals which are adapted to life in water-carrying sand or gravel, as far as their body structure and nutrition etc. are concerned ("Fauna des Mesopsammals", *Husmann* [1968]).

According to former views, the colonisation extended to a depth of about 20 cm (8"). Newer investigations by *Husmann* [1958, 1960] on the mesopsammon fauna in the slow sand filters of the Bremen water works showed that the best conditions for life and high counts of individuals were found in samples from 20–42 cm below the sand surface. He was able to identify 20 animal species, made up of 11 polychaetes and nematodes, 3 annelids, 5 different copepods and 1 water mite (Hydracarina). In one litre of filter sand Husmann found 6000 individuals of the above listed species.

The extent of an animal colonisation by certain species was shown especially clearly in investigations of the sand filters of Osaka (Japan). As *Rössner* reported [1966], there were 300.000 annelids of the species *Nais communis* and 500.000 of the species *Pristina rosea* per square metre surface of sand. Since these worm species belong to the mesopsammon fauna, there is no general fear of their migrating into the pure water.

The many species of protozoa are significant to some extent for restricting the growth of bacteria. Colourless flagellates, naked and shell-bearing rhizopodes of the genera *Amoeba, Nebela, Hyalodiscus,* and ciliates (*Stylonychia, Glaucoma, Aspidisca* and *Vorticella*) participate equally in forming the filter covering. Flagellates, ciliates and rotifers (*Rotaria, Keratella, Callidina,* etc.) predominate in the dense veil of sludge straight above the covering.

The free water space is the real living domain of many small crustaceans which, when the ecological conditions are suitable, develop massively here. The odour and taste of the filtrate is impaired and the bacterial numbers increase if these animals are killed off through sudden fluctuations of temperature or treatments such as with copper or chlorine, just as when algal "water bloom" dies off.

Similar phenomena are encountered also when larger "wads" of algae die off and, when the filter covering is lifted, through excessive oxygen development.

←

Fig. 9 Types of iron bacteria. 1. *Crenothrix polyspora*, a. habitus, b. mineralised type (1500). 2. *Gallionella ferruginea* a. habitus, part of a dichomic branched thread (1200), b. typically developed cell, c. dividing stage (b & c, 2500), d. cross-section of thread (5000). 3. *Clonothrix fusca*, a. habitus (1100), b. cross-section of thread (5000). 4. *Leptothrix ochracea* (1000). 5. *Leptothrix lopholea* (ca. 700). 6. *Leptothrix trichogenes* (650). 7. *Siderocapsa treubii* (500). 8. *Siderocapsa monoeca* (4000). 9. *Siderocystis vulgaris* (2000). 10. *Siderocapsa botryoides* (2000). 11. *Leptothrix discophora*, a) free-swimming thread, b) non-motile thread, adhering disc at the broader end (900).
(Magnifications given in brackets)

Drawings: *Kowalsky*, after *Beger, Dorff, Molisch*

2. Rapid Filters

Modern rapid filters, built as open or closed (so-called pressure filters) plants also are not free from interferences by plant and animal life. Enough examples are known in practice. E.g., in one case the alga *Oscillatoria rubescens*, was only half retained by the uppermost gravel layer; the remainder entered the lower layers and was rinsed out with the filtrate. In another case, the small desmids of the genus *Cosmarium* developed enormously and passed through the rapid filters without difficulty. The filtrate was a distinct green! May these few examples indicate that, despite the use of flocculants, many species of organisms are fully capable of passing through the filters, not only passively but also actively (e.g., worms, small crustaceans).

3. Activated Carbon Filters

The ciliates, nematodes (threadworms), small crustaceans (copepods, etc.), frequently observed in active carbon filters, find living space here similar to that of the mesopsammon fauna in the slow filters. Massive development of bacteria is often a nuisance which showers microorganisms into the pure water.

Thanks to the help of modern water conditioning techniques, organisms now appear far less frequently in the pure water. However, inferior processing can, for example, lead to accumulations of sludge in the clean water chambers and favourable conditions are established for development of iron- and sulphur bacteria and of ciliates, worms, water-lice, small crustaceans and other organisms. Small flies get in through inadequately secured vents and any cracks in the walls facilitate entry of ground water animals.

4. Water Distribution System

Despite the extreme conditions for life which prevail in the water distribution system (darkness, low oxygen content, meagre nutrient supply), it has its permanent residents such as protozoa, some species of worm, rotifers, water-lice, small crustaceans and even blood-worms (Chironomus larvae); these may multiply considerably and can be combated only with difficulty. *Rössner* [1966] has stressed that all the reasons for not having organisms in drinking water can apply only to the pipe system. They are indispensable in the individual conditioning stages on account of their powers of degradation, which are not to be underestimated. An eventual control of these creatures in the various processing plants might displace their region of colonisation to the pipe system.

IV. Methods of Combating Organisms in Drinking Water Supply Plants

No general or standard directions for treatment can be laid down for a biologically effected interference. As is shown in practice, every case is a special one and is associated with a different chemical and biological situation.

Precautionary measures include frequent checks, cleanliness in the works and careful construction. The routes taken by many intruding organisms can be recognised in due time only through regular inspection. This includes, further, frequent de-sludging of dug wells, cisterns and tappings of springs; intact sieves in source chambers to hinder entrance of ground water animal life; fly-proof windows and vents in all works buildings; the use of unsmoothed cement and of Eternit (asbestos cement), both of which hinder colonisation of plant and animal organisms which grow on surfaces. Reference may be made too to the use of coloured window glass to limit excessive algae production, and to the action of aluminium sulphate and other compounds for flocculating plankton.

A direct control with chemical destruction agents depends principally on the specific sensitivity of the various species and the chemical nature of the water. The necessary precautionary measures must be strictly respected in order to rule out the possibility of harm to drinking water consumers.

The toxic chemicals which are most used in water works operation are: Chlorine and some chlorine-containing compounds; copper sulphate; potassium permanganate; quaternary ammonium compounds; organic zinc compounds; antibiotics; insect powders (pyrethrum and derris preparations); and, recently, also insecticides, herbicides and algicides. *Beger* [1966] gives a detailed compilation of the control agents, including the necessary amounts for use.

The use of chlorine is sufficiently well-known. It is applied in the liquid or gas state or in the form of chlorine compounds, e.g. Caporit (calcium hypochlorite with 70% active chlorine) or sodium hypochlorite solution. The so-called residual or excess chlorine amount in the water determines how much chlorine is needed for the treatment. This is the amount of chlorine which is not consumed by the chlorine-combining material in the water. Many protozoa and bacteria species are generally destroyed in a short time with 0.2–0.5 mg/l of excess chlorine. Smaller amounts of chlorine are usually ineffective or scarcely damage the organisms concerned. Some lower forms of life, embedded in slimy or gelatinous envelopes (e.g. *Crenothrix polyspora*, certain slime-forming bacteria and some diatoms) are more resistant to chlorine, because chlorine in smaller amounts possesses no marked profound action. Animal forms living in the sludge (small crustaceans, protozoa, *Asellus cavaticus*) are only insignificantly harmed. In all these cases, *"intermittent chlorination"* is used, in which the combated living matter is subjected continuously to chlorinated and unchlorinated water alternately. Maximum success is achieved when a 2 minute chlorination wave is applied every 10 minutes. Another way of chlorinating more effectively is the so-called "superchlorination", i.e., one works with larger amounts of chlorine. Any dechlorination necessary after a longer chlorination period is performed with sodium thiosulphate, sodium sulphite or active carbon. *Rössner* [1966] pointed out that in older pipes which had been occupied by sulphate-reducing bacteria, hydrogen sulphide was formed after 7–10 mg/l of chlorine had been used; this persisted a long time and was intolerable for the consumer.

Chlorine amounts of 3 mg/l and more are needed to destroy *Synura uvella,* rotifers, small crustaceans and larvae of may-flies. *Chang* [1961] recommends 0.4–0.5 mg/l of residual chlorine with 6 hours time of action to remove thread worms *(Nematodes)* which often occur in filters and can multiply considerably there.

2 mg chlorine/l are enough to destroy 95% of the worms of the Oligochaeta class (genus *Nais* and related species) within 15 minutes. Common salt (1% solution, acting for 24 hours) is a simple but effective agent for removing *Nais* species (*Learner* and *Edwards* [1963]).

5 mg residual chlorine/l often do not suffice for the chlorine-resistant algae of the genera *Chlorella, Scenedesmus, Pedaistrum, Ankistrodesmus* etc. The water-louse (*Asellus aquaticus),* which frequently becomes a nuisance, likewise tolerates amounts of chlorine up to 15 mg/l. At least 25 mg chlorine/l with action times of 20 hours and more are necessary to combat Chironomidae larvae. The well-known filamentous alga, *Cladophora,* is killed off only by treatment with 150 mg active chlorine/l for 1 hour or 30 mg/l for 48 hours.

Ammonia has been finding use recently in filter chambers and water towers especially for removing winged insects (e.g. may flies, Chironomidae). The parts of the buildings which are plagued are sealed and treated with the gas. 8 mg NH_3/l are highly toxic to small crustaceans (Cyclopidae) within a few hours, and 10 mg/l for water-lice.

Chloramines have the reputation of being the best agents for destroying slimy bacterial coatings *(Crenothrix)* or animal growths (fresh water sponges, Bryozoa, *Dreissena polymorpha* etc.) in pipes. The amounts of chloramine needed are small but very effective. There is, moreover, no danger of the unpleasant taste to the water from chlorophenols which ordinary chlorination yields when phenols are present. The toxic action of chloramine begins more slowly than that of chlorine but persists longer. According to *Beger* [1966], chloramine concentrations corresponding to 1 mg/l of combined available chlorine are as active as 100 mg/l of free, available chlorine. Treatment consists of first passing ammonia into the water and then, immediately afterwards, the chlorine (or chlorine water), in the ratio of 1:2. Periods of contact of 30 minutes to 2 hours are enough, according to experience so far. Even chlorine- and copper-resistant species can be removed fairly certainly through intermittent application. The use of copper, as copper sulphate, has various advantages. It is commercially available in solid form which can be easily despatched and transported without essential change of properties. Since it dissolves in water relatively quickly, it may be employed in standing and also slowly moving waters. In the amounts in which it is used, it is completely harmless to human beings. Copper sulphate has become known as an excellent agent above all for combating plankton organisms and is applied with success to destroy water bloom in reservoirs, impounding basins, storage tanks and swimming baths. It is known in practice that certain species, such as Bryozoa, water-lice, blood-worms and leeches

can accustom themselves to copper to a greater or lesser degree. The species, already quoted, resistant to chlorine, are also resistant to copper; regular additions of 5 mg Cu/l are thus said to be without effect on them.
The copper sulphate amounts (mg/l) required for some of the important algae in water works operation are given below:

Anabaena	0.12	*Ankistrodesmus*	1.0
Microcystis	0.12–0.25	*Scenedesmus*	1.0
Oscillatoria	0.2 –0.5	*Eudorina & Pandorina*	10.0
Euglena	0.55	*Synura*	0.12–0.25
Staurastrum	1.5	*Melosira*	0.33
Peridinium	0.5 –2.0	*Fragilaria*	0.25
Enteromorpha	0.5	*Cladophora*	0.5
Ulothrix	0.2	*Tribonema*	0.35
Spirogyra	0.12–2.5	*Chara*	0.1 –0.5

The following amounts are needed for bacteria and fungi:

Sphaerotilus	0.2	*Beggiatoa*	5.0
Crenothrix	0.33	*Leptomitus*	0.4

Apart from copper sulphate, the industrial product "Grünkupfermittel" (green copper agent) has been used with success for control of excessive algal growth.
Among protozoa, the ciliates are particularly sensitive to copper. *Paramaecium, Stylonychia, Glaucoma* and related species die off at concentrations of 0.2 to 0.6 mg/l of copper sulphate. Rotifers and small crustaceans are more resistant; they require amounts exceeding 1 mg/l. For combating gammarids successfully, up to 10 mg/l must be used and up to 50 mg/l for mud snails *(Limnaea stagnalis)*. 1 mg/l of copper sulphate is enough in hard waters to kill off worms (*Nais* species) in less than 6 hours. Copper sulphate concentrations of this same value, 1 mg/l, act more strongly in soft waters at neutral pH-value (*Learner* and *Edwards,* [1963]).
If water is pumped away after such a pest control action, one must bear in mind that these toxic waters should not be introduced into the nearest receiving water without having been sufficiently diluted. They would destroy fish extensively, since they are much more sensitive. Thus 0.1 mg/l of copper sulphate is fatal to salmon-type fish, 0.35 to carp and 5 to white fish (see p. 191 for removal of copper).
The well-known insect powder is available as a reliable contact nerve poison for insect larvae and water-lice. This powder is obtained by grinding the flower heads of Dalmatian, Spanish, Caucasian and, above all, African chrysanthemum species. The active components are two pyrethrins which are contained in the oily resin of the powder. The compounds are not really soluble in water but are 90–95% present in colloidal state after 24 hours.
Insect powder is utterly unsuitable for combating protozoa since this animal group has no nerves. Rotifers and worms are killed off with certainty only with amounts of 100–250 mg/l. Pyrethrum powder is at its most successful in the control of blood worms (Chironomidae larvae), for which 2 mg/l are already enough; they die off after 1–1½ days. 1 mg/l is fatal to

water-lice within 12–18 hours, and 5 mg/l for small crustaceans (Daphnia) within 20–48 hours. Great care must be taken when insect powder is used, that these treated waters on no account enter the complete supply network, so that no sensitive consumers suffer harm. The possibility of death of fish in aquaria, zoological concerns, fish shops etc. must be considered in this connection. If the waters which have been treated with insect powder are to be introduced into the nearest receiving water, their content of insect powder must not exceed 5 mg/l. No damage to fish and fish food organisms is then to be expected.

Many synthetic toxic materials of high activity have been brought on to the market by the chemical industry in recent times. These are the generally known contact insecticides. The so-called Hexa (BHC, HCH, Gammexan), phosphate ester preparations and chlorinated hydrocarbons do not come into consideration at all for water works operation on account of their high toxicity to man. On the other hand, the DDT-preparations (Dichloro*di*phenyl*t*richloreoethane) have proved to be usable, provided that the necessary precautionary measures are carefully adhered to. Investigations on an impounding basin in N. America have shown that the amounts of DDT needed to control insect larvae (0.1 mg/l) caused no damage to man or animal which drank this water. Moreover, after a control treatment, 80–98% of the residual active DDT agent can be removed by subsequent flocculation, sedimentation and filtration. The remaining few percent can be reduced considerably by treatment with active carbon.

A 5% suspension in water (1–2 g per 10 square feet of wall) is enough to destroy winged insects in water towers, filter plants etc. Some necessary amounts of DDT used in practice for successful control of some animal species are quoted here:

Simulium larvae	1.0 mg/l
Sticklebacks	0.1 mg/l
Gammarids	0.02 mg/l
Fly larvae (Culex, Aedes)	0.05 mg/l
Chironomidae larvae	0.001 mg/l
Daphnids	0.001 mg/l

An interesting fact must be mentioned in this connection. In the whole wide-ranging literature on the application of insecticides of very widely varying chemical composition, there is scarcely a hint to be found about their use in water works operation. It must be said after many years of personal experience that DDT preparations and other less toxic contact insecticides have not made the breakthrough that was expected many years ago. The old established agents (chlorine, copper sulphate) are still being used today in cases of internal interferences in order to take no unnecessary risk.

Herbicides, algicides and related materials are being used increasingly for combating water bloom and undesired plant growth on riparian stretches. Without wishing to discuss in more detail the chemical properties and the toxic activities of these numerous preparations, the following problems may be referred to here briefly: Little has become known so far about the behaviour of herbicides, algicides etc during passage through the soil (bank filtered surface water), in drinking water tanks and during water conditioning itself. In my opinion, one ought therefore to consider whether water bloom or algae in drinking water reservoirs should not be combated with other, less harmful agents than the highly toxic compounds

about which we really do not know how they ultimately affect humans. It is known that certain parts of these agricultural pesticides can get into the raw water during pumpage and are not held back adequately or at all during the usual conditioning.
For the same reasons, the use of colouring matters for protective coatings to prevent growth of algae in drinking water supply plants cannot be permitted (*Reimann* [1961]).

V. Bibliography of the Biological Part

The translated titles are given after the bibliography in German, as was done for the bibliographies of the chemical and bacteriological parts:

Literatur zum biologischen Teil

Beger, Herbert: Die Zielsetzungen der Wasserwerksbiologie. Ibid. *93,* 88 (1952).

– Tierische Schädlinge in der Wasserwerkspraxis. Ztschr. f. angew. Zool. *41,* H. 1, 1–18 (1954).

– Leitfaden der Trink- und Brauchwasserbiologie. 2. Aufl. (1966).

Bursche, E. M.: Beitrag zum Phytoplanktonhaushalt des Müggelsees mit einer Betrachtung über die Verschiedenartigkeit von Netz-, Sieb- und Schöpfplantkon. Ztschr. f. Fischerei und deren Hilfswissenschaften II, N. F., H. 3/4, 209–226 (1953).

Chang, Sh. L.: Viruses, Amebas and Nematodes and Public Water Supplies. JAWWA 53, Nr. 3, 288–296 (1961).

Cohn, F.: Über lebende Organismen im Trinkwasser. Günsberg's Ztschr. f. klin. Mediz. *4,* 229 (1853). ibid. Jahresber. Schles. Ges. f. Vaterl. Kultur *31,* 91.

Helfer, H.: Geschichte der biologischen Wasseranalyse. Arch. f. Hydrobiol. *11,* 565 (1916).

– Das Saprobiensystem. Die makroskopisch-biologische Wasseranalyse nach ihrem heutigen Stande. Kl. Mitt. Mitgl. Ver. f. Wasser-, Boden- und Lufthygiene, Berlin-Dahlem 7, 138 (1931).

Heynig, H., und *Ockert, G.:* Beitrag zur Methodik biologisch-hygienischer Trinkwasseruntersuchungen. Ztschr. f. d. ges. Hygiene u. ihre Grenzgebiete *8,* H. 7, 486–507 (1962).

Husmann, S.: Untersuchungen über die Sandlückenfauna der bremischen Langsamfilter. Abh. d. Braunschweigischen Wissenschaftl. Ges. X, 93–116 (1958).

– Über eine blinde Arthropodenart aus dem Mesopsammal bremischer Langsamfilter. Abh. naturw. Ver., Bremen *35,* Nr. 3, 321–437 (1960).

– Langsamfilter als Biotopmodell der experimentalökologischen Grundwasserforschung. Gewässer und Abwässer, H. 46, 20–49 (1968).

Kolkwitz, R.: Entnahme- und Beobachtungsinstrumente für die biologische Wasseruntersuchung. Mitt. K. Prüfanst. f. Wasservers. u. Abwasserbes., Berlin-Dahlem *9,* 111 (1970).

– Plankton-Membranfilter. Ber. Deutsch. Botan. Ges. *42,* 205 (1924).

– Ökologie der Saprobien. Über die Beziehungen der Wasserorganismen zur Umwelt. Schriftenreihe Ver. f. Wasser-, Boden- und Lufthygiene. Berlin-Dahlem *4* (1950).

– und *Marrson, M.:* Grundsätze für die biologische Beurteilung des Wassers nach seiner Flora und Fauna. Mitt. K. Prüfanst. f. Wasservers. u. Abwasserbes., Berlin-Dahlem *1,* 33 (1902).

– und *Marrson, M.:* Ökologie der pflanzlichen Saprobien. Ber. Deutsch. Botan. Ges. *26a,* 55 (1908).

– und *Marrson, M.:* Ökologie der tierischen Saprobien. Int. Revue f. d. ges. Hydrobiol. *2* (1909).

Kothé, P.: Der "Artenfehlbetrag", ein einfaches Gütekriterium und seine Anwendung bei biologischen Vorfluteruntersuchungen. Dtsch. Gewässerkundl. Mitt. *6,* H. 3, 60–65 (1962).

Knöpp, H.: Abwasserlastplan und biologischer Gütelängsschnitt – ein kritischer Vergleich. Die Wasserwirtschaft *46*, H. 3, 57–59 (1955).
– Grundsätzliches zur Frage biologischer Untersuchungen, erläutert an einem Gütelängsschnitt des Maines, Arch. f. Hydrobiol., Suppl. XXII. H. 3/4, 363–368 (1955).
Learner, M. A., AND *Edwards, R. W.:* The toxicity of some substances to *Nais (Oligochaete). Proc.* of the Society for Water Treatment and Examination 12, Nr. 3, 161–168 (1963).
Liebmann, H.: Die Notwendigkeit einer Revision des Saprobiensystems und deren Bedeutung für die Wasserbeurteilung. Ges.-Ing. *68*, 33 (1947).
– Handbuch der Frischwasser- und Abwasserbiologie. Band I. Verlag R. Oldenbourg, München (1951).
– Methodik und Auswertung der biologischen Wassergütekartierung. Münchn. Beiträge zur Abwasser-, Fischerei- und Flußwasserbiologie *6*, 134–156 (1959).
– und *Reimann, K.:* Hydrobiologische Prüfungsmethoden. Föderation Europäischer Gewässerschutz, Zürich (Schweiz), Informationsblatt Nr. *11*, 43–53 (1964).
Marvan, P.: Notes on the application of membrane filtration for quantitative phytoplancton research. Sbornik, Scientific papers from the Inst. of chem. Technology. Prague *4*, Teil 1, 399–418 (1960).
Mez, C.: Mikroskopische Wasseranalyse. Berlin (1898).
Pantle, R.: Biologische Flußüberwachung. Die Wasserwirtschaft *46*, H. 8, 206–209 (1956).
– und *Buck, H.:* Die biologische Überwachung der Gewässer und die Darstellung der Ergebnisse. Gas- und Wasserfach *96*, H. 18, 604 (1955).
– und *Buck, H.:* Biologische Flußüberwachung. Sonderschrift d. Reg. Präs. Nordwürttemberg (1959).
Reimann, K.: Verhinderung der Algenansiedlung durch Schutzanstriche. Münchn. Beiträge zur Abwasser-, Fischerei- u. Flußbiologie *8*, 135–143 (1961).
Rößner, F. X.: Organismen in den technischen Anlagen der Wasserversorgung. Neue Deliwa-Ztschr. H. 3, 93–102 (1966).
Schneider, H.: Brunnenbauüberwachung mittels Unterwasserkameras. Bohrtechnik – Brunnenbau *13*, H. 1, 17–26 (1962).
Schultz, K.: Eine neue Subminiatur-Fernsehkamera für Unterwasserbeobachtungen und zum Einsatz in explosionsgefährdeten Räumen. VDI-Zeitschr. *105*, Nr. 12, 489–492 (1963).
Schwoerbel, H.: Methoden der Hydrobiologie (Süßwasserbiologie). Franckh'sche Verlagshandlung, Stuttgart (1966).
Thienemann, A.: Die Binnengewässer Mitteleuropas. In *Thienemann,* Die Binnengewässer 1 (1926).
Whipple, G. C.: The microscopy of drinking-water. 4. Aufl. New York (1927).

Translated Titles

Beger, H.: Aims of water works biology
– Animal pests in water works practice
– Guiding principles in the biology of drinking and service water
Bursche, E. M.: Contribution to the phytoplankton budget of the Müggelsee, considering the variety of the net, sieve and drawn plankton
Cohn, F.: Concerning living organisms in drinking water
Helfer, H.: History of biological water analysis
– The saprobic system; macroscopic-biological water analysis today
Heynig, H., and *Ockert, G.:* Contribution to the methods of biological-hygienic examination of drinking water
Husmann, S.: Investigations of the mesopsammon fauna of the slow filters in Bremen
– A blind Anthropod species from the mesopsammon fauna of the slow filters in Bremen
– Slow filters as biotopic models in experimental ecological research on ground water

Kolkwitz, R.: Equipment for sampling and observation in biological examination of water
– Plankton-membrane filters
– Ecology of saprobic species. Concerning the relation of water organisms to environment
– and *Marsson, M.:* Basic principles of biological assessment of water according to its flora and fauna
–– Ecology of plant saprobic species
–– Ecology of animal saprobic species
Kothé, P.: The "species deficit", a simple criterion of quality, and its application to biological investigations of receiving waters
Knöpp, P.: Wastewater load plan and biological longitudinal section of quality–a critical comparison
– Basic comments on the question of biological studies of receiving waters, illustrated on a longitudinal section of quality of the R. Main
Liebmann, H.: The necessity of revising the saprobic system and its significance in the assessment of water
– Handbook of the biology of fresh water and wastewater
– Methods and evaluation of maps of water quality
– and *Reimann, K.:* Hydrobiological methods of testing
Mez, C.: Microscopic water analysis
Pantle, R.: Biological river supervision
– and *Buck, H.:* Biological supervision of waters and presentation of the results
–– Biological supervision of rivers
Reimann, K.: Prevention of algae colonisation through protective paint coating
Rössner, F.: Organisms in the technical plants for water supply
Schneider, H.: Supervision of well construction by means of underwater cameras
Schultz, K.: A new sub-miniature television camera for underwater observations and use in places exposed to explosion danger
Schwoerbel, H.: Methods of hydrobiology (fresh water biology)
Thienemann, A.: The inland waters of central Europe

Index

A
Abyssinian wells 1, 366
Accelator process 53, 225, 226, 368
Acid-base titration 90–93
Acid consumption 91–92
Acidulous waters 5, 84, 159, 164, 237, 244, 367
Activated carbon filters 53–56, 145, 147, 372
– sludge process 225
Aeration and aeration processes 166–168, 194, 195, 369
Aerobacter 297–300, 303, 330
Agar nutrient media 294–295
Aggressive carbon dioxide 86, 172, 182, 209, 211
Aggressivity, evaluation 83–108
Akdolit filtration 54, 169–171, 176–178, 191
Albuminoid nitrogen, determination 30
Algae, algal growth 234, 347, 358, 362, 367–369, 371–377
Alkalinity 91
Aluminate, flocculation in water decoloration 53
Aluminium, corrosion of 158
–, determination 130
–, sanitary significance 192
Aluminium sulphate, flocculation with 52–53, 205
Amino acids, determination 207
Ammonia, albuminoid 30
–, for combating organisms 374
–, protein 30
Ammonium ion, determination 27–30
– –, pollution indicator 68, 204
Analyses of medicinal water 241–254
– – tapwater 82–147
– – –, minor 27–50
Anthracite filters 62
Antimony in water, significance 193
Anti-rust protective layer 142, 148, 151, 152, 163–166, 170, 172–174, 181, 194
– – –, prevention of formation by carbon dioxide 97, 165
Arsenic, determination 137
–, occurrence 137, 191
–, removal 191
–, significance 191
Asbestos cement pipes 3, 156, 159–160, 162, 373
Assessment of drinking water 51–81
Associated carbon dioxide 65, 86, 94, 97, 114, 160, 163–166, 182
Atomisation jets 167
Azide-containing nutrients 311, 322, 331

B
Bacterial count see Bacterial number
Bacterial counting apparatus 277, 293–294
Bacterial number 288–289, 318–320, 324–325, 331
– –, determination 290–296
Bacteriological investigations 6, 273–332
– working place 276–277
Bacteriophages 314–316
Bacterium coli 56, 58, 67, 297–308
Balance, carbonate see Carbonate balance
Bank filtration 59
Base consumption 92–93
Berkefeld filters 66–67
Beryllium, determination 253
Bibliography, bacteriological 333–341

–, biological 378–380
–, chemical 264–269
Biochemical oxygen demand 105, 215–216, 223
Biological purification, procedures 225–227
Biological water investigation 345–377
Biomass titration 216
Blending of waters 152–153
Blood agar plates 326
BOD 105, 215–216, 223
Boiler feed water 142, 168, 194, 230–233
Bore wells 1, 2
Boric acid, determination 247
Bottom materials 347, 353–355
Breakpoint chlorination 64, 128
Brewing, wastewater from 228
–, water for 151, 174, 234–236
Brilliant green bile broth 305, 307, 330
Brines 41, 238
Bromide, determination 248
Bücher process (softening) 183
Building purposes, water for 208–212

C
Cadmium in water, significance 193
Calcium, determination 117–118, 119
Calcium hydroxide process (deacidification) 171
Candlot's salt 151, 209
Carbonate balance (equilibrium) 64, 88, 89, 96, 97, 149, 151, 160, 161, 163, 164, 172, 175, 182, 205
Carbonate hardness 59, 64, 88, 95, 112–114, 152, 162, 165, 168, 170, 172, 176, 179, 180, 183, 184, 205, 211
Carbon dioxide, aggressive 86, 172, 182, 209, 211
– –, associated see Associated carbon dioxide
– –, combined 162
– –, determination 86–90
– –, excess 94, 95, 97, 161, 164–165, 172
– –, free 15, 86–90, 94, 96, 97, 101, 113, 152, 161, 162–166, 170–172, 176
– –, lead-aggressive 97, 165–166
– –, lime-aggressive 94, 96–97, 165–166
– –, preventing formation of anti-rust protective layer 97, 155, 165
Carcinogenic materials in water 78–79, 146, 282
Caustic embrittlement 232
Cell formation (local elements) in corrosion 157
Chemical oxygen depletion 105
– purification of wastewater 225
– sterilisation 62–67
Chloramines for combating organisms 374
Chloride ion, corrosion due to 74, 155–158, 192, 211
– –, determination 40–43, 246
– –, as pollution indicator 73–74
Chlorination, intermittent 373–374
–, as sterilisation 63–64, 66, 203, 373–374
– tablets 66
Chlorine, combined available 126–127
– consumption 128
– demand 128
– dioxide, sterilisation with 64–65, 204
– –, determination 128
–, free, determination 126–127
– number, determination 40
– –, as pollution indicator 72–73
–, total 126–127
Chlorophenols 63–65, 205, 215, 374
Cholera 278, 331
Chromium, determination 136
–, significance 193
Cisterns 1, 2
Citrate test (IMViC) 299, 303, 304
Clarification of water 54–55
Clarity of the water 17–18
Clostridia, detection 310–313, 332
–, growth 323
Coagulant aids 150
Coliform bacteria 56, 318
– –, detection 297–308
– –, in swimming bath water 324–325
Coli strains, pathogenic 309

– titre and determination 301–308, 325
Collector wells 3
Colorimetric methods 23–26
Colorimetry 22–26
Colour as pollution indicator 72
– of the water 19, 52–54
– – – –, removal and improvement 52–54
Combined available chlorine 126–127
– carbon dioxide 162
Complexone 50, 115, 116
Concrete, damage to 74, 76, 101, 151, 161, 162, 165, 204, 208–211, 223
– mixing, water for 211
Conditioning of drinking water, regulations 148–150
Conductivity, electrolytic 121–122, 230
Conservation of water samples 14, 220
Contamination of ground water by chemicals 147
– – – – – mineral oil 143–147
– – – – – radioactivity 147
Cooling water 233–234
– –, softening of 189
Copper, corrosion of 157
–, determination 133, 250
–, occurrence 133, 190
– pipes, attack by sulphur water 195
–, removal 191
–, sanitary significance 190
–, sterilisation with 65, 205, 374–375
Core samplers 353
Corrosion 151–160
–, protective layer against 142, 148, 151–156, 163–166, 170, 172–174, 181, 194
–, – – –, prevention of formation by carbon dioxide 97, 165
Cumasina process 65
Cyanide 228–229
Cyclator process 226

D

Dairies, water for 150, 174, 175
DDT 376
Deacidification 59, 166–174
–, chemical 168–174
–, mechanical 166–168
Decarbolith process 171
Decontamination, radioactive 125, 147
Deep wells 2, 3
Demineralisation 74–75, 121, 230
–, total 187
Density, determination 122
Desalination 187
Detergents 78, 183, 217–218, 282
Dithizone methods 248–253
Dolomite filters and filtration 169–171, 176
Dredges 353–355
Drilled wells 1, 2, 366
Drilling procedures 3
Drinking water analyses, examples 80–81
– – –, hygiene-chemical 27–50
– – –, major 82–147
– – –, minor 27–50
– –, assessment 51–81
– –, biological investigation 365–377
– –, regulations for conditioning 148–150
– –, requirements of 51–52
– –, sterilisation 60–67
Dug wells 1, 2, 9
Duolites 231, 232
Dye test 9
Dysentery 56, 278–279
Dyspepsia coli 279

E

Ecological procedures of biological water analysis 355–364
EDTA 50, 115, 116
Effervescent waters 238
Eijkman procedure 301, 305
Endo nutrient media 302, 304–306, 321, 331
Enterobacteria 297–298, 300, 332
Enterococci 297, 332
–, detection 310–313, 322, 323
– in swimming bath water 324–325
Eosin-methylene blue nutrient media 302, 306

Epilimnion 59, 193
Escherichia coli 278, 279, 318, 321, 331
– –, detection 297–308
– – in swimming bath water 324–326
Eternite pipes 3, 156, 159–160, 162, 371, 373
Excess carbon dioxide 94, 95, 97, 161, 164–165, 172
Exchange filters 186
Expendible culture media 330–331

F

Faecal reaction 47
Ferric chloride flocculation 54, 79, 225
Flame photometry 244–245
Flocculation accelerators 53
– tanks (in iron removal) 175
– in water decoloration 52–54
Fluoridation of drinking water 140–141
Fluoride, determination 138–140
–, significance for health 140–141, 281–282
Free available chlorine 126–127
– carbon dioxide see Carbon dioxide, free
– chlorine, determination 126–127
Fur (= scale) 172, 183

G

Gelatine liquefying bacteria 294, 295, 299, 300
– nutrient media 291–292, 294
Geothermal gradient 14, 15, 240
Gram staining 274, 307–308
Ground water supply 58
Gypsum-containing scale 183

H

Hardness 178–189
–, apparent carbonate 179
–, carbonate see Carbonate hardness
–, concept of 178
–, degree of 178–179
–, determination 112–119
–, formers 57, 180, 182, 183, 186, 187, 188, 189, 230, 234
–, non-carbonate 64, 117, 165, 179, 180, 183
–, permanent 17, 97, 117
–, temporary 179
–, total, determination 112–119
Hehner cylinders, work with 26
Hepatitis 279, 314
Heyer test 15, 88–89
High pressure boiler water 189
Hollow ram pump wells 1
Horizontal wells 3–4
Humins, determination 40
–, removal 52–54, 176–177
Hydrazine, determination 131, 232
Hydrogen ion concentration 160–161
– peroxide, determination 129
– sulphide, corrosion by 157, 195, 209
– –, determination 119–120, 222
– –, occurrence 194, 369
– –, removal 55, 195, 369
Hypolimnion 175, 193

I

Ignition, loss on 77, 121
–, residue on 120–121, 221
IMViC tests 299, 300, 302–305
Incubators 276–277
Indican test 207
Indicator organisms 358–364
Indole detection test (IMViC) 299, 302, 304
Infants, damage to health by nitrates 70–71, 281
Insecticides 376
Inspection of the site 6–8, 283–285
Interferometer value 122
Iodide, determination 141–142, 248
–, occurrence 141
Ion exchange procedures, boiler feed water 230
– – –, decontamination of radioactivity 147
– – –, fluoride removal 140
– – –, potassium determination 135

– – –, in water softening 185–187
– specific electrodes 22, 36, 43, 139, 230
Iron bacteria 175, 366–370
–, corrosion of 74, 97, 101, 151–152, 155, 211
–, determination 108–110, 246
–, occurrence 108, 174
–, removal 175–177
Irradiation with ultraviolet light (in water sterilisation) 66–67

K
Katadyn process 66, 67
Klebsiella species 297–298
Koser's citrate test (IMViC) 299, 303–304

L
Lactose-peptone solution 302–305
Laundries 151, 174, 182, 183
–, wastewater from 228
"Layer" wells 3
Lead-attacking carbon dioxide 97, 165–166
Lead, corrosion 83, 97, 101, 151–152, 154–155, 158, 165–166, 172
–, determination 131–132, 250–251
–, occurrence 131, 189
– pipes 83, 97, 101, 154, 173, 181, 195, 281
– poisoning 154, 189, 195, 281
–, removal 190
–, sanitary significance 189
– solubility test 101
Leifson's (deoxycholate) agar 329, 331
Lewatit 185, 187, 230, 231
Lime-aggressive carbon dioxide 94, 96, 97, 165–166
Lime-aggressivity of waters 90, 94
Lime-soda process 184
Limiting value (of bacteria in drinking water) 295–296
Lithium, determination 245
Littoral 193
Loss on ignition 77, 121

M
MacConkey's agar (broth) 305
Magnesite filtration (deacidification) 169
Magnesium, determination 118, 119
Magnetic water treatment 187
Magno filtration 54, 169–171, 176–178, 191
Major drinking water analysis 82–147
Manganese bacteria 175, 178, 367–370
–, determination 110–112
–, occurrence 110, 174
–, removal 177–178
Maps of water quality 360, 363–364
Marble filtration 168
Mechanical filters 79
Medicinal waters 238–254
Membrane filter procedures 57, 315, 317–323, 327–331, 351–352
– – – , staining in 320–321
Mercury, determination 252–253
Mesopsammon fauna 371
Methaemoglobinaemia 71, 281
Methylene blue as antidote 71, 281
– – test 221
Methyl red test (IMViC) 299, 303–305
Microsieves in water clarification 55
Mineral oil, carcinogenic properties 78, 146–147
– –, contamination of ground water 143–146
– –, detection 79, 143
– –, removal 79, 144, 146–147
– –, seepage 145
Mineral waters 77, 237–238
Minor drinking water analysis 27–50
Mixing bulbs 27–50
Mixing (blending) of waters 152–153
Mortar, attack on 74, 76, 101, 151, 162, 165, 208–210
"Most probable numer" test 305–307, 325
Multiple layer filtration 62
m-Value 89–93

N
Nannoplankton net (Höll) 77, 350, 366,

Nitrate ion, determination 33–36
– –, harm due to 71, 192, 211, 281
– – as pollution indicator 69–71
– –, removal 71
Nitrite ion as pollution indicator 69
– –, determination 31–32
Non-carbonate hardness 64, 117, 165, 179, 180, 183
Nutrient agar media 294
– cardboard discs 330–331

O

Odour 15–17, 55
Oil-binding materials 146
Oligodynamic action of silver 65
Organisms in drinking water supply plants, combating of 373–377
Organoleptic examination 16
Oxygen in boiler feed water 105, 232
– bottle 12
–, consumption 105, 213
– deficit or deficiency 105, 106
– depletion, chemical 105
–, determination 13, 101–106, 214, 222
–, occurrence 193
–, saturation deficit 105, 106
– – index 15, 106, 107
–, significance 193–194, 213–216
Ozone, determination 128
–, sterilisation with 54, 56, 62–63, 204

P

Palisade scraper 351, 352, 366
Paper industry, hardness of water for 182
– –, wastewater from 229
Paratyphoid 56, 278, 328
Pathogenic bacteria, determination in water 327, 332
– coli strains 309
Pelagial 12, 162
Percolation filters 225
Permanent hardness 17, 97, 117
Permanganate consumption see Permanganate number
– number 37–40, 176, 216, 223, 234
– – as pollution indicator 71–73
Permutit 185, 187, 230, 231
Pesticides 219
Phages 314–316
Phenol(s), determination 217, 227–228
Phosphate dosage 113, 154–155, 164, 166, 170, 172–173, 177, 178, 184, 187–189, 230–231
– ion, determination 43–45
– – as pollution indicator 75–76
pH-value 93
–, determination 15, 83–86
Phytoplankton 59
Plankton 161
– nets 347, 349–351, 366, 369
–, removal 59, 60, 234
–, sampling 347
Plant growth sampling 347, 351, 352
Plastic pipes 156, 158–159
Plate count (number) see Bacterial number
Pollution indicators 67–79
Polyacrylamide-based flocculating agents 53
Polyphosphate dosage see Phosphate dosage
–, determination 45–46, 130
Potassium ion, determination 134–136, 244
– – as pollution indicator 77
– permanganate consumption see Permanganate number
Preservation of water samples 14, 220
Profundal 12, 60, 193
Protective zones of water supply 7
– layer formation against corrosion see Corrosion, protective layer against
Protein nitrogen, determination 30
Proteus species 298, 330
Protozoa contaminants 280
– species 366, 371
Pumping test 4–6
Purification stage, third 173, 227
p-Value 89–93

Pyrethrum powder for combating organisms 375
Pyrogen test 58

Q

Qc-Value 89, 90, 94

R

Radioactivity and its determination 123–126
–, decontamination 125
–, fallout 60
Radon 125
Rainwater supply 60
Rapid filters 61, 372
Reagents for investigating drinking water 256–261
Reaktivator process 53, 225, 226
Redox potential 5, 86, 293
Reed knives 351, 352
Referrisation 156, 157
Refuse dumps, tips 180–181, 191
Regeneration of exchange filters 186
Removal of arsenic 191
– – iron 175–177
– – lead 190
– – manganese 177–178
Reservoir water supply 59
Residue on evaporation 120–121, 244
– – – as pollution indicator 77
– – ignition 120–121, 221
River water supply 59

S

Salmonella 56, 278–280, 298, 315–316, 327–331
Salmonella-Shigella (S-S) agar 329–331
Salt lenses 5, 77
Sample bottles, filling 5, 11
– size 13, 287
Sampling 9
– for bacteriological analysis 286–289
– – biological analysis 346–347
– bottles 10, 287
– of plant growth 347, 351–353
Sand filters 125, 369, 371
Saprobic system 356–364
Saturation index 15, 94, 96–97
Scale 172, 183
Scirpus lacustris 225, 228
Screenable material 77, 347
– – as pollution indicator 77
Sedipur process 228
Seepage of oil 145–147
Selenium, determination 138
Sent-in samples 14
Settleability of the water 18
Settleable matter 213, 220
Sewage 219–230
–, clarification 224
– farms 225
– lagoons 225
–, sampling 219–220
–, testing 220–222
Shallow wells 2, 3
Shigella 278, 298, 315, 316
Sievable material 347
Sighting disc 19, 350
Silica, in boiler feed water 142, 231–232
–, determination 142–143
– gel nutrient media 295
–, occurrence 142
–, removal 231
–, significance 142
Silicate treatment in protection against corrosion 142, 154, 174
Silicophos process 174
Silver, determination 129, 251–252
–, oligodynamic action 65
–, sterilisation with 65–66
Site, inspection and investigation 6–8, 14
Slow filters 61, 369
Sludge-blanket filters 225
Sludge core samplers 353
– cup 353, 354
– investigation 353–355
– sieve 354, 355
Soda regenerative process 184
Sodium aluminate flocculation 53

– ion, determination 136, 244
–– as pollution indicator 77
Softening processes 183–189
Species deficit Kothé 363
Spraying (deacidification) 167
Spring water supply 60
Staphylococci 324–326
Sterilisation 60–67, 276
Stirring sphere 25, 44, 111, 132
Streptococci 310
Strohecker-Langelier equation 96, 100
Submerged funnel in sampling 11
Sulphadur cement 210
Sulphate ion content and corrosion 76, 151, 156, 158, 159, 192, 209–211
– – , determination 47–50, 247
– – as pollution indicator 76–77
Sulphate-resistant cements 210
Sulphides, determination 119–120
– , corrosion by 209
Sulphite, determination 131, 229
– liquor 228
Sulphur bacteria 369
Surface water 212–219
Suspended matter 213, 221
Swimming bath water, bacteriological examination 324–326
– – –, chemical examination 202–208

T

Table waters 237–238
– Water Decree 237, 238
Tanning industry 182
Tap water, major analysis of 82–147
– –, minor analysis of 27–50
– –, normal and typical values 80–82
Taste 15, 17, 55
Tellurite-containing broth 311
Temperature, water, determination 14
– –, significance 14–15
Temporary hardness 179
Tensides 78, 183, 217–218, 282
Tetrathionate-crystal violet digestion broth 329–330
"Thermal contamination" 215
Thermal waters 240
Thermocline 12, 60, 102, 161
Third purification stage 173, 227
Threshold odour number 16–17
Tin, corrosion 192
–, determination 138
–, occurrence 138
Total chlorine 126–127
– hardness 180
Trade wastewaters 227–230
Transparency of the water 17–18
Trickling (deacidification) 167, 175
Triphenyltetrazolium chloride 331
Tridodium phosphate process (softening) 184, 230–231
Tube wells 1
Tularaemia 279
Turbidity and degree of turbidity 18, 54–55
Typhoid 56, 278, 280, 315–316, 327

U

Ultraviolet radiation in sterilisation 66–67
Urea, detection and determination 206
Urinary components in swimming bath water 206
Urochrome 46, 57, 282

V

Vacuum degassing 168
Viruses 280, 314–316
Visibility disc 19, 350
Voges-Proskauer reaction (IMViC) 299, 303–305

W

Wastewater 219–230
–, chemical purification 225
–, sampling 219–220
–, trade 227–230
Water bloom 39, 43, 214, 358, 364, 371, 374, 376
– for brewing 151, 174, 234–236

– – building purposes 208–212
– – concrete mixing 211
–, decoloration 52–54
– and disease 278–282
– drawers 12–13, 288, 349–350
– illness 56, 280
– as a material 254
– quality classes 358–364
–, requirements of a 51–52
– softening 183–189
–, sterilisation 60–67
Wells 1–4, 366
Wilson-Blair culture medium 322, 328–331
Wisprofloc process 53
Wofatit process 185, 231
Wrought iron pipes 156, 162

Z

Zinc, corrosion of 97, 157
–, determination 133, 250–251
–, occurrence 133, 192
–, significance 192

Bei Fragen zur Produktsicherheit wenden Sie sich bitte an:
If you have any questions regarding product safety,
please contact:

Walter de Gruyter GmbH
Genthiner Straße 13
10785 Berlin
productsafety@degruyterbrill.com